Ergebnisse der Mathematik und ihrer Grenzgebiete

3. Folge · Band 16

A Series of Modern Surveys in Mathematics

Gerard van der Geer

Hilbert Modular Surfaces

With 39 Figures

Springer-Verlag
Berlin Heidelberg New York
London Paris Tokyo

Gerard van der Geer
Mathematisch Instituut
Universiteit van Amsterdam
Roetersstraat 15
1018 WB Amsterdam
The Netherlands

Mathematics Subject Classification (1980): 11F41, 11G18, 11I20

ISBN 3-540-17601-2 Springer-Verlag Berlin Heidelberg New York
ISBN 0-387-17601-2 Springer-Verlag New York Berlin Heidelberg

Library of Congress Cataloging-in-Publication Data
Geer, Gerard van der. Hilbert modular surfaces.
(Ergebnisse der Mathematik und ihrer Grenzgebiete ; 3. Folge, Bd. 16)
Bibliography: p. Includes index.
1. Hilbert modular surfaces. I. Title. II. Series. QA573.G44 1987 512'.7 87-20634

Printed in Germany

Typesetting: Asco Trade Typesetting Ltd., Hong Kong
Printing and bookbinding: Konrad Triltsch, Würzburg
2141/3140-543210

Preface

Work on this book started in 1981 following a suggestion of F. Hirzebruch. I would like to thank him heartily for this suggestion and for his constant support, in particular for the invitations to visit the Max-Planck-Institut für Mathematik in Bonn. I profited a lot from the stimulating atmosphere at this institute. The manuscript was completed during a stay at the Mathematical Sciences Research Institute in Berkeley. I acknowledge with gratitude the support of both research institutes.

During the preparation of the manuscript suggestions and corrections were made by C. Faber, J. Koehl, S. Maurmann and R. Schoof. J. Koehl kindly provided the table of numerical invariants at the end of this book. I thank them for their help. I also would like to thank B. Gross and B. Mazur for their comments on the last two chapters, and J. Lagerberg for drawing the figures. Thanks are also due to the publishers for their cooperation.

Finally, I would like to express my gratitude to D. Zagier. Over the years I learned a lot from him in the course of numerous discussions we had on the subject of this book and on other parts of mathematics. Moreover, he read a large part of the manuscript and his comments were most helpful. For all this and the fact that he contributed a section, VII.5 to be precise, I thank him very much.

Cambridge, Mass., June 1987 Gerard van der Geer

Table of Contents

Introduction

If K is a totally real algebraic number field then $\mathrm{PSL}(2, \mathfrak{o}_K)$, with $\mathfrak{o}_K$ the ring of integers of K, is called the Hilbert modular group (of K). If the degree of K over $\mathbb{Q}$ is n the n embeddings $K \hookrightarrow \mathbb{R}$ define an embedding of $\mathrm{PSL}(2, K)$ into $\mathrm{PSL}(2, \mathbb{R})^n$. Since $\mathrm{PSL}(2, \mathbb{R})$ is the holomorphic automorphism group of the complex upper half plane $\mathfrak{H}$ the Hilbert modular group $\mathrm{PSL}(2, \mathfrak{o}_K)$, or more generally $\mathrm{PSL}(\mathscr{L})$ with $\mathscr{L} \subset K^2$ a projective $\mathfrak{o}_K$-module of rank 2, acts on $\mathfrak{H}^n$ and we can form the quotient. This is quasi-projective algebraic variety with finitely many isolated singularities and it is called a Hilbert modular variety. For $n = 1$ ($K = \mathbb{Q}$) one finds the classical modular group and the j-line $\mathrm{PSL}(2, \mathbb{Z})\backslash\mathfrak{H}$. This book deals with the case where the field K is a real quadratic field.

Hilbert explained his interest for this group action in his lecture at the International Congress at Paris in 1900 when he formulated his twelfth problem. His aim, he said, was to find analytic functions playing the same role for arbitrary algebraic number fields as the exponential function does in Kronecker's theorem about abelian extensions of $\mathbb{Q}$ and as the modular function j does for imaginary quadratic number fields in the theory of complex multiplication. He expected that research in this direction would stimulate the development of the theory of analytic functions of several complex variables. Hilbert worked on this problem around 1893 and later proposed to his student Blumenthal to study the function theory of the quotients $\mathrm{PSL}(2, \mathfrak{o}_K)\backslash\mathfrak{H}^n$. Several authors (Blumenthal, Hecke, Kloosterman) dealt with the subject, but none of them made it flourish. Contrary to Hilbert's expectation progress on Hilbert modular varieties required a mature theory of analytic functions of several complex variables and of algebraic geometry.

The quotients $\mathrm{PSL}(\mathscr{L})\backslash\mathfrak{H}^n$ admit a natural compactification obtained by adding finitely many points, one for each ideal class of K. These points, the cusps, are highly singular. In 1970 Hirzebruch succeeded in resolving these cusp singularities for real quadratic fields. This made it possible to apply the knowledge of non-singular projective algebraic surfaces to these Hilbert modular surfaces and thus paved the way to a rapid development of the subject. It is the purpose of this book to give a survey of the state of the art in this field.

The surfaces we study here are rather special, but they are nevertheless important in arithmetic algebraic geometry. On the one hand, Hilbert modular varieties are the moduli spaces of abelian varieties whose endomorphism ring contains a given order from a totally real field. Thus the points of a Hilbert modular variety have a natural meaning and this makes these varieties much more accessible than arbitrary varieties. In particular, besides providing beautiful examples of algebraic varieties, they are an ideal testing ground for the conjectures of (arithmetic) algebraic geometry. On the other hand, a complex of conjectures in arithmetic algebraic geometry due to Weil, Tate, Langlands and others predicts, roughly speaking, that certain summands of the cohomology of Hilbert modular varieties constitute building blocks for the cohomology of varieties defined over number fields. To give one example, it is conjectured that if E is an elliptic curve defined over a totally real number field K of degree n over $\mathbb{Q}$, then there is a summand H of the n-th cohomology group $H^n(X)$ of a Hilbert modular variety X associated to K such that $H \cong \bigotimes_i H^1(E^{(i)})$, where the tensor product is taken over the conjugates $E^{(i)}$ of E with respect to the field extension $K/\mathbb{Q}$. (This isomorphism should be interpreted in the widest possible sense: as "motives".)

Some words about this book. I have tried to make this book as self-contained as possible, requiring from the reader only a general education in algebraic geometry and number theory. I tried to choose that approach that brings out the underlying ideas. The book is divided in eleven chapters, each of which is preceded by a short introduction that usually contains remarks on the history of the subject. I have tried to give complete proofs, except in part of the last two chapters, where also the level of prerequisites is higher. Finally, a remark about the terminology. Some recent authors use the term Hilbert-Blumenthal modular variety instead of Hilbert modular variety. Since I found no compelling historical nor linguistic reason to do this I shall stick to the old name.

Notations and Conventions Concerning Quadratic Number Fields

An integer D is called a fundamental discriminant if it is the discriminant of (the maximal order of) a quadratic number field. For a fundamental discriminant we denote by χ_D the primitive Dirichlet character mod $|D|$ satisfying $\chi_D(p) = \left(\frac{D}{p}\right)$ for any odd prime, $\chi_D(2) = +1$ (resp. -1) if $D \equiv 1 \pmod 8$ (resp. $D \equiv 5 \pmod 8$)) and $\chi_D(-1) = \operatorname{sgn} D$. A fundamental discriminant is called a prime discriminant if it is divisible by only one prime (i.e. $D = -4, +8, -8, p$ with p a prime $\equiv 1 \pmod 4$, $-p$ with p a prime $\equiv 3 \pmod 4$). Every fundamental discriminant is a product of prime discriminants:

$$D = \prod_{p|D} D(p)\,.$$

If K is a real quadratic field we denote the action of its non-trivial Galois automorphism by $x \to x'$. We denote by N its norm map over $\mathbb{Q}$, by $\mathcal{C}\ell(K)$ (or simply $\mathcal{C}\ell$) its class group and by h_K or h its class number. We say that two non-zero fractional ideals $\mathfrak{a}$ and $\mathfrak{b}$ are equivalent in the narrow sense (or strictly equivalent) if there exists a $\lambda \in K$ with $\mathsf{N}(\lambda) > 0$ and $\lambda\mathfrak{a} = \mathfrak{b}$. The equivalence classes form a group $\mathcal{C}\ell^+$ of order h^+.

Let $\mathrm{Sq}^+ : \mathscr{C}\ell^+ \to \mathscr{C}\ell^+$ be the homomorphism defined by squaring. Then $\mathscr{C}\ell^+/\mathrm{Sq}^+(\mathscr{C}\ell^+)$ is called the genus group; a coset of $\mathrm{Sq}^+(\mathscr{C}\ell^+)$ is called a genus and $\mathrm{Sq}^+(\mathscr{C}\ell^+)$ is called the principal genus. Then two non-zero fractional ideals $\mathfrak{a}$ and $\mathfrak{b}$ belong to the same genus if and only if there exists a $\lambda \in K$ with $\mathsf{N}(\lambda) > 0$ and $\mathsf{N}(\lambda)\mathsf{N}(\mathfrak{a}) = \mathsf{N}(\mathfrak{b})$. The real characters $\chi: \mathscr{C}\ell^+ \to \mathbb{R}$ are called genus characters. There is bijective correspondence between the genus characters of the real quadratic field K and the decompositions of its discriminant $D = D_1 D_2$ as a product of two fundamental discriminants. The genus character $\chi_{D_1 D_2}$ corresponding to $D = D_1 D_2$ is defined on prime ideals $\mathfrak{p}$ by

$$\chi_{D_1 D_2}(\mathfrak{p}) = \begin{cases} \chi_{D_1}(\mathsf{N}(\mathfrak{p})) & \text{if } (\mathsf{N}(\mathfrak{p}), D_1) = 1 \\ \chi_{D_2}(\mathsf{N}(\mathfrak{p})) & \text{if } (\mathsf{N}(\mathfrak{p}), D_2) = 1 \end{cases}$$

It follows that the genus group of K is isomorphic to $(\mathbb{Z}/2\mathbb{Z})^{t-1}$, where t is the number of (distinct) primes dividing D. Let $\mathfrak{a}$ be an integral ideal of K. Then a natural number n is the norm of an ideal from the genus of $\mathfrak{a}$ if and only if $n\mathsf{N}(\mathfrak{a})$ is the norm of an element of K.

We shall use the following notation for the genera of K. We list the t primes dividing D in increasing order. Then the genus γ of $\mathfrak{a}$ is denoted by t signs: $\gamma = (\ldots, \gamma_p, \ldots)_{p|D}$ with $\gamma_p =$ sign of $(D, \mathsf{N}(\mathfrak{a}))_p$ with $(\ ,\ )_p$ the Hilbert symbol. If $\mathsf{N}(\mathfrak{a})$ is prime to D (which we may assume after replacing $\mathfrak{a}$ if necessary by another ideal in the same ideal class), then $(D, \mathsf{N}(\mathfrak{a}))_p = \chi_{D_1 D_2}(\mathfrak{a})$, where $D_1 = D(p)$, $D_2 = D/D(p)$.

I. Hilbert's Modular Group

In a paper in Mathematische Annalen [13] Blumenthal did the first pioneering work in a program outlined by Hilbert with the aim of creating a theory of modular functions of several variables that should be just as important in number theory and geometry as the theory of modular functions of one variable was at the beginning of this century. Since no general theory of complex spaces was available this was by no means an easy task. Blumenthal had at his disposal a manuscript by Hilbert from 1893/94 on the action of the modular group Γ_K of a totally real field K of degree n over $\mathbb{Q}$ on the product $\mathfrak{H}^n$ of n upper half planes. According to Blumenthal it gave a sketchy description of general properties such as properly discontinuous action and fundamental domain but it contained precise information on the construction of modular functions by means of theta functions. Blumenthal gave a detailed account of the function theory involved but his construction of a fundamental domain had a flaw: he obtained a fundamental domain with only one cusp as in the case of the classical modular group. This mistake was corrected many years later by Maass [110] who showed that the number of cusps equals the class number h of K.

In his famous lectures on analytic number theory [156] Siegel gave a detailed exposition of the construction of a fundamental domain for the action of Hilbert's modular group. Such a fundamental domain is in a natural way the union of h pieces, each piece corresponding to an ideal class of K. In fact, such a piece consists of the orbits of the points which are nearest with respect to a certain metric to one cusp. The interior points of each piece come from points with a trivial isotropy group in $\mathfrak{H}^n$. The boundary, however, may contain orbits coming from elliptic fixed points and these give rise to quotient singularities on $\Gamma_K \backslash \mathfrak{H}^n$. This quotient space will turn out to be a quasi-projective variety.

The Hilbert modular group $\mathrm{PSL}(2, \mathfrak{o}_K)$ with $\mathfrak{o}_K$ the ring of integers of K arises from the automorphism group of the $\mathfrak{o}_K$-module $\mathfrak{o}_K \oplus \mathfrak{o}_K$. In order to obtain a satisfactory theory one is forced to treat at the same time and on equal footing the discontinuous groups defined by the automorphism groups of all projective $\mathfrak{o}_K$-modules of rank 2 (up to isomorphism). They appear in a natural way at the same time when one uses the adelic language. This has big advantages when one treats arithmetic problems on Hilbert modular varieties, but for geometric problems it is often better to treat them separately.

This chapter contains the basic definitions of the Hilbert modular group, its companion groups and their extensions, the construction of a fundamental domain and a paragraph on elliptic fixed points. Though this book essentially deals with modular surfaces no inconvenience results from treating the general case at this point. The restriction $n = 2$ will therefore be imposed only after this chapter.

1. The Action of the Hilbert Modular Group

Let K be an algebraic number field of degree n over $\mathbb{Q}$ which is totally real, that is, admitting n different embeddings into the field of real numbers

$$K \to \mathbb{R} \qquad \alpha \mapsto \alpha^{(i)} \qquad i = 1, \ldots, n\,.$$

We shall often assume that an embedding is given and that $\alpha = \alpha^{(1)}$. An element $\alpha \in K$ with $\alpha^{(i)} > 0$ $i = 1, \ldots, n$ is called *totally positive*. Notation: $\alpha \gg 0$.

The group

$$\mathrm{GL}^+(2, K) = \left\{\begin{pmatrix} a & b \\ c & d \end{pmatrix}: a, b, c, d \in K, ad - bc \gg 0\right\}$$

can be embedded in $\mathrm{GL}(2, \mathbb{R}) \times \cdots \times \mathrm{GL}(2, \mathbb{R})$ (n times) using the n embeddings $K \hookrightarrow \mathbb{R}$.

Put

$$\mathrm{PGL}^+(2, K) = \mathrm{GL}^+(2, K) \Big/ \left\{\begin{pmatrix} a & 0 \\ 0 & a \end{pmatrix}: a \in K^*\right\}.$$

Let $\mathfrak{o}_K$ be the ring of algebraic integers of K. The *Hilbert modular group* for the field K is the group $\mathrm{PSL}(2, \mathfrak{o}_K) = \mathrm{SL}(2, \mathfrak{o}_K)/\{+1, -1\}$ contained in $\mathrm{PGL}^+(2, K)$. We shall denote it by Γ_K. The classical modular group $\mathrm{PSL}(2, \mathbb{Z})$ is just the special case $n = 1$, $K = \mathbb{Q}$.

We denote by $\mathfrak{H}$ be upper half plane of $\mathbb{C}$:

$$\mathfrak{H} = \{z \in \mathbb{C}: \operatorname{Im} z > 0\}\,.$$

The group $\mathrm{GL}^+(2, \mathbb{R}) = \left\{\begin{pmatrix} a & b \\ c & d \end{pmatrix}: a, b, c, d \in \mathbb{R}, ad - bc > 0\right\}$ acts on $\mathfrak{H}$ by fractional linear transformations

$$z \mapsto \frac{az + b}{cz + d}\,.$$

This action factors through an action of $\mathrm{PGL}^+(2, \mathbb{R}) = \mathrm{GL}^+(2, \mathbb{R}) \Big/ \left\{\begin{pmatrix} a & 0 \\ 0 & a \end{pmatrix}: a \neq 0\right\}$ which is the group of all biholomorphic maps of $\mathfrak{H}$ onto itself. In particular, $\mathrm{PGL}^+(2, \mathbb{R})$ acts effectively.

The group $\mathscr{G}$ of all biholomorphic maps of the n-fold Cartesian product $\mathfrak{H}^n$ to itself contains as connected component of the identity the n-fold direct product of $\mathrm{PGL}^+(2, \mathbb{R})$ with itself. In fact, $\mathscr{G}$ is the semi-direct product of $\mathfrak{S}_n$, the symmetric group of permutations of the n factors of $\mathfrak{H}^n$, with this product.

The Hilbert modular group Γ_K is a discrete subgroup of $\mathrm{PGL}^+(2,\mathbb{R})^n$ (by means of the n embeddings $K \hookrightarrow \mathbb{R}$) and therefore Γ_K acts properly discontinuously on $\mathfrak{H}^n$. This generalizes the well-known action of the classical modular group on $\mathfrak{H}$. The action of Γ_K and also of $\mathrm{PGL}^+(2,K)$ on $\mathfrak{H}^n$ is *irreducible* in the following sense: if $\gamma = (\gamma^{(1)},\dots,\gamma^{(n)})$ and if $\gamma^{(i)}$ acts on the i-th factor of $\mathfrak{H}^n$ as the identity then $\gamma = \mathrm{id}$.

The domain $\mathfrak{H}^n$ is contained in $\mathbb{P}^1(\mathbb{C}) \times \cdots \times \mathbb{P}^1(\mathbb{C})$ (n times). The n embeddings of K into the reals induce an embedding

$$\mathbb{P}^1(K) \to \mathbb{P}^1(\mathbb{R})^n \subset \mathbb{P}^1(\mathbb{C})^n .$$

$\mathrm{PGL}^+(2,\mathbb{R})$ acts on $\mathbb{P}^1(\mathbb{R})$ by fractional linear transformations and so does $\mathrm{PGL}^+(2,K)$ on $\mathbb{P}^1(K) \subset \mathbb{P}^1(\mathbb{R})^n$. The orbits of $\mathbb{P}^1(K)$ under Γ_K or any group $\Gamma \subset \mathrm{PGL}^+(2,K)$ which is discrete in $\mathrm{PGL}^+(2,\mathbb{R})^n$ are called the *cusps* of Γ_K or Γ. Very often also the representatives are called cusps.

Let $\sigma = (\alpha:\beta) \in \mathbb{P}^1(K)$. We may assume that α and β are integral. To σ we associate the (well-defined) class of the ideal (α,β) of $\mathfrak{o}_K$ generated by α and β in the ideal class group $\mathcal{C}\ell(K)$ of K.

(1.1) Proposition. *The association $(\alpha:\beta)\mapsto$ class of (α,β) sets up a bijective correspondence between the set of cusps of Γ_K and the ideal class group $\mathcal{C}\ell(K)$ of K. In particular, the number of cusps of Γ_K equals the class number h of K.*

Proof. If $\tau = \dfrac{a\sigma + b}{c\sigma + d} \sim \sigma$ then the ideal class of τ is that of the ideal $(a\alpha + b\beta, c\alpha + d\beta)$ which equals the ideal (α,β) since $\begin{pmatrix} a & b \\ c & d\end{pmatrix}$ is unimodular. Suppose now that $\sigma = (\alpha:\beta)$ and $\tau = (\gamma:\delta)$ are two cusps with the same associated ideal class. After multiplication by a suitable number from K we may assume $(\alpha,\beta) = (\gamma,\delta) = \mathfrak{a}$. Then $1 \in \mathfrak{o}_K = \mathfrak{a}\mathfrak{a}^{-1} = \alpha\mathfrak{a}^{-1} + \beta\mathfrak{a}^{-1}$, i.e. there are α^*, $\beta^* \in \mathfrak{a}^{-1}$ such that $\alpha\beta^* - \alpha^*\beta = 1$ and similarly γ^*, $\delta^* \in \mathfrak{a}^{-1}$ such that $\gamma\delta^* - \gamma^*\delta = 1$.

That is, the matrix $M_\sigma = \begin{pmatrix} \alpha & \alpha^* \\ \beta & \beta^*\end{pmatrix}$ $\left(\text{resp. } M_\tau = \begin{pmatrix} \gamma & \gamma^* \\ \delta & \delta^*\end{pmatrix}\right)$ has determinant 1 and transforms the cusp $\infty = (1:0)$ to σ (resp. τ). The matrix $M_\sigma M_\tau^{-1}$ is contained in $\mathrm{SL}(2,\mathfrak{o}_K)$ and transforms τ to σ. The observation that every ideal class is associated to a cusp (since every ideal of $\mathfrak{o}_K$ can be generated by two elements) finishes the proof. □

The cusp given by $(1:0) \in \mathbb{P}^1(K)$ is called the *cusp at* ∞ and often denoted by $\infty = (\infty,\dots,\infty)$. As remarked in the proof of the preceding proposition an element $\sigma = (\alpha:\beta)$ of $\mathbb{P}^1(K)$ ($\alpha,\beta \in \mathfrak{o}_K$) can be transformed to ∞ by the inverse of a unimodular matrix $M_\sigma = \begin{pmatrix} \alpha & \alpha^* \\ \beta & \beta^*\end{pmatrix}$ with α^*, $\beta^* \in \mathfrak{a}^{-1}$, $\mathfrak{a} = (\alpha,\beta)$. Instead of looking at the cusp represented by σ for Γ_K we may as well look at the cusp ∞ for $M_\sigma^{-1}\mathrm{SL}(2,\mathfrak{o}_K)M_\sigma/\{+1,-1\}$. This group equals $\mathrm{PSL}(\mathfrak{o}_K \oplus \mathfrak{a}^2)$, where for an invertible $\mathfrak{o}_K$-module $\mathfrak{b}$ of rank 1 we set

$$\mathrm{SL}(\mathfrak{o}_K \oplus \mathfrak{b}) = \left\{\begin{pmatrix} a & b \\ c & d\end{pmatrix} \in \mathrm{SL}(2,K) \colon a,d \in \mathfrak{o}_K,\, b \in \mathfrak{b}^{-1},\, c \in \mathfrak{b}\right\}.$$

Let U_K be the unit group of $\mathfrak{o}_K$. Then the isotropy group of the cusp ∞ in $\mathrm{PSL}(\mathfrak{o}_K \oplus \mathfrak{a}^2)$ equals

$$\left\{\begin{pmatrix} \varepsilon & \mu \\ 0 & \varepsilon^{-1} \end{pmatrix} : \varepsilon \in U_K, \mu \in \mathfrak{a}^{-2}\right\} \Big/ \{+1, -1\}$$

which equals

$$\left\{\begin{pmatrix} \varepsilon^2 & \mu \\ 0 & 1 \end{pmatrix} : \varepsilon \in U_K, \mu \in \mathfrak{a}^{-2}\right\}$$

as a transformation group.

We call the pair $(\mathfrak{a}^{-2}, U_K^2)$ *the type of the cusp* σ. In general the type of a cusp is of the form (M, V), where M is a $\mathbb{Z}$-module of rank n in K and V is a group of totally positive units of rank $n-1$ preserving M (cf. Ch. II.1).

2. The Distance to the Cusps

Let $\sigma = \dfrac{\alpha}{\beta} \in \mathbb{P}^1(K)$. For $z \in \mathfrak{H}^n$ we define

$$\begin{aligned}\mu(\sigma, z) &= \frac{\mathrm{N}((\alpha,\beta))^2 y_1 \cdot \ldots \cdot y_n}{|(-\beta^{(1)} z_1 + \alpha^{(1)}) \cdot \ldots \cdot (-\beta^{(n)} z_n + \alpha^{(n)})|^2} \\ &= \frac{\mathrm{N}(\mathfrak{a})^2 \mathrm{N}(y)}{|\mathrm{N}(-\beta z + \alpha)|^2}, \qquad \mathfrak{a} = (\alpha, \beta), \qquad y_i = \operatorname{Im} z_i .\end{aligned}$$

This is independent of the choice of α and $\beta \in \mathfrak{o}_K$ and for any $\gamma \in \Gamma_K$ we have the invariance property

$$\mu(\gamma\sigma, \gamma z) = \mu(\sigma, z) .$$

(2.1) Lemma. *There exists a positive real number r depending only on K such that for* $z \in \mathfrak{H}^n$ *the inequalities* $\mu(\sigma, z) > r$ and $\mu(\tau, z) > r$ *for* $\sigma, \tau \in \mathbb{P}^1(K)$ *imply* $\sigma = \tau$.

Proof. By Proposition 1.1 and the invariance property just mentioned we can write $\sigma = \dfrac{\alpha}{\beta}, \tau = \dfrac{\gamma}{\delta}$ with integers $\alpha, \beta, \gamma, \delta$ such that the ideals $\mathfrak{a} = (\alpha, \beta)$ and $\mathfrak{b} = (\gamma, \delta)$ have norm less than a constant C depending only on K.

For every n-tuple $(t_1, \ldots, t_n)$ of non-zero real numbers there exists a unit $\varepsilon \in K$ such that

$$|t_i \varepsilon^{(i)}| \leqslant c \sqrt[n]{|\mathrm{N}(t)|}, \qquad \mathrm{N}(t) = \prod_{i=1}^{n} t_i$$

with c a constant depending only on K. This follows from Dirichlet's unit theorem. Hence, after multiplying α and β by a suitable unit we have

$$y_i^{-1}|-\beta^{(i)}z_i + \alpha^{(i)}|^2 \leq c\mu(\sigma, z)^{-1/n}C^{2/n} < cr^{-1/n}C^{2/n} \qquad i = 1, \ldots, n\,.$$

This gives

$$|-\beta^{(i)}x_i + \alpha^{(i)}|\, y_i^{-1/2} < c^{1/2}r^{-1/2n}C^{1/n} \qquad (x_i = \operatorname{Re} z_i)$$

and

$$|\beta^{(i)}|\, y_i^{1/2} < c^{1/2}r^{-1/2n}C^{1/n}\,.$$

For γ, δ we obtain similar inequalities. But now

$$\alpha^{(i)}\delta^{(i)} - \beta^{(i)}\gamma^{(i)} = (-\beta^{(i)}x_i + \alpha^{(i)})y_i^{-1/2}\delta^{(i)}y_i^{1/2} - (-\delta^{(i)}x_i + \gamma^{(i)})y_i^{-1/2}\beta^{(i)}y_i^{1/2}$$

and consequently

$$|\mathsf{N}(\alpha\delta - \beta\gamma)| < (2c)^n r^{-1} C^2\,.$$

So for $r > (2c)^n C^2$ the norm of the algebraic integer $\alpha\delta - \beta\gamma$ has absolute value less than 1, that is $\alpha\delta - \beta\gamma = 0$ and this says $\sigma = \tau$. □

We can consider $\dfrac{1}{\mu(\sigma, z)^{1/2}}$ as the "distance" of z to the cusp σ. For $\sigma = \infty$ the distance is just $\dfrac{1}{\mathsf{N}(y)^{1/2}}$. This distance is universally bounded in the following sense.

(2.2) Lemma. *There exists a positive real number s depending only on K such that for $z \in \mathfrak{H}^n$ there exists a cusp σ such that $\mu(\sigma, z) > s$.*

Proof. Since $N(\mathfrak{a}) \geq 1$, it is sufficient to satisfy the inequality

$$|N(-\beta z + \alpha)|^2 \mathsf{N}(y)^{-1} < s^{-1}$$

with a solution (α, β) in $\mathfrak{o}_K$. Consider the inequalities

$$|-\beta^{(i)}x_i + \alpha^{(i)}|\, y_i^{-1/2} \leqslant c_i \qquad i = 1, \ldots, n\,,$$

$$|\beta^{(i)}|\, y_i^{1/2} \leq d_i \qquad i = 1, \ldots, n\,,$$

which we may write using a basis $\omega_1, \ldots, \omega_n$ of $\mathfrak{o}_K$ as a system of $2n$ linear inequalities. According to a theorem of Minkowski [15] we can find a solution $\alpha = \sum a_i\omega_i$, $\beta = \sum b_i\omega_i$ with $a_i, b_i \in \mathbb{Z}$ if $c_1 \ldots c_n d_1 \ldots d_n \geqslant$ determinant of this system, which is $|(\omega_i^{(j)})|^2 = D_K$, the discriminant of K. So we can take $c_i = d_j = D_K^{1/2n}$, and hence $s = 2^n D_K$. □

Let $\sigma \in \mathbb{P}^1(K)$. We define the "*sphere of influence*" of σ:

$$F_\sigma = \{z \in \mathfrak{H}^n \colon \mu(\sigma, z) \geqslant \mu(\tau, z) \text{ for all } \tau \in \mathbb{P}^1(K)\}\,.$$

The invariance condition $\mu(\gamma\sigma, \gamma z) = \mu(\sigma, z)$ implies $F_{\gamma\sigma} = \gamma F_\sigma$ $(\gamma \in \Gamma_K)$.

(2.3) Proposition. *The action of Γ_K in the interior F_σ^0 of F_σ reduces to that of the isotropy group Γ_σ of σ, i.e., if z and γz both belong to F_σ^0, then $\gamma\sigma = \sigma$.*

Proof. We have

$$\begin{array}{ccc} \mu(\gamma^{-1}\sigma, z) & \leqslant & \mu(\sigma, z) \\ \| & & \| \\ \mu(\sigma, \gamma z) & \geqslant & \mu(\gamma\sigma, \gamma z)\,, \end{array}$$

for $z, \gamma z \in F_\sigma^0$ and the inequalities are strict if $\gamma\sigma \neq \sigma$. □

Example. Let $K = \mathbb{Q}$. In this case $h = 1$ and if we take $\sigma = \infty$ we get for F_σ the region consisting of the $\mathbb{Z}$-translates of $\{z \in \mathfrak{H}: |z| \geqslant 1, |\mathrm{Re}\, z| \leqslant \frac{1}{2}\}$.

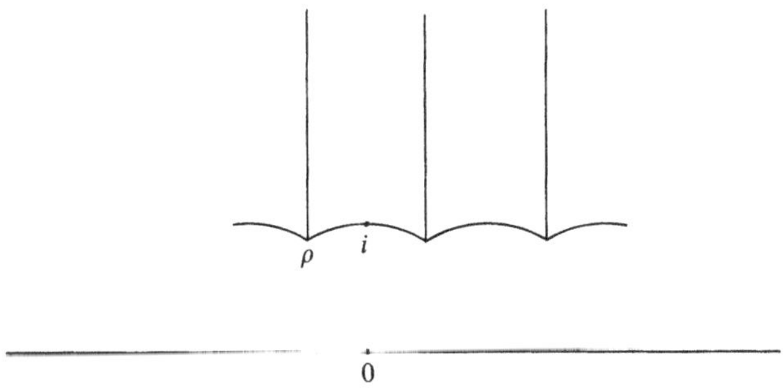

Sphere of influence of the cusp at ∞

The boundary of F_σ consists of pieces of "generalized isometric circles" given by equalities $\mu(\sigma, z) = \mu(\tau, z)$ with $\tau \neq \sigma$.

Using Lemma (2.2) and the Proposition we arrive at the following way to decompose the orbit space $\Gamma_K \backslash \mathfrak{H}^n$ into h pieces glued in some way along parts of their boundary. Let $i_\sigma\colon \Gamma_\sigma \backslash F_\sigma \to \Gamma_K \backslash \mathfrak{H}^n$ be the natural map. Then

$$\Gamma_K \backslash \mathfrak{H}^n = \bigcup_\sigma i_\sigma(\Gamma_\sigma \backslash F_\sigma)$$

where the union is taken over a set of h cusps representing the ideal classes of K. Each piece corresponds to an ideal class of K.

Observe that the decomposition just given is not the only possible one. If we replace the distance function $\mu(\sigma, z)$ by $c(\sigma)\mu(\sigma, z)$, where $c(\sigma) \in \mathbb{R}_{>0}$ depends only on the class of σ in $\Gamma_K \backslash \mathbb{P}^1(K)$, then $\mu(\gamma\sigma, \gamma z) = \mu(\sigma, z)$ still holds and we can use the new μ to define "spheres of influence" as before. In this way we get a $(h-1)$-dimensional family of decompositions of $\Gamma_k \backslash \mathfrak{H}^n$ into h pieces.

Note that the action of Γ_σ on $\mathfrak{H}^n$ is free. Thus we get:

(2.4) Corollary. *All fixed points of Γ_K on $\mathfrak{H}^n$ lie on the boundaries of the F_σ.*

3. A Fundamental Domain

We now proceed to the construction of a fundamental domain for the action of Γ_K on $\mathfrak{H}^n$. First we construct a fundamental domain for the action of Γ_σ on $\mathfrak{H}^n$.

As in Proposition (1.1) we take an element $g \in \mathrm{SL}(2, K)$ such that $g\sigma = \infty$ and $g\mathrm{SL}(2, \mathfrak{o}_K)g^{-1} = \mathrm{SL}(\mathfrak{o}_K \oplus \mathfrak{a}^2)$. Note that the isotropy group of ∞ in $g\Gamma_K g^{-1}$ is generated by translations

$$z \mapsto z + \mu, \qquad \mu \in \mathfrak{a}^{-2},$$

and by dilations

$$z \mapsto \varepsilon z,$$

where ε runs through the group $V = U_K^2$.

Consider the map

$$\begin{aligned} \mathrm{Im}\colon \mathfrak{H}^n &\to (\mathbb{R}_+)^n \\ z &\mapsto \mathrm{Im}\, z := (\mathrm{Im}\, z_1, \ldots, \mathrm{Im}\, z_n) =: y\,. \end{aligned}$$

It induces a map $g\Gamma_\sigma g^{-1}\backslash\mathfrak{H}^n \to V\backslash(\mathbb{R}_+)^n$ which exhibits $g\Gamma_\sigma g^{-1}\backslash\mathfrak{H}^n$ as a torus bundle over $V\backslash(\mathbb{R}_+)^n$ with fibre the n-dimensional torus $\mathbb{R}^n/\mathfrak{a}^{-2}$. Having factored out the action of the translations we only have to construct a fundamental domain for the action of V on $(\mathbb{R}_+)^n$. We look first at the action of V on $B = \{y \in (\mathbb{R}_+)^n\colon \mathrm{N}(y) = 1\}$. By taking logarithms it is transformed bijectively into an $\mathbb{R}^{n-1}$:

$$\begin{aligned} B &\overset{\log}{\to} \mathbb{R}^{n-1} = \{(r_1, \ldots, r_n) \in \mathbb{R}^n\colon \textstyle\sum r_i = 0\} \\ y &\mapsto (\log y_1, \ldots, \log y_n)\,, \end{aligned}$$

while the action of V on B is carried over into an action on $\mathbb{R}^{n-1}$ by translations: $a_i \mapsto a_i + \log \varepsilon^{(i)}$. By Dirichlet's theorem on units the logarithm transforms V into a lattice in $\mathbb{R}^{n-1}$. The exponentional transforms a fundamental domain (e.g. a fundamental parallelopiped) for this action back into a fundamental domain B_V for the action of V on B. The cone over B_V, that is $\mathbb{R}_+ \cdot B_V \subset (\mathbb{R}_+)^n$, is a fundamental domain for the action of V on $(\mathbb{R}_+)^n$. If we denote by T a fundamental domain for the action of the translations by elements of $\mathfrak{a}^{-2}$ on $\mathbb{R}^n$, then

$$E = \{z \in \mathfrak{H}^n\colon \mathrm{Re}\, z \in T, \mathrm{Im}\, z \in \mathbb{R}_+ \cdot B_V\}$$

is a fundamental domain for the action of $g\Gamma_\sigma g^{-1}$ on $\mathfrak{H}^n$.

One now easily checks that the intersection of E with $g(F_\sigma)$ is connected. So $g^{-1}E \cap F_\sigma$ is a fundamental domain for the action of Γ_σ on F_σ which we call D_σ.

In this way we obtain h domains D_{σ_i} $(i = 1, \ldots, h)$, one for each cusp. It is easy to see that there exist $\alpha_1, \ldots, \alpha_h \in \Gamma_K$ such that $\bigcup_{i=1}^h \alpha_i(D_{\sigma_i})$ is connected and hence a fundamental domain for Γ_K.

Finally, we define a compact manifold with boundary:

$$S_r = \Gamma_K\backslash\{z \in \mathfrak{H}^n\colon \mu(\sigma, z) \leqslant r \text{ for all cusps } \sigma\}$$

where r is so large that for all cusps σ $\{z \in \mathfrak{H}^n\colon \mu(\sigma, z) > r\}$ is contained in F_σ. The boundary ∂S_r consists of h compact manifolds $i_\sigma(\Gamma_\sigma\backslash\partial W(\sigma, r))$ of dimension $2n - 1$ with $W(\sigma, r) = \{z \in \mathfrak{H}^n\colon \mu(\sigma, z) \geqslant r\}$.

In order to describe these, let $\Lambda = \{(\lambda_1, \ldots, \lambda_n)\colon \lambda_i \in \mathbb{R}_+, \mathrm{N}(\lambda) = 1\}$ act on $\mathbb{R}^n$ by component-wise multiplication. The semi-direct product $E = \mathbb{R}^n \rtimes \Lambda$ acts on $\mathfrak{H}^n$ by

$$z_i \mapsto \lambda_i z_i + \mu_i\,.$$

The boundary $\partial W(\infty,r)$ is a principal homogeneous space for this semi-direct product. We view $g\Gamma_\sigma g^{-1}\backslash\partial W(\infty,r)$ as the quotient of E by the discrete subgroup $g\Gamma_\sigma g^{-1}$. It is a n-torus bundle over $V\backslash\Lambda$ with fibre $\mathbb{R}^n$ modulo the translations in $g\Gamma_\sigma g^{-1}$.

The manifold with boundary S_r is homotopically equivalent to $\Gamma_K\backslash\mathfrak{H}^n$. It will serve us later.

We have $\Gamma_K\backslash\mathfrak{H}^n = S_r \bigcup_{\partial S_r}(\partial S_r \times [0,\infty))$, i.e. $\Gamma_K\backslash\mathfrak{H}^n$ is topologically a manifold with h "ends" of the form (T^n-bundle over T^{n-1}) $\times [0,\infty)$.

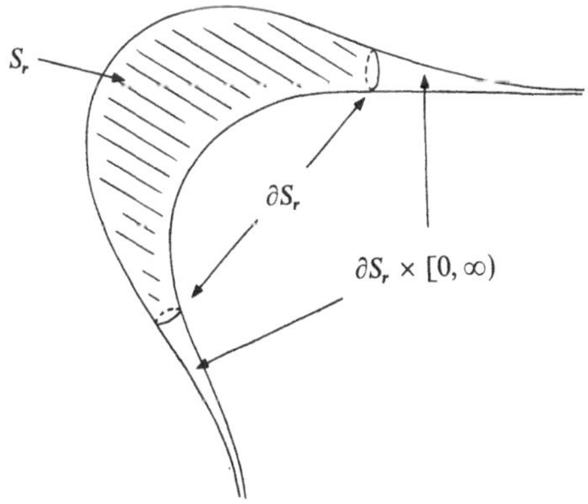

4. The Hurwitz-Maass Extension

Though the full Hilbert modular group Γ_K will play a central role in our book there are some related groups of equal interest. According to a theorem of *A*. Selberg (see [142] for a sketch of proof) a discrete irreducible subgroup Γ of $\mathrm{PGL}^+(2,\mathbb{R})^n$ for which $\Gamma\backslash\mathfrak{H}^n$ is not compact but has finite volume is conjugate inside $\mathrm{PGL}^+(2,\mathbb{R})^n$ to a group which is commensurable with a Hilbert modular group Γ_K (i.e. there exists a g such that $g\Gamma g^{-1}\cap\Gamma_K$ is of finite index in both Γ and Γ_K.) The class of subgroups of $\mathrm{PGL}^+(2,\mathbb{R})^n$ commensurable with Γ_K was determined by Helling [69].

Let $U_K = \mathfrak{o}_K^*$ be the group of units and U_K^+ the subgroup of totally positive units.

We define the *extended Hilbert modular group* $\hat{\Gamma}_K$ by

$$\hat{\Gamma}_K = \{\alpha\in\mathrm{GL}(2,\mathfrak{o}_K)\colon \det\alpha\in U_K^+\}\Big/\left\{\begin{pmatrix}\varepsilon & 0\\ 0 & \varepsilon\end{pmatrix}\colon \varepsilon\in U_K\right\}.$$

The map $\alpha\mapsto\det\alpha$ gives rise to an exact sequence

$$1\to\Gamma_K\to\hat{\Gamma}_K\to U_K^+/U_K^2\to 1\,.$$

For the special case of a real quadratic field one has

$$\#(U_K^+/U_K^2) = \begin{cases} 1 & K \text{ possesses a unit of norm } -1\,.\\ 2 & \text{otherwise}\,.\end{cases}$$

A direct generalization of the Hilbert modular group Γ_K is obtained by taking the automorphism group of a projective $\mathfrak{o}_K$-module of rank 2 modulo its center. Any projective $\mathfrak{o}_K$-module of rank 2 is isomorphic to a module of the form $\mathfrak{a} \oplus \mathfrak{b}$ with $\mathfrak{a}$ and $\mathfrak{b}$ fractional ideals of K. Its isomorphism class only depends on the ideal class of $\mathfrak{a}\mathfrak{b}$ in $\mathscr{C}\ell(K)$. In particular $\mathfrak{a} \oplus \mathfrak{b} \cong \mathfrak{o}_K \oplus \mathfrak{a}\mathfrak{b}$. So it is natural to consider the groups $\mathrm{PSL}(\mathfrak{o}_K \oplus \mathfrak{a})$ and $\mathrm{PGL}^+(\mathfrak{o}_K \oplus \mathfrak{a})$. To be explicit, consider the maximal order

$$\begin{pmatrix} \mathfrak{o}_K & \mathfrak{a}^{-1} \\ \mathfrak{a} & \mathfrak{o}_K \end{pmatrix} \tag{1}$$

in the matrix algebra $M_2(K)$. Then $\mathrm{SL}(\mathfrak{o}_K \oplus \mathfrak{a})$ (resp. $\mathrm{GL}^+(\mathfrak{o}_K \oplus \mathfrak{a})$) consists of the matrices from this order with determinant 1 (resp. determinant in U_K^+). Their images in $\mathrm{PGL}^+(2, K)$ are denoted by $\Gamma(\mathfrak{o}_K \oplus \mathfrak{a})$ (resp. $\hat{\Gamma}(\mathfrak{o}_K \oplus \mathfrak{a})$). In particular, $\Gamma_K = \Gamma(\mathfrak{o}_K \oplus \mathfrak{o}_K)$, $\hat{\Gamma}_K = \hat{\Gamma}(\mathfrak{o}_K \oplus \mathfrak{o}_K)$. In the sequel we often shall delete the index K of $\mathfrak{o}_K$ and just write $\Gamma(\mathfrak{o} \oplus \mathfrak{a})$, etc.

Two maximal orders of the form (1) in $M_2(K)$ are conjugate (by an inner automorphism) if and only if the corresponding ideals, $\mathfrak{a}_1$ and $\mathfrak{a}_2$ say, belong to the same genus (that is, $\mathfrak{a}_2 = \lambda\mathfrak{b}^2\mathfrak{a}_1$ for some totally positive $\lambda \in K$ and an ideal $\mathfrak{b}$). In fact, let M be an element of

$$\begin{pmatrix} \mathfrak{b} & \mathfrak{a}_1^{-1}\mathfrak{b}^{-1} \\ \mathfrak{a}_1\mathfrak{b} & \mathfrak{b}^{-1} \end{pmatrix}$$

with determinant 1. Then

$$\begin{pmatrix} \lambda^{-1} & 0 \\ 0 & 1 \end{pmatrix} M^{-1}\mathrm{SL}(\mathfrak{o} \oplus \mathfrak{a}_1)M \begin{pmatrix} \lambda & 0 \\ 0 & 1 \end{pmatrix} = \mathrm{SL}(\mathfrak{o} \oplus \mathfrak{a}_2)\,.$$

This gives explicitly the equivalence of $\Gamma(\mathfrak{o} \oplus \mathfrak{a}_1)$ and $\Gamma(\mathfrak{o} \oplus \mathfrak{a}_2)$ as transformation groups of $\mathfrak{H}^n$. We may therefore restrict ourselves to the groups $\Gamma(\mathfrak{o} \oplus \mathfrak{a})$ where $\mathfrak{a}$ runs through a set of representatives of all genera of K. We stress that the groups $\Gamma(\mathfrak{o} \oplus \mathfrak{a})$ and $\hat{\Gamma}(\mathfrak{o} \oplus \mathfrak{a})$ are of equal importance with the usual Hilbert modular group Γ_K and its extension $\hat{\Gamma}_K$ (which is the case: $\mathfrak{a}$ in the principal genus), see Sec. 7.

We now describe the maximal discrete subgroups of $\mathrm{PGL}^+(2, \mathbb{R})^n$ containing the generalized Hilbert modular groups $\Gamma(\mathfrak{o} \oplus \mathfrak{a})$.

Let $\mathscr{C}\ell^+(K)$ be the *narrow ideal class group* of K. Recall that two ideals $\mathfrak{a}_1$ and $\mathfrak{a}_2$ of K are said to be *equivalent in the narrow sense* (or *strictly equivalent*) if there exists a totally positive element $\lambda \in K$ such that $\lambda\mathfrak{a}_1 = \mathfrak{a}_2$. Notation: $\mathfrak{a}_1 \approx \mathfrak{a}_2$.

Let $\mathfrak{g}_1, \ldots, \mathfrak{g}_r$ be integral ideals in $\mathfrak{o}$ representing the r elements of ker Sq, where

$$\mathrm{Sq}\colon \mathscr{C}\ell(K) \to \mathscr{C}\ell^+(K)$$

is the homomorphism given by squaring. Consider then the subset of $M_2(K)$

$$\bigcup_{i=1}^{r} \left\{\alpha \in \mathfrak{g}_i \begin{pmatrix} \mathfrak{o}_K & \mathfrak{a}^{-1} \\ \mathfrak{a} & \mathfrak{o}_K \end{pmatrix} : \det\alpha \gg 0,\ (\det\alpha) = \mathfrak{g}_i^2 \right\}.$$

The image of this set of matrices in $\mathrm{PGL}^+(2, K)$ is a discrete subgroup of $\mathrm{PGL}^+(2, \mathbb{R})^n$, called the *Hurwitz-Maass extension* of $\Gamma(\mathfrak{o} \oplus \mathfrak{a})$. *Notation*: $\Gamma_m(\mathfrak{o} \oplus \mathfrak{a})$. It contains $\Gamma(\mathfrak{o} \oplus \mathfrak{a})$ as a normal subgroup.

By associating to α the ideal class $[\mathfrak{g}_i]$ we get a surjective homomorphism

$$\pi\colon \Gamma_m(\mathfrak{o}\oplus\mathfrak{a})/\Gamma(\mathfrak{o}\oplus\mathfrak{a})\to\ker\mathrm{Sq}\subset\mathscr{C}\ell(K)$$

with kernel $\hat{\Gamma}(\mathfrak{o}\oplus\mathfrak{a})/\Gamma(\mathfrak{o}\oplus\mathfrak{a})$. So the index is given by

$$[\Gamma_m:\Gamma] = \#U_K^+/U_K^2\cdot\#\ker\mathrm{Sq}\{\mathscr{C}\ell(K)\to\mathscr{C}\ell^+(K)\}$$

Note that Γ_m/Γ is a group of type $(2,\ldots,2)$. Since

$$\begin{aligned}\#\ker\{\mathscr{C}\ell^+\to\mathscr{C}\ell\} &= \#(K^*/K_+^*U_K) = 2^n/[U_K:U_K^+] = [U_K:U_K^2]/[U_K:U_K^+]\\ &= [U_K^+:U_K^2]\end{aligned}$$

we find

$$[\Gamma_m:\Gamma] = \#\ker\{\mathrm{Sq}^+\colon\mathscr{C}\ell^+(K)\to\mathscr{C}\ell^+(K)\}\,.$$

The Hurwitz-Maass extension $\Gamma_m(\mathfrak{o}\oplus\mathfrak{a})$ is the *maximal discrete subgroup of* $\mathrm{PGL}^+(2,\mathbb{R})^n$ *containing* $\hat{\Gamma}(\mathfrak{o}\oplus\mathfrak{a})$. This was proved by Maass [111] for $\mathfrak{a}=\mathfrak{o}$ (cf. also Hurwitz's Werke) and by Helling in the general case [69].

The Hurwitz-Maass extension $\Gamma_m(\mathfrak{o}\oplus\mathfrak{a})$ acts properly discontinuously on $\mathfrak{H}^n$ and this action induces an action of Γ_m/Γ on $\Gamma(\mathfrak{o}\oplus\mathfrak{a})\backslash\mathfrak{H}^n$.

In the particular case that K is a real quadratic field with discriminant D we have $\#\ker\mathrm{Sq}^+ = 2^{t-1}$ with t the number of distinct primes dividing D, by genus theory, hence $[\Gamma_m:\Gamma] = 2^{t-1}$.

The subgroups of order 2 of Γ_m/Γ arise then in the following way. Let $D = D_1\cdot D_2$ be a decomposition of D into two fundamental discriminants. Let $\Gamma_{D_1,D_2}(\mathfrak{o}_K\oplus\mathfrak{a})$ be the subgroup of $\Gamma_m(\mathfrak{o}_K\oplus\mathfrak{a})$ arising from the set of matrices

$$\left\{\alpha\in\sqrt{(D_1)}\begin{pmatrix}\mathfrak{o}_K & \mathfrak{a}^{-1}\\ \mathfrak{a} & \mathfrak{o}_K\end{pmatrix}\colon(\det\alpha) = (D_1)\right\},$$

where $\sqrt{(D_1)}$ is an ideal whose square equals (D_1). Interchanging D_1 and D_2 does not affect the definition of Γ_{D_1,D_2}. Then $\Gamma_{D_1,D_2}/\Gamma$ is cyclic of order 2 unless the decomposition is $D = 1\cdot D$ or $D = -4\cdot -d$ in which case $\Gamma_{D_1,D_2} = \Gamma$.

We shall use the following notation. $\Gamma_{D_1,D_2}(\mathfrak{o}_K\oplus\mathfrak{a})$ is denoted by $\Gamma_w(\mathfrak{o}_K\oplus\mathfrak{a})$ or simply Γ_w, where w is the square free part of $|D_1|$. If $\tilde{w}$ denotes the square-free part of D/w, then

$$\Gamma_{w_1} = \Gamma_{w_2}\text{ if and only if }w_1 = w_2\text{ or }w_1 = \tilde{w}_2\,.$$

Note that

$$\Gamma_m = \bigcup_{w|D}\Gamma_w\,,\qquad \Gamma_m/\Gamma = \bigcup_{w|D}\Gamma_w/\Gamma\,.$$

If $\mathfrak{a}_1$ and $\mathfrak{a}_2$ are two ideals from the same genus, then the choice of a $\lambda\in K$, $\lambda\gg 0$, an ideal $\mathfrak{b}$ such that $\mathfrak{a}_2 = \lambda\mathfrak{b}^2\mathfrak{a}_1$ and of a determinant 1 matrix M from $\begin{pmatrix}\mathfrak{b} & \mathfrak{a}_1^{-1}\mathfrak{b}^{-1}\\ \mathfrak{a}_1\mathfrak{b} & \mathfrak{b}^{-1}\end{pmatrix}$ determines an isomorphism

$$\begin{aligned}\Gamma(\mathfrak{o}\oplus\mathfrak{a}_2)\backslash\mathfrak{H}^n &\xrightarrow{\sim}\Gamma(\mathfrak{o}\oplus\mathfrak{a}_1)\backslash\mathfrak{H}^n\\ z&\mapsto M(\lambda z)\,.\end{aligned}$$

Two such isomorphisms differ by an element of Γ_m/Γ.

The group $\Gamma_m(\mathfrak{o} \oplus \mathfrak{a})$ also acts on $\mathbb{P}^1(K) = K \cup \{\infty\} \subset \mathbb{P}^1(\mathbb{C})^n$. Under the action of $\Gamma(\mathfrak{o} \oplus \mathfrak{a})$ there are h orbits, the *cusps* ($h = \#\mathscr{C}\ell(K)$). Each cusp corresponds in a natural way to an ideal class of K: if $(\alpha : \beta)$ represents a cusp then the corresponding ideal class is the class of $\alpha\mathfrak{o} + \beta\mathfrak{a}^{-1}$. The group $\Gamma_m(\mathfrak{o} \oplus \mathfrak{a})/\Gamma(\mathfrak{o} \oplus \mathfrak{a})$ acts on the set of cusps of $\Gamma(\mathfrak{o} \oplus \mathfrak{a})$. An element corresponding to $\mathfrak{g} \in \ker \mathrm{Sq}$ sends a cusp belonging to the ideal class of $\mathfrak{c}$ to the class of $\mathfrak{g}\mathfrak{c}$. These facts can be checked directly. The group $\hat{\Gamma}(\mathfrak{o} \oplus \mathfrak{a})$ fixes the cusps of $\Gamma(\mathfrak{o} \oplus \mathfrak{a})$. Hence $\Gamma_m(\mathfrak{o} \oplus \mathfrak{a})$ possesses $\# \operatorname{Im}\{\mathrm{Sq}: \mathscr{C}\ell(K) \to \mathscr{C}\ell^+(K)\}$ cusps.

If K is a real quadratic field then $\Gamma_m(\mathfrak{o} \oplus \mathfrak{a})$ has $h/2^{t-1}$ (resp. $h/2^{t-2}$) cusps with t the number of distinct primes dividing the discriminant D of K if K has a unit of norm -1 (resp. if K does not have such a unit), so the number of cusps is $h^+/2^{t-1}$ with $h^+ = \#\mathscr{C}\ell^+$. Each cusp corresponds to a strict ideal class in the genus of $\mathfrak{a}$.

The isotropy group of the cusp of $\Gamma(\mathfrak{o} \oplus \mathfrak{a})$ given by $(\alpha : \beta) \in \mathbb{P}^1(K)$ with α, β integral and coprime equals (set $\mathfrak{c} = \alpha\mathfrak{o} + \beta\mathfrak{a}^{-1}$)

$$\left\{\begin{pmatrix} \varepsilon & \mu \\ 0 & \varepsilon^{-1} \end{pmatrix} : \varepsilon \in U_K, \mu \in \mathfrak{a}^{-1}\mathfrak{c}^{-2}\right\} \Big/ \{+1, -1\}.$$

For $\hat{\Gamma}(\mathfrak{o} \oplus \mathfrak{a})$ it equals

$$\left\{\begin{pmatrix} \varepsilon & \mu \\ 0 & 1 \end{pmatrix} : \varepsilon \in U_K^+, \mu \in \mathfrak{a}^{-1}\mathfrak{c}^{-2}\right\}.$$

We call the pair $(\mathfrak{a}^{-1}\mathfrak{c}^{-2}, U_K^2)$ (resp. $(\mathfrak{a}^{-1}\mathfrak{c}^{-2}, U_k^+)$) the *type of the cusp* as we did for Γ_k.

The full biholomorphic automorphism group $\mathscr{G}$ of $\mathfrak{H}^n$ is a semi-direct product of $\mathfrak{S}_n$, the symmetric group on the n factors of $\mathfrak{H}^n$ with $\mathrm{PGL}^+(2, \mathbb{R})$. If the field K is Galois over $\mathbb{Q}$ we have an injective homomorphism $\mathrm{Gal}(K/\mathbb{Q}) \to \mathfrak{S}_n$ determined by the action of $\mathrm{Gal}(K/\mathbb{Q})$ on the n real embeddings of K. In this case $\mathfrak{S}_n$ acts as a group of automorphisms on $\Gamma_K \backslash \mathfrak{H}^n$.

If K is a real quadratic field and $\Gamma = \Gamma(\mathfrak{o} \oplus \mathfrak{a})$ with $\mathrm{N}(\mathfrak{a}) = A$ we define an involution $\tau: \Gamma\backslash\mathfrak{H}^2 \to \Gamma\backslash\mathfrak{H}^2$ by the action on $\mathfrak{H}^2$:

$$(z_1, z_2) \mapsto (-1/Az_2, -1/Az_1).$$

This extends to an action of τ on the cusps $\Gamma\backslash\mathbb{P}^1(K)$. The group $\Gamma \cup \tau\Gamma$ is a discrete extension of Γ inside $\mathscr{G}$. We call it a *symmetric Hilbert modular group*. The group $\Gamma_m \cup \tau\Gamma_m$ with $\Gamma_m = \Gamma_m(\mathfrak{o} \oplus \mathfrak{a})$ is the maximal discrete extension of $\Gamma(\mathfrak{o} \oplus \mathfrak{a})$ inside $\mathscr{G}$.

In Section 7 we give the adelic description of the quotient of the Hilbert modular group. The companion groups $\Gamma(\mathfrak{o} \oplus \mathfrak{a})$ then all appear in a natural way at the same time.

5. Elliptic Fixed Points

The Hilbert modular groups for a totally real field K of degree n over $\mathbb{Q}$ in general do not act freely on $\mathfrak{H}^n$ and this section deals with this phenomenon.

As we saw in Section 3 the action of $\Gamma(\mathfrak{o}_K \oplus \mathfrak{a})$ is free on the interior of F_σ for each cusp σ. Therefore, the points $z \in \mathfrak{H}^n$ with a non-trivial isotropy group must lie on the boundary of F_σ.

An element $(\alpha_1, \ldots, \alpha_n) \in \mathrm{GL}^+(2, \mathbb{R})^n$ is called *elliptic* if

$$\mathrm{tr}(\alpha_i)^2 - 4\det(\alpha_i) < 0 \qquad i = 1, \ldots, n \,. \tag{1}$$

An elliptic element of $\mathrm{GL}^+(2, \mathbb{R})^n$ has precisely one fixed point in $\mathfrak{H}^n$.

A point $z \in \mathfrak{H}^n$ is called an elliptic point for a discrete subgroup G of $\mathrm{GL}^+(2, \mathbb{R})^n$ if it occurs as a fixed point of an elliptic element of G. For $z \in \mathfrak{H}^n$ let

$$G_z = \{g \in G\colon g(z) = z\} \,,$$

the isotropy group of z.

Let us now fix a group G, say $G = GL^+(\mathfrak{o}_K \oplus \mathfrak{a})$. Assume $\alpha \in G$. If the polynomial $X^2 - \mathrm{tr}(\alpha)X + \det\alpha \in K[X]$ is irreducible then we can consider $K(\alpha)$ as a quadratic field extension of K, where we consider K to be embedded diagonally in the algebra $M_2(K)$ of 2×2 matrices over K.

If α is elliptic then (1) implies that $L = K(\alpha)$ is a totally imaginary quadratic extension of K. To the fixed point $z \in \mathfrak{H}^n$ of α we associate an order $\mathfrak{O}_z$ of this field L:

$$\mathfrak{O}_z = L \cap \begin{pmatrix} \mathfrak{o}_K & \mathfrak{a}^{-1} \\ \mathfrak{a} & \mathfrak{o}_K \end{pmatrix}.$$

The isotropy group of z in $\bar{G} = \mathrm{PGL}^+(\mathfrak{o}_K \oplus \mathfrak{a})$ is

$$\{\gamma \in \mathfrak{O}_z^*\colon \det\gamma \gg 0\}/U_K \,. \tag{2}$$

It is finite since G acts properly discontinuously and it is cyclic since any finite subgroup of the multiplicative group of a field is cyclic.

Any order $\mathfrak{O}$ in a totally imaginary quadratic extension of K contained in $M_2(K)$ for which $\{\gamma \in \mathfrak{O}^*\colon \det\gamma \gg 0\}/U_K$ is non-trivial determines an elliptic fixed point of $\mathfrak{H}^n$. Two elliptic points $z, z' \in \mathfrak{H}^n$ define the same point of $G\backslash\mathfrak{H}^n$ if and only if there exists a $g \in G$ such that

$$g\mathfrak{O}_{z'}g^{-1} = \mathfrak{O}_z \,.$$

It is well-known that if $\mathfrak{O}_1$ and $\mathfrak{O}_2$ are two isomorphic orders contained in $M_2(K)$ then there exists an invertible element $\beta \in M_2(K)$ such that $\beta\mathfrak{O}_1\beta^{-1} = \mathfrak{O}_2$. It follows that the number $a(\mathfrak{O})$ of G-inequivalent elliptic fixed points z of $\mathfrak{H}^n$ whose order $\mathfrak{O}_z$ is isomorphic to a given order $\mathfrak{O}$ is the number of G-inequivalent orders of the form $\beta\mathfrak{O}\beta^{-1}$, provided that such an $\mathfrak{O}$ occurs. A formula for this number was given by Shimizu [153].

$$a(\mathfrak{O}) = \frac{2^{n-1}}{[U_K^+ : U_K^2][\mathfrak{O}^* : \mathfrak{O}_+^*]} \cdot \frac{h(\mathfrak{O})}{h}$$

with $\mathfrak{O}_+^* = \{\gamma \in \mathfrak{O}^*\colon \det\gamma \gg 0\}$, $h(\mathfrak{O})$ the class number of $\mathfrak{O}$ and h the class number of K. We refer to Prestel [129, 130] for a detailed proof and to Vignéras [167] for results concerning related groups. Prestel also calculates the type of action of the isotropy groups in the following sense: let $\alpha = \begin{pmatrix} a & b \\ c & d \end{pmatrix} \in \mathrm{GL}^+(2, \mathbb{R})^n$ be an elliptic element with fixed point $z = (z_1, \ldots, z_n)$

$$z_i = \frac{a_i - d_i}{2c_i} + \frac{1}{2|c_i|}\sqrt{\operatorname{tr}(\alpha_i)^2 - 4\det(\alpha_i)}\,.$$

By the transformation $\zeta_i \to \dfrac{\zeta_i - z_i}{\zeta_i - \bar{z}_i}$ the isotropy group $\bar{G}_z$ goes over into a group of rotations around the origin of an n-fold product of unit discs. Each element is determined by its *rotation factor* $r = (r_1, \ldots, r_n)$ with

$$r_i = e^{2i\theta_j}\,, \qquad \cos\theta_j = \frac{\operatorname{tr}(\alpha_j)}{2\sqrt{\det(\alpha_j)}}\,, \qquad c_j \sin\theta_j > 0\,.$$

According to Prestel the number of G-inequivalent elliptic fixed points with order isomorphic to $\mathfrak{O}$ and a given rotation factor is the same for all rotation factors that occur. Moreover, note that r is determined by $t = \operatorname{tr}(\alpha)$ and $d = \det(\alpha)$ up to the choice of signs

$$r \to (r_1^{\varepsilon_1}, \ldots, r_n^{\varepsilon_n})\,, \qquad \varepsilon_j = \pm 1\,, \qquad j = 1, \ldots, n\,.$$

Prestel shows that all signs ε_j occur if $K(\alpha)/K$ is ramified, while for $K(\alpha)/K$ unramified the signs ε_j that occur satisfy

$$\prod \varepsilon_i = \left(\frac{K(\alpha)}{\mathfrak{a}}\right).$$

with $\left(\dfrac{K(\alpha)}{\mathfrak{a}}\right)$ the Artin-symbol.

If K is a real quadratic field and $G = \mathrm{SL}(\mathfrak{o}_K \oplus \mathfrak{a})$ then α is elliptic only if $\operatorname{tr}(\alpha) \ll 4\det\alpha = 4$, so trace α is restricted to the following possibilities

$$0,\ \pm 1,\ \pm\sqrt{2},\ \pm\sqrt{3},\ \pm\frac{1 \pm \sqrt{5}}{2}$$

and the order of α with such a trace as element of $\mathrm{PSL}(\mathfrak{o}_K \oplus \mathfrak{a})$ equals 2, 3, 4, 6, 5 respectively. We see that if the discriminant D of K is different from 5, 8 and 12 then only order 2 and order 3 occur. For $D = 5$, 8 and 12 elliptic elements with order 5, 4 and 6 dȯ occur. The elliptic fixed points with isotropy group in $\mathrm{PSL}(\mathfrak{o}_K \oplus \mathfrak{a})$ of order 3 fall into two types: the isotropy group is generated by an element α with rotation factor (ζ_3, ζ_3) or (ζ_3, ζ_3^{-1}), where ζ_3 is a primitive third root of 1. Those of order 2, of course, have rotation factor $(-1, -1)$. Prestel gives formulas for the numbers of inequivalent fixed points of all three kinds; these formulas involve the class numbers of $\mathbb{Q}(\sqrt{-D})$ and $\mathbb{Q}(\sqrt{-3D})$. One can construct tables using these formulas (cf. [129, 130, 102]). In Chapter V we shall describe a different method. We shall give formulas and tables for $\mathrm{PSL}(\mathfrak{o}_K \oplus \mathfrak{a})$, $\hat{\Gamma}(\mathfrak{o}_K \oplus \mathfrak{a})$ and $\Gamma_m(\mathfrak{o}_K \oplus \mathfrak{a})$ on page 267.

We give a few formulas to illustrate the results. Let $h(\Delta)$ be the class number of the (imaginary) quadratic field with discriminant Δ. Then we have for $K = \mathbb{Q}(\sqrt{D})$,

$D > 12$ and $\Gamma = \Gamma_K$:

$$a_2(\Gamma) = \begin{cases} h(-4D) & D \equiv 1 \bmod 4 \\ 3h(-D) & D/4 \equiv 2 \bmod 4 \\ 10h(-D/4) & D/4 \equiv 3 \bmod 8 \\ 4h(-D/4) & D/4 \equiv 7 \bmod 8 \end{cases}$$

If a_3^+ (resp. a_3^-) denotes the number of quotient singularities of order 3 with rotation factor (ζ_3, ζ_3) (resp. (ζ_3, ζ_3^{-1})) then we have

$$a_3^+(\Gamma) = \begin{cases} \frac{1}{2}h(-3D) & D \not\equiv 0 \bmod 3 \\ 4h(-D/3) & D \equiv 3 \bmod 9 \\ 3h(-D/3) & D \equiv 6 \bmod 9 \end{cases}$$

$$a_3^-(\Gamma) = \begin{cases} \frac{1}{2}h(-3D) & D \not\equiv 0 \bmod 3 \\ h(-D/3) & D \equiv 3 \bmod 9 \\ 0 & D \equiv 6 \bmod 9\,. \end{cases}$$

6. Hilbert Modular Forms

In this section K denotes a totally real number field of degree n over $\mathbb{Q}$ and Γ a subgroup of finite index of $\Gamma_m(\mathfrak{o} \oplus \mathfrak{a})$.

(6.1) Definition. A holomorphic function $f\colon \mathfrak{H}^n \to \mathbb{C}$ $(n > 1)$ is called a *Hilbert modular form* of weight $\mathbf{k} = (k_1, \ldots, k_n) \in \mathbb{Z}^n$ on Γ if for all $\gamma = \begin{pmatrix} a & b \\ c & d \end{pmatrix} \in \Gamma$ one has

$$f(\gamma z) = \left(\prod_{i=1}^{n} (c^{(i)} z_i + d^{(i)})^{k_i} (\det \gamma^{(i)})^{-k_i/2} \right) f(z)\,. \tag{1}$$

For $n = 1$ one has to add the condition that f is holomorphic at the cusps, see [150]. Instead of $\prod_{i=1}^n (c^{(i)} z_i + d^{(i)})^{k_i}$ we shall often write $(cz + d)^{\mathbf{k}}$. If we introduce for $\gamma \in \mathrm{GL}^+(K)$ the operator $\cdot|_{\mathbf{k}}\gamma$

$$f|_{\mathbf{k}}\gamma = \left(\prod_{i=1}^{n} (\det \gamma^{(i)})^{k_i/2} (c^{(i)} z_i + d^{(i)})^{-k_i} \right) f(\gamma z)$$

then (1) says $f|_{\mathbf{k}}\gamma = f$ for all $\gamma \in \Gamma$. If $\mathbf{k}$ has the form $(k, \ldots, k)$ for some $k \in \mathbb{Z}$, we speak of a Hilbert modular form of weight k.

Consider now the cusp ∞ at infinity. A Hilbert modular form f of weight $\mathbf{k}$ on Γ is invariant under all translations $z \to z + \mu$ in the isotropy group of this cusp. Therefore, if the type of the cusp is (M, V), then f has a Fourier expansion:

$$f = \sum_{\nu \in M^\vee} a_\nu \exp(2\pi i \operatorname{Tr} \nu z)\,, \qquad (\operatorname{Tr} \nu z = \sum \nu^{(i)} z_i) \tag{2}$$

where $M^\vee = \{\lambda \in K\colon \operatorname{Tr} \lambda\mu \in \mathbb{Z}\ \forall \mu \in M\}$. Note that if $\mathbf{k} = (k,\ldots,k)$ then $a_\nu = a_{\nu\varepsilon}$ for all $\varepsilon \in V$. We claim (for $n > 1$)

$$a_\nu \neq 0 \Rightarrow \nu = 0 \quad \text{or} \quad \nu \gg 0\,.$$

Indeed, suppose $a_\nu \neq 0$ and $\nu^{(1)} < 0$ (say). We can find a unit $\varepsilon \gg 0$ with $\varepsilon^{(1)} > 1$, $\varepsilon^{(i)} < 1\ i = 2,\ldots,n$ and with $\operatorname{Tr} \varepsilon\nu < 0$. But then the terms in (2) with $\nu = \nu_0 \varepsilon^m$ ($m \in \mathbb{Z}$) and $z = (i,\ldots,i)$ tend to infinity as $m \to \infty$, contradicting the convergence of the series. This shows that the condition of holomorphicity at the cusps is superfluous for $n > 1$, in contrast with the case $n = 1$ (*Koecher principle*).

Now let $\sigma \in \mathbb{P}^1(K)$ be an arbitrary cusp and choose $\gamma \in \mathrm{GL}^+(2,K)$ with $\gamma\sigma = \infty$. We then can expand $f|_{\mathbf{k}}\gamma$ in a Fourier series (2).

(6.2) Definition. A Hilbert modular form f of weight $\mathbf{k}$ is a *cusp form* if the constant term a_0 vanishes in the Fourier series of $f|_{\mathbf{k}}\gamma$ for all $\gamma \in \mathrm{GL}^+(2,K)$.

The requirements in the definition in fact amount to a finite number of conditions. We denote by $M_{\mathbf{k}}(\Gamma)$ (resp. $S_{\mathbf{k}}(\Gamma)$) the vector space of all Hilbert modular forms of weight $\mathbf{k}$ (resp. cusp forms of weight $\mathbf{k}$) on Γ.

(6.3) Lemma. *If $M_{\mathbf{k}}(\Gamma) \neq (0)$, then either $\mathbf{k} = (0,\ldots,0)$ or all k_i are positive. Moreover, $M_{\mathbf{0}}(\Gamma) = \mathbb{C}$, $S_{\mathbf{0}}(\Gamma) = (0)$.*

Proof. (i) If $f \in M_{\mathbf{k}}(\Gamma)$, $f \neq 0$, then $(f|_{\mathbf{k}}\gamma)(z,\ldots,z)$ with $z \in \mathfrak{H}$, $\gamma \in \mathrm{GL}^+(2,K)$, is an elliptic modular form of weight $\sum k_i$, and it is non-zero for suitable γ (since the images of the diagonal under $\mathrm{GL}^+(2,K)$ are dense in $\mathfrak{H}^n$). Hence $\sum k_i \geqslant 0$ and $\Sigma\, k_i = 0$ if and only if f is constant.

(ii) Now assume $\sum k_i > 0$ and that some k_i, say k_1, is zero. If ε is a unit we have $f(\varepsilon z) = (\prod_{i=2}^n (\varepsilon^{(i)})^{k_i}) f(z)$, hence $|a_0| = (\prod_{i=2}^n (\varepsilon^{(i)})^{k_i})|a_0|$. Therefore, (choosing ε suitably) we find that $a_0 = 0$. This works for all cusps. Let $g(z) = (\prod_{i=2}^n y_i^{k_i/2}) f(z)$. Then g is holomorphic in z_1 and $|g|$ is invariant under Γ. Moreover, $|g|$ vanishes in the cusps.

Hence $|g|$ is bounded on $\mathfrak{H}^n$, so by Liouville's theorem g must be constant as a function of z_1, so $|g(z_1, \alpha^{(2)} z_2, \ldots, \alpha^{(n)} z_n)| = |g(z)|\ \forall \alpha \in \Gamma$. But this action of Γ on $\mathfrak{H}^{n-1}$ (last $n-1$ coordinates) is no longer discontinuous; in fact, the orbit of each point $z' \in \mathfrak{H}^{n-1}$ under Γ is dense. Hence $|g|$ is constant, even zero because it is zero at the cusps. This shows that f is zero.

(iii) Finally, assume $\sum k_i > 0$ and that some k_i, say k_1, is negative. As we shall see shortly there are non-zero modular forms with all weights positive. Replacing f by a power of f times a non-zero form with positive weights we find a form with $k_1 = 0$ and $\sum k_i > 0$. This contradicts (ii). □

If $f, g \in M_{\mathbf{k}}(\Gamma)$ and one of them is a cusp form then we put

$$\langle f,g \rangle = \int_{\Gamma\backslash\mathfrak{H}^n} f(z)\overline{g(z)}\, y^{\mathbf{k}}\, d\mu \qquad (y^{\mathbf{k}} = \prod y_i^{k_i})$$

with $d\mu = \prod_{i=1}^{n} \frac{dx_i \wedge dy_i}{y_i^2}$. This defines an inner product on $S_{\mathbf{k}}(\Gamma)$, the *Petersson product*. It is clear that $\langle f, f \rangle > 0$ for $f \neq 0$, so this makes $S_{\mathbf{k}}(\Gamma)$ into a Hilbert space (finite-dimensional as we shall see later).

Eisenstein series. Let B be a wide ideal class, $\mathfrak{b} \in B$ an ideal. We set for $k > 2$, k even

$$G_{k,B}(z) = \mathrm{N}(\mathfrak{b})^k \sum_{\substack{(\alpha,\beta) \in \mathfrak{ba} \times \mathfrak{b} \\ (\alpha,\beta) \neq (0,0)}}' \mathrm{N}(\alpha z + \beta)^{-k}.$$

The summation here is taken over all non-associated points $(\alpha, \beta) \neq (0, 0)$ in $\mathfrak{ba} \times \mathfrak{b}$. One checks that $G_{k,B}$ defines a holomorphic function on $\mathfrak{H}^n$ which does not depend on the choice of $\mathfrak{b} \in B$ and is a modular form of weight k on $\Gamma(\mathfrak{o} \oplus \mathfrak{a})$. If $K = \mathbb{Q}$ then $\frac{1}{2} G_{k,B}$ gives the usual Eisenstein series.

If $\alpha \in \mathrm{GL}^+(2, K)$ with $\alpha(\infty) = \sigma$ for a cusp σ associated to the ideal $\mathfrak{c}$ then $G_{k,B}|_{\mathbf{k}}\alpha$ for $G_{k,B} \in M_{\mathbf{k}}(\Gamma(\mathfrak{o} \oplus \mathfrak{a}))$ is a multiple of $G_{k,BC^{-1}} \in M_{\mathbf{k}}(\Gamma(\mathfrak{o} \oplus \mathfrak{ac}^2))$ with C the ideal class of $\mathfrak{c}$. Hence, to know the Fourier expansions of $G_{k,B} \in M_{\mathbf{k}}(\Gamma(\mathfrak{o} \oplus \mathfrak{a}))$ at the various cusps it suffices to know the Fourier expansions at ∞ of the $G_{k,B}$ on the various $\Gamma(\mathfrak{o} \oplus \mathfrak{a})$ where $\mathfrak{a}$ runs through the genera.

To compute the Fourier series one applies the Poisson summation formula. This gives

$$\sum_{\beta \in \mathfrak{b}} \mathrm{N}(\alpha z + \beta)^{-k} = \frac{(2\pi i)^{kn}}{(k-1)!^n} D^{1/2-k} \mathrm{N}(\mathfrak{b}^{-1}) \operatorname{sgn} \mathrm{N}(\alpha^k) \sum_{\substack{\lambda \in \mathfrak{b}^\vee \\ \alpha\lambda \gg 0}} \mathrm{N}(\lambda \mathfrak{d})^{k-1} \exp(2\pi i \operatorname{Tr} \lambda \alpha z)$$

$$(\alpha \neq 0, k \geqslant 2)$$

with $\mathfrak{d}$ the different. Splitting the sum defining $G_{k,B}$ as $(\sum_{\alpha=0} + \sum_{\alpha \neq 0}) \sum_\beta$ then gives the following proposition.

(6.4) Proposition. *The Eisenstein series $G_{k,B}(z)$ on $\Gamma(\mathfrak{o} \oplus \mathfrak{a})$ for $k \geqslant 2$, k even has the following Fourier series at ∞:*

$$G_{k,B}(z) = \zeta_{B^{-1}}(k) + \frac{(2\pi i)^{kn}}{(k-1)!^n} D^{1/2-k} \sum_{\substack{\nu \in \mathfrak{ad}^{-1} \\ \nu \gg 0}} \sigma_{k-1,\mathfrak{d}B}(\mathfrak{a}^{-1}\mathfrak{d}(\nu)) e^{2\pi i \operatorname{Tr} \nu z},$$

where for an ideal class A we have set

$$\zeta_A(k) = \sum_{\substack{\mathfrak{c} \in A \\ \mathfrak{c} \text{ integral}}} \mathrm{N}(\mathfrak{c})^{-k}$$

and

$$\sigma_{k-1,A}(\mathfrak{e}) = \sum_{\substack{\mathfrak{c} \in A \\ \mathfrak{c} \text{ integral} \\ \mathfrak{c} | \mathfrak{e}}} \mathrm{N}(\mathfrak{c})^{k-1} \qquad (\mathfrak{e} \textit{ integral}).$$

Using the functional equation of $\zeta_B(s)$ we can rewrite this as

$$G_{k,B}(z) = \frac{(2\pi i)^{kn}}{(k-1)!^n} D^{1/2-k} [2^{-n}\zeta_{\mathfrak{d}B}(1-k) + \sum \sigma_{k-1,\mathfrak{d}B}(\mathfrak{a}^{-1}\mathfrak{d}(\nu)) e^{2\pi i \operatorname{Tr}(\nu z)}],$$

where the Fourier coefficients of the expression in the square brackets are now all rational numbers (in fact, all rational integers except for the constant term).

Remarks. 1. We can also define $G_{k,B}$ for k odd. However, the definition is no longer independent of $\mathfrak{b} \in B$: replacing $\mathfrak{b}$ by $(\lambda)\mathfrak{b}$ with $\mathrm{N}(\lambda) < 0$ changes the sign of $G_{k,B}$. This function is zero if n is odd, or more generally if K has a unit of norm -1. In general, however, it is not zero.

2. By restricting to the diagonal one gets an explicit elliptic modular form of weight kn. Siegel [159] observed that this can be used to obtain explicit formulas for $\zeta_A(1-k)$. For $\mathfrak{a} = \mathfrak{o}$, $k = 2$, $n = 2$, for instance, we find that

$$G_{2,B}(z,z) = \zeta_{B^{-1}}(2) + \frac{(2\pi i)^4}{D^{3/2}} \sum_{\substack{\nu \in \mathfrak{d}^{-1} \\ \nu \gg 0}} \sigma_{1,\mathfrak{d}B}(\mathfrak{d}(\nu)) e^{2\pi i \operatorname{Tr}(\nu z)}$$

is a modular form of weight 4 on $\mathrm{SL}(2,\mathbb{Z})$, hence a multiple of the Eisenstein series

$$G_4(z) = \zeta(4) + \frac{(2\pi i)^4}{3!} \sum_{n \geq 1} \sigma_3(n) e^{2\pi i n z} \qquad \left(\sigma_r(n) = \sum_{d|n} d^r\right).$$

Comparing coefficients of $e^{2\pi i z}$ and using $\zeta(4) = \pi^4/90$, we find

$$\zeta_{B^{-1}}(2) = \frac{\pi^4}{15 D^{3/2}} \sum_{\substack{\nu \in \mathfrak{d}^{-1} \\ \nu \gg 0, \operatorname{Tr}(\nu) = 1}} \sigma_{1,\mathfrak{d}B}(\mathfrak{d}(\nu))$$

or (replacing B by $B\mathfrak{d}^{-1}$ and using the functional equation)

$$\zeta_B(-1) = \frac{1}{60} \sum_{\substack{\nu \in \mathfrak{d}^{-1} \\ \nu \gg 0, \operatorname{Tr}(\nu) = 1}} \sigma_{1,B}(\mathfrak{d}(\nu))$$

Summing over the ideal classes and using the easy identity

$$\sum_{B \in \mathscr{C}\ell(K)} \sigma_{m,B}(\mathfrak{a}) = \sum_{d|\mathfrak{a}} \left(\frac{D}{d}\right) d^m \sigma_m\left(\frac{\mathrm{N}(\mathfrak{a})}{d^2}\right)$$

and writing ν as $(1 + x/\sqrt{D})/2$ we obtain the following reformulation of a theorem of Siegel ([159], cf. [18, 173]):

(6.5) Theorem (Siegel). *Let K be a real quadratic field with discriminant D. Then*

$$\zeta_K(-1) = \frac{1}{60} \sum_{x \in \mathbb{Z}} \sigma_1\left(\frac{D - x^2}{4}\right),$$

where $\sigma_1(x) = 0$ if $x \notin \mathbb{Z}_{\geq 1}$ and $\sigma_1(x) = \sum_{d|x} d$ if $x \in \mathbb{Z}_{\geq 1}$.

3. For $k = 2$ the series defining $G_{k,B}$ is not absolutely convergent and it is therefore not obvious that $G_{k,B}$ is a Hilbert modular form. Following Hecke, we

replace the definition of $G_{2,B}$ by

$$G_{2,B} = \lim_{s\to 0} \mathsf{N}(\mathfrak{b})^{2+s} \sum \mathsf{N}(\alpha z + \beta)^{-2} |\mathsf{N}(\alpha z + \beta)|^{-s}$$

in this case, cf. [65] and [5]. Then if $n > 1$ the function $G_{2,B}$ is a Hilbert modular form of weight 2 and Proposition (6.4) still holds. For $n = 1$, the function

$$G_{2,[\mathbb{Z}]} = \frac{\pi^2}{6} - \frac{\pi}{2y} - 4\pi^2 \sum_{n=1}^{\infty} \sigma_1(n) e^{2\pi i n z}$$

is not holomorphic.

4. As in the one-dimensional case one can impose a primitivity condition:

$$G^*_{k,B}(z) = \mathsf{N}(\mathfrak{b})^k \sum_{\substack{(\alpha,\beta)\\ \alpha\mathfrak{a}^{-1}+\beta\mathfrak{o}=\mathfrak{b}}}' \mathsf{N}(\alpha z + \beta)^{-k}\,.$$

One thus obtains h Eisenstein series, each of which is non-zero at precisely one cusp. One has $G_{k,B}(z) = \sum_A \zeta_A(k) G^*_{k,AB}(z)$. The functions $G^*_{k,B}$ have rational Fourier coefficients at every cusp.

(6.6) Proposition. *The Fourier coefficients a_ν of a Hilbert modular form of weight $k > 2$ satisfy the growth condition*

$$a_\nu = O(\mathsf{N}(\nu)^{k-1}) \qquad \textit{as } \mathsf{N}(\nu) \to \infty\,.$$

Proof. For the Eisenstein series $G_{k,B}$ this follows from (6.4). By remark 4 we can subtract from any Hilbert modular form a linear combination of Eisenstein series to get a cusp form. If f is a cusp form, then the function

$$\phi(z) = \mathsf{N}(y)^{k/2} |f(z)|$$

is invariant under Γ and tends to 0 as z tends to any cusp, so $\phi(z) \leqslant C$ in $\mathfrak{H}^n$ for some $C \in \mathbb{R}_{>0}$. The ν-th Fourier coefficient of f is given by the integral

$$a_\nu = \frac{e^{-2\pi\,\mathrm{Tr}(\nu y)}}{\mathrm{vol}(\mathbb{R}^n/M)} \int_{\mathbb{R}^n/M} \cdots \int f(x + iy) e^{-2\pi i\,\mathrm{Tr}\,\nu x}\, dx$$

($y \in \mathbb{R}^n_{>0}$ arbitrary), and we find

$$|a_\nu| \leqslant C e^{-2\pi\,\mathrm{Tr}(\nu y)} \mathsf{N}(y)^{-k/2}\,.$$

Choosing $y = \left(\frac{1}{\nu^{(1)}}, \ldots, \frac{1}{\nu^{(n)}}\right)$ gives

$$a_\nu = O(\mathsf{N}(\nu)^{k/2})\,. \qquad \square$$

This argument is due to Hecke. For $k = 2$ the same holds for cusp forms, but we have only

$$a_\nu = O(\mathsf{N}(\nu) \log \mathsf{N}(\nu))$$

for the Eisenstein series.

In the following we consider the L-function of a Hilbert modular form. We make the assumption that the strict class number h^+ of K is one. Without this assumption things become more complicated. To avoid these technicalities one should use the adelic language (cf. I.7, VI.1 or [43]).

If f is a cusp form of weight $\mathbf{k} = (k, \ldots, k)$ on $\Gamma = \mathrm{SL}(2, \mathfrak{o})$ then it has a Fourier expansion

$$f = \sum_{\substack{\nu \in \mathfrak{d}^{-1} \\ \nu \gg 0}} a_\nu e^{2\pi i \operatorname{Tr} \nu z} .$$

We define

$$L(f, s) = \sum_{\substack{\nu \in \mathfrak{d}^{-1}/U^+ \\ \nu \gg 0}} a_\nu \mathrm{N}(\nu\mathfrak{d})^{-s} .$$

Note that $a_{\varepsilon\nu} = a_\nu$ ($\varepsilon \in U^+$), so a_ν depends only on the ideal $(\nu\mathfrak{d})$ of $\mathfrak{o}$. We sometimes write $a(\nu\mathfrak{d})$ instead of a_ν. Then $L(f, s) = \sum_{\mathfrak{m}} a(\mathfrak{m}) N(\mathfrak{m})^{-s}$ where the sum is over all integer non-zero ideals of $\mathfrak{o}$. We put

$$\Lambda(f, s) = D^s (2\pi)^{-ns} \Gamma(s)^n L(f, s) .$$

We have:

$$\mathrm{N}(\nu\mathfrak{d})^{-s} = D^{-s} \left(\frac{(2\pi)^s}{\Gamma(s)} \right)^n \int_0^\infty \cdots \int_0^\infty e^{-2\pi \operatorname{Tr} \nu y} \mathrm{N}(y)^{s-1} \mathrm{N}(dy) ,$$

(where $y_i = \operatorname{Im} z_i$, $i = 1, \ldots, n$, $\mathrm{N}(y) = y_1 \ldots y_n$, $\mathrm{N}(dy) = dy_1 \ldots dy_n$ and $\operatorname{Tr} \nu y = \sum_{i=1}^n y_i \nu^{(i)}$). Hence

$$\Lambda(f, s) = D^{-s} (2\pi)^{ns} \Gamma(s)^{-n} \int_{\mathbb{R}_+^n/U^+} f(iy) \mathrm{N}(y)^{s-1} \mathrm{N}(dy)$$

with $f(iy) = f(iy_1, \ldots, iy_n)$.

The last formula gives the analytic continuation of $\Lambda(f, s)$ to all of the s-plane and the functional equation

$$\Lambda(f, s) = \Lambda(f, k - s)$$

using the transformation $z \to -1/z$.

Still under the assumption $h^+ = 1$ we consider

$$E(z, s) = \sum_{\substack{(\gamma, \delta) \in \mathfrak{o}^2 \bmod U \\ (\gamma, \delta) = 1}} \mathrm{N}(y^s / |\gamma z + \delta|^{2s}) , \qquad (\operatorname{Re}(s) > 1)$$

a non-holomorphic Eisenstein series. We have

$$\begin{aligned} \zeta_K(2s) E(z, s) &= \sum_{(\gamma, \delta) \in \mathfrak{o}^2 \bmod U} \mathrm{N}(y^s / |\gamma z + \delta|^{2s}) \\ &= \zeta_K(2s) \mathrm{N}(y)^s + \mathrm{N}(y)^s \sum_{\substack{\gamma \in \mathfrak{o} \bmod U \\ \gamma \neq 0}} \sum_{\delta \in \mathfrak{o}} \mathrm{N}(|\gamma z + \delta|)^{-2s} \end{aligned}$$

where we separated $\gamma = 0$ and $\gamma \neq 0$ and used the fact that $\{(\gamma, \delta) \in \mathfrak{o}^2 : \gamma \neq 0\} \bmod U = (\{\gamma \in \mathfrak{o} : \gamma \neq 0\} \bmod U) \times \mathfrak{o}$.

(6.7) Lemma. *Let* $K_s(t) = \int_0^\infty e^{-t\cosh x}\cosh(sx)\,dx$. *Then for* $t \neq 0$ *and* $\mathrm{Re}(s) > \frac{1}{2}$

$$\int_{-\infty}^{+\infty} e^{itx}(x^2+1)^{-s}\,dx = 2\sqrt{\pi}\,\Gamma(s)^{-1}|t|/2^{s-1/2}K_{s-1/2}(|t|)\,.$$

Proof. We have

$$\begin{aligned}
\Gamma(s)(|t|/2)^{-s}&\int_{-\infty}^{+\infty} e^{itx}(x^2+1)^{-s}\,dx \\
&= \int_{-\infty}^{+\infty} e^{itx}\left(\int_0^\infty u^{s-1}e^{-|t|(x^2+1)u/2}\,du\right)dx \\
&= \int_0^\infty u^{s-1}\left[\sqrt{\frac{2\pi}{tu}}\,e^{-t(u+u^{-1})/2}\right]du\,.
\end{aligned}$$

Now write $u = e^x$. □

(6.8) Lemma. *Let* $z \in (\mathbb{C}\backslash\mathbb{R})^n$. *Then*

$$\sum_{\mu\in\mathfrak{o}} \mathsf{N}(|z+\mu|)^{-2s} = \sum_{\nu\in\mathfrak{o}^\vee} c_\nu(y)e^{2\pi i\,\mathrm{Tr}\,\nu x}$$

with

$$c_\nu = \begin{cases} D^{-1/2}|\mathsf{N}(y)|^{1/2-s}\Gamma(s)^{-n}2^n\pi^{ns}|\mathsf{N}(\nu)|^{s-1/2}\prod_{i=1}^n K_{s-1/2}(2\pi|\nu^{(i)}|y_i) & \text{if } \nu \neq 0\,, \\ D^{-1/2}|\mathsf{N}(y)|^{1-2s}\pi^{n/2}\Gamma(s)^{-n}\Gamma(s-\frac{1}{2})^n & \text{if } \nu = 0\,. \end{cases}$$

Proof. The left hand side is periodic, so has an expansion

$$\sum_{\nu\in\mathfrak{o}^\vee} c_\nu(y)e^{2\pi i\,\mathrm{Tr}\,\nu y}$$

with

$$\begin{aligned}
c_\nu(y) &= \int_{\mathbb{R}^n/\mathfrak{o}}\cdots\int (\mathrm{LHS})e^{-2\pi i\,\mathrm{Tr}\,\nu x}\mathsf{N}(dx) \\
&= D^{-1/2}\int_{\mathbb{R}^n}\cdots\int e^{-2\pi i\,\mathrm{Tr}\,\nu x}\left(\prod_{j=1}^n (x_j^2+y_j^2)^{-s}\right)dx_1\ldots dx_n \\
&= D^{-1/2}\Gamma(s)^{-n}|\mathsf{N}(y)|^{1/2-s}\pi^{ns}|\mathsf{N}(\nu)|^{s-1/2}\prod_{i=1}^n K_{s-1/2}(2\pi|\nu|^{(i)}y_i)
\end{aligned}$$

if $\nu \neq 0$. If $\nu = 0$ we use

$$\int_{-\infty}^{+\infty}\frac{dx}{(x^2+y^2)^s} = \frac{\pi^{1/2}\Gamma(s-\frac{1}{2})}{\Gamma(s)}|y|^{1-2s}\,.$$ □

Let $\zeta_K^*(s) = \pi^{-ns/2}D^{s/2}\Gamma\left(\frac{s}{2}\right)^n\zeta_K(s)$ (so that the functional equation of ζ_K takes the form $\zeta_K^*(1-s) = \zeta_K^*(s)$) and set $E^*(z,s) = \zeta_K^*(2s)E(z,s)$.

Then we have:

(6.9) Proposition. *The function $E^*(z,s)$ has a meromorphic continuation over the s-plane. Its only poles are simple ones at $s=0$ and $s=1$ with residues independent of z.*

Proof. Substituting (6.8) in the equation preceding (6.7) we find the Fourier expansion

$$E^*(z,s)=\zeta_K^*(2s)y^s+\zeta_K^*(2s-1)y^{1-s}$$

$$+\,2^n\mathrm{N}(y)^{1/2}\sum_{\substack{\mu\in\mathfrak{d}^{-1}/U\\ \mu\neq 0}}\mathrm{N}((\mu)\mathfrak{d})^{s-1/2}\sigma_{1-2s}((\mu)\mathfrak{d})\prod_{i=1}^{n}K_{s-1/2}(2\pi|\mu^{(i)}|y_i)e^{2\pi i\,\mathrm{Tr}\,\mu x},$$

where as usual $\sigma_r(\mathfrak{a})=\sum\mathrm{N}(\mathfrak{b})^r$ (sum over the divisors of an integral ideal). This Fourier expansion converges absolutely for all s and gives the meromorphic continuation. The functional equation also follows because $\mathrm{N}(\mathfrak{a})^{s-1/2}\sigma_{1-2s}(\mathfrak{a})$ for any integral ideal $\mathfrak{a}$ and $K_{s-1/2}(t)$ for any $t>0$ are even functions of $s-1/2$ and because of the functional equation of ζ_K^*. The only poles come from the pole of $\zeta_K^*(s)$ at $s=1$. We have

$$\mathrm{res}_{s=1}\,E^*(z,s)=\tfrac{1}{2}\,\mathrm{res}_{s=1}\,\zeta_K^*(s)=\tfrac{1}{2}D^{1/2}\,\mathrm{res}_{s=1}\,\zeta_K(s)$$

and hence

$$\mathrm{res}_{s=1}\,E(z,s)=\frac{1}{\zeta_K^*(2)}\,\mathrm{res}_{s=1}\,E^*(z,s)=\frac{\pi^n}{2\sqrt{D}}\,\mathrm{res}_{s=1}\frac{\zeta_K(s)}{\zeta_K(2s)}.\qquad\square$$

Now we treat the so-called *Rankin trick*. Suppose $f,g\in S_k(\Gamma)$ and consider the integral

$$\int_{\Gamma\backslash\mathfrak{H}^n}E(z,s)f(z)\overline{g(z)}\mathrm{N}(y^{k-2}\,dx\,dy).$$

Let $f=\sum a_\nu e^{2\pi i\,\mathrm{Tr}\,\nu z}$, $g=\sum b_\nu e^{2\pi i\,\mathrm{Tr}\,\nu z}$. Writing $E(z,s)$ as

$$\sum_{\Gamma_\infty\backslash\Gamma}\mathrm{N}(\mathrm{Im}(\gamma z)^s)$$

where $\Gamma_\infty=G(\mathfrak{o},U^+)$ is the stabilizer of the cusp at infinity in $\Gamma=\mathrm{SL}(2,\mathfrak{o})$ and using the modular properties of f and g we find that it equals

$$\begin{aligned}R(s)&=\int_{\Gamma\backslash\mathfrak{H}^n}\mathrm{N}(y)^sf(z)\overline{g(z)}\mathrm{N}(y^{k-2}\,dx\,dy)\\&=\int_{\mathbb{R}_+^n/U^+}\left(\int_{\mathbb{R}^n/\mathfrak{o}}f(x+iy)\overline{g(x+iy)}\mathrm{N}(dx)\right)\mathrm{N}(y^{s+k-2}\,dy)\\&=\int_{\mathbb{R}_+^n/U^+}(\mathrm{vol}(\mathbb{R}^n/\mathfrak{o}))\sum_{\substack{\nu\in\mathfrak{d}^{-1}\\ \nu\gg 0}}a_\nu\bar{b}_\nu e^{-4\pi\,\mathrm{Tr}\,\nu y}\mathrm{N}(y^{k-2+s}\,dy)\\&=D^{1/2}\sum_{\substack{\nu\in\mathfrak{d}^{-1}/U^+\\ \nu\gg 0}}a_\nu\bar{b}_\nu\int_{\mathbb{R}_+^n}\mathrm{N}(y)^{k-2+s}e^{-4\pi\,\mathrm{Tr}\,\nu y}\mathrm{N}(dy)\\&=D^{1/2}\left(\frac{\Gamma(k-1+s)}{(4\pi)^{k-1+s}}\right)^n\sum_{\nu\in\mathfrak{d}^{-1}/U^+}\frac{a_\nu\bar{b}_\nu}{\mathrm{N}(\nu)^{k-1+s}}.\end{aligned}$$

Write

$$L_{f,g}(s) = \sum_{\substack{\mathfrak{m} \subset \mathfrak{o} \\ \mathfrak{m} \neq (0)}} a(\mathfrak{m})\overline{b(\mathfrak{m})}\mathrm{N}(\mathfrak{m})^{-s},$$

with $a(\mathfrak{m}) = a_\nu$, $b(\mathfrak{m}) = b_\nu$ for $\mathfrak{m} = \nu\mathfrak{d}^{-1}$. Here the sum is taken over all non-zero integral ideals of K. We set

$$\Lambda_{f,g}(s) = (2\pi)^{-2ns}D^{2s}\Gamma(s)^n\Gamma(s-k+1)^n\zeta_K(2s-2k+2)L_{f,g}(s).$$

Then our integral identity becomes

$$\int_{\Gamma\backslash\mathfrak{H}^n} E^*(z, s-k+1)f(z)\overline{g(z)}\mathrm{N}(y^{k-2}\,dx\,dy) = D^{-k+3/2}\pi^{n(k-1)}\Lambda_{f,g}(s).$$

Proposition (6.9) now implies:

(6.10) Proposition. i) *$\Lambda_{f,g}(s)$ has an analytic continuation to the whole s-plane except for poles of order one at $s = k$ and $s = k - 1$ and satisfies the functional equation $\Lambda_{f,g}(s) = \Lambda_{f,g}(2k-1-s)$.*

ii) $\mathrm{res}_{s=k}\,\Lambda_{f,g}(s) = D^{k-3/2}\pi^{-n(k-1)}\kappa\langle f, g\rangle$, *where $\kappa \neq 0$ is the residue of $E^*(z,s)$ at $s = 1$.*

7. The Adelic Version

We give the adelic description of Hilbert modular surfaces. This is more natural; for example, all the companion groups $\Gamma(\mathfrak{o} \oplus \mathfrak{a})$ introduced in Section 4 will appear at the same time.

The set of finite (resp. infinite) places of a totally real field K of degree n over $\mathbb{Q}$ is denoted by S_f (resp. S_∞). We write $\mathbb{A}_K$ (resp. $\mathbb{A}_K^*$) for the adelic ring (resp. idèles) of K. The idèle group contains K^* and the quotient $\mathbb{A}_K^*/K^*$ is the idèle class group of K. For a place v of K we denote by K_v the completion of K at v and by $\mathfrak{o}_v$ the valuation ring of K_v. If we identify K_v with $\mathbb{R}$ for $v \in S_\infty$ we have

$$\mathbb{A}_K^*/K^*\left(\prod_{v\in S_f}\mathfrak{o}_v^*\right)\left(\prod_{v\in S_\infty}\mathbb{R}_+\right) \simeq \mathscr{C}\ell^+(K) \tag{1}$$

Let $\mathsf{G} = \mathsf{R}_{K/\mathbb{Q}}\mathrm{GL}(2)$ be the reductive algebraic group defined over $\mathbb{Q}$ obtained by restriction of scalars from K to $\mathbb{Q}$. Then $\mathsf{G}(\mathbb{R}) = \mathrm{GL}(2,\mathbb{R})^n$, one copy for each infinite place. Let

$$h_0\colon \mathbb{C}^* \to \mathsf{G}(\mathbb{R})$$

be the homomorphism defined by

$$x + iy \mapsto \left(\begin{pmatrix} x & -y \\ y & x \end{pmatrix}, \ldots, \begin{pmatrix} x & -y \\ y & x \end{pmatrix}\right).$$

The centralizer of h_0 in $\mathsf{G}(\mathbb{R})$ is

$$\mathsf{K}_\infty = \left\{ \begin{pmatrix} x_1 & -y_1 \\ y_1 & x_1 \end{pmatrix}, \ldots, \begin{pmatrix} x_n & -y_n \\ y_n & x_n \end{pmatrix} : x_i, y_i \in \mathbb{R} \right\}$$

Note that K_∞ is connected. The quotient $\mathsf{G}(\mathbb{R})/\mathsf{K}_\infty$ is in a natural way a manifold of real dimension $2n$ and consists of 2^n connected components. These components are permuted by

$$\pi_0(\mathsf{G}(\mathbb{R})) = \mathsf{G}(\mathbb{R})/\mathsf{G}(\mathbb{R})^0 \cong (\mathbb{Z}/2\mathbb{Z})^n .$$

We can put a complex structure on $\mathsf{G}(\mathbb{R})/\mathsf{K}_\infty$ via the identification

$$\mathsf{G}(\mathbb{R})/\mathsf{K}_\infty \leftrightarrow (\mathbb{C} - \mathbb{R})^n \qquad g \mapsto (g_1(i), \ldots, g_n(i)) .$$

The action of $\varepsilon_j = \left(1, \ldots, \begin{pmatrix} -1 & 0 \\ 0 & 1 \end{pmatrix}, \ldots, 1\right) \in \pi_0(\mathsf{G}(\mathbb{R}))$ is then complex conjugation of the j-th coordinate. We shall consider the quotient

$$\mathsf{G}(\mathbb{A})/\mathsf{K}_\infty = (\mathsf{G}(\mathbb{R})/\mathsf{K}_\infty) \times \mathsf{G}(\mathbb{A}^f) ,$$

($\mathbb{A}^f$: finite adèles) which is also provided with this complex structure. The group $\mathsf{G}_f = \mathsf{G}(\mathbb{A}^f)$ acts on the left on $\mathsf{G}(\mathbb{A})/\mathsf{K}_\infty$.

The group $\mathsf{G}(\mathbb{A})$ is a topological group with the topology determined by the fact that $(\prod_{v \in S_f} \mathrm{GL}(2, \mathfrak{o}_v)) \times \mathsf{G}(\mathbb{R})^0$ is an open subgroup. Now we choose a compact open subgroup K_f of G_f and consider the quotient

$$X'_{\mathsf{K}_f} := \mathsf{G}(\mathbb{Q})\backslash\mathsf{G}(\mathbb{A})/\mathsf{K}_\infty\mathsf{K}_f .$$

To analyze it we use the map $\mathsf{G}(\mathbb{A}) \to \mathbb{A}_K^*$ defined by the determinant. This induces

$$\mathsf{G}(\mathbb{Q})\backslash\mathsf{G}(\mathbb{A})/\mathsf{K}_f\mathsf{K}_\infty \to \mathbb{A}_K^*/K^* \det(\mathsf{G}(\mathbb{R})^0\mathsf{K}_f) \tag{2}$$

Choose $g_1, \ldots, g_m \in \mathsf{G}_f$ such that $\det g_i$ $i = 1, \ldots, m$ form a complete set of representatives for the right-hand side of (2).

(7.1) Proposition (Strong approximation for GL(2)).

$$\mathsf{G}(\mathbb{A}) = \bigcup_{j=1}^{m} \mathsf{G}(\mathbb{Q})g_j\mathsf{G}(\mathbb{R})^0\mathsf{K}_f .$$

Proof. See [128].

(7.2) Proposition. *There is an identification*

$$\mathsf{G}(\mathbb{Q})\backslash\mathsf{G}(\mathbb{A})/\mathsf{K}_\infty\mathsf{K}_f = \bigcup_{j=1}^{m} \Gamma_j\backslash\mathfrak{H}^n$$

with $\Gamma_j = g_j\mathsf{G}(\mathbb{R})^0\mathsf{K}_f g_j^{-1} \cap \mathsf{G}(\mathbb{Q})$.

Proof. $$\mathsf{G}(\mathbb{Q})\backslash\mathsf{G}(\mathbb{A})/\mathsf{K}_\infty\mathsf{K}_f = \mathsf{G}(\mathbb{Q})\Big\backslash\bigcup_{j=1}^{m}\mathsf{G}(\mathbb{Q})g_j\mathsf{G}(\mathbb{R})^0\mathsf{K}_f/\mathsf{K}_\infty\mathsf{K}_f$$
$$= \bigcup_{j=1}^{m} g_j\mathsf{G}(\mathbb{R})^0\mathsf{K}_f g_j^{-1}\cap\mathsf{G}(\mathbb{Q})\backslash(\mathsf{G}(\mathbb{R})^0/\mathsf{K}_\infty)$$
$$= \bigcup_{j=1}^{m}\Gamma_j\backslash\mathfrak{H}^n\,. \qquad\square$$

If we do this for the compact open $\mathsf{K}_0 := \prod_{v\in S_f}\mathrm{GL}(2,\mathfrak{o}_v)$ then we find:

(7.3) Corollary. $\mathsf{G}(\mathbb{Q})\backslash\mathsf{G}(\mathbb{A})/\mathsf{K}_\infty\mathsf{K}_0$ *can be identified with* $\bigcup_{\mathfrak{a}}\hat{\Gamma}(\mathfrak{o}\oplus\mathfrak{a})\backslash\mathfrak{H}^n$, *where* $\mathfrak{a}$ *runs through a complete set of representatives of* $\mathscr{C}\ell^+(K)$.

Proof. The components are in $1-1$ correspondence with the elements of $\mathbb{A}_K^*/K^*\det(\mathsf{G}(\mathbb{R})^0\mathsf{K}_f)\cong\mathscr{C}\ell^+(K)$ via (1). The result follows now from (7.2). $\square$

This explains why we need to consider all $\Gamma(\mathfrak{o}\oplus\mathfrak{a})\backslash\mathfrak{H}^2$. In a large part of this book we consider geometric properties of the components of $\mathsf{G}(\mathbb{Q})\backslash\mathsf{G}(\mathbb{A})/\mathsf{K}_\infty\mathsf{K}_0$ and there we do not use the adelic language. When we deal with arithmetic aspects it often becomes imperative to use the adelic language.

We close this chapter with the following remark. Let K_1, K_2 be compact open subgroups of $\mathsf{G}(\mathbb{A}^f)$ and let $g\in\mathsf{G}(\mathbb{A}^f)$ such that $g^{-1}\mathsf{K}_1 g\subset\mathsf{K}_2$. Then there is a natural morphism $I_{\mathsf{K}_2\mathsf{K}_1}(g)\colon X'_{\mathsf{K}_1}\to X'_{\mathsf{K}_2}$ given by right multiplication by g. We have

i) $I_{\mathsf{K}_3\mathsf{K}_2}(h)I_{\mathsf{K}_2\mathsf{K}_1}(g) = I_{\mathsf{K}_3\mathsf{K}_1}(gh)$
$I_{\mathsf{K}_1\mathsf{K}_1}(g) = Id.$

ii) If $\mathsf{K}_1\subset\mathsf{K}_2$ is a normal subgroup then $\mathsf{K}_2/\mathsf{K}_1$ acts on X_{K_1} and $I_{\mathsf{K}_2\mathsf{K}_1}(1)$ induces an isomorphism

$$X'_{\mathsf{K}_1}/(\mathsf{K}_2/\mathsf{K}_1)\xrightarrow{\sim}X_{\mathsf{K}_2}\,.$$

II. Resolution of the Cusp Singularities

In January 1971 Hirzebruch received a letter from Serre in which he was asked whether he knew how to resolve the cusp singularities of Hilbert modular surfaces. Hirzebruch's answer consisted in a long letter (dated 18 January 1971) in which he explained the resolution process discovered by him just a few days before Serre's letter arrived. The resolution of cusp singularities is like that of 2-dimensional quotient singularities related to continued fractions, but in contrast to these, the curves occurring in a resolution form a cyclic configuration. By the way, Hirzebruch thus obtained interesting counter examples to a claim made at the end of his own thesis [71] (in which he resolved two-dimensional quotient singularities) saying that such cyclic configurations should not occur.

The resolution opened the way to the geometric study of Hilbert modular surfaces and as a result our knowledge about these surfaces has been greatly enlarged. But apart from this it stimulated developments in other directions. Together with work done by Igusa on the resolutions of singularities of Siegel modular varieties it formed the motivation for the theory of toroidal compactification as developed by Mumford and his school [89, 8]. On the other hand it led to the Zagier-Shintani method of computing values at non-positive integers of zeta-functions of totally real algebraic number fields [175, 154]. It is a remarkable fact that this—in a sense elementary—method was not discovered much earlier.

In this chapter I explain Hirzebruch's resolution of two-dimensional cusp singularities. I have added a section which describes the resolution of two-dimensional quotient singularities in a somewhat similar way.

1. The Local Ring at Infinity

Let K be a real quadratic field and let M be an additive subgroup of K which is a free abelian group of rank 2. Such a group M is called a *complete $\mathbb{Z}$-module* of K.

Let U_M^+ be the group of totally positive elements ε of K satisfying $\varepsilon M = M$. Such an ε is always a unit.

Two modules M_1, M_2 are called (*strictly*) *equivalent* if there exists a (totally positive) element $\lambda \in K$ with $\lambda M_1 = M_2$. For equivalent modules one has $U_{M_1}^+ = U_{M_2}^+$.

For a complete $\mathbb{Z}$-module $M \subset K$ the set

$$\{x \in K: xM \subseteq M\}$$

is an order in K, i.e. a subring of the ring $\mathfrak{o}_K$ of algebraic integers of K containing 1. The norm restricted to this order defines a primitive binary integral quadratic form of discriminant $\Delta = Df^2$, with D the discriminant of K and f the index of the order in $\mathfrak{o}_K$. We call this order $\mathfrak{o}_\Delta$. (Note that $\mathfrak{o}_D = \mathfrak{o}_K$.)

Let M be a complete $\mathbb{Z}$-module in K and $V \subset U_M^+$ a subgroup of U_M^+ of finite index. We define

$$\begin{aligned} G(M, V) &= \left\{\begin{pmatrix} \varepsilon & \mu \\ 0 & 1 \end{pmatrix} \in \mathrm{GL}^+(2, K): \varepsilon \in V, \mu \in M\right\} \\ &= M \rtimes V \text{ (\textit{semi-direct product})}. \end{aligned}$$

The isotropy group of a cusp of the Hilbert modular group for a real quadratic field or one of its extensions is a group of this type (as transformation group). The group $G(M, V)$ acts freely and properly discontinuously on $\mathfrak{H}^2$.

If M and N are strictly equivalent modules, say $N = \lambda M$ with $\lambda \gg 0$, then the map $\mathfrak{H}^2 \to \mathfrak{H}^2$ given by $(z_1, z_2) \mapsto (\lambda z_1, \lambda' z_2)$ induces an equivalence between the actions of $G(M, V)$ and $G(N, V)$ on $\mathfrak{H}^2$.

The equivalence class of a module M contains one or two strict equivalence classes, depending on whether or not there exists a unit ε of norm -1 with $\varepsilon M = M$. In the latter case, the two strict equivalence classes contained in the equivalence class of M are represented by M and λM with $\lambda \in K$ an element of negative norm.

Let $M' = \{\mu' \in K: \mu \in M\}$ be the conjugate of M. The interchange of the two factors of $\mathfrak{H}^2$: $(z_1, z_2) \mapsto (z_2, z_1)$ induces an equivalence between the actions of $G(M, V)$ and $G(M', V)$ on $\mathfrak{H}^2$.

We now add a point to the complex manifold $G(M, V)\backslash\mathfrak{H}^2$ which we call ∞:

$$\overline{G(M, V)\backslash\mathfrak{H}^2} = G(M, V)\backslash\mathfrak{H}^2 \cup \{\infty\}.$$

A fundamental system of neighbourhoods of ∞ in $\overline{G(M, V)\backslash\mathfrak{H}^2}$ is defined by the sets (see I.3 for notation)

$$(G(M, V)\backslash\mathring{W}(\infty, r)) \cup \{\infty\} \qquad r > 0.$$

Define as local ring at ∞ the ring of continuous functions on $G(M, V)\backslash\mathfrak{H}^2$ which are holomorphic when restricted to some $G(M, V)\backslash\mathring{W}(\infty, r)$.

Let f be a holomorphic function on $G(M, V)\backslash\mathring{W}(\infty, r)$ The pull-back of f to $W(\infty, r)$ is invariant under translations $z \mapsto z + \mu$ with $\mu \in M$ and therefore has an absolutely convergent Fourier series

$$\sum_{\nu \in M^\vee} a_\nu \mathbf{e}(\operatorname{Tr} \nu z),$$

where $\mathbf{e}(-) = e^{2\pi i(-)}$ and $M^\vee$ is the dual of M w.r.t the trace:

$$M^\vee = \{\nu \in K: \operatorname{Tr} \mu\nu \in \mathbb{Z}\ \forall \mu \in M\}.$$

Moreover, $a_{\varepsilon\nu} = a_\nu$ for all $\varepsilon \in V$.

(1.1) Lemma. *The local ring at ∞ is the ring of all Fourier series*

$$a_0 + \sum_{\substack{\nu \in M^\vee \\ \nu \gg 0}} a_\nu \mathbf{e}(\operatorname{Tr} \nu z) \tag{1}$$

converging in some $\mathring{W}(\infty, r)$ and satisfying $a_{\varepsilon\nu} = a_\nu$ for all $\varepsilon \in V$.

Proof. Everything is clear, except that a Fourier series in the local ring satisfies: $a_\nu = 0$ if $\nu \neq 0$ and ν is not totally positive. Suppose, say, $\nu < 0$. Since $\lim_{n\to\infty} \operatorname{Tr}(\eta^n \nu) = -\infty$, with $\eta > 1$ the fundamental unit, we obtain a divergent subseries

$$\sum_n |a_\nu| \exp(-2\pi \operatorname{Tr} \eta^n \nu)$$

for $z = (i, i)$ unless $a_\nu = 0$. This proves the lemma. □

The functions which have a non-vanishing constant term a_0 in their expansion are units in the ring. This justifies the term local ring.

(1.2) *Remark 1.* The condition that (1) converges on some $W(\infty, r)$ is equivalent to the growth condition $a_\nu = O(e^{c\mathsf{N}(\nu)})$ for some c. The Fourier expansions coming from functions defined on all of $G(M, V)\backslash\mathfrak{H}^2$ satisfy this condition for all $c > 0$; those coming from Hilbert modular forms even satisfy $a_\nu = O(\mathsf{N}(\nu)^c)$ for some c, as we saw in I.6.

Remark 2. Lemma (1.1) is a special case of the so-called "Koecher-principle".

The space $\overline{G(M, V)\backslash\mathfrak{H}^2}$ with the local ring at ∞ as defined above is a normal complex space as we shall see later.

2. Glueing

The module M acts on $\mathbb{C}^2$ by translations $(z_1, z_2) \mapsto (z_1 + \mu^{(1)}, z_2 + \mu^{(2)})$ and the quotient $M\backslash\mathbb{C}^2$ is an algebraic torus. A choice of a basis μ_1, μ_2 of M determines an isomorphism

$$\begin{array}{c} \varphi_{\mu_1,\mu_2}\colon M\backslash\mathbb{C}^2 \to \mathbb{C}^* \times \mathbb{C}^* \\ \quad\quad\quad \cup\!\!\!\!\!\!\shortmid \quad\quad\quad \cup\!\!\!\!\!\!\shortmid \\ z \bmod M \mapsto (u, v)\,, \end{array}$$

where u, v are determined by

$$\mathbf{e}(z_j) = u^{\mu_1^{(j)}} v^{\mu_2^{(j)}} \qquad (j = 1, 2)\,,$$

or equivalently

$$2\pi i z_j \equiv \mu_1^{(j)} \log u + \mu_2^{(j)} \log v \qquad (\operatorname{mod} 2\pi i M)\,.$$

A change of basis of M yields a commutative diagram

$$\begin{array}{ccc} M\backslash\mathbb{C}^2 & \xrightarrow{\varphi_{\mu_1,\mu_2}} & \mathbb{C}^* \times \mathbb{C}^* \\ \| & & \downarrow \psi \\ M\backslash\mathbb{C}^2 & \xrightarrow{\varphi_{\rho_1,\rho_2}} & \mathbb{C}^* \times \mathbb{C}^*, \end{array} \tag{1}$$

with ρ_1, ρ_2 the second basis and $\psi = \varphi_{\rho_1,\rho_2} \circ \varphi_{\mu_1,\mu_2}^{-1}$. If $g \in \mathrm{GL}(2,\mathbb{Z})$ transforms (μ_1,μ_2) into (ρ_1,ρ_2) and if $g = \begin{pmatrix} a & b \\ c & d \end{pmatrix}$, then ψ can be given explicitly by

$$(u,v) \mapsto (u^a v^b, u^c v^d) \tag{2}$$

Now $M\backslash\mathbb{C}^2$ contains $M\backslash\mathfrak{H}^2$ as an open subset, and if for $z \in \mathfrak{H}^2$ $\operatorname{Im} z_1 \operatorname{Im} z_2$ tends to infinity, then either $|u|$ or $|v|$ (or both) tends to zero. Therefore, it is natural to look at the inclusion $\mathbb{C}^* \times \mathbb{C}^* \subset \mathbb{C}^2$. We can extend the map ψ in (1) to those $w = (u,v) \in \mathbb{C}^2$ satisfying: $u \neq 0$ if a or c is negative, $v \neq 0$ if b or d is negative. For such w the extension $\tilde{\psi}$ is given by (2).

This extension $\tilde{\psi}$ of ψ has the property that its graph is closed in $\mathbb{C}^2 \times \mathbb{C}^2$, for it is defined by (2) whenever this makes sense. Therefore, the topological space $Y_{\tilde{\psi}}$ obtained by glueing the two copies of $\mathbb{C}^2$ by $\tilde{\psi}$ has the Hausdorff separation property. Note that $\tilde{\psi}$ is a biholomorphic map since $\begin{pmatrix} a & b \\ c & d \end{pmatrix}$ is invertible. Therefore, $Y_{\tilde{\psi}}$ is a complex manifold.

The module M does not have a canonical basis. However, there is a natural infinite set of bases. One is led to these by studying the behaviour of elements of the maximal ideal of the local ring at ∞.

Let f be an element of the local ring at ∞ of $\overline{G(M,V)\backslash\mathfrak{H}^2}$ with Fourier expansion

$$f = \sum_{\substack{\nu \in M^\vee \\ \nu \gg 0}} a_\nu \mathbf{e}(\operatorname{Tr} \nu z).$$

The dominant terms in this series when $\operatorname{Im} z_1 \operatorname{Im} z_2 \to \infty$ are those for which $\operatorname{Tr} \nu y$ is minimal ($\nu \neq 0$, $y = \operatorname{Im} z$). Therefore, we look at the map

$$M\backslash\mathfrak{H}^2 \to (\mathbb{R}_+)^2$$

$$(z_1, z_2) \bmod M \to (y_1, y_2) \qquad y_i = \operatorname{Im} z_i,$$

and the behaviour of the function

$$t(y) = \min_{\substack{\nu \in M^\vee \\ \nu \gg 0}} \operatorname{Tr} \nu y$$

on $\mathbb{R}_+^2$. Obviously, $t(ry) = rt(y)$ for $r \in \mathbb{R}_+$.

In order to describe the behaviour of $t(y)$ on $(\mathbb{R}_+)^2$ we consider the embedding of $M_+ = \{\mu \in M : \mu \gg 0\}$ in $(\mathbb{R}_+)^2$ given by the two embeddings of the field K into $\mathbb{R}$:

$$M_+ \to \mathbb{R}_+^2$$

$$\mu \mapsto (\mu^{(1)}, \mu^{(2)}).$$

Take the convex hull of M_+ in $(\mathbb{R}_+)^2$. Its boundary is a polygon consisting of infinitely many line segments (see fig. on this page).

Denote by A_k, $k \in \mathbb{Z}$, the points of M_+ lying on the boundary. We assume that $A_{k+1}^{(1)} < A_k^{(1)}$ for all k.

Put

$$\sigma_k = \{sA_{k-1} + tA_k : s, t \in \mathbb{R}_+\} .$$

(2.1) Lemma. *Two consecutive boundary points A_{k-1}, A_k form a basis of the group M. The function $t(y)$ takes the value 1 on each A_k and is linear on the cone σ_k spanned by A_{k-1} and A_k. Moreover, $t(y)$ is linear on $\sigma_k \cup \sigma_{k+1}$ if and only if A_{k-1}, A_k and A_{k+1} lie on a straight line.*

Proof. We only prove the first statement, leaving the other two to the reader. Two consecutive boundary points clearly form a $\mathbb{Q}$-basis of M. If they do not form a $\mathbb{Z}$-basis, we can find $\mu = \alpha A_{k-1} + \beta A_k \in M$ with $\alpha, \beta \in \mathbb{Q}$, $(\alpha, \beta) \notin \mathbb{Z}^2$. Replacing μ by $-\mu$ if necessary and subtracting a suitable integral linear combination of A_{k-1} and A_k from μ we can suppose $0 \leqslant \alpha, \beta < 1$, $\alpha + \beta \leqslant 1$, $(\alpha, \beta) \neq (0,0)$. But then μ lies in or on the boundary of the triangle with vertices 0, A_{k-1} and A_k, contradicting the definition of the A_i. □

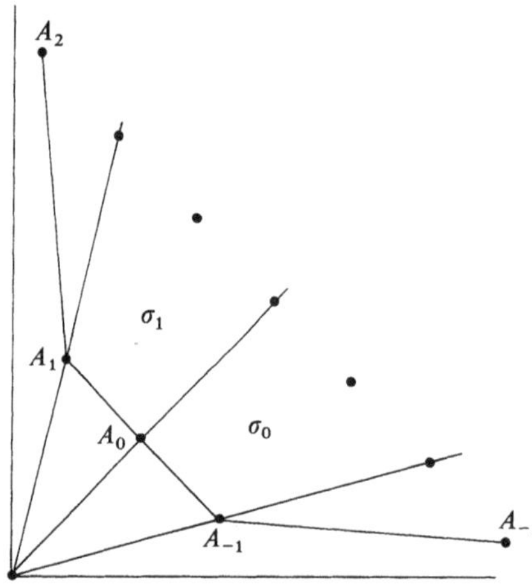

$$M = \mathbb{Z}1 + \mathbb{Z}\frac{3+\sqrt{3}}{3}$$
$$\ldots, b_{-1} = 3, b_0 = 2, b_1 = 3, \ldots$$

The boundary points A_k provide us with a natural set of bases of M. Take for each 2-dimensional cone σ_k a copy $(\mathbb{C}^2)_{\sigma_k}$ of $\mathbb{C}^2$, say with coordinates u_k, v_k. We can glue these copies by the procedure described above and thus obtain a complex manifold Y.

From Lemma 2.1 it follows that $A_{k-1} + A_{k+1}$ is an integral multiple of A_k, say

$$A_{k-1} + A_{k+1} = b_k A_k \qquad (b_k \in \mathbb{Z}) , \tag{3}$$

and that $b_k \geqslant 2$. The coordinate transformation between $(\mathbb{C}^2)_{\sigma_k}$ and $(\mathbb{C}^2)_{\sigma_{k+1}}$ is given by (cf. (2))

$$(u_k, v_k) \mapsto (u_k^{b_k} v_k, 1/u_k) \qquad u_k \neq 0 .$$

The coordinate axes of the $(\mathbb{C}^2)_{\sigma_k}$ give rise to a string of non-singular rational curves S_k in Y. The curve S_k is defined by $v_k = 0$ in $(\mathbb{C}^2)_{\sigma_k}$ and by $u_{k+1} = 0$ in $(\mathbb{C}^2)_{\sigma_k+1}$. The curves S_{k-1} and S_k intersect transversally in one point, the origin of $(\mathbb{C}^2)_{\sigma_k}$. We denote this point by p_k. If $|k-l| > 1$, then the curves S_k and S_l do not intersect.

Now we construct meromorphic functions on Y. Let $\ell: M \to \mathbb{Z}$ be a linear map. Define f_ℓ on $(\mathbb{C}^2)_{\sigma_k}$ by

$$f_\ell(u_k, v_k) = u_k^{\ell(A_{k-1})} v_k^{\ell(A_k)} .$$

This definition is compatible with the glueing (by the linearity) and hence we get a meromorphic function on Y. The divisor of f_ℓ equals

$$\sum_{k \in \mathbb{Z}} \ell(A_k) S_k .$$

Since the divisor of f_ℓ is homologous to zero, we obtain

$$\begin{aligned} 0 = (f_\ell) \cdot S_k &= \ell(A_{k-1}) S_{k-1} \cdot S_k + \ell(A_k) S_k \cdot S_k + \ell(A_{k+1}) S_{k+1} \cdot S_k \\ &= \ell(A_k)(S_k^2 + b_k) . \end{aligned}$$

Choosing an ℓ with $\ell(A_k) \neq 0$ we find

$$S_k^2 = -b_k .$$

Note that since $b_k \geqslant 2$ all intersection numbers S_k^2 are negative and different from -1. In particular, there is *no exceptional curve among the* S_k.

3. Dividing by the Units

We now construct the resolution of the singularity of $\overline{G(M, V)\backslash \mathfrak{H}^2}$. To do this, we define a neighborhood of $\bigcup_k S_k$ in Y by

$$Y^+ = \Phi(M \backslash \mathfrak{H}^2) \cup \bigcup_k S_k ,$$

where $\Phi: M \backslash \mathbb{C}^2 \to Y$ is the embedding defined by φ_{A_{k-1}, A_k}:

$$M \backslash \mathbb{C}^2 \to \mathbb{C}^* \times \mathbb{C}^* \hookrightarrow Y .$$

(This is independent of k, see (1) of the preceding section.)

The exact sequence

$$0 \to M \to G(M, V) \to V \to 0$$

implies that V acts on $M \backslash \mathfrak{H}^2$. Let ε be a generator of the cyclic group V. Then there exists an integer $r = r(M, V)$ such that

$$\varepsilon A_k = A_{k+r} \quad \text{for all } k , \tag{1}$$

since ε must carry boundary points to boundary points. Define an action of V on Y as follows: ε^n sends a point with coordinates u_k, v_k in $(\mathbb{C}^2)_{\sigma_k}$ to the point with the same coordinates in $(\mathbb{C}^2)_{\sigma_{k+nr}}$. This is compatible with the identifications made, since obviously $b_k = b_{k+nr}$ by (1) and Section 2 (3). We leave it to the reader to check that the actions of V on $M\backslash\mathbb{C}^2$ and on Y are compatible under the biholomorphic map $\Phi: M\backslash\mathbb{C}^2 \to Y$ and that Y^+ is invariant under V.

(3.1) Lemma. *The action of V on Y^+ is free and properly discontinuous.*

Proof. The action of V on $M\backslash\mathfrak{H}^2$ is free and properly discontinuous, so the action of V on $Y^+ - \bigcup_{k\in\mathbb{Z}} S_k$ also is. An element ε^n of V sends S_k to S_{k+nr} and one sees that the action is free on $\bigcup S_k$.

Now assume that $p, q \in Y^+$. We must exhibit neighbourhoods U_p, U_q such that $gU_p \cap U_q \neq \varnothing$ for only finitely many $g \in V$. Since the action is properly discontinuous on the open set $Y^+ - \bigcup_{k\in\mathbb{Z}} S_k$ this is clear if p and q do not belong to $\bigcup_{k\in\mathbb{Z}} S_k$. If $p \in \bigcup_{k\in\mathbb{Z}} S_k$ and $q \notin \bigcup_{k\in\mathbb{Z}} S_k$ then we take

$$U_q = \left\{u \in Y^+ - \bigcup_{k\in\mathbb{Z}} S_k : u \notin \Phi(W(\infty, \rho))\right\}$$

with $\rho \in \mathbb{R}_+$ sufficiently large that $q \notin \Phi(W(\infty,\rho))$. (For $W(\infty,\rho)$ see I.3.) Since V preserves U_q we can take for U_p the complement of $\bar{U}_q$ in Y^+.

Now suppose $p \in S_k \cap (\mathbb{C}^2)_{\sigma_k}$, $q \in S_l \cap (\mathbb{C}^2)_{\sigma_l}$. Take neighbourhoods

$$U_p = \{(u_k, v_k) \in (\mathbb{C}^2)_{\sigma_k} : |u_k| < \rho_1, |v_k| < \rho_2\}$$

$$U_q = \{(u_l, v_l) \in (\mathbb{C}^2)_{\sigma_l} : |u_l| < \rho_1, |v_l| < \rho_2\}$$

for suitable ρ_1 and sufficiently small ρ_2. Now note that for any pair of open sets V_1, V_2 in Y^+ we have $V_1 \cap V_2 = \varnothing$ if and only if $V_1 \cap V_2 \cap (Y^+ - \bigcup_{k\in\mathbb{Z}} S_k) = \varnothing$. The map

$$\mathrm{Im}: Y^+ - \bigcup_{k\in\mathbb{Z}} S_k \cong M\backslash\mathfrak{H}^2 \to \mathbb{R}^2_+$$

is equivariant with respect to the action of V. The image of U_p (resp. U_q) is contained in the cone C spanned by σ_k and by $A_{k-1}(-\log\rho_1) + A_k(-\log\rho_2)$ (resp. σ_l and $A_{l-1}(-\log\rho_1) + A_l(-\log\rho_2)$). So it is obvious that $g(C) \cap C \neq \varnothing$ for only finitely many $g \in V$ and also $g(U_p) \cap U_q \cap (Y^+ - \bigcup S_k) \neq \varnothing$ for only finitely many $g \in V$. □

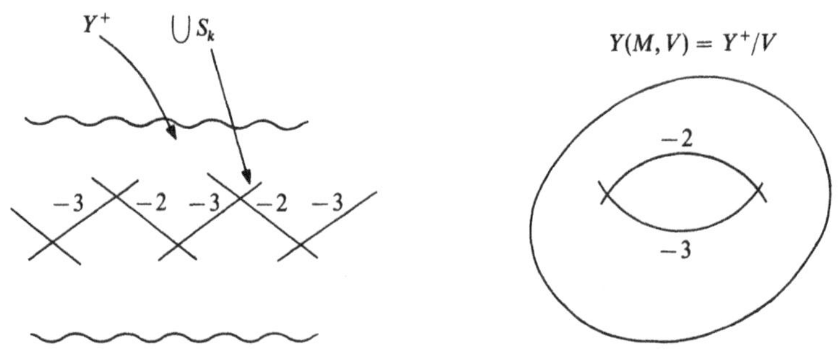

Schematic diagram of $\bigcup S_k$ and $\bigcup S_k/V$ for $M = \mathbb{Z} + \mathbb{Z}\frac{3+\sqrt{3}}{3}$

$V = U_M^+$ (cf. Fig. on p. 32)

We form the orbit space $Y(M, V) = Y^+/V$. By Lemma 3.1 $Y(M, V)$ is a complex manifold. The curve S_k in Y^+ is identified with S_{k+r} by the generator of V; we shall denote the image of S_k in $Y(M, V)$ under the quotient map again by S_k, but with k now in $\mathbb{Z}/r\mathbb{Z}$.

If $r \geqslant 3$, then the curves S_k ($k \in \mathbb{Z}/r\mathbb{Z}$) in $Y(M, V)$ form a cycle of non-singular rational curves with intersection numbers given by

$$S_k \cdot S_l = \begin{cases} -b_k & \text{if } k = l \text{ in } \mathbb{Z}/r\mathbb{Z} \\ 1 & \text{if } k = l \pm 1 \text{ in } \mathbb{Z}/r\mathbb{Z} \\ 0 & \text{otherwise} \end{cases}$$

with the b_k determined by (3) of Section 2. Note that $b_k = b_{k+r}$.

If $r = 2$, then S_0, S_1 in $Y(M, V)$ are again non-singular rational curves with self-intersections $-b_0$ and $-b_1$, but they have two intersection points: $S_0 \cdot S_1 = 2$, and the configuration looks like the one above.

If $r = 1$, the curve S_0 in $Y(M, V)$ is singular. It is a rational curve with one ordinary double point. The self-intersection of S_0 in $Y(M, V)$ is given by the following lemma.

(3.2) Lemma. *If $r = 1$ the self-intersection number of S_0 in $Y(M, V)$ equals $-b_0 + 2$.*

Proof. Let $\tilde{c}_1$ (resp. c_1) be the first Chern class of Y^+ (resp. $Y(M, V)$) and let $\rho\colon Y^+ \to Y(M, V)$ be the natural map. Then $\rho^*(c_1) = \tilde{c}_1$ and if we denote S_k in Y^+ by $\tilde{S}_k$ in order to avoid confusion, we have $\rho(\tilde{S}_k) = S_k$ and hence

$$\tilde{c}_1(\tilde{S}_0) = c_1(S_0)\,.$$

By the adjunction formula

$$\tilde{c}_1(\tilde{S}_0) - \tilde{S}_0 \cdot \tilde{S}_0 = 2$$

$$c_1(S_0) - S_0 \cdot S_0 = 0\,,$$

since the virtual genus of the singular curve S_0 is one. Combining these yields

$$S_0 \cdot S_0 = \tilde{S}_0 \cdot \tilde{S}_0 + 2 = -b_0 + 2\,. \qquad \square$$

The intersection matrix $(S_k \cdot S_l)$, $k, l \in \mathbb{Z}/r\mathbb{Z}$ is negative definite. This follows from the shape of the intersection matrix and the fact that at least one of the b_k is greater than 2 (which is equivalent to the fact that not all A_k lie on a straight line in $\mathbb{R}^2$).

Now we invoke a theorem of Grauert [47] which says that the negative definiteness of $(S_k \cdot S_l)$ implies that $\bigcup S_k$ can be blown down to a normal isolated singular point, that is, there exists a normal complex surface $Y^*(M, V)$ with one isolated singular point p and a holomorphic map $\varphi\colon Y(M, V) \to Y^*(M, V)$ with $\varphi(\bigcup S_k) = p$ and such that φ induces an isomorphism between $Y(M, V) - \bigcup S_k$ and $Y^*(M, V) - p$. Since then $Y^*(M, V) - p$ is isomorphic to $G(M, V)\backslash\mathfrak{H}^2$ the local ring of $Y^*(M, V)$ at p is easily seen to be the ring described in Lemma (1.1). Thus $G(M, V)\backslash\mathfrak{H}^2$ can be identified with $Y^*(M, V)$ and hence is a normal complex surface. Since the cycle consisting of the S_k does not contain exceptional curves (of the first kind) it yields the minimal resolution of the singular point ∞. A theorem of Laufer says that the minimal resolution of a normal isolated surface singularity is unique, cf. [108].

(3.3) Theorem. *The space $\overline{G(M, V)\backslash\mathfrak{H}^2}$ with the local ring at ∞ defined as above is a normal complex space and $Y(M, V)$ is the unique minimal resolution of the singular point ∞ of $\overline{G(M, V)\backslash\mathfrak{H}^2}$.*

The proof of Grauert's theorem uses a fundamental system of strongly pseudo-convex neighbourhoods. The sets $\Phi(W(\infty, r) \cup (\bigcup S_k))$ form such a fundamental system. In fact, $\mu(\infty, z)^{-1} = (y_1 y_2)^{-1}$ defines a strongly pluri-subharmonic function on $Y(M, V) - \bigcup S_k$.

In the higher dimensional case, i.e. if $[K:\mathbb{Q}] > 2$, we can construct in a similar way resolutions of singularities $Y(M, V) \to \overline{G(M, V)\backslash\mathfrak{H}^2}$. Indeed, take a pair (M, V) with M a complete $\mathbb{Z}$-module in K and $V \subset U_K^+$ a subgroup of finite index preserving M. As in Section 2 one considers the algebraic torus $T = M\backslash\mathbb{C}^n$. A basis $\mu_1, \ldots, \mu_n$ of M defines an isomorphism $M\backslash\mathbb{C}^n \to \mathbb{C}^{*n}$ and by composition with $\mathbb{C}^{*n} \to \mathbb{C}^n$ a map $M\backslash\mathbb{C}^n \to \mathbb{C}^n$. A different choice of basis gives a diagram as in (1) of Section 2. Then if we have a suitable cone decomposition $\{\sigma_\alpha\}$ of $(M \otimes \mathbb{R})_+ = \mathbb{R}_+^n$ with cones $\sigma_\alpha = \mu_1\mathbb{R}_+ + \cdots + \mu_n\mathbb{R}_+$ as in Section 2 we can construct as in Section 2 a complex manifold Y containing $M\backslash\mathfrak{H}^n$. If the cone decomposition of $(M \otimes \mathbb{R})_+$ is invariant under the natural action of V then V acts on the interior Y^+ of the closure of $M\backslash\mathfrak{H}^n$ in Y. If V acts freely then we can form the quotient $Y^+/V = Y(M, V)$. It is a complex manifold containing $G(M, V)\backslash\mathfrak{H}^n$. Note that it depends on the chosen decomposition of $(M \otimes \mathbb{R})_+$. Now using the theorem of Grauert again [47] one shows that the divisor $Y(M, V) - G(M, V)\backslash\mathfrak{H}^n$ can be blown down to the point ∞ of a complex space $\overline{G(M, V)\backslash\mathfrak{H}^n}$ (defined as in Section (1)). So $\overline{G(M, V)\backslash\mathfrak{H}^n}$ is a normal complex space and $Y(M, V) \to \overline{G(M, V)\backslash\mathfrak{H}^n}$ is a resolution of singularities. That a V-invariant decomposition of $(M \otimes \mathbb{R})_+$ with the required properties exists was shown by Ehlers [28]. In the case $n = 2$ there is a canonical decomposition. It is given by the boundary points of the convex hull of M_+ inside $(M \otimes \mathbb{R})_+$ and it was used in Section 2. In contrast to this there is no such canonical decomposition for $n \geqslant 3$. This is related to the fact that in the theory of algebraic varieties of dimension $\geqslant 3$ there is not a theory of minimal models as for surfaces. A class of special examples with $n = 3$ was studied by Vasquez and Thomas [164].

The constructions described for $n = 2$ and for $n > 2$ are special cases of the theory of toroidal embeddings, cf. [8].

4. Digression: the Elliptic r-gon

Now take $\mathbb{Q} \oplus \mathbb{Q}$ instead of K and $\mathbb{Z} \oplus \mathbb{Z}$ instead of M. Again M acts on $\mathbb{C}^2$ by translations and we have a map $\mathbb{C}^2/M \to \mathbb{R}^2$ by taking the imaginary parts. Consider in the half plane $\mathbb{R}_+ \times \mathbb{R}$ the set $M_+ = \{(a, b) \in \mathbb{Z} \oplus \mathbb{Z} \colon a \geqslant 1\}$. The boundary points $(1, k)$, $k \in \mathbb{Z}$, all lie on a straight line. Two consecutive points $A_k = (1, k)$, $A_{k+1} = (1, k+1)$ form a basis of M and we can carry out the construction of Section 2. Since $A_{k-1} + A_{k+1} = 2A_k$ we obtain a complex manifold Y on which there is an infinite string of non-singular rational curves S_k with self-intersection numbers -2.

Let V be the infinite cyclic group generated by the transformation of $M\backslash\mathfrak{H} \times \mathbb{C}$

$$(z, \zeta) \operatorname{mod} M \mapsto (z, \zeta + rz) \operatorname{mod} M$$

for some $r \in \mathbb{Z}$. The induced action of V on $\mathbb{R}_+ \times \mathbb{R}$ sends A_k to A_{k+r}. Define Y^+ as the union of the image of $M\backslash\mathfrak{H} \times \mathbb{C}$ with $\bigcup_{k \in \mathbb{Z}} S_k$. We can extend the action of V on $M\backslash\mathfrak{H} \times \mathbb{C}$ to a free and properly discontinuous action on Y^+ and we can form the quotient space $V\backslash Y^+$.

Put $X = M\backslash\mathfrak{H} \times \mathbb{C}$. The projection $V\backslash X \to \mathbb{Z}\backslash\mathfrak{H} \subset \mathbb{C}^*$ induced by $(z, \zeta) \mapsto z$ is well-defined and the fibre over $z \operatorname{mod} \mathbb{Z}$ is the elliptic curve $\mathbb{C}/(\mathbb{Z} + \mathbb{Z}z)$.

Let ℓ be the linear map $M \to \mathbb{Z}$ defined by $\ell(A_k) = 1$ (all k). The corresponding function f_ℓ (see Sec. 2) on Y^+ has as its divisor $\sum S_k$ and hence is holomorphic in Y^+. Moreover, f_ℓ is invariant under V and it defines a holomorphic function $f_\ell'\colon V\backslash Y^+ \to \mathbb{C}$. This function extends the map $V\backslash X \to \mathbb{C}^* \subset \mathbb{C}$ and the fibre over $0 \in \mathbb{C}$ is a generalized elliptic curve (singular fibre of type ${}_1I_r$ in Kodaira's terminology). The intersection matrix $(S_k \cdot S_l)$ is only negative *semi*-definite in this case.

5. Continued Fractions

There is a classical algorithm for finding the boundary points of the convex hull of M_+, which we describe here together with the relation between the boundary points of the convex hull of M_+ and of $(\lambda M)_+$ for λ with negative norm.

For a complete $\mathbb{Z}$-module M in K we choose a basis α, β satisfying $\alpha\beta' - \alpha'\beta > 0$. (We call this an *oriented basis.*) We put

$$Q(x, y) = \frac{1}{\mathsf{N}(M)} \mathsf{N}(x\alpha + y\beta),$$

where $\mathsf{N}(x\alpha + y\beta)$ is the norm of $x\alpha + y\beta$, $(x, y \in \mathbb{Z})$ and where $\mathsf{N}(M)$ is the unique rational number >0 which makes Q an integral primitive quadratic form. The primitive integral binary quadratic form Q is indefinite and has discriminant $\Delta = Df^2$ with D the discriminant of K and f a natural number. The $\mathrm{SL}(2, \mathbb{Z})$-equivalence class of Q does not depend on the choice of an oriented basis and does not change if we replace M by a strictly equivalent module ρM, $\rho \gg 0$.

Conversely, given a primitive binary integral quadratic form $ax^2 + bxy + cy^2$, we associate to it the module $\mathbb{Z}1 + \mathbb{Z}\dfrac{b + \sqrt{\Delta}}{2c}$, where $\Delta = b^2 - 4ac$ is the discriminant of the form.

We thus obtain a bijection: strict equivalence classes of $\mathbb{Z}$-modules of rank 2 in real quadratic fields $\leftrightarrow$ $\mathrm{SL}(2, \mathbb{Z})$-equivalence classes of integral primitive indefinite binary quadratic forms of non-square discriminant.

A quadratic form $Q(x,y) = ax^2 + bxy + cy^2$ is called *reduced* if

$$\frac{b+\sqrt{\Delta}}{2c} > 1 > \frac{b-\sqrt{\Delta}}{2c} > 0\,. \tag{1}$$

Every module M is strictly equivalent to a module of the form $\mathbb{Z}1 + \mathbb{Z}w_0$ with $w_0 > 1 > w_0' > 0$. Equivalently every $\mathrm{SL}(2,\mathbb{Z})$-equivalence class of quadratic forms of the type described above contains a reduced form. To see this, choose the A_k (boundary points of the convex hull of M_+) as in Section 2, and set $w_0 = A_{-1}/A_0$. Since $\{A_0, A_{-1}\}$ is a basis of M we find that $\{1, w_0\}$ is a basis of the strictly equivalent module $A_0^{-1}M$. For convenience we replace M by this module, so

$$A_0 = 1\,, \qquad A_{-1} = w_0\,.$$

If we set

$$w_k = \frac{A_{k-1}}{A_k}\,,$$

then the relation

$$A_{k-1} + A_{k+1} = b_k A_k \tag{2}$$

is equivalent to

$$w_k = b_k - \frac{1}{w_{k+1}}\,. \tag{3}$$

Here b_k is the smallest integer greater than w_k. So (3) leads to an infinite continued fraction with negative signs

$$w_k = b_k - \cfrac{1}{b_{k+1} - \cfrac{1}{\ddots}}$$

which we denote by

$$w_k = [[b_k, b_{k+1}, \ldots]]\,.$$

Since a generator ε of U_M^+ transforms boundary points of the convex hull into boundary points and since it preserves the order of the A_k we have $\varepsilon A_k = A_{k+s}$ for some s. Then $b_{k+s} = b_k$ and consequently $w_{k+s} = w_k$. Since $(w_1 w_2 \ldots w_s)^{-1}$ transforms A_0 into A_s, it equals ε. Moreover, if $b_{k+t} = b_k$ for all k, then $(w_1 w_2 \ldots w_t)^{-1}$ is a unit hence a power of ε and thus $t \equiv 0 \bmod s$, i.e. *s is the minimal period* and $w_1 w_2 \ldots w_s$ is a generator of U_M^+.

From the relation (3) one obtains

$$0 < w_k' < 1 < w_k\,,$$

i.e. the quadratic form associated to $\mathbb{Z}1 + \mathbb{Z}w_k$ is reduced.

Put

$$w_k = \frac{M_k + \sqrt{\Delta}}{2N_k}\,.$$

Then the associated quadratic form $Q_k(x, y)$ equals

$$Q_k(x, y) = N_{k-1}x^2 + M_kxy + N_ky^2$$

with $M_k^2 - 4N_{k-1}N_k = \Delta$.

The forms $Q_k(x, y)$, $k \bmod s$, are exactly the reduced quadratic forms lying in the $SL(2, \mathbb{Z})$-equivalence class determined by M.

In order to explain the relation with ordinary continued fractions, we consider the lattice points of M lying in $\mathbb{R}_+ \times \mathbb{R}$. We now take the support polygons of the convex hull of M_+ in $(\mathbb{R}_+)^2$ and of $M_{+-} = \{\mu \in M: \mu > 0, \mu' < 0\}$ in $\mathbb{R}_+ \times \mathbb{R}_-$. We are interested in the *vertices* B_k, $k \in \mathbb{Z}$, of the union of the two polygons. We assume $B_k < B_{k-1}$. To fix a numbering, let $M = \mathbb{Z}1 + \mathbb{Z}w_0$ with $0 < w_0' < 1$, $w_0 > 2$. (That such a w_0 exists is easy to see.) We set

$$B_0 = 1\,, \qquad B_{-1} = z_0 - w_0 - 1\,.$$

It is not difficult to see that B_{k+1} lies on the half-line $B_{k-1} - tB_k$, $t \in \mathbb{R}_+$. Indeed,

$$B_{k-1} = a_kB_k + B_{k+1} \tag{4}$$

with $a_k = [B_{k-1}/B_k] \in \mathbb{Z}_{\geqslant 1}$. Two consecutive B_k form a basis of M satisfying $B_{k-1}B_k' - B_{k-1}'B_k = (-1)^k\sqrt{\Delta}$ as follows by induction using (4). Moreover, $B_{2l} \in M_+$, $B_{2l+1} \in M_{+-}$.

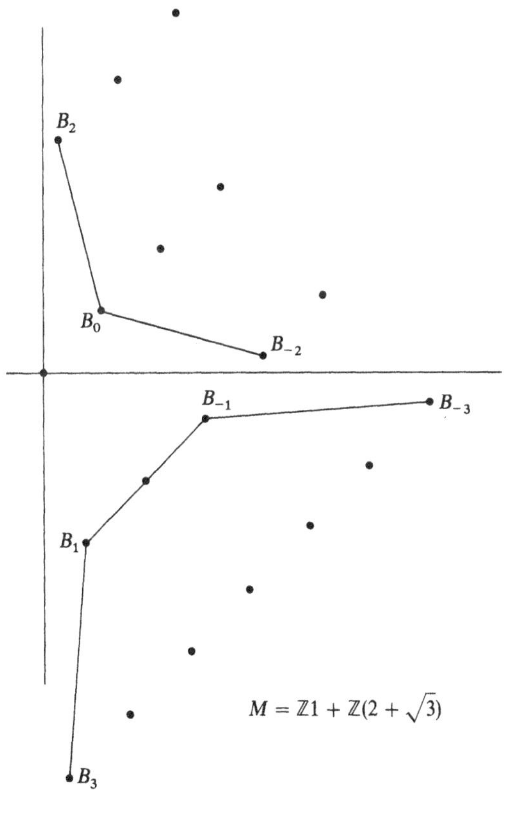

$M = \mathbb{Z}1 + \mathbb{Z}(2 + \sqrt{3})$

If we put

$$z_k = \frac{B_{k-1}}{B_k},$$

then we obtain a continued fraction with positive signs

$$z_0 = a_0 + \cfrac{1}{a_1 + \cfrac{1}{\ddots}}, \tag{5}$$

which we denote by $z_0 = [a_0, a_1, \ldots]$. A generator ε of U_M^+ sends B_k to B_{k+2l} for some l; hence $a_{k+2l} = a_k$, $z_{k+2l} = z_k$. In case there exists a unit η of norm -1 with $\eta M = M$, then $\eta B_k = B_{k\pm l}$, with l odd and $(a_1, \ldots, a_{2l})$ is twice the primitive period of the continued fraction (5).

The relation between the vertices B_{2k} of the support polygon of the convex hull of M_+ and the boundary points A_k on it is as follows. Suppose $A_m = B_{2k}$ for some k and m. Then A_{m-1} and A_{m+1} are given by $B_{2k} + B_{2k-1}$ and $B_{2k} - B_{2k+1}$. Hence using (4)

$$\begin{aligned} A_{m-1} + A_{m+1} &= 2B_{2k} + B_{2k-1} - B_{2k+1} \\ &= (2 + a_{2k})A_m . \end{aligned}$$

i.e. $b_m = a_{2k} + 2$.

If A_m is not a vertex, but lies on the line segment joining B_{2k} and B_{2k+2}, then its neighbours $A_{m\pm 1}$ are $A_m \pm B_{2k+1}$, i.e. $b_m = 2$.

Therefore, the period (or cycle) $((b_1, \ldots, b_r))$ of the continued fraction with negative signs belonging to M_+ is (up to cyclic permutation) given by

$$((\underbrace{2, 2, \ldots, 2}_{a_1 - 1},\; a_2 + 2, \underbrace{2, \ldots, 2}_{a_3 - 1},\; a_4 + 2, \underbrace{2, \ldots, 2}_{a_5 - 1}, \ldots,\; a_{2l} + 2)), \tag{6}$$

whereas the period of the continued fraction with negative signs associated to M_{+-} is given by

$$((\underbrace{2, 2, \ldots, 2}_{a_2 - 1},\; a_3 + 2, \underbrace{2, \ldots, 2}_{a_4 - 1},\; a_5 + 2, 2, \ldots, 2, \ldots,\; \underbrace{2, \ldots, 2}_{a_{2k} - 1},\; a_1 + 2)) \tag{7}$$

Note that (6) and (7) coincide if and only if l is odd, i.e. if and only if there exists a unit η with $\mathrm{N}(\eta) = -1$, $\eta M = M$. The periods (6) and (7) represents the periods of M and λM with $\lambda \in K$ of negative norm.

(5.1) *Examples.* For discriminants D as in the left column the cycles of $\mathfrak{o}_K$ and $\sqrt{D}\,\mathfrak{o}_K$ with $K = \mathbb{Q}(\sqrt{D})$ are as follows.

D	cycle of $\mathfrak{o}_K$	cycle of $\sqrt{D}\mathfrak{o}_K$
$\alpha^2 - 4$	$((\alpha))$	$((3, \underbrace{2, \ldots, 2}_{\alpha-3}))$
$4(\alpha^2 - 2)$	$((\alpha, 2\alpha))$	$((3, \underbrace{2, \ldots, 2}_{\alpha-3}, 3, \underbrace{2, \ldots, 2}_{2\alpha-3}))$
$4\alpha^2 + 1$	$((2\alpha + 1, 3, \underbrace{2, \ldots, 2}_{2\alpha-2}, 3))$	same as for $\mathfrak{o}_K$
$\alpha^2 + 4$	$((\alpha + 2, \underbrace{2, \ldots, 2}_{2\alpha-1}))$	same as for $\mathfrak{o}_K$
$4(\alpha^2 + 2)$	$((2\alpha + 2, \underbrace{2, \ldots, 2}_{\alpha-1}))$	$((\alpha + 2, \underbrace{2, \ldots, 2}_{2\alpha-1}))$

More generally, suppose that $d = 4(\alpha^2 + \beta)$ with $\alpha, \beta > 0$ and $\beta | 2\alpha$. Then the two cycles are respectively $((2\alpha + 2, \underbrace{2, \ldots, 2}_{\frac{2\alpha}{\beta}-1}))$ and $\left(\left(\frac{2\alpha}{\beta} + 2, \underbrace{2, \ldots, 2}_{2\alpha-1}\right)\right)$.

If $\alpha^2 + 1$ is even then the cycle $((\alpha + 1, 3, \underbrace{2, \ldots, 2}_{\alpha-2}, 3))$ represents the cycle of a module not equivalent to $\mathfrak{o}_K$. (Note that in this case $\alpha^2 + 1$ has at least two prime factors and $\mathsf{N}(\varepsilon) = -1$, hence $h > 1$.)

The cycles given in (5.1) cover most of the cycles that occur for small discriminants.

(5.2) *Remark.* If x is an isolated surface singularity which is resolved by a cycle of r non-singular rational curves or by a rational curve with one ordinary double point, then x is called a *cusp singularity*. Suppose $r > 1$. The self-intersection numbers $-b_k$ $k = 1, \ldots, r$ determine a quadratic irrationality

$$w = [[b_1, \ldots, b_{k+r}]]$$

and hence a module $M = \mathbb{Z}1 + \mathbb{Z}w$ in the field $K = \mathbb{Q}(w)$. Moreover, if the A_k are the boundary points of the convex hull of M_+ in $\mathbb{R}_+^2$ and ε is an element of K with $\varepsilon A_k = A_{k+r}$ then ε generates a cyclic group of totally positive units preserving M. The singularity of $\overline{G(M, V)\backslash\mathfrak{H}^2}$ is resolved by a cycle of r non-singular rational curves with self-intersection numbers $-b_k$, $k = 1, \ldots, r$. If $r = 1$ one must take $-S_1^2 + 2$ for b_1. According to a theorem of Laufer [108] the resolution cycle determines the isomorphism class of the singularity. Therefore, the singularity x is isomorphic to the singularity of $\overline{G(M, V)\backslash\mathfrak{H}^2}$.

6. Resolution of Cyclic Quotient Singularities

Let G be a finite cyclic group of order n. We define an action of G on $\mathbb{C}^2$ by letting a generator of G act by

$$(z_1, z_2) \mapsto (\zeta^{a_1} z_1, \zeta^{a_2} z_2), \qquad a_i \in \mathbb{Z}$$

with ζ a primitive n-th root of 1, $a_i \not\equiv 0 \bmod n$. The quotient space $G \backslash \mathbb{C}^2$ is a normal analytic space with a singularity at the origin. Let us assume moreover that $(a_1, n) = (a_2, n) = 1$. Then this singularity is an isolated singularity. Denote by $\pi\colon \mathbb{C}^2 \to G\backslash\mathbb{C}^2$ the natural quotient map. The local ring at $\bar{0} \in G\backslash\mathbb{C}^2$ is the ring of invariants of the local ring at $0 \in \mathbb{C}^2$. An element f of this local ring yields by pullback under π a holomorphic function in a neighbourhood of $0 \in \mathbb{C}^2$ with expansion

$$\sum a_\lambda z_1^{\lambda_1} z_2^{\lambda_2},$$

where the sum is over all $\lambda \in \mathbb{Z}^2_{\geqslant 0}$ satisfying

$$\lambda_1 a_1 + \lambda_2 a_2 \equiv 0 \pmod{n}.$$

We set

$$L = \{(\lambda_1, \lambda_2) \in \mathbb{Z}^2 \colon \lambda_1 a_1 + \lambda_2 a_2 \equiv 0 \pmod{n}\}$$

and define the dual by

$$M = \{(\mu_1, \mu_2) \in \mathbb{Q}^2 \colon \langle \lambda, \mu \rangle := \lambda_1 \mu_1 + \lambda_2 \mu_2 \in \mathbb{Z}, \forall \lambda \in L\}.$$

We can view $G\backslash\mathbb{C}^2$ as the analytic space of the scheme Spec(R) with R the ring of invariants $R = \mathbb{C}[z_1, z_2]^G$.

Consider the semi-group of exponents $\lambda \in L$ of monomials occurring in R. Let us assume (as we may) that $a_1 = 1$, $a_2 = q$. Choose a basis e_1, e_2 of L and a dual basis f_1, f_2 of M, say $e_1 = (n, 0)$, $e_2 = (-q, 1)$ and $f_1 = (1/n, q/n)$, $f_2 = (0, 1)$. The elements of the semi-group are the $\lambda \in L$ satisfying

$$\langle \lambda, \mu \rangle \geqslant 0 \quad \text{for all } \mu \in \mathbb{R}_+(nf_1 - qf_2) + \mathbb{R}_+ f_2.$$

So we are led to the cone $\mathscr{C} \subset M \otimes \mathbb{R}$ spanned by $nf_1 - qf_2$ and f_2. We take the convex hull of $M \cap \mathscr{C}$. Its boundary is a finite polygon. Let $C_d = f_2 = (0, 1)$, $C_{d-1} = \left(\frac{1}{n}, \frac{q}{n}\right)$, …, $C_1 = \left(\frac{q'}{n}, \frac{1}{n}\right)$, $C_0 = (1, 0)$ (where q' $(0 < q' < n)$ is the inverse of q $(\bmod n)$: $qq' = 1 \pmod{n}$) be the successive boundary points. Two consecutive boundary points C_{i-1}, C_i form a basis of M. We have $C_{i-1} + C_{i+1} = c_i C_i$ with $c_i \in \mathbb{Z}_{\geqslant 2}$.

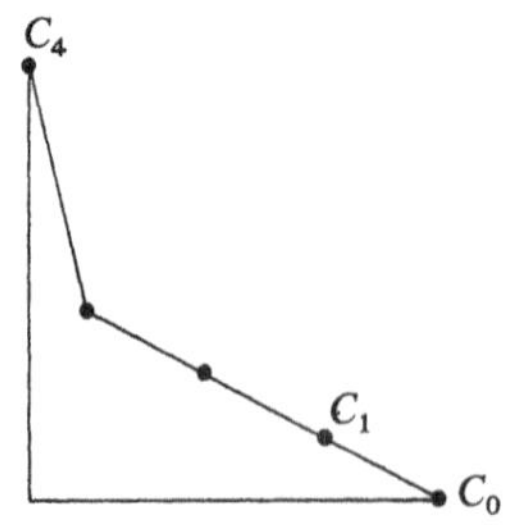

$n = 7, q = 3$

We construct the resolution of the singularity of $G\backslash\mathbb{C}^2$. Take for each basis $\sigma_i = (C_{i-1}, C_i)$ a copy $(\mathbb{C}^2)_i$ of $\mathbb{C}^2$ with coordinates u_i, v_i and glue these as follows.

Let $g \in \mathrm{GL}(2, \mathbb{Z})$ with $g({}^tC_{i-1}, {}^tC_i) = ({}^tC_{j-1}, {}^tC_j)$ and set $g = \begin{pmatrix} \alpha & \beta \\ \gamma & \delta \end{pmatrix}$. Then glue $(\mathbb{C}^2)_i$ with $(\mathbb{C}^2)_j$ by

$$(u_i, v_i) \mapsto (u_i^\alpha v_i^\beta, u_i^\gamma v_i^\delta)$$

whenever this makes sense: $u_i \neq 0$ if α or γ is negative, $v_i \neq 0$ if β or δ is negative. Or in a more invariant way, let $(\mathbb{C}^2)_i$ be the analytic space of $\mathrm{Spec}(R_i)$ with $R_i = \mathbb{C}[z^\lambda]_{\langle \lambda, \sigma_i \rangle \geqslant 0}$ the ring generated by $z_1^{\lambda_1} z_2^{\lambda_2}$ for $\lambda = (\lambda_1, \lambda_2)$ with $\langle \lambda, \mu \rangle \geqslant 0$ for all $\mu \in \mathbb{R}_+ C_{i-1} + \mathbb{R}_+ C_i$. Then the glueing is the obvious map obtained by identifying elements of R_i and R_j. The identification is biholomorphic on $(\mathbb{C}^{*2})_i$. By glueing the d copies of $\mathbb{C}^2$ we obtain a complex manifold V. The inclusion $R \subset R_i$ induces a holomorphic map

$$\varphi_i\colon (\mathbb{C}^2)_i \to G \backslash \mathbb{C}^2, \qquad z^\lambda = u_i^{\langle \lambda, C_{i-1} \rangle} v_i^{\langle \lambda, C_i \rangle}$$

and the φ_i are compatible with the identifications $v_{k+1} = 1/u_k$, $u_{k+1} = u_k^{c_k} v_k$. So we get a holomorphic map

$$\varphi\colon V \to G \backslash \mathbb{C}^2$$

which is biholomorphic on $V - \varphi^{-1}(0)$. This yields a resolution of the singularity of $G\backslash\mathbb{C}^2$. The inverse image of $\bar{0} \in G\backslash\mathbb{C}^2$ consists of $d-1$ copies of $\mathbb{P}^1$, $S_1, \ldots, S_{d-1}$ with S_j given by $u_j = 0$ in $(\mathbb{C}^2)_j$ and by $v_{j+1} = 0$ in $(\mathbb{C}^2)_{j+1}$. Moreover, the "coordinate axes" of $G\backslash\mathbb{C}^2$: $z_1 = 0$, $z_2 = 0$ correspond to $u_d = 0$ and $v_1 = 0$, respectively. These are denoted by S_d and S_0, resp..

Using meromorphic functions as in Section 2, one shows that

$$S_i^2 = -c_i\,.$$

Since all $c_i \geqslant 2$ there are no exceptional curves in the resolution and therefore φ gives the minimal resolution. One finds the c_i as follows: from $C_0 + C_2 = c_1 C_1$ we see $C_2 = c_1 C_1 - C_0$ with minimal c_1 such that C_2 lies in $\mathscr{C}$, so $c_1 = \left[\dfrac{n}{q'}\right] + 1$, etc. We get

$$\frac{n}{q'} = c_1 - \cfrac{1}{c_2 - \cfrac{}{\ddots \, - \cfrac{1}{c_{d-1}}}}\,,$$

the continued fraction of $\dfrac{n}{q'}$. Note that the continued fraction of $\dfrac{n}{q}$ is just

$$c_{d-1} - \cfrac{1}{c_{d-2} - \cfrac{}{\ddots \, - \cfrac{1}{c_1}}}\,.$$

7. The Baily-Borel Compactification

Let now K be a totally real field and let $\Gamma \subset \Gamma_m(\mathfrak{o} \oplus \mathfrak{a})$ of finite index. We define two compactifications of the quotient space $\Gamma\backslash\mathfrak{H}^n$.

First, enlarge $\Gamma\backslash\mathfrak{H}^n$ by adding the cusps

$$\overline{\Gamma\backslash\mathfrak{H}^n} = \Gamma\backslash\mathfrak{H}^n \cup \Gamma\backslash\mathbb{P}^1(K).$$

We extend the topology on $\Gamma\backslash\mathfrak{H}^n$ by taking as a fundamental system of neighbourhoods of a cusp $\sigma \in \Gamma\backslash\mathbb{P}^1(K)$ the sets

$$\sigma \cup (\Gamma_\sigma\backslash \mathring{W}(\sigma, r)) = \sigma \cup \Gamma_\sigma\backslash\{z \in \mathfrak{H}^n\colon \mu(\sigma, z) > r\}$$

with r sufficiently big (i.e. such that this is contained in $\sigma \cup F_\sigma$, cf. I.2).

We showed in Sections 1-3 that $\sigma \cup (\Gamma_\sigma\backslash W(\sigma, r))$ carries a natural structure of a complex analytic space compatible with that of $\Gamma\backslash\mathfrak{H}^n$. By glueing we obtain a compact complex analytic space again denoted $\overline{\Gamma\backslash\mathfrak{H}^n}$, called the *Satake compactification* of $\Gamma\backslash\mathfrak{H}^n$.

Hilbert modular forms now define an algebraic structure on $\Gamma\backslash\mathfrak{H}^n$ and $\overline{\Gamma\backslash\mathfrak{H}^n}$:

(7.1) Theorem (Baily-Borel [9]). *The complex analytic space* $\overline{\Gamma\backslash\mathfrak{H}^n}$ *is the normal complex analytic space of* Proj($\mathscr{M}(\Gamma)$), where $\mathscr{M}(\Gamma)$ *is the graded ring* $\oplus_{\mathbf{k}} M_{\mathbf{k}}(\Gamma)$ *of modular forms of* Γ.

In other words, modular forms of a sufficiently big weight on Γ define an embedding of $\Gamma\backslash\mathfrak{H}^n$ into some projective space and the closure of the image of $\Gamma\backslash\mathfrak{H}^n$ is isomorphic to $\overline{\Gamma\backslash\mathfrak{H}^n}$. The idea of the proof consists in exhibiting sufficiently many modular forms to separate points and tangents. These can be constructed explicitly via Poincaré series, see [9]. In this way $\overline{\Gamma\backslash\mathfrak{H}^n}$ becomes a complex projective algebraic variety denoted by X_Γ. It is called the *Baily-Borel compactification*. If $n = 2$ we showed how to resolve the singular points (the quotient singularities and the cusps) of X_Γ. By doing this in a minimal way we obtain a non-singular algebraic surface Y_Γ. The surface obtained by resolving only the cusps is denoted by Z_Γ. The variety X_Γ (n arbitrary) is called a *Hilbert modular variety*. For $n = 2$ the surfaces $\Gamma\backslash\mathfrak{H}^2$, X_Γ, Z_Γ and Y_Γ are called *Hilbert modular surfaces.*

Finally, for later use we define two line bundles L_i ($i = 1, 2$) on Z_Γ corresponding to the automorphy factors $j_1(\gamma, z) = (c^{(i)}z_i + d^{(i)})^2 \cdot (\det \gamma^{(i)})$, $i = 1, 2$ (for $n = 2$).

Define an action of Γ on $\mathfrak{H}^2 \times \mathbb{C}$ by

$$(z, \zeta) \mapsto (\gamma z, j_i(\gamma, z)\zeta) \qquad i = 1, 2.$$

The quotient spaces L_i^0 are line bundles on $\Gamma\backslash\mathfrak{H}^2$. In a "neighbourhood" $W(\sigma, r)$ of σ, σ a cusp of type (M, V), we have

$$(W(\sigma, r) \times \mathbb{C})/M = (W(\sigma, r)/M) \times \mathbb{C},$$

so we find a trivial bundle here. We extend it to the trivial bundle over $(W(\sigma, r)/M) \cup \bigcup_k S_k$. The group V operates naturally on this extension and the quotient is a line bundle over a neighbourhood of the resolution of σ. In this way we obtain two line bundles L_i, $i = 1, 2$ on Z_Γ. We set $L = L_1 \otimes L_2$.

III. Local Invariants

This chapter deals with some local invariants and properties of the cusps and quotient singularities. Since a cusp is associated to a purely arithmetic object, one naturally expects that the geometric invariants of the cusp singularity and its resolution correspond to arithmetic invariants. We shall see that the contribution $\chi(M_\sigma, V_\sigma)$ of a cusp σ to the arithmetic genus is essentially equal to the value at $s = 0$ of an L-series $L_{M_\sigma, V_\sigma}(s)$ associated to (M_σ, V_σ). In the quadratic case that can be checked directly since Meyer expressed such special L-values in terms of Dedekind sums, but for higher degree fields it turned out to be quite difficult to prove such identities. The relation between $\chi(M_\sigma, V_\sigma)$ and $L_{M_\sigma, V_\sigma}(0)$ stimulated Zagier and Shintani to compute the values at negative integers of L-series of totally real algebraic number fields directly using the cone decompositions occurring in the resolution of cusp singularities (the first in the quadratic case, the latter in the general case). It remains an interesting question to express the values at negative integers as geometric invariants.

In the first section we introduce and compute the invariant $\chi(M, V)$. The relation with the values of L-series in the second section is given without proofs. In the third section we study Hilbert modular forms near cusp and quotient singularities. We also study the extension of holomorphic differential forms over the resolution of cusp and quotient singularities.

1. Local Chern Classes

In Chapter II we associated to a pair (M, V) a complex manifold $Y(M, V)$. It contains a cyclic configuration of curves S_k, $k \in \mathbb{Z}/r\mathbb{Z}$. Let D be the divisor $\sum_{k \in \mathbb{Z}/r\mathbb{Z}} S_k$ on $Y(M, V)$.

We define a sheaf $\mathscr{F}$ on $Y(M, V)$ by defining $\mathscr{F}(U)$ for an open subset U of $Y(M, V)$ as the set of all meromorphic one-forms φ on U satisfying the two properties

i) φ is holomorphic on $U \cap (Y(M, V) - D)$

ii) φ and $d\varphi$ have at most simple poles along D.

The notation commonly used for this sheaf is $\Omega^1(\log D)$.

Denote by

$$i_k\colon S_k \to Y(M, V) \qquad (k \bmod r)$$

the inclusion of S_k in $Y(M, V)$. We can compare $\mathscr{F}$ with the sheaf $\Omega^1_{Y(M,V)}$ of germs of holomorphic 1-forms on $Y(M, V)$ by means of the exact sequence

$$0 \to \Omega^1_{Y(M,V)} \to \mathscr{F} \xrightarrow{\rho} \bigoplus_{k \bmod r} (i_k)_* \mathcal{O}_{S_k} \to 0\,, \tag{1}$$

where ρ is the Poincaré residue map, defined as follows. Suppose the intersection of D with the open set U is given by $\prod \zeta_k = 0$. We can write $\varphi \in \mathscr{F}(U)$ as

$$\varphi = \sum \alpha_k \frac{d\zeta_k}{\zeta_k} + \psi\,,$$

with ψ holomorphic. This way of writing φ is not unique, but the restriction of α_k (which is a section of $\mathcal{O}_U$) to S_k is well-defined and it gives the k-th component of $\rho(\phi)$.

(1.1) Proposition. *The sheaf $\Omega^2(\log D) = \bigwedge^2 \Omega^1(\log D)$ is trivial on $Y(M, V)$.*

Proof. This sheaf is locally free of rank one and has a non-vanishing section. Indeed, the form $dz_1 \wedge dz_2$ on $\mathfrak{H}^2$ is $G(M, V)$-invariant and defines a form

$$\frac{\sqrt{\Delta(M)}\, du_k \wedge dv_k}{4\pi^2 u_k v_k}$$

which clearly gives a non-vanishing section. □

Combining this proposition with the exact sequence (1) one finds the first Chern class of $Y(M, V)$:

$$c_1 = \sum_{k \bmod r} s_k\,,$$

where $s_k \in H^2_c(Y(M, V), \mathbb{Z})$ is the Poincaré dual of the homology class (with compact support) of S_k.

For a compact algebraic variety V the arithmetic genus[1])

$$\chi(V)\,(= \textstyle\sum (-1)^i \dim H^i(V, \mathcal{O}_V))$$

can be expressed as a polynomial (the Todd polynomial) in the Chern classes of V evaluated at the fundamental class $[V]$ of V (cf. [72]). For a surface S this formula reads

$$\chi(S) = \tfrac{1}{12}(c_1^2 + c_2)[S]\,,$$

called *Noether's formula.* Here $c_2[S]$ is the Euler number of S.

The manifold $Y(M, V)$ is not compact, but it makes sense to consider the expression $\frac{1}{12}(c_1^2 + c_2)$, where we interpret c_2 as the Euler number e, that is, define

$$12\chi(M, V) = \Big(\sum_{k \bmod r} S_k\Big)^2 + e(Y(M, V))\,.$$

---

[1]) This terminology conflicts with that of some standard textbooks in Algebraic Geometry, where the arithmetic genus is defined as $(-1)\dim V(\chi(\mathcal{O}_V) - 1)$. Our χ however is multiplicative.

The following theorem expresses $\chi(M, V)$ in terms of the integers b_k defined in Chapter II. Let $V = \langle \varepsilon \rangle$ and assume that $\varepsilon A_k = A_{k+r}$.

(1.2) Theorem. *Let M be a complete $\mathbb{Z}$-module in a real quadratic field K and let $V \subset U_M^+$ be a subgroup of finite index. Then*

$$\chi(M, V) = \frac{1}{12} \sum_{k \bmod r} (-b_k + 3).$$

Proof. We have

$$\left(\sum_{k \bmod r} S_k\right)^2 = \sum_{k \bmod r} (-b_k + 2).$$

Since $Y(M, V)$ has $\bigcup_{k \bmod r} S_k$ as a deformation retract the Euler number of $Y(M, V)$ equals r. □

Note the trivial identity $\chi(M, V) = [U_M^+ : V]\chi(M, U_M^+)$. The expression $\sum(-b_k + 3)$ occurring in the right hand side of (1.2) for $V = U_M^+$ can be rewritten in terms of the period cycle of continued fractions with *positive* signs as follows (conventions as in Sec. 5 of Ch. II):

$$\sum(-b_k + 3) = -\sum_{i=1}^{2l} (-1)^i a_i. \tag{2}$$

If we replace M by λM with λ of negative norm, then the period (6) of Sec. 5, Ch. II is replaced by (7) there and we then get $\sum_{i=1}^{2l}(-1)^i a_i$ instead of $-\sum_{i=1}^{2l}(-1)^i a_i$. This proves the following proposition.

(1.3) Proposition. *If $\lambda \in K$ is an element with negative norm then $\chi(M, V) = -\chi(\lambda M, V)$. In particular, the existence of a unit ε of norm -1 with $\varepsilon M = M$ implies $\chi(M, V) = 0$.*

Another expression for $\chi(M, V)$ is obtained as follows. The dual $M^\vee$ of M w.r.t the trace is strictly equivalent to $\lambda M'$ with λ of negative norm. For any complete $\mathbb{Z}$-module M the cycle of M' is the cycle of M with its orientation reversed (cf. Ch. II.5). So the cycles given in (6) and (7) in II.5 are up to orientation the cycles of (M, V) and $(M^\vee, V)$. Now the length of (6) and (7) is given by $\sum_{\substack{i=1 \\ i \text{ odd}}}^{2l-1} a_i$ and $\sum_{\substack{i=2 \\ i \text{ even}}}^{2l} a_i$, respectively.

Denoting the length of the cycle associated to (M, V) by $\ell(M, V)$ (so $\ell(M, V) = r$ in our earlier notation), we get the identity

$$-12\chi(M, V) = \ell(M^\vee, V) - \ell(M, V). \tag{3}$$

Note the trivial identity

$$\tfrac{1}{12}(c_1^2 + c_2) = \tfrac{1}{4}(\tfrac{1}{3}(c_1^2 - 2c_2) + c_2). \tag{4}$$

For a compact algebraic surface S the expression $\frac{1}{3}(c_1^2 - 2c_2)$ gives the signature of the underlying 4-manifold. This is the signature theorem of [72] for surfaces. For non-compact manifolds the intersection form on the second homology with compact support is not necessarily non-degenerate (actually, it is for $Y(M, V)$) but its signature (the number of positive entries minus the number of negative ones in a diagonalized version of the intersection matrix) is well-defined. For $Y(M, V)$ it equals $-r$ since the S_k generate H_2^c and since $(S_k \cdot S_l)_{kl}$ is negative definite. The Euler number $e(Y(M, V))$ (which replaces c_2 here) equals r, so (4) says

$$4\chi(M, V) = \tfrac{1}{3}(c_1^2 - 2e(Y(M, V))) - \operatorname{sign} Y(M, V),$$

i.e. the invariant χ measures the extent to which the signature formula fails to hold for the non-compact manifold $Y(M, V)$.

2. Meyer's Theorem

In the preceding section we associated to a pair (M, V) an invariant $\chi(M, V)$ defined in a geometric way. It is natural to expect that this invariant can be defined in arithmetic terms.

We define for a pair (M, V) an L-series as follows

$$L_{M,V}(s) := \sum_{\substack{\mu \in M/V \\ \mu \neq 0}} \frac{\operatorname{sign} \mathrm{N}(\mu)}{|\mathrm{N}(\mu)|^s} \qquad s \in \mathbb{C}, \qquad \mathrm{Re}(s) > 1,$$

where the summation is taken over all non-zero $\mu \in M$ modulo multiplication by units from V. Using the definition of the Γ-function we can rewrite a general term as

$$\frac{\mathrm{N}(\mu)}{|\mathrm{N}(\mu)|^{s+1}} = \Gamma\left(\frac{s+1}{2}\right)^{-2} \int_0^\infty \int_0^\infty \mathrm{N}(\mu) e^{-(\mu^2 t + \mu'^2 t')} (tt')^{(s+1)/2} \, d\log t \, d\log t'. \quad (1)$$

If we use the definition of the Γ-function again we see that this equals

$$= 2\Gamma(s+1)\Gamma\left(\frac{s+1}{2}\right)^{-2} \int_0^\infty \frac{\mathrm{N}(\mu)}{(\mu^2 v + \mu'^2 v^{-1})^{s+1}} \, d\log v. \quad (2)$$

We can view (1) as an integral over $\mathbb{R}_+^2 = (M \otimes \mathbb{R})_+$. If we do this then (2) becomes an integral over a branch of a hyperbola. Substituting (2) into the definition of $L_{M,V}$ gives

$$L_{M,V}(s) = 2\Gamma(s+1)\Gamma\left(\frac{s+1}{2}\right)^{-2} \sum_{\substack{\mu \in M/V \\ \mu \neq 0}} \int_0^\infty \frac{\mathrm{N}(\mu)}{(\mu^2 v + \mu'^2 v^{-1})^{s+1}} \, d\log v.$$

We now use a trick of Hecke: the last expression equals

$$2\Gamma(s+1)\Gamma\left(\frac{s+1}{2}\right)^{-2} \int_r^{\varepsilon^2 r} \sum_{\substack{\mu \in M \\ \mu \neq 0}} \frac{\mathrm{N}(\mu)}{(\mu^2 v + \mu'^2 v^{-1})^{s+1}} \, d\log v,$$

where $r \in \mathbb{R}_+$ is arbitrary and ε generates V. Choose an oriented basis ξ_1, ξ_2 of M and define discriminant and norm by

$$\Delta_M = (\xi_1\xi_2' - \xi_1'\xi_2)^2, \qquad \mathsf{N}(M) = \gcd\{\mathsf{N}(\xi): \xi \in M\} \in \mathbb{Q}.$$

The quadratic form $\mu^2 v + \mu'^2 v^{-1}$ on M is given with respect to ξ_1, ξ_2 by a matrix

$$\begin{pmatrix} \xi_1^2 v + \xi_1'^2 v^{-1} & \xi_1\xi_2 v + \xi_1'\xi_2' v^{-1} \\ \xi_1\xi_2 v + \xi_1'\xi_2' v^{-1} & \xi_2^2 v + \xi_2'^2 v^{-1} \end{pmatrix}$$

Let C be the cone of real 2×2 symmetric matrices which are positive definite. We have maps

$$(M \otimes \mathbb{R})_+ \hookrightarrow C \leftarrow \mathfrak{H}$$

$$(v, v') \mapsto \frac{1}{\sqrt{\Delta_M}} \begin{pmatrix} \operatorname{Tr}\xi_1^2 v & \operatorname{Tr}\xi_1\xi_2 v \\ \operatorname{Tr}\xi_1\xi_2 v & \operatorname{Tr}\xi_2^2 v \end{pmatrix}$$

$$\frac{1}{y}\begin{pmatrix} x^2 + y^2 & x \\ x & 1 \end{pmatrix} \leftarrow\!\dashv z = x + iy.$$

The action of the generator $\varepsilon \in V$ is carried over into an action of an element $T_\varepsilon \in \mathrm{SL}(2, \mathbb{Z})$ on $\mathfrak{H}$. Using these maps our integral (2) becomes an integral over part of a geodesic arc in the upper half plane $\mathfrak{H}$. Working out the transformations one finds (see [177])

$$L_{M,V}(s) \cdot \Delta_M = 2i\Gamma(s)\Gamma\left(\frac{s+1}{2}\right)^{-2} \int_p^{T_\varepsilon p} \frac{\partial}{\partial z} E(z, s)\, dz \tag{3}$$

where $E(z, s)$ is the non-holomorphic (in z) Eisenstein series[1])

$$E(z, s) = \sum_{\substack{m,n \in \mathbb{Z} \\ (m,n) \neq (0,0)}} \frac{y^s}{|mz + n|^{2s}} \qquad (z \in \mathfrak{H}, \operatorname{Re}(s) > 1)$$

and $p \in \mathfrak{H}$ is an arbitrary point. It is well-known that $E(z, s)$ can be extended to a meromorphic function (of s) on $\mathbb{C}$ with a (simple) pole at $s = 1$ whose residue equals π, see I (6.9). So (3) implies that $L_{M,V}(s)$ can be extended to a holomorphic function on all of $\mathbb{C}$. Moreover, the functional equation

$$\pi^{-s}\Gamma(s)E(z, s) = \pi^{s-1}\Gamma(1 - s)E(z, 1 - s)$$

implies a functional equation for $L_{M,V}(s)$ as follows. Let

$$\Lambda_M(s) = \pi^{-s}\Gamma\left(\frac{s+1}{2}\right)^2 \frac{\Delta_M^{s/2}}{[U_M^+ : V]} L_{M,V}(s);$$

then $\Lambda_M(s)$ only depends on the strict equivalence class of M and satisfies the functional equation

$$\Lambda_M(1 - s) = \Lambda_{M^\vee}(s).$$

Meyer computed in his book [112] the right-hand side of (3) for $s = 1$ using the Kronecker limit formula and the transformation formula of the Dedekind η-function, which occurs in the limit formula. He expressed his result in Dedekind

---

[1]) The notation differs slightly from the one used in I.6.

sums, but it can be translated into a formula involving the numbers b_k associated to the pair (M, V) in Chapter II (cf. [177])

$$L_{M,V}(1) = \frac{\pi^2}{3\sqrt{\Delta_M}} \sum_{i=1}^{r} (b_i - 3).$$

Using the functional equation and (1.2) one can reformulate Meyer's theorem as follows.

(2.1) Theorem. *The geometric invariant $\chi(M, V)$ and the arithmetic invariant $L_{M,V}(0)$ of the pair (M, V) are related by*

$$L_{M,V}(0) = -4\chi(M, V).$$

In view of the shape of (3) one might ask whether it is possible to compute $\chi(M, V)$ directly as an integral over the boundary of a neighbourhood of the resolution cycle on $Y(M, V)$.

Theorem (2.1) was the starting point for the computation of the values of L-functions of totally real algebraic number fields at negative integers by means of decompositions of $(M \otimes \mathbb{R})_+$ as constructed in Chapter II. This was done by Zagier in the quadratic case [173] and by Shintani in the general case [154]. Define

$$Z_M(s) = \mathrm{N}(M)^s \sum_{\substack{\mu \in M/U_M^+ \\ \mu \gg 0}} \mathrm{N}(\mu)^{-s}.$$

Note that $\mathrm{N}(M)^s L_{M,V}(s) = 2[U_M^+ : V](Z_M(s) - Z_{\lambda M}(s))$ for $\lambda \in K$, $\mathrm{N}(\lambda) < 0$. If M is a fractional ideal of K then $Z_M(s)$ equals

$$\sum_{\mathfrak{a} \in A} \mathrm{N}(\mathfrak{a})^{-s}$$

where A is the strict ideal class of M. Recall the cone decomposition of $(M \otimes \mathbb{R})_+$. Every $\mu \in M_+$ can be written as

$$\mu = pA_{k-1} + qA_k \qquad p, q \in \mathbb{Z}_{\geqslant 0}$$

for some k and k is unique if $p, q \in \mathbb{Z}_{>0}$. Then A_{k-1}, A_k define a quadratic form Q_k by

$$\mathrm{N}(pA_{k-1} + qA_k) = \mathrm{N}(M) Q_k(p, q).$$

Using the decomposition of $(M \otimes \mathbb{R})_+$ the following proposition is more or less obvious.

(2.2) Proposition. *Let M be a complete $\mathbb{Z}$-module in K. Then the zeta-function $Z_M(s)$ can be decomposed as*

$$Z_M(s) = \sum_{Q} \left(\sum_{p,q}' Q(p,q)^{-s} \right),$$

where Q runs over the reduced quadratic forms in the class of forms corresponding to M and $\sum'_{p,q}$ denotes $\sum_{p,q \in \mathbb{Z}_{>0}} + \frac{1}{2} \sum_{\substack{p=0,\, q \in \mathbb{Z}_{>0} \\ q=0,\, p \in \mathbb{Z}_{>0}}}$.

Let $Q(p,q) = ap^2 + bpq + cq^2$ with $a, b, c > 0$, $D = b^2 - 4ac > 0$ and consider its zeta-function

$$Z_Q(s) = \sum_{p,q}{}' Q(p,q)^{-s} \qquad \mathrm{Re}(s) > 1 .$$

Zagier proved in [175] the following theorem.

(2.3) Theorem. *The function $Z_Q(s)$ possesses an analytic continuation as a meromorphic function on $\mathbb{C}$ with one (simple) pole at $s = 1$ with residue $-\frac{1}{2\sqrt{D}}\log\frac{w}{w'}$, $w = \frac{b+\sqrt{D}}{2c}$. Moreover, for $n \in \mathbb{Z}_{\geqslant 0}$ its value at $-n$ equals*

$$Z_Q(-n) = \frac{(-1)^n n!}{2}\,\frac{B_{2n+2}}{2n+2}\left(\frac{1}{a^{n+1}} + \frac{1}{c^{n+1}}\right)\sum_{r=0}^{n}(-1)^r\frac{(n-r)!}{r!(2n+1-2r)!}a^r b^{2n+1-2r}c^r$$

$$+ \sum_{r=0}^{2n}\frac{B_{r+1}B_{2n-r+1}}{(r+1)(2n-r+1)}d_{r,n} ,$$

where B_k is the k-th Bernoulli number and $d_{r,n}$ is the coefficient of X^rY^{2n-r} in $(aX^2 - bXY + cY^2)^n$.

If $Q_k = N_{k-1}p^2 + M_k pq + N_k q^2$ the theorem gives for $n = 0$

$$Z_M(0) = \frac{1}{24}\sum_{k=1}^{r}\left(\frac{M_{k+1}}{N_k} + \frac{M_{k+1}}{N_{k+1}} - 6\right) .$$

Using $M_{k+1} + M_k = 2b_k N_k$ one gets again

$$\frac{1}{4}L_{M,U_M^+}(0) = Z_M(0) = \frac{1}{12}\sum_{k=1}^{r}(b_k - 3) = -\chi(M, U_M^+) .$$

It is not unlikely that the values at negative integers also can be interpreted as local geometric invariants of some variety.

Theorem (2.1) was generalized to the case where $[K:\mathbb{Q}] > 2$ by Atiyah, Donelly and Singer [7]. That

$$(-1)^{n/2}2^n\sum_\sigma \chi(M_\sigma, V_\sigma) = \sum_\sigma L_{M_\sigma,V_\sigma}(0) ,$$

where σ runs over all cusps and χ is the Todd polynomial in the local Chern classes of a resolution of σ, is relatively easy, see [81], but proving the equality of the corresponding terms is much harder.

3. Extension of Differential Forms

A Hilbert modular form f of weight $2m$ on Γ defines a Γ-invariant m-tensor $f(dz_1 \wedge dz_2)^{\otimes m}$ on $\mathfrak{H}^2$. This gives a section of $\omega_{X_\Gamma^{\mathrm{reg}}}^{\otimes m}$, where X_Γ^{reg} is the smooth part of X_Γ and where for a surface S we put ω_S for the relative dualizing sheaf (= sheaf of germs of holomorphic 2-forms if S is smooth). We introduce for $m \geqslant 1$

$$\omega_{X_\Gamma}^{[m]} = i_* \omega_{X_\Gamma^{\text{reg}}}^{\otimes m},$$

where $i\colon X_\Gamma^{\text{reg}} \hookrightarrow X_\Gamma$ is the natural inclusion of the smooth locus.

(3.1) Lemma. *By associating $f(dz_1 \wedge dz_2)^{\otimes m}$ to f one obtains an isomorphism $M_{2m}(\Gamma) \xrightarrow{\sim} H^0(X_\Gamma, \omega_{X_\Gamma}^{[m]})$.*

Proof. As we just saw a modular form of weight $2m$ defines a section of $\omega_{X_\Gamma}^{[m]}$. Conversely, pulling back a section of $\omega_{X_\Gamma}^{[m]}$ to $\mathfrak{H}^2$ gives a Γ-invariant form $f(dz_1 \wedge dz_2)^{\otimes m}$ which is holomorphic outside a discrete subset of $\mathfrak{H}^2$, hence holomorphic everywhere. The Γ-invariance now implies that f is a modular form of weight $2m$ on Γ. □

We now wish to compute $\omega_{X_\Gamma}^{[m]}$ near the cusps and near the quotient singularities of X_Γ. We start with the cusps.

Near a cusp singularity $\omega_{X_\Gamma}^{[m]}$ is invertible (assume the cusp is at ∞; then $(dz_1 \wedge dz_2)^{\otimes m}$ defines a nowhere vanishing section). We can also look at the resolution $Y(M, V)$ of a cusp of type (M, V) to see what happens.

We have a diagram

$$\begin{array}{ccc} G(M,V)\backslash\mathfrak{H}^2 & \xrightarrow{\ j\ } & Y(M,V) \\ & {\scriptstyle i}\searrow & \downarrow{\scriptstyle \pi} \\ & & \overline{G(M,V)\backslash\mathfrak{H}^2} \end{array}$$

(3.2) Lemma. $j_*\omega_{G(M,V)\backslash\mathfrak{H}^2} = \omega_{Y(M,V)}(\sum_k S_k) \cong \mathcal{O}_{Y(M,V)}$

Proof. In the k-th coordinate system (u_k, v_k) of the resolution $Y(M, V)$ we have

$$4\pi^2\, dz_1 \wedge dz_2 = \sqrt{\Delta_M}\,\frac{du_k \wedge dv_k}{u_k v_k}\,. \tag{1}$$

Since $dz_1 \wedge dz_2$ is a nowhere vanishing section of $\omega_{G(M,V)\backslash\mathfrak{H}^2}$ this proves the first equality. The second results from the fact that $dz_1 \wedge dz_2$ extends to a nowhere vanishing section of $\omega_{Y(M,V)}(\sum_k S_k)$ by (1). (Note that this checks with our computation of c_1 for $Y(M, V)$ in the first section of this chapter.) □

Next we look at 2-dimensional cyclic quotient singularities. As in Ch. II, such a singularity is given as the origin of $W = G\backslash\mathbb{C}^2$, where G is the cyclic group of order n generated by $(z_1, z_2) \mapsto (\zeta z_1, \zeta^q z_2)$. Recall that such a singularity was resolved by a series of curves S_i, $i = 1, \ldots, d-1$, corresponding to elements C_i of M (notation as in II.6).

Now let

$$\phi = \Big(\sum_\lambda c_\lambda z^\lambda\Big)(dz_1 \wedge dz_2)^{\otimes m}$$

be a G-invariant m-canonical form on $\mathbb{C}^2 - \{0\}$. The transition of coordinates implies

$$n\frac{dz_1 \wedge dz_2}{z_1 z_2} = \frac{du_i \wedge dv_i}{u_i v_i} \qquad i = 1, \ldots, d\,, \tag{2}$$

and consequently

$$\begin{aligned}\phi &= \left(\sum c_\lambda z^{\lambda+m(1,1)}\right)\left(\frac{dz_1 \wedge dz_2}{z_1 z_2}\right)^{\otimes m} \\ &= \left(\sum c_{\gamma_\lambda, \delta_\lambda} u_i^{\gamma_\lambda} v_i^{\delta_\lambda}\right)\left(\frac{du_i \wedge dv_i}{u_i v_i}\right)^{\otimes m},\end{aligned}$$

where

$$\gamma_\lambda = \langle \lambda + m(1,1), C_{i-1} \rangle\,.$$

Hence, ϕ can be extended holomorphically over $S_i\,(u_i = 0)$ if and only if the following holds for all λ

$$c_\lambda = 0 \quad \text{or} \quad \langle \lambda + m(1,1), C_{i-1} \rangle \geqslant m\,.$$

Note that $\lambda \geqslant 0$, hence $\langle \lambda + m(1,1), C_i \rangle \geqslant m(l_i + m_i)$ if $C_i = (l_i, m_i)$ and a sufficient condition for the extendability is that $l_i + m_i \geqslant 1$ for all C_i, $i = 1, \ldots, d-1$. By the convexity all C_i lie in the lower half plane bounded by the straight line through C_0 and C_d, i.e. $l_i + m_i \leqslant 1$. This gives the special case:

(3.3) Lemma. *If all C_i lie on the line through C_0 and C_d (i.e. $c_i = 2$ for $i = 1, \ldots, d-1$) then every m-tensor on W minus the singularity gives rise to an m-tensor on the resolution of the singularity.*

The condition expressed in the lemma implies that the singularity is a rational double point.

In particular, we find that every 2-form ($m = 1$) extends over the resolution of a quotient singularity, a result due to Freitag [33]. (Use the G-invariance of ϕ: if $c_\lambda \neq 0$ then $(\lambda_1 + 1) + q(\lambda_2 + 1) \equiv 0 \bmod q$.)

Let

$$i\colon W - \{s\} \hookrightarrow W = G\backslash\mathbb{C}^2$$

be the natural inclusion. We define for $m \geqslant 1$

$$\omega_W^{[m]} = i_*(\omega_{W-\{s\}}^{\otimes m}) \qquad (\text{then } \omega_W = \omega_W^{[1]})\,.$$

Let $\pi\colon Y \to W$ is the minimal resolution of singularities of W.

(3.4) Proposition. *With the notations as above we have:*

i) $\omega_Y = \mathcal{O}(-\sum_{i=0}^d S_i)$

ii) $\omega_W = \pi_* \omega_Y$

iii) $\pi^* \omega_W \cong \omega_Y$ (mod *torsion*)

iv) $\omega_W^{[m]} \cong \pi_* \mathcal{O}_Y(m' \sum_{i=0}^d S_i)$ *with $m' \in \mathbb{Z}$ such that $m + m' \equiv 0 \pmod n$.*

Proof. i) From the equalities

$$n\frac{dz_1 \wedge dz_2}{z_1 z_2} = \frac{du_1 \wedge dv_i}{u_i v_i} \qquad j = 1, \ldots, d$$

it follows that the left-hand side defines a trivialization of $\omega_Y(\sum S_i)$, hence $\omega_Y \cong \mathcal{O}(-\sum S_i)$.

ii) This follows since every 2-form extends as we just showed.

iii) A section of ω_W. is given by $\varphi \cdot \left(\frac{dz_1 \wedge dz_2}{z_1 z_2}\right)$, regular on $\mathbb{C}^2$, hence $\pi^*\varphi$ has to vanish on S_0 and S_d. This means that if $\varphi = \sum \gamma_\lambda z^\lambda$

$$\gamma_\lambda = 0 \quad \text{or} \quad \begin{cases} \langle \lambda, C_0 \rangle \geqslant 1\,, & \text{and} \\ \langle \lambda, C_d \rangle \geqslant 1 \end{cases}$$

and this implies $\langle \lambda, C_i \rangle > 0$ if $\gamma_\lambda \neq 0$ $(i = 0, \ldots, d)$. By i) this gives $\pi^*\omega_W \subseteq \omega_Y$. The sections $u_i v_i$ of $\pi^*\omega_W$ generate ω_Y, hence iii).

iv) Note that the left-hand side only depends on $m(\bmod n)$, since $\omega_W^{[n]}$ has a nowhere vanishing section $(dz_1 \wedge dz_2)^{\otimes n}$. Let $m' \in \mathbb{Z}_{\geqslant 0}$ such that $m + m' = en$. A section of $\omega_W^{[m]}$ is given by

$$\psi'(dz_1 \wedge dz_2)^{\otimes m} = \psi(dz_1 \wedge dz_2)^{\otimes en}\left(\frac{dz_1 \wedge dz_2}{z_1 z_2}\right)^{\otimes -m'} \tag{3}$$

regular on $\mathbb{C}^2 - \{0\}$, i.e. $\pi^*\psi$ has poles of order $\leqslant m'$ on S_0 and S_d. But then the poles of $\pi^*\psi$ on S_i are of order $m'(l_i + m_i) \leqslant m'$. Conversely, any section χ of $\mathcal{O}_Y(m' \sum S_i)$ can be written as $\pi^*\psi$ with a ψ such that (3) is regular. Together with ii) this proves iv) for $m' \geqslant 0$. The case $m' < 0$ is similar. □

If $x \in S$ is an isolated surface singularity we now can define an invariant which measures the obstruction for extending holomorphic m-tensors over the resolution of x. Let $\pi\colon \tilde{S} \to S$ give a resolution of x and let $\omega = \omega_{\tilde{S}}$. We put for $m \in \mathbb{Z}_{\geqslant 1}$

$$d_m = \dim \varinjlim_{x \in U} \Gamma(\pi^{-1}(U - x), \omega^{\otimes m}) / \Gamma(\pi^{-1}(U), \omega^{\otimes m})\,,$$

where U runs through a fundamental system of neighbourhoods of x. It follows from general theorems that d_m is finite, does not depend on the resolution of singularities and is a quadratic polynomial in m.

Let us compute d_m for two-dimensional cusp singularities, that is, singularities resolved by a cycle of non-singular rational curves or one rational curve with an ordinary double point. Such a singularity occurs as the singular point ∞ of $\overline{G(M, V)\backslash\mathfrak{H}^2}$ for suitable M and V.

A holomorphic m-tensor f on $U - \infty$ is of the form

$$g(dz_1 \wedge dz_2)^{\otimes m}$$

with g holomorphic on some $W(\infty, r)$; so g has a Fourier expansion

$$g = a_0 + \sum_{\nu \in M_+^\vee} a_\nu \mathbf{e}(\operatorname{Tr} \nu z) . \qquad M_+^\vee = \{\nu \in M^\vee : \nu \gg 0\} .$$

In the k-th coordinate system (u_k, v_k) of $Y(M, V)$ we can rewrite it as

$$a_0 + \sum a_\nu u_k^{\operatorname{Tr} \nu A_{k-1}} v_k^{\operatorname{Tr} \nu A_k}$$

while

$$4\pi^2 \, dz_1 \wedge dz_2 = \sqrt{\Delta_M} \frac{du_k \wedge dv_k}{u_k v_k} .$$

We see: f can be extended holomorphically over the coordinate axes in the k-th coordinate system if and only if $a_\nu = 0$ for $\nu = 0$ and for those $\nu \in M_+^\vee$ with $\operatorname{Tr} \nu A_{k-1} < m$ or $\operatorname{Tr} \nu A_k < m$. Therefore, f can be extended over the complete resolution cycle if and only if

$$a_\nu = 0 \quad \text{for all } \nu \in M_+^\vee \cup \{0\} \text{ with } \min_{y \in M_+} \operatorname{Tr} \nu y < m .$$

Since any element g of the local ring at ∞ defines a holomorphic m-tensor $g(dz_1 \wedge dz_2)^{\otimes m}$ we get

$$d_m = \# \left\{ \nu \in M_+^\vee : \min_{y \in M_+} \operatorname{Tr} \nu y < m \right\} \cup \{0\} \bmod V .$$

Clearly $d_1 = 1$. We now compute d_m for $m \geqslant 2$ following Tai [8]. Define points D_k of $M_+^\vee$ as follows. Let $\sqrt{\Delta_M} = A_k A'_{k+1} - A'_k A_{k+1}$ be the determinant of the lattice M and put

$$D_k = \frac{A'_{k+1} - A'_k}{\sqrt{\Delta_M}} .$$

One has

$$\operatorname{Tr} D_k A_{k+1} = \operatorname{Tr} D_k A_k = 1 , \qquad \operatorname{Tr} D_k A_l \geqslant 1 \qquad (\forall l) . \tag{4}$$

Note that

$$\begin{aligned} D_{k-1} = D_k &\Leftrightarrow \operatorname{Tr} D_k(A_{k-1} + A_{k+1}) = 2 \\ &\Leftrightarrow \operatorname{Tr} D_k b_k A_k = 2 \\ &\Leftrightarrow b_k = 2 . \end{aligned}$$

We define a polyhedral cone (degenerate if $b_k = 2$) by

$$E_k = \mathbb{R}_{>0} D_{k-1} + \mathbb{R}_{\geqslant 0} D_k .$$

The E_k with $b_k \geqslant 3$ cover $\mathbb{R}_+^2$. Their meaning is explained by the following lemma.

(3.5) Lemma. $\bar{E}_k = \{\alpha \in M_+^\vee : \min_{x \in M_+} \operatorname{Tr} \alpha x = \operatorname{Tr} \alpha A_k\} \cup \{0\}$.

Proof. The inclusion $\subseteq$ is obvious. For $\supseteq$, let α be an interior point of E_i and suppose $i \neq k$. If $\operatorname{Tr} \alpha A_k = \min \operatorname{Tr} \alpha x$ then $\operatorname{Tr} D_i A_k = \operatorname{Tr} D_{i-1} A_k = 1$. Then (4) with k replaced by i implies that A_k lies on the straight line through A_{i+1} and A_i and on the straight line through A_i and A_{i-1}. Since $A_k \neq A_i$ we conclude that A_{i-1}, A_i and A_{i+1} are collinear, hence $D_i = D_{i-1}$. This contradicts the fact that α is an interior point. If $\alpha \in \mathbb{R}_+ D_i$ then $\operatorname{Tr} D_i A_k = 1$ and again A_k lies on the line through A_{i+1} and A_i. If $k < i$ then the collinearity of A_k, A_{i+1} and A_i implies $D_k = D_{k+1} = \cdots = D_i$ and if $k > i$ $D_i = D_{i+1} = \cdots = D_{k-1}$. This proves the lemma. □

Suppose that a generator ε of V sends A_k to A_{k+r}. Then $\overline{\mathbb{R}_+ D_k + \mathbb{R}_+ D_{k+r}}$ is a fundamental domain for the action of V on $(M^\vee \otimes \mathbb{R})_+$.

We now have to count

$$\left\{ v \in M_+^\vee : \min_{y \in M_+} \operatorname{Tr} vy < m \right\} \cup \{0\} \bmod V .$$

If $D_k \neq D_{k+1}$ then $|D_k D'_{k+1} - D'_k D_{k+1}|$ is the number of lattice points in the parallelogram spanned by D_k and D_{k+1}. Therefore we get

$$\begin{aligned} d_m &= \sum_{k=1}^{r} |D_k D'_{k+1} - D'_k D_{k+1}|(1 + 2 + \cdots + (m-1)) \\ &= \frac{1}{2} m(m-1) \sum_{k=1}^{r} (b_k - 2) . \end{aligned}$$

From Ch. II we know that $\sum (b_k - 2) = -\mathrm{K} \cdot \sum S_k$, hence

$$-d_m = \tfrac{1}{2} m(m-1) \mathrm{K} \cdot \sum S_k \quad (\mathrm{K}\text{: canonical divisor})$$

which shows that $-d_m$ is the contribution from the cycle $\sum S_k$ in the Riemann-Roch formula for $\mathrm{K}^{\otimes m}$, cf. Ch. IV.

Similarly, we find for a quotient singularity

$$d_m = \#\{\lambda \in L : \lambda - m(1,1) \geqslant 0, \langle \lambda, C_i \rangle < m\} .$$

In particular, $d_1 = 0$ for all cyclic quotient singularities.

By counting as above one finds

$$\lim_{m \to \infty} \frac{d_m}{m^2} = \frac{1}{2} \sum_{i=1}^{d-1} (c_i - 2) - (n - q - 1)/2n - (n - q' - 1)/2n .$$

The last two terms are correction terms coming from the boundary of the cone in $M \otimes \mathbb{R}$. We summarize the results as follows.

(3.6) Proposition. *For a two-dimensional cusp singularity the obstruction d_m for extending holomorphic m-tensors over the resolution of the singularity is given by $d_m = \frac{1}{2} m(m-1) \sum (b_k - 2)$. For a quotient singularity one has $\lim_{m \to \infty} d_m/m^2 = \frac{1}{2} \sum (c_i - 2) - (2n - q - q' - 2)/2n$.*

For future refence we state

(3.7) Proposition. *By sending $f \in S_2(\Gamma)$ to $f\,dz_1 \wedge dz_2$ one obtains an isomorphism $S_2(\Gamma) \to H^0(Y_\Gamma, \omega_{Y_\Gamma})$.*

Proof. As we saw $d_1 = 0$ for cyclic quotient singularities. So for every cusp form of weight 2 the differential form extends over the resolution of all quotient singularities and it also extends over the cusp resolutions since a_0 is zero in the Fourier expansion for every cusp. So the result follows from (3.1). □

IV. Global Invariants

In Chapter I we proved that for a Hilbert modular group Γ of a real quadratic field K the compactified quotient $\overline{\Gamma\backslash\mathfrak{H}^2}$ is a (singular) projective algebraic surface X_Γ.

By resolving the singularities in a minimal way as explained in Chapter II we obtain a non-singular algebraic surface Y_Γ. It is the purpose of this Chapter to compute some of the invariants of these non-singular models Y_Γ and to show that they are defined over a number field.

Let S be a non-singular connected algebraic surface over $\mathbb{C}$. Regarding S as a closed oriented 4-manifold we have the topological invariants: the Betti numbers $b_i = \operatorname{rk} H_i(S, \mathbb{Z})$ satisfying $b_0 = b_4 = 1, b_1 = b_3$ and the Euler number $e = \sum(-1)^i b_i$. As a complex (Kähler) manifold S gives rise to the Hodge numbers $h^{p,q} = \dim_{\mathbb{C}} H^q(S, \Omega^p)$ with Ω^p the sheaf of holomorphic p-forms on S, and they satisfy $h^{p,q} = h^{q,p}$. It is customary in the theory of algebraic surfaces to call $h^{0,2}$ the *geometric genus* and $h^{0,1}$ the *irregularity* of S, denoted by p_g and q respectively. The alternating sum $\chi = h^{0,0} - h^{0,1} + h^{0,2}$ is called the *arithmetic genus* of S. It is related to the Chern numbers c_1^2 and $c_2 = e$ of S by Noether's formula: $12\chi = c_1^2 + c_2$.

It will be shown in this Chapter that the surfaces Y_Γ are regular, i.e. $b_1 = 2q = 0$. Therefore, the knowledge of c_1^2 and c_2 suffices to find b_2, p_g and χ. The Chern numbers of Y_Γ can be computed by means of Hirzebruch's proportionality principle which expresses them in terms of the volume of $\Gamma\backslash\mathfrak{H}^2$ plus certain correction terms due to the non-compactness and the quotient singularities of $\Gamma\backslash\mathfrak{H}^2$. The volume of $\Gamma\backslash\mathfrak{H}^2$ in the natural invariant metric was computed by Siegel and it can be expressed in terms of the zeta-function of the field K. We use the invariants thus obtained to calculate dimension formulas for spaces of Hilbert modular cusp forms and to compute representations of groups on certain spaces of cusp forms. Finally, in the last section we show that X_Γ is defined over a number field. For the theory of algebraic surfaces we refer to the standard text [10].

1. The Volume of $\Gamma\backslash\mathfrak{H}^2$

The upper half plane carries the Riemannian metric ($z = x + iy$)

$$\frac{(dx)^2 + (dy)^2}{y^2},$$

and this metric is invariant under the automorphism group $\mathrm{PGL}^+(2,\mathbb{R})$ of $\mathfrak{H}$. Similarly, on $\mathfrak{H}^n$ we have a metric invariant under the full automorphism group G of $\mathfrak{H}^n$

$$\sum_{j=1}^{n}\frac{(dx_j)^2+(dy_j)^2}{y_j^2}$$

and the corresponding normalized volume form (exterior form of type (n,n)) equals

$$\omega=(-1)^n\frac{1}{(2\pi)^n}\frac{dx_1\wedge dy_1}{y_1^2}\wedge\cdots\wedge\frac{dx_n\wedge dy_n}{y_n^2}. \tag{1}$$

It turns out that for groups $\Gamma\subset\mathrm{PGL}^+(2,\mathbb{R})^n$ commensurable with the Hilbert modular group the volume of $\Gamma\backslash\mathfrak{H}^n$ is finite. It thus yields a basic invariant of the quotient space $\Gamma\backslash\mathfrak{H}^n$.

Let $\zeta_K(s)$ be the Dedekind zeta-function of a totally real number field K, defined for $s\in\mathbb{C}$, $\mathrm{Re}(s)>1$ by

$$\zeta_K(s)=\sum \mathrm{N}(\mathfrak{a})^{-s}, \tag{2}$$

where the sum is extended over all non-zero integral ideals $\mathfrak{a}$ of $\mathfrak{o}_K$. It is a meromorphic function of s, holomorphic except for a simple pole at $s=1$. It satisfies a functional equation: let

$$\xi_K(s)=\left(\frac{D_K}{\pi^n}\right)^{s/2}\Gamma\left(\frac{s}{2}\right)^n\zeta_K(s),$$

where $n=[K:\mathbb{Q}]$ and D_K is the discriminant of K, then

$$\xi_K(s)=\xi_K(1-s).$$

(1.1) Theorem (Siegel). *The volume of $\Gamma_K\backslash\mathfrak{H}^n$ is given by the formula*

$$\int_{\Gamma_K\backslash\mathfrak{H}^n}\omega=2\zeta_K(-1).$$

Siegel expressed the volume in terms of $\zeta_K(2)$ which relates by the functional equation to $\zeta_K(-1)$. He gave two proofs for this result. His first proof is based on the embedding of $\mathfrak{H}^n$ into the space of positive quadratic forms in $2n$ variables, see [157] and the correction. He gave a second proof in [158], which uses the definition of $\zeta_K(2)$ as an infinite sum. The modern approach, based on the fact that the Tamagawa number of $\mathrm{SL}(2,\mathfrak{o}_K)$ equals 1, uses instead the Euler product decomposition of $\zeta_K(2)$.

One can give a fairly elementary proof by adapting the argument of Zagier in [176] (which uses a variant of Rankin's method for a non-holomorphic Eisenstein series). This gives

$$\mathrm{vol}(S_t)=(-1)^n\pi^{-2n}D_K^{3/2}\zeta_K(2)+O(1/t^2),$$

where $S_t=\Gamma\backslash\{z\in\mathfrak{H}^n\colon \mu(\sigma,z)\leqslant t \text{ for all cusps } \sigma\}$.

If $\Gamma\subset\mathrm{PGL}^+(2,\mathbb{R})^n$ is commensurable with the Hilbert modular group Γ_K one finds for the volume of $\Gamma\backslash\mathfrak{H}^n$

$$\operatorname{vol}(\Gamma\backslash\mathfrak{H}^n) = [\Gamma_K : \Gamma] 2\zeta_K(-1)\,,$$

where the index $[\Gamma_K : \Gamma] \in \mathbb{Q}$ is defined as $[\Gamma_K : \Gamma_K \cap \Gamma]/[\Gamma : \Gamma_K \cap \Gamma]$.

We saw in Chapter I that the orbit space $\Gamma_K\backslash\mathfrak{H}^n$ admits a natural decomposition into h pieces. Each piece consists of the points nearest to one cusp and therefore corresponds to an ideal class of K. On the other hand, also the zeta function $\zeta_K(s)$ admits a natural decomposition into h pieces defined by

$$\zeta_K = \sum_{C \in \mathscr{Cl}(K)} \zeta_{K,C}$$

with

$$\zeta_{K,C}(s) = \sum_{\substack{\mathfrak{a} \in C \\ \mathfrak{a} \neq (0)}} \mathsf{N}(\mathfrak{a})^{-s}\,.$$

Zagier raised the question whether $(-1)^n \pi^{-2n} D_K^{3/2} \zeta_{K,C}(2)$ can be interpreted in a natural way as the volume of a piece of $\Gamma_K\backslash\mathfrak{H}^2$, perhaps $\Gamma_\sigma\backslash F_\sigma$, where σ is the cusp corresponding to the ideal class C and F_σ is defined by some "distance to the cusps" function, cf. I.2.

Siegel also provided formulas for $\zeta_K(-1)$. In the case of a real quadratic field the result is given in I.6.

The form ω in (1) is the Gauss-Bonnet form and so differential geometry tells us that the integral of ω over a fundamental domain $\Gamma\backslash\mathfrak{H}^n$ for a freely acting group Γ with compact quotient gives the Euler number of the manifold $\Gamma\backslash\mathfrak{H}^n$. If $\Gamma\backslash\mathfrak{H}^n$ is not compact, but has a finite volume this still holds. To be precise:

(1.2) Theorem. *Let $\Gamma \subset \mathrm{PGL}^+(2,\mathbb{R})^n$ be a discrete subgroup commensurable with a Hilbert modular group. Then the Euler number $e(\Gamma\backslash\mathfrak{H}^n)$ is given by*

$$e(\Gamma\backslash\mathfrak{H}^n) = \int_{\Gamma\backslash\mathfrak{H}^n} \omega + \sum_{r>1} a_r(\Gamma)\frac{r-1}{r}\,,$$

where a_r is the number of points in $\Gamma\backslash\mathfrak{H}^n$ coming from points in $\mathfrak{H}^n$ with isotropy group of order r.

This theorem is (up to the correction term for the elliptic fixed points) a special case of a theorem of Harder [60]. We give a sketch of proof.

Let $S_t^0 = \Gamma\backslash\{z \in \mathfrak{H}^n : \mu(\sigma, z) < t \text{ for all cusps } \sigma\}$ (cf. Ch. I) with t sufficiently large. The generalization of the classical Gauss-Bonnet formula by Allendoerfer and Weil as given by Chern [17] tells us that if Γ acts freely

$$e(S_t) = \int_{S_t} \omega + \sum_\sigma \int_{\Gamma_\sigma\backslash\partial W(\sigma,t)} \Pi\,,$$

where Π is the pull-back of a $(2n-1)$-form on the sphere bundle B of unit tangent vectors of S_t under the section $\partial S_t \to B$ defined by the unit outward normal vectors of ∂S_t. If $t \to \infty$ then $\int_{S_t} \omega$ tends to $\int_{\Gamma\backslash\mathfrak{H}^n} \omega$.

Hence we have to show

$$\lim_{t\to\infty}\int_{\Gamma_\sigma\backslash\partial W(\sigma,t)} \Pi = 0 \quad \text{for all } \sigma .$$

Now $\Gamma_\sigma\backslash\partial W(\sigma,t)$ is a homogeneous space (for $\mathbb{R} \rtimes \Lambda$, see Ch. I), from which we infer that

$$\int_{\Gamma_\sigma\backslash\partial W(\sigma,t)} \Pi = c\cdot \operatorname{vol} \partial W(\sigma,t) ,$$

where c is a constant independent of t since $\partial W(\sigma,t)$ and $\partial W(\sigma,t')$ are isometric (for t and t' sufficiently large). We know that $\Gamma_\sigma\backslash\partial W(\sigma,t)$ is a torus bundle over $V\backslash\Lambda$ whose fibre F is $\mathbb{R}^n$ modulo the "translations" of Γ_σ. A simple computation yields

$$\operatorname{vol}\partial W(\sigma,t) = -\frac{\sqrt{2}\log\varepsilon}{2\pi t}\operatorname{vol}(F) ,$$

where $\operatorname{vol}(F)$ is the Euclidian volume of F, and this goes to zero for $t\to\infty$. Since $\Gamma\backslash\mathfrak{H}^n$ is homotopy equivalent to S_t the Euler number of $\Gamma\backslash\mathfrak{H}^n$ equals $\int_{\Gamma\backslash\mathfrak{H}^n}\omega$.

The formula for groups Γ which are not torsion-free can be deduced from the torsion-free case by noting that such a Γ contains with finite index a torsion-free normal subgroup Γ'. The relation between the Euler numbers $e(\Gamma\backslash\mathfrak{H}^n)$ and $e(\Gamma'\backslash\mathfrak{H}^n)$ yields the desired formula. Alternately one may remove suitable neighbourhoods of the elliptic fixed points and compute the integral $\int\Pi$ over their boundaries, cf. Formula (16) of [17].

2. Chern Numbers of Y_Γ

We now turn to the Chern numbers of Y_Γ. The basic idea for calculating them is the following principle, formulated by Hirzebruch [73] who was inspired by a paper of Igusa.

(2.1) Hirzebruch's Proportionality Principle. *Let $\mathcal{D}$ be a bounded symmetric domain and let Γ be a discrete torsion-free group of automorphisms of $\mathcal{D}$ such that $\Gamma\backslash\mathcal{D}$ is compact. Then the Chern numbers of $\Gamma\backslash\mathcal{D}$ are proportional to those of the compact dual $\check{\mathcal{D}}$ of $\mathcal{D}$ and the factor of proportionality equals the volume of $\Gamma\backslash\mathcal{D}$ (suitably normalized).*

The compact dual $\check{\mathcal{D}}$ is a rational algebraic variety which contains $\mathcal{D}$ in a natural way. It is a flag variety and its Chern numbers are easily calculated. For example, if $\mathcal{D}$ is the unit disc in $\mathbb{C}$ then $\check{\mathcal{D}} = \mathbb{P}^1$. The domain $\mathfrak{H}$ is equivalent to the unit disc in $\mathbb{C}$. In the case of $\mathfrak{H}^n$ that we are dealing with the principle is more or less obvious. In fact, on $\mathfrak{H}$ the form $-\frac{1}{2\pi}\frac{dx\wedge dy}{y^2}$ represents the first Chern class in terms of the invariant metric as we saw in the preceding section. The l-th elementary symmetric function of the forms $w_j = -\frac{1}{2\pi}\frac{dx_j \wedge dy_j}{y^2}$ represents the l-th Chern class of $\mathfrak{H}^n$. Note that our normalization of the volume differs from the normalization in the principle above by a factor 2^{-n}.

In our case where Γ is a Hilbert modular group acting on $\mathfrak{H}^n$ we cannot apply (2.1) directly since Γ is not co-compact and often not torsion-free. One expects that there should be corrections due to the torsion and to the cusps and that the principle should be valid for suitably modified Chern classes. In [117] Mumford proved an extension of the proportionality principle for the non-compact case which shows that this is indeed the case. Herȩ we shall give a direct proof.

Let us assume first that Γ is a *torsion-free* Hilbert modular group and let $Y = Y_\Gamma$ be the minimal non-singular model defined in the preceding section. We denote by $D = \sum S_k$ the divisor of the curves occuring in the cusp resolutions of Y (with multiplicity 1). It will turn out in this situation that $\Omega_Y^1(\log D)$ and not Ω_Y^1 (or its dual) is the object for whose Chern numbers a proportionality principle holds.

If $F = \Omega^1(\log D)$ defined as in Section 1 of Chapter III, then F is the locally free sheaf of rank two associated to the vectorbundle $L_1 \oplus L_2$ with L_j corresponding to the factor of automorphy $(cz_j + d)^2$, see Ch. II.7.

Let $\tilde{c} = 1 + \tilde{c}_1 + \tilde{c}_2$ be the total Chern class of F. In order to compute $\tilde{c}_i$ we define an hermitian metric on $L_{j|_{\Gamma\backslash\mathfrak{H}^2}}$ as follows. Pull L_j back to $\mathfrak{H} \times \mathfrak{H}$ and trivialize it there. Then the norm of a local section s is given by

$$h_j(s(z), s(z)) = y_j^{-2}|s(z)|^2 \qquad (j = 1, 2).$$

The Chern form of L_j with respect to this hermitian metric is then

$$-w_j = \frac{1}{2\pi}\frac{dx_j \wedge dy_j}{y_j^2} \qquad (j = 1, 2).$$

One easily checks that by $\eta \mapsto \lim_{t\to\infty} \int_{S_t} \eta \wedge w_j$ w_j defines a current (functional on forms) on Y_Γ. (Use the behaviour at ∞: Poincaré growth cf. [117].)

(2.2) Lemma. *The current $-w_j$ represents the first Chern class of L_j, $j = 1, 2$.*

Proof. Take any C^∞ hermitian metric h_j' on Y_Γ of L_j. Then $\partial\bar\partial \log h_j' - \partial\bar\partial \log h_j$ again defines a current on Y_Γ. Checking whether this current is a boundary on Y_Γ can be done on $\Gamma\backslash\mathfrak{H}^2$ because of the Poincaré growth at the cusps, cf. Proposition 1.2 of [117]. But on $\Gamma\backslash\mathfrak{H}^2$ it is a boundary. □

(2.3) Corollary. *The Chern classes $\tilde{c}_1$ and $\tilde{c}_2$ of $\Omega^1(\log D)$ are represented by the currents $-(w_1 + w_2)$ and $w_1 \wedge w_2$.*

Lemma (2.2) shows that the Chern numbers of F are proportional to the Chern numbers of $\Omega^1_{\mathbb{P}^1}$: (the volume is given in (1.1))

$$\begin{aligned} \tilde{c}_1^2 &= 2\,\mathrm{vol}(\Gamma\backslash\mathfrak{H}^2) \\ \tilde{c}_2 &= \mathrm{vol}(\Gamma\backslash\mathfrak{H}^2)\,. \end{aligned} \tag{1}$$

For torsion-free groups Γ we thus find the Chern numbers of Y_Γ from III.1. (1), (1) and the local Chern numbers:

$$\begin{aligned} c_1^2 &= 2\operatorname{vol}(\Gamma\backslash\mathfrak{H}^2) + \sum_k (2 - b_k) \\ c_2 &= \operatorname{vol}(\Gamma\backslash\mathfrak{H}^2) + \sum_k 1 \end{aligned} \tag{2}$$

Now we introduce Chern classes and Chern numbers for singular algebraic surfaces in order to derive from (2) the Chern numbers for the case in which Γ is allowed to have torsion. Chern classes for singular algebraic varieties were first introduced by MacPherson. We only deal with the case of compact normal complex algebraic surfaces.

For a compact normal surface X we have the Chern classes $d_i = d_i(X)$, $i = 1, 2$. These are homology classes lying in $H_2(X, \mathbb{Z})$, resp. $H_0(X, \mathbb{Z})$ and possess the following properties. Under the identification of $H_0(X, \mathbb{Z})$ with $\mathbb{Z}$ the class d_2 corresponds to the Euler number of X. If $\pi\colon Y \to X$ is a desingularization of X, then

$$d_1(X) = \pi_* d_1(Y) = \pi_* P_Y c_1(Y), \tag{3}$$

where $P_Y c_1(Y)$ is the Poincaré dual of $c_1(Y)$. In particular, if X is non-singular then d_1 is the Poincaré dual of $c_1(X)$. Moreover, if $f\colon X \to Z$ is a finite morphism of compact normal algebraic surfaces with only isolated ramification points we have

$$f_* d_1(X) = \deg(f) d_1(Z). \tag{4}$$

By P_X we denote the Poincaré homomorphism from $H^2(X)$ to $H_2(X)$. Then

$$f_* P_X f^* = \deg(f) P_Z \tag{5}$$

and if we take (co)homology with rational coefficients rank $\pi_* =$ rank P_X. So $d_1 = P_X e$ for some e in $H^2(X, \mathbb{Q})$. We now define the rational characteristic number $d_1^2 = d_1^2(X)$ by

$$d_1^2 = e^2 = (\pi^* e)^2 .$$

This is well-defined because $d_1 = P_X e = P_X e'$ implies $e - e' \in \ker P_X = \ker \pi^*$. One can compute d_1^2 using a non-singular model as follows. Since X is normal it has only isolated singular points, say $p_1, \ldots, p_k$. If p is such a singular point and if $C_1, \ldots, C_m$ are the irreducible curves on Y forming the resolution of p there exists a unique divisor $Z = \sum_{i=1}^m a_i C_i$ with rational numbers a_i such that the adjunction formula holds:

$$ZC_i - C_iC_i = 2 - 2p_a(C_i) \qquad (p_a = \text{classical arithmetic genus}).$$

Since the intersection matrix $(C_iC_j)_{ij}$ is negative definite the divisor Z is unique and it is called the *local Chern cycle* of p on Y. If Z_i $(i = 1, \ldots, k)$ are the local Chern cycles of $p_1, \ldots, p_k$ and if $d_1 = P_X e$, then $c_1(Y) - \pi^* e = \sum Z_i$. Hence

$$d_1^2(X) = c_1^2(X) - \sum_{i=1}^k Z_i^2 . \tag{6}$$

(2.4) Proposition (Hammond [57]). *Let G be a finite group of automorphisms of a compact normal surface X. If the non-trivial elements of G have only isolated fixed points then*

$$d_1^2(X) = \#G \cdot d_1^2(G\backslash X)\,.$$

Proof. Denote by $f\colon X \to G\backslash X$ the quotient map. Since $d_1(X)$ is invariant under G we can find a G-invariant cohomology class e with $d_1(X) = P_X e$. Let e be the image of g under the isomorphism $f^*\colon H^2(G\backslash X, \mathbb{Q}) \to H^2(X, \mathbb{Q})^G$. By (4) and (5) we have

$$\deg(f) P_{G\backslash X}\, g = f_* P_X f^* g = f_* d_1(X) = \#G \cdot d_1(G\backslash X)\,,$$

so $P_{G\backslash X}\, g = d_1(G\backslash X)$. We find

$$d_1^2(X) = e^2 = \#G \cdot g^2 = \#G \cdot d_1^2(G\backslash X)\,. \qquad \square$$

For a cusp singularity whose resolution cycle is $\{S_1, \ldots, S_r\}$ the local Chern cycle is (see III.1)

$$Z = \sum_{i=1}^{r} S_i\,,$$

hence $Z^2 = \sum_{i=1}^{r} (2 - b_k)$. Combination of (2) and (6) gives us for torsion-free groups Γ

$$d_1^2(\Gamma\backslash\mathfrak{H}^2) = 2\,\mathrm{vol}(\Gamma\backslash\mathfrak{H}^2)\,.$$

Let Γ now be a effectively operating group, possibly with torsion. Then Γ contains with finite index a normal subgroup Γ' which is torsion-free. Then Γ/Γ' acts on $\Gamma'\backslash\mathfrak{H}^2$ with quotient $\Gamma\backslash\mathfrak{H}^2$ and the last proposition yields

$$d_1^2(\Gamma\backslash\mathfrak{H}^2) = [\Gamma : \Gamma']^{-1}\, d_1^2(\Gamma') = 2\,\mathrm{vol}(\Gamma\backslash\mathfrak{H}^2)\,.$$

Finally, by applying (6) again one finds the desired Chern numbers.

(2.5) Theorem. *Let $\Gamma \subset \mathrm{PGL}^+(2, \mathbb{R})^2$ be commensurable with the Hilbert modular group and let Y_Γ be the minimal desingularization of $\overline{\Gamma\backslash\mathfrak{H}^2}$. Then*

$$c_1^2(Y_\Gamma) = 2\,\mathrm{vol}(\Gamma\backslash\mathfrak{H}^2) + \sum (2 - b_k) + \sum a(\Gamma; n; a, b) c(n; a, b)\,,$$

$$c_2(Y_\Gamma) = \mathrm{vol}(\Gamma\backslash\mathfrak{H}^2) + \ell + \sum a(\Gamma; n; a, b)\left(\ell(n; a, b) + \frac{n-1}{n}\right),$$

where

$a(\Gamma; n; a, b) =$ *# quotient singularities of $\Gamma\backslash\mathfrak{H}^2$ of type $(n; a, b)$,*

$c(n; a, b) =$ *self-intersection number of the local Chern cycle of a quotient singularity of type $(n; a, b)$,*

$\ell(n; a, b) =$ *# curves in the resolution of a quotient singularity of type $(n; a, b)$,*

$\ell =$ *# curves in the resolution of the cusps.*

Note that for a cyclic quotient singularity the Chern cycle equals

$$\sum_{i=0}^{d} (l_i + m_i - 1)S_i\,, \tag{7}$$

where the notation is as in II.6.

In a table at the end of this book we shall give formulas for the numbers $a(\Gamma; n; a, b)$ for the standard groups. We add here a table listing the numbers $c(n; a, b)$ that we shall need later.

Table

type $(n; a, b)$	$c(n; a, b)$	resolution
$(a; 1, 1)$	$4 - a - 4/a$	(a)
$(a; a-1, 1)$	0	$\underbrace{(2, \ldots, 2)}_{a-1}$
$(5; 2, 1)$	$-\frac{2}{5}$	$(2, 3)$
$(5; 3, 1)$	$-\frac{2}{5}$	$(3, 2)$
$(12; 5, 1)$	$-\frac{1}{2}$	$(3, 2, 3)$

(2.6) Corollary. *The arithmetic genus χ of Y_Γ is given by the formula*

$$\chi = \frac{1}{4}\operatorname{vol}(\Gamma\backslash\mathfrak{H}^2) + \sum_\sigma \chi(M_\sigma, V_\sigma) + \sum_x \chi(x)$$

where the last sum is over all quotient singularities x of $\Gamma\backslash\mathfrak{H}^2$ and $\chi(x) = \frac{1}{12}\left(c(n; a, b) + \ell(n; a, b) + \frac{n-1}{n}\right)$.

We can express this differently. We have

$$\begin{aligned}\chi(Y_\Gamma) &= \tfrac{1}{12}(c_1^2 + c_2)(Y_\Gamma)\\ &= \tfrac{1}{4}(\tfrac{1}{3}(c_1^2 - 2c_2) + c_2)(Y_\Gamma)\\ &= \tfrac{1}{4}(e + s)(Y_\Gamma)\,,\end{aligned}$$

with e the Euler number and $s = \operatorname{sign}$ the signature. Both the Euler number and the signature behave "additively". So if U_x (resp. U_σ) denotes a suitable neighbourhood of a quotient singularity x (resp. a cusp σ) then

$$\begin{aligned}4\chi(Y_\Gamma) &= (e + s)\left(\Gamma\backslash\mathfrak{H}^2 - \bigcup_x U_x - \bigcup_\sigma U_\sigma\right) + \sum_x (e + s)(U_x) + \sum_\sigma (e + s)(U_\sigma)\\ &= \operatorname{vol}(\Gamma\backslash\mathfrak{H}^2) + \sum_x (n_x - 1)/n_x + \operatorname{sign}(\Gamma\backslash\mathfrak{H}^2)\,.\end{aligned}$$

(we have written n_x for the order of a quotient singularity x.) Note $(e + s)(U_x) = (e + s)(U_\sigma) = 0$. One finds using (2.6):

$$\operatorname{sign}(\Gamma\backslash\mathfrak{H}^2) = \sum_x \left(4\chi(x) - \frac{n_x - 1}{n_x}\right) + \sum_\sigma 4\chi(M_\sigma, V_\sigma)\,. \tag{8}$$

Let now $\mathfrak{H}_- = \{z \in \mathbb{C}: \operatorname{Im} z < 0\}$. The map $(z_1, z_2) \to (z_1, \bar{z}_2)$ induces an orientation reversing homeomorphism $\Gamma\backslash\mathfrak{H}^2 \to \Gamma\backslash\mathfrak{H} \times \mathfrak{H}_-$, hence

$$\operatorname{sign}(\Gamma\backslash\mathfrak{H}^2) = -\operatorname{sign}(\Gamma\backslash\mathfrak{H} \times \mathfrak{H}_-).$$

A cusp of type (M_σ, V_σ) of $\Gamma\backslash\mathfrak{H}^2$ corresponds to a cusp of type $(M_\sigma^\vee, V_\sigma)$ for $\Gamma\backslash\mathfrak{H} \times \mathfrak{H}_-$ and $\chi(M_\sigma, V_\sigma) = -\chi(M_\sigma^\vee, V_\sigma)$. In fact all terms change sign in (8) when one replaces $\mathfrak{H}^2$ by $\mathfrak{H} \times \mathfrak{H}_-$. In particular we have:

(2.7) Proposition. *Let* $G = \mathrm{PGL}(\mathfrak{o} \oplus \mathfrak{a})$ *and let* X_G *be the Baily-Borel compactification of* $G\backslash(\mathbb{C}\backslash\mathbb{R})^2$. *Then we have*

$$\chi(X_G) = \tfrac{1}{4}e(G\backslash(\mathbb{C}\backslash\mathbb{R})^2).$$

We end this section with a description of the first Chern class of a Hilbert modular surface.

(2.8) Proposition. *Let* Y_Γ *be a Hilbert modular surface. Then its first Chern class* $c_1 \in H^1(Y_\Gamma, \mathbb{Q})$ *is given by*

$$c_1 = \gamma + \sum_x Z_x,$$

where γ *is the cohomology class in* $H^2(Y_\Gamma, \mathbb{Q})$ *defined by* $w_1 + w_2$, Z_x *is the cohomology class of the local Chern cycle of a resolution of a singularity* x *and where* x *runs over all quotient and cusp singularities of* X_Γ.

Proof. If Γ acts freely we know by (2.3) that $c_1(\Omega^1(\log D))$ is represented by $-(w_1 + w_2)$. The proposition follows immediately in this case. If Γ does not act freely, we consider a normal subgroup Γ' of finite index and blow up $Y_{\Gamma'}$ in the cusps and in the points lying over the quotient singularities of X_Γ to get a morphism $Y_{\Gamma'} \to Y_\Gamma$. Then one uses the Hurwitz formula for the canonical class. We leave the details to the reader. □

3. Inequalities for χ and c_1^2

We shall now prove inequalities for the invariants c_1^2 and χ of the surfaces Y_Γ, $\Gamma = \mathrm{PSL}(\mathfrak{o} \oplus \mathfrak{a})$, when $\mathfrak{a}$ lies in different genera. For this we first need an expression for the value $\sum_\sigma \chi(M_\sigma, V_\sigma)$, where the sum is over the cusps of Γ.

For the moment, let K be a totally real algebraic number field. Its ideal class group (resp. narrow ideal class group) is denoted by $\mathcal{Cl}$ (resp. $\mathcal{Cl}^+$). The kernel of the natural map $\mathcal{Cl}^+ \to \mathcal{Cl}$ is denoted by $\mathcal{P}$. Note that if $A \in \mathcal{Cl}$, then A^2 naturally lives in $\mathcal{Cl}^+$.

We assume that all units of $\mathfrak{o}_K$ have norm $+1$. Then the map $\lambda \mapsto \operatorname{sgn} \mathrm{N}(\lambda)$ from $K^* \to \{\pm 1\}$ depends only on the principal ideal (λ) and defines a homomorphism $\tau: \mathcal{P} \to \{\pm 1\}$.

A character χ of $\mathscr{C}\ell^+$ is called a *norm-signature character* if $\chi|\mathscr{P} = \tau$. For such a character we define the L-series.

$$L(s,\chi) = \sum_{\substack{\mathfrak{a} \\ \mathfrak{a} \subset \mathfrak{o}}} \chi(\mathfrak{a}) \mathrm{N}(\mathfrak{a})^{-s} .$$

For a narrow ideal class $B \in \mathscr{C}\ell^+$ we define the zeta function

$$\zeta_B(s) = \sum_{\mathfrak{b} \in B} \mathrm{N}(\mathfrak{b})^{-s}$$

and the L-function

$$L_B(s) = \sum_{E \in \mathscr{P}} \tau(E) \zeta_{EB}(s) ,$$

a sum over all $B' \in \mathscr{C}\ell^+$ having the same image in $\mathscr{C}\ell$ as B. For χ of norm-signature type the L-series $L(s,\chi)$ can be written as

$$L(s,\chi) = \frac{\#\mathscr{C}\ell}{\#\mathscr{C}\ell^+} \sum_{C \in \mathscr{C}\ell^+} \chi(C) L_C(s) ,$$

which by abuse of notation we can write as

$$L(s,\chi) = \sum_{C \in \mathscr{C}\ell} \chi(C) L_C(s) .$$

(Note that for $C \in \mathscr{C}\ell^+$ the expression $\chi(C)L_C(s)$ only depends on the wide class of C.)

The following result generalizes Theorem 1.2 of [59]:

(3.1) Proposition. *Let $\chi_1, \ldots, \chi_r$ be the real (i.e. order 2) characters of norm-signature type. For any $B \in \mathscr{C}\ell^+$ one has*

$$\sum_{A \in \mathscr{C}\ell} L_{A^2B}(s) = \begin{cases} 0 & \textit{if } r = 0 \\ \sum_{j=1}^{r} \chi_j(B) L(s,\chi_j) & \textit{otherwise} . \end{cases}$$

Proof. (i) Suppose first that $r = 0$. Then all order 2 characters of $\mathscr{C}\ell^+$ take the value $+1$ on $\mathscr{P}$, hence by duality of finite abelian groups $\mathscr{P} \subset SQ^+ = \{A^2 \in \mathscr{C}\ell^+ : A \in \mathscr{C}\ell^+\}$. But then

$$\{A^2 \in \mathscr{C}\ell^+ : A \in \mathscr{C}\ell, A^2 = 1 \in \mathscr{C}\ell\} = \mathscr{P} .$$

Hence we have

$$\sum_{A \in \mathscr{C}\ell} L_{A^2B}(s) = \sum_{A \in \mathscr{C}\ell/\mathscr{C}\ell[2]} \sum_{C \in \mathscr{P}} L_{A^2BC}(s) ,$$

where $\mathscr{C}\ell[2] = \{A \in \mathscr{C}\ell : A^2 = 1 \in \mathscr{C}\ell\}$. But for any $D \in \mathscr{C}\ell^+$ we have

$$\sum_{C \in \mathscr{P}} L_{CD}(s) = 0$$

because it is a sum of terms $\sum_{C \in \mathscr{P}} \tau(C) \zeta_G$.

(ii) Now we may assume $r > 0$. Observe that r is the number of elements of order $\leqslant 2$ in $\mathscr{C}\ell$. So the left-hand side equals

$$\sum_{C \in B \cdot SQ} L_C(s) \sum_{\substack{A \\ A^2B=C}} 1 = r \sum_{C \in B \cdot SQ} L_C(s), \tag{1}$$

where $SQ = \{A^2 \in \mathscr{C}\ell^+ : A \in \mathscr{C}\ell\}$. The right-hand side can be written as

$$\sum_{A \in \mathscr{C}\ell} \sum_{j=1}^{r} \chi_j(B)\chi_j(A)L_A(s).$$

We claim

$$\sum_{j=1}^{r} \chi_j(B)\chi_j(A) = 0 \quad \text{if } A \notin SQ \cdot B.$$

In fact, let $A' \in \mathscr{C}\ell^+$ map to A under $\mathscr{C}\ell^+ \to \mathscr{C}\ell$. Then

$$\sum_{\chi} \chi(A')\chi(B) = 0,$$

where we sum over all real characters of $\mathscr{C}\ell^+$ since the narrow class of $A'B$ is not a square. But AB is also not a square in $\mathscr{C}\ell$. So if we sum over the real characters of $\mathscr{C}\ell$ we also get zero. What we are left with in the right-hand side is

$$\sum_{j=1}^{r} \sum_{A \in SQ} \chi_j(A)L_{AB}(s).$$

Since the expression in the inner sum equals $\sum_A L_{AB}(s)$ we get (1). □

If K is as before a real quadratic field with discriminant D the real norm signature characters are just the genus characters χ_{D_1,D_2} corresponding to the decompositions $D = D_1 \cdot D_2$ of D as a product of discriminants of imaginary quadratic fields, see [156]. The L-series $L(s,\chi)$ is decomposed accordingly

$$L(s,\chi) = L(s,\chi_{D_1})L(s,\chi_{D_2})$$

and classical number theory provides the formula

$$L(1,\chi_{D_i}) = \frac{2\pi h(D_i)}{\sqrt{|D_i|}\, w_i}$$

with w_i the number of roots of 1 contained in $\mathbb{Q}(\sqrt{D_i})$. We set $h'(D_i) = \frac{2}{w_i} h(D_i)$, the modified class number (so $h'(-3) = \frac{1}{3}$, $h'(-4) = \frac{1}{2}$, $h' = h$ for $D_i < -4$). Thus we obtain the following expression for the right-hand side in Proposition (3.1)

$$\sum_{\substack{D=D_1 \cdot D_2 \\ D_i<0}} \frac{\pi^2}{\sqrt{D}} \chi_{D_1,D_2}(B)h'(D_1)h'(D_2) \tag{2}$$

On the other hand, the contribution to the arithmetic genus of a cusp σ equals (see Ch. III.2).

$$\chi(M_\sigma, V_\sigma) = \frac{-\sqrt{\Delta_M}}{4\pi^2} L_{M_\sigma,V_\sigma}(1).$$

Let $\mathfrak{b}$ be an ideal representing cusp σ with $\mathfrak{b}^{-2}\mathfrak{a} = M = M_\sigma$. By

$$\mu \leftrightarrow \mu\mathfrak{b}^2\mathfrak{a}^{-1}$$

we get a correspondence between the non-zero elements μ of M mod U_M^+ and the integral ideals in the class of M^{-1}. (Note that we may suppose that K has no unit of norm -1 since otherwise $\chi(M_\sigma, V_\sigma) = 0$.) In this way we get

$$\begin{aligned} \mathrm{N}(M)^s L_{M,U_M^+}(s) &= \sum_{\substack{0 \neq \mu \in M \\ \mathrm{mod}\ U_M^+}} \frac{\mathrm{N}(M)^s \operatorname{sgn} \mathrm{N}(\mu)}{|\mathrm{N}(\mu)|^s} \\ &= \sum \frac{\chi((\mu))}{\mathrm{N}(\mu\mathfrak{b}^2\mathfrak{a}^{-1})^s}, \end{aligned}$$

where χ is any real norm signature character which induces $\operatorname{sgn} \mathrm{N}(\mu)$ on K^*. It follows that

$$\sum_\sigma \mathrm{N}(M_\sigma) L_{M_\sigma, U_{M_\sigma}^+}(1) = 2 \sum_{B \in \mathscr{C}\ell} L_{B^2A}(s)$$

where B (resp. A) is the class (resp. narrow class) of $\mathfrak{b}$ (resp. $\mathfrak{a}^{-1}$). Combination of (2), and this last formula with Proposition (3.1) gives the following theorem. (Note that we can assume $[U^+ : V_\sigma] = 2$.)

(3.2) Theorem. *Let $\Gamma = \mathrm{PSL}(\mathfrak{o} \oplus \mathfrak{a})$. The contribution of the cusps to the arithmetic genus $\chi(Y_\Gamma)$ satisfies:*

$$\sum_\sigma \chi(M_\sigma, V_\sigma) = -\sum \chi_{D_1,D_2}(\mathfrak{a}) h'(D_1) \cdot h'(D_2).$$

Here the right-hand sum is over all decompositions of D as a product of discriminants of imaginary quadratic fields with $D_1 < D_2$ and χ_{D_1,D_2} is the genus character corresponding to the decomposition $D = D_1 \cdot D_2$. Moreover, h' denotes the modified class number with $h'(-3) = \frac{1}{3}$, $h'(-4) = \frac{1}{2}$.

As a corollary we obtain estimates on χ.

(3.3) Corollary. *Let $D > 12$ and let $\mathfrak{a}$ be an integral ideal. Then*

$$\chi(X_{\mathrm{PSL}(\mathfrak{o}\oplus\mathfrak{a})}) - \chi(X_{\mathrm{PSL}(\mathfrak{o}\oplus\mathfrak{o})}) = \sum_{\substack{D = D_1 D_2 \\ D_1 < -4}} \frac{1}{2}(1 - \chi_{D_1,D_2}(\mathfrak{a})) h(D_1) h(D_2) \geqslant 0,$$

where the sum is over all decompositions of D as a product of discriminants of imaginary quadratic number fields and $D_1 < -4$. (Note: h is the usual class number.)

Proof. The numbers a_2, $a_3^+ + a_3^-$ do not depend on $\mathfrak{a}$, so the contribution of the elliptic fixed points to the left hand side is

$$\tfrac{1}{9}(a_3^-(\Gamma(\mathfrak{o} \oplus \mathfrak{a})) - a_3^-(\Gamma(\mathfrak{o} \oplus \mathfrak{o}))) = \tfrac{1}{6} h(-D/3)(1 - \chi_{-3,-D/3}(\mathfrak{a}))$$

if $D \equiv 0 \pmod 3$ and zero otherwise, cf. the tables on p. 267. □

Now we wish to estimate c_1^2. To do this we use Theorem (2.5). Recall a definition: $\ell(M, V)$ = number of curves in the resolution cycle of $G(M, V)\backslash\mathfrak{H}^2 \cup \{\infty\}$ and $\ell(M) = \ell(M, U_M^+)$. Since $\ell(M)$ depends only on the narrow equivalence class of M it makes sense to define

$$\ell(D) = \sum_{A \in \mathcal{C}\ell^+} \ell(A).$$

By Chapter II we know that $\ell(D)$ equals the number of all reduced quadratic irrationalities w of discriminant D. (i.e. w with $w > 1 > w' > 0$). If w satisfies $aw^2 + bw + c = 0$ with $b^2 - 4ac = D$ then necessarily $\gcd(a, b, c) = 1$. Hence

$$\ell(D) = \#\left\{(a, b, c) \in \mathbb{Z}^3 \colon b^2 - 4ac = D, \frac{b + \sqrt{D}}{2a} > 1 > \frac{b - \sqrt{D}}{2a} > 0\right\}$$

Put $x = b - 2a$. As c is determined by a and x we get

$$\ell(D) = \#\{(a, x) \in \mathbb{Z}^2 \colon a > 0, x^2 < D, x^2 \equiv D \pmod{4a}, x + 2a > \sqrt{D}\}.$$

We claim

$$2\ell(D) = \#\{(a, x) \in \mathbb{Z}^2 \colon a > 0, x^2 < D, x^2 \equiv D \pmod{4a}\}$$

Indeed, the latter set is in $1:1$ correspondence with the set S of triples $(x, a, a') \in \mathbb{Z}^3$ with a, $a' > 0$, $x^2 + 4aa' = D$. Since $x + 2a > x > -\sqrt{D}$ the condition $x + 2a > \sqrt{D}$ is equivalent to $(x + 2a)^2 > D$ or to $x > a' - a$ and applying the involution $(x, a, a') \to (-x, a', a)$ we see that exactly half of the elements of S satisfy this condition. Hence

$$\ell(D) = \frac{1}{2} \sum_{\substack{x^2 < D \\ x^2 \equiv D \pmod 4}} \sum_{\substack{a > 0 \\ a \mid (D - x^2)/4}} 1 = \frac{1}{2} \sum_{\substack{x^2 < D \\ x^2 \equiv D \pmod 4}} \sigma_0\left(\frac{D - x^2}{4}\right). \tag{3}$$

The reader will not fail to notice the resemblance of this formula with the formula of Siegel for $\zeta_K(2)$, see I.6.

Now define for a genus γ (viewed as a collection of elements of $\mathcal{C}\ell^+$)

$$\ell_\gamma(D) = \sum_{A \in \gamma} \ell(A), \qquad \ell_\gamma^\vee(D) = \sum_{A \in \gamma} \ell(A^\vee),$$

where $A^\vee$ is the dual of A.

To get the inequality for c_1^2 we now estimate $\ell_\gamma(D) + \ell_\gamma^\vee(D)$. Note that the reduced quadratic form $aX^2 + bXY + cY^2$ with $x + 2a > \sqrt{D}$ and $b^2 - 4ac = D$ belongs to the genus γ of $\mathfrak{a}$ if and only if

$$\prod_{p \mid D} \frac{1}{2}(1 + (\mathrm{N}(\mathfrak{a})a, D)_p) \neq 0.$$

Therefore we have

$$\ell_\gamma(D) + \ell_\gamma^\vee(D) = \frac{1}{2} \sum_{\substack{x^2 < D \\ x^2 \equiv D \pmod 4}} \sum_{\substack{a > 0 \\ a \mid (D - x^2)/4}} \left\{\prod_{p \mid D} \frac{1}{2}(1 + (\mathrm{N}(\mathfrak{a})a, D)_p) + \prod_{p \mid D} \frac{1}{2}(1 + (-\mathrm{N}(\mathfrak{a})a, D)_p)\right\}$$

$$\left(\text{since if } (a,x) \text{ belongs to } M \text{ then } \left(\frac{D-x^2}{4a}, -x\right) \text{ belongs to } M^\vee\right)$$

$$= \frac{1}{2}\sum_x\sum_a \sum_{\substack{w|D \\ w \text{ square free}}} \chi_{D,w}(\mathrm{N}(\mathfrak{a})a) + \chi_{D,w}(-\mathrm{N}(\mathfrak{a})a) \qquad \left(\chi_{D,w} = \prod_{p|w}(\ ,D)_p\right) \tag{4}$$

$$= \frac{1}{2}\sum_{w|D} \chi_{D,w}(\mathrm{N}(\mathfrak{a}))\sum_x\sum_a (\chi_{D,w}(a) + \chi_{D,w}(-a)).$$

Observe that the inner sum is non-negative. Clearly $\ell_\gamma(D) + \ell_\gamma^\vee(D)$ is maximal if γ is the principal genus. Note that we cannot separate $\ell_\gamma(D)$ and $\ell_\gamma^\vee(D)$ since we left out the condition $x + 2a > \sqrt{D}$. Using (4) we now show:

(3.4) Proposition. *We have*

$$c_1^2(Y_{\mathrm{PSL}(\mathfrak{o}\oplus\mathfrak{a})}) \geqslant c_1^2(Y_{\mathrm{PSL}(\mathfrak{o}\oplus\mathfrak{o})})$$

Proof. Recall that $12\chi(M,V) = \ell(M,V) - \ell(M^\vee,V)$. By Theorem (5.9) $\sum_\sigma(\ell(M_\sigma^\vee) - \ell(M_\sigma))$, where σ runs over the cusps of $\Gamma = \mathrm{PSL}(\mathfrak{o}\oplus\mathfrak{a})$, is maximal if $\mathfrak{a}$ is in the principal genus. We just saw that this is also true for $\sum_\sigma(\ell(M_\sigma^\vee) + \ell(M_\sigma))$. Hence $\sum_\sigma \ell(M_\sigma^\vee)$ is maximal for $\mathfrak{a}$ in the principal genus. If $D \geqslant 13$ only quotient singularities of order 2 and 3 occur and we have

$$c_1^2 = 4\zeta_K(-1) - \frac{1}{3}a_3^+ + \sum(2-b_k)$$

$$= 4\zeta_K(-1) - \frac{1}{3}a_3^+ + \sum_\sigma \ell(M_\sigma^\vee)$$

since $\sum_\sigma(\ell(M_\sigma^\vee) - \ell(M_\sigma)) = \sum(b_k - 3)$. Since a_3^+ is also maximal for the principal genus the result follows, except for $D \leqslant 12$, where it can be checked directly. □

4. Dimensions of Spaces of Cusp Forms

As an application of the results obtained so far we now compute the dimension of the space of Hilbert modular cusp forms of weight $2m$, $m \in \mathbb{Z}_{\geqslant 1}$ for a Hilbert modular group Γ.

We begin by assuming that Γ *is torsion-free.* As we saw in Chapter III in this case a Hilbert modular form of weight $2m$ on Γ can be viewed as a section over Y_Γ of $L^{\otimes m}$ with $L = \Omega^2(\log D)$, where $D = \sum S_i$ is the sum of all curves occurring in the cusp resolutions. By applying the Riemann-Roch theorem to $L^{\otimes m}$ we get

$$\chi(L^{\otimes m}) = \sum(-1)^i \dim_{\mathbb{C}} H^i(Y_\Gamma, L^{\otimes m})$$

$$= \tfrac{1}{2}(mc_1(L))(mc_1(L) + c_1) + \chi(Y_\Gamma).$$

From III.1 and (2.8) (or by the isomorphism $L \cong \mathcal{O}(\mathrm{K} + D)$ on Y_Γ with K the canonical divisor class) we find

$$c_1(L) = \tilde{c}_1 = -c_1 + \sum s_i \,, \tag{1}$$

where $c_i = c_i(Y_\Gamma)$ and where the s_i are the Poincaré duals of the homology classes of the S_i. Substituting (1) in the Riemann-Roch formula and using (2.3) and $\tilde{c}_1 \cdot s_i = 0$ we get

$$\begin{aligned}\chi(L^{\otimes m}) &= \tfrac{1}{2}(m\tilde{c}_1)\left((m-1)\tilde{c}_1 + \sum s_i\right) + \chi(Y_\Gamma)\\ &= m(m-1)\operatorname{vol}(\Gamma\backslash\mathfrak{H}^2) + \chi(Y_\Gamma)\,.\end{aligned}$$

Serre duality says

$$\chi(L^{\otimes m}) = \chi(L^{-m} \otimes \Omega^2_{Y_\Gamma})$$

and so we obtain

$$\chi(L^{\otimes m} \otimes \Omega^2_{Y_\Gamma}) = m(m+1)\operatorname{vol}(\Gamma\backslash\mathfrak{H}^2) + \chi(Y_\Gamma)\,.$$

Since we learned in Chapter II that the sections of $L^{\otimes m}$ embed Y_Γ into a projective space for large m we may apply Kodaira's Vanishing Theorem:

$$\dim H^i(L^{\otimes m} \otimes \Omega^2_{Y_\Gamma}) = 0 \qquad \text{for } i > 0, m \geqslant 1\,.$$

For $m > 1$ we thus find

$$\dim H^0(L^{\otimes(m-1)} \otimes \Omega^2_{Y_\Gamma}) = \chi(L^{\otimes(m-1)} \otimes \Omega^2_{Y_\Gamma})\,.$$

But the left hand side equals

$$\dim S_{2m}(\Gamma)\,,$$

the dimension of the space of cusp forms of weight $2m$ on Γ. Moreover by (2.6)

$$\chi(Y_\Gamma) = \frac{1}{4}\operatorname{vol}(\Gamma\backslash\mathfrak{H}^2) + \sum_\sigma \chi(M_\sigma, V_\sigma)\,,$$

where σ runs through the cusps of Γ. Combining these things yields:

(4.1) Proposition. *Let Γ be a torsion-free Hilbert modular group. The dimension of the vector space of cusp forms of even weight $k > 2$ on Γ is given by*

$$\dim S_k(\Gamma) = \frac{(k-1)^2}{4}\operatorname{vol}(\Gamma\backslash\mathfrak{H}^2) + \sum_\sigma \chi(M_\sigma, V_\sigma)\,.$$

In order to get a formula in the general case where Γ has torsion we shall apply the holomorphic Lefschetz formula, cf. [84]. If X is a compact complex manifold, V a holomorphic vector bundle on X and g an automorphism of X then one defines

$$\chi(X, V; g) = \sum (-1)^i \operatorname{Tr}(g^* | H^i(X, \mathcal{O}(V)))\,.$$

If G is a finite group of automorphisms of X then

$$\chi(X,V)^G \stackrel{\text{def}}{=} \sum (-1)^i \dim H^i(X,\mathcal{O}(V))^G$$
$$= \frac{1}{\# G} \sum_{g\in G} \chi(X,V;g) .$$

The holomorphic Lefschetz formula generalizes the Riemann-Roch formula $\chi(V) = (\mathrm{ch}(V)\,\mathrm{td}(X))[X]$ by a formula

$$\chi(X,V;g) = (\mathrm{ch}(V,g)\,\mathrm{td}(X,g))[X^g] ,$$

the notation of which we now explain. By X^g we mean the fixed point manifold of g and by $[X^g]$ its fundamental class. Let

$$c(T_{X^g}) = \prod_y (1+y)$$

be a formal factorization of the total Chern class of the tangent bundle (in the sense of [72]) and let

$$N^g = \bigoplus_\alpha N^g_\alpha$$

be the decomposition of the normal bundle of X^g in X in subbundles such that g acts on N^g_α by $e^{i\alpha}$. Moreover, let $c(N^g_\alpha) = \prod (1+y_\alpha)$. We thus have

$$c(T_X|X^g) = c(T_{X^g})c(N^g)$$
$$= \prod_y (1+y) \prod_\alpha \left(\prod_{y_\alpha} (1+y_\alpha) \right) .$$

Then the Todd class $\mathrm{td}(X,g) \in H^*(X^g,\mathbb{C})$ is defined by

$$\mathrm{td}(X,g) = \prod \frac{y}{1-\exp(-y)} \prod \frac{1}{1-\exp(-y_\alpha - i\alpha)} ,$$

where the products run over the same indices as in the last formula. The Chern character $\mathrm{ch}(X,V;g)$ is defined as follows. Let V be a complex bundle with G-action and let $R(G)$ be the complex representation ring of G. We then can write

$$[V] = \sum a_i \otimes \chi_i , \qquad a_i \in K(X) , \qquad \chi_i \in R(G) ,$$

where $K(X)$ is the Grothendieck group of X and χ_i runs through the characters of the irreducible representations of G. Then

$$\mathrm{ch}(X,V;g) = \sum \chi_i(g)\,\mathrm{ch}(a_i) \in H^*(X,\mathbb{C})$$

with ch the usual Chern character. For more details we refer to [84].

We apply this in the case where Γ is a Hilbert modular group, Γ' is a torsion-free normal subgroup of finite index of Γ and $G = \Gamma/\Gamma'$, $X = Y_{\Gamma'}$ and $V = \Omega^2 \otimes L^{(m-1)}$.

In the "finite part" $\Gamma'\backslash\mathfrak{H}^2$ of $Y_{\Gamma'}$ the fixed point set of $g \in G$, $g \neq \mathrm{id}$ consists of finitely many points corresponding to quotient singularities on $\Gamma\backslash\mathfrak{H}^2$ under $\Gamma'\backslash\mathfrak{H}^2 \to \Gamma\backslash\mathfrak{H}^2$. To a quotient singularity of $\Gamma\backslash\mathfrak{H}^2$ of type $(n;q,1)$ there corresponds a contribution

$$\frac{\# G}{n} \sum_{\substack{\zeta \neq 1 \\ \zeta^n = 1}} \frac{\zeta^{(q+1)m}}{(1-\zeta)(1-\zeta^q)} \qquad (\in \mathbb{Q})$$

to $\sum_{\mathrm{id} \neq g \in G} \chi(Y_{\Gamma'}, V; g)$, since for an isolated fixed point with rotation factor (ζ, ζ^q) one has $\mathrm{td}(X, g) = \dfrac{1}{(1-\zeta)(1-\zeta^q)}$.

The other components of the fixed point set lie on the cusp resolutions. Note that L is trivial in a neighbourhood of a cusp resolution (transfer the cusp to ∞, then $dz_1 \wedge dz_2$ yields a section at ∞). So here we may replace V by Ω^2. An element $g \in \Gamma/\Gamma'$ can have fixed points in the cusp resolution of cusp σ only if $g\sigma = \sigma$. In this case g is represented by an element of $\Gamma_\sigma/\Gamma'_\sigma$. We assume $\sigma = \infty$ and let $\Gamma_\infty = G(M, V)$ and $\Gamma'_\infty = G(M', V')$ with $V' \subseteq V$, $M' \subset M$.

Let $\mu \in M$. Note that V' fixes the residue class of $\mu \cdot (\mathrm{mod}\, M')$ since $\Gamma'_\sigma = (g\Gamma' g^{-1})_\sigma = g\Gamma'_\sigma g^{-1}$ for all $g \in \Gamma_\sigma$. Therefore, the translation $z \mapsto z + \mu$ on $\mathfrak{H}^2$ induces an action T_μ on $Y(M', V')$. If $\mu = \alpha_k A_{k-1} + \beta_k A_k$ with $\alpha_k, \beta_k \in \mathbb{Q}$ then T_μ is given in the local coordinates (u_k, v_k) on $Y(M', V')$ by

$$u_k \to u_k \exp(2\pi i \alpha_k), \qquad v_k \to v_k \exp(2\pi i \beta_k) \tag{2}$$

in particular, T_μ is the identity if $\mu \in M'$. Clearly T_μ fixes the origin of every coordinate system u_k, v_k, i.e. all the intersection points of the resolution curves. Moreover,

$$T_\mu | S_k = \mathrm{id} \Leftrightarrow (\mu + M) \cap \mathbb{R}_+ A_k \neq \emptyset .$$

The contribution of the automorphism T_μ to $\chi(Y(M', V'), \Omega^2; T_\mu)$ was computed by Sczech [141].

Let $\mu + M'_{be}$ a V-invariant residue class. Define

$$\chi(M' + \mu, V) := \mathrm{td}(Y, T_\mu)\,\mathrm{ch}(\Omega_Y, T_\mu)[Y^{T_\mu}] ,$$

where $Y = Y(M', V)$. If $\mu \in M'$ then $T_\mu = \mathrm{id}$ and by a suitable interpretation or just by definition one has

$$\chi(M' + 0, V) = \chi(M', V)$$

with the right hand side defined as in Chapter III.

Let us define

$$\text{for } x \in \mathbb{C} - \mathbb{Z} \quad c_0(x) = 0 , \quad c_1(x) = \cot \pi x , \quad c_2(x) = \mathrm{cosec}^2\, \pi x$$

$$\text{for } x \in \mathbb{Z} \quad c_0(x) = 1 , \quad c_1(x) = 0 , \quad c_2(x) = 1/3 .$$

(In general $c_k(x)$ is defined as $\pi^{-k} \sum_{m \in x + \mathbb{Z}} m^{-k}$, but the series is not absolutely convergent for $k = 0$ or 1. The functions $c_k(x)$ satisfy the distribution relation

$$\sum_{j \,(\mathrm{mod}\, n)} c_k\left(\frac{x+j}{n}\right) = n^k c_k(x) .)$$

(4.2) Proposition (Sczech, [141]). *Let $\mu = \alpha_k A_{k-1} + \beta_k A_k$ with $\alpha_k, \beta_k \in \mathbb{Q}$, where the A_k ($k \bmod r$) form the boundary points* (mod V) *of the convex hull of M'_+. Then*

$$\chi(M' + \mu, V) = -\frac{1}{4} \sum_{k \bmod r} \{b_k c_0(\alpha_k) c_2(\beta_k) + c_1(\alpha_k) c_1(\beta_k) - 1\} .$$

Proof. By working out td and ch one finds for an isolated fixed point $(u_k, v_k) = (0,0)$ the contribution

$$-\tfrac{1}{4}(c_1(\alpha_k) - i)(c_1(\beta_k) - i),$$

whereas a non-singular rational curve S_k fixed by T_μ (i.e. $\alpha_k \in \mathbb{Z}$, $\beta_k \notin \mathbb{Z}$) yields a contribution

$$-\tfrac{1}{4}\{b_k c_2(\beta_k) - 1 - 2ic_1(\beta_k)\}$$

(The c_k enter by the relation

$$\frac{X}{1 - \exp(-X - 2\pi i\alpha)} = c_0(\alpha) - (c_1(\alpha) + i)\frac{iX}{2} - \sum_{k=2}^{\infty} c_k(\alpha)\left(\frac{iX}{2}\right)^k.)$$

The imaginary terms disappear if we sum over k because

$$\sum_{\alpha_k, \beta_k \notin \mathbb{Z}} \{c_1(\alpha_k) + c_1(\beta_k)\} + 2 \sum_{\substack{\alpha_k \in \mathbb{Z} \\ \beta_k \notin \mathbb{Z}}} c_1(\beta_k) = \sum_{\alpha_k \notin \mathbb{Z}} c_1(\alpha_k) + \sum_{\beta_k \notin \mathbb{Z}} c_1(\beta_k)$$
$$= 0$$

(since $\beta_{k+1} = -\alpha_k$ for all k and $\alpha_{k+1} = \beta_k$ if $\alpha_k \in \mathbb{Z}$) and one obtains the desired formula. □

Note that for $\mu \in M'$ we get the formula for $\chi(M', V)$ given in Chapter III.

We saw at the end of Section 1 of Chapter III that $4\chi(M, V)$ equals the signature defect for the manifold with boundary $Y = Y(M, V)$. Something similar is true for $\chi(M' + \mu, V)$. Suppose that X is a 4-manifold with G-action such that G preserves the orientation and acts without fixed points on ∂X. One defines the α-invariant (introduced by Atiyah and Singer) for $g \in G$, by

$$\alpha(Y, g) := \operatorname{sign}(Y, g) - \mathscr{L}(Y, g),$$

that is, the deviation from the equivariant signature theorem: $\operatorname{sign}(Y, g) = \mathscr{L}(Y, g)$ which holds if X has no boundary. We refer to [84] for the definition of $\operatorname{sign}(Y, g)$ and $\mathscr{L}(Y, g)$. A computation for $X = Y(M', V)$, $g = T_\mu$ (cf. [113]) yields

$$\mathscr{L}(Y, T_\mu) = \sum_{k \bmod r} \{b_k c_0(\alpha_k) c_2(\beta_k) - c_1(\alpha_k) c_1(\beta_k)\} \qquad \mu \notin M'.$$

Since $H_2(Y, \mathbb{Q})$ is generated by the classes of the S_k and $T_\mu S_k = S_k$ (orientation preserved) we have $\operatorname{sign}(Y, T_\mu) = -r$.

We thus find

$$-4\chi(M' + \mu, V) = \alpha(Y, T_\mu).$$

(4.3) Lemma. *The invariant $\chi(M' + \mu, V)$ satisfies the distribution formula*

$$\sum_{\mu \in M/M'} \chi(M' + \mu, V) = [M : M']\chi(M, V)$$

for any $M' \subset M$ of finite index such that $V \subset U_M^+$ (a subgroup of finite index) preserves the residue classes $\mu + M' \bmod M$.

Proof. We use the equality $-4\chi(M+\mu', V) = \alpha(Y, T_\mu)$. The desired identity then results directly from a similar identity for α:

$$\sum_{g \in G} \alpha(Y, g) = \# G \cdot \alpha(Y/G, 1)$$

cf. [84], p. 75. □

Presumably one can give a more elementary proof by using (4.2) and the distribution relations of the $c_k(x)$.

(4.4) Theorem. *Let Γ be a Hilbert modular group. The dimension of the space of cusp forms of weight $2m \geqslant 4$ on Γ is given by*

$$\dim S_{2m}(\Gamma) = \left(\frac{2m-1}{2}\right)^2 \operatorname{vol}(\Gamma \backslash \mathfrak{H}^2) + \sum \sum_{\substack{\zeta^n = 1 \\ \zeta \neq 1}} \frac{\zeta^{(q+1)m}}{n(1-\zeta)(1-\zeta^q)} + \sum_{\sigma} \chi(M_\sigma, V_\sigma),$$

where the first sum is over all quotient singularities of Y_Γ (their type is given by $(n; q, 1)$) and where the last sum extends over all cusps of Γ.

Proof. Take a torsion-free normal subgroup Γ' of finite index in Γ. Apply the holomorphic Lefschetz formula to $Y_{\Gamma'}$, $G = \Gamma/\Gamma'$ and $V = \Omega^2 \otimes L^{(m-1)}$ on $Y_{\Gamma'}$. Adding the contributions and using Proposition (4.1) and Lemma (4.3) one arrives at the formula above. □

Since the defect d_1 vanishes for all cyclic quotient singularities (see Ch. III) a cusp form of weight 2 is a section of Ω^2. So

$$\dim S_2(\Gamma) = p_g(Y_\Gamma).$$

In Chapter III we saw that $-\chi(M, V) = \dfrac{d(M')}{4\pi^2} L(M, V; 1)$. This suggests that

$$-\chi(M'+\mu, V) = \frac{d(M')}{4\pi^2} L(M', V, \mu; 1), \tag{3}$$

where

$$L(M', V, \mu; s) = \sum_{\lambda \in M'+\mu/V} \frac{\operatorname{sign} \mathsf{N}(\lambda)}{|\mathsf{N}(\lambda)|^s} \qquad \operatorname{Re} s > 1.$$

Here the summation is over non-zero non-associated representatives of $M' + \mu$. This L-series has an analytic continuation to all of $\mathbb{C}$ and satisfies a functional equation. Following an idea of Siegel Sczech proved that one can rewrite the summation of this L-series in such a way that

$$\frac{d(M')}{\pi^2} L(M', V; 1) = \sum_{k \bmod r} \{c_0(\alpha_k)c_2(\beta_k) - c_1(\alpha_k)c_1(\beta_k) - 1\}$$

Combining this with (4.2) shows that (3) is true.

Anyway, using the relation $-\chi(M, V) = \frac{d(M)}{\pi^2} L(M, V; 1)$ one gets from (4.4) Shimizu's dimension formula for even weight. In [153] Shimizu proved a general formula for the dimensions of spaces of Hilbert modular forms using the Selberg Trace Formula, see also [2].

5. Representations on Spaces of Cusp Forms

Let $\mathfrak{c}$ be an integral ideal of K and let $\Gamma[\mathfrak{c}]$ be the image in $\Gamma = \mathrm{PSL}(\mathfrak{o} \oplus \mathfrak{a})$ of the group

$$\left\{ \begin{pmatrix} \alpha & \beta \\ \gamma & \delta \end{pmatrix} \in \mathrm{SL}(\mathfrak{o} \oplus \mathfrak{a}) \colon \alpha \equiv \delta \equiv 1 \ (\mathrm{mod}\ \mathfrak{c}), \gamma \in \mathfrak{c}\mathfrak{a}, \beta \in \mathfrak{c}\mathfrak{a}^{-1} \right\} . \tag{1}$$

It is called the *principal congruence subgroup* of Γ of level $\mathfrak{c}$. The natural map $X_{\Gamma[\mathfrak{c}]} \to X_\Gamma$ is a Galois covering with Galois group $\Gamma/\Gamma[\mathfrak{c}] \cong \mathrm{PSL}(2, \mathfrak{o}/\mathfrak{c})$. This group also acts on $Y_{\Gamma[\mathfrak{c}]}$ and on $S_k(\Gamma[\mathfrak{c}])$.

In this section we prove a theorem on the representation r of $\mathrm{PSL}(2, \mathfrak{o}/\mathfrak{c})$ on $S_k(\Gamma[\mathfrak{c}])$ for $\mathfrak{c}$ a prime ideal. This theorem generalizes a theorem of Hecke [66] for the classical modular group. This generalization was proved by Yoshida and Saito [139] using the Selberg trace formula for even weight $k \geqslant 4$. It was proved by Meyer and Sczech [113] for $k \geqslant 2$ using the holomorphic Lefschetz Theorem of Atiyah and Singer.

We assume that $\mathfrak{c} = \mathfrak{p}$ is a prime ideal. It is no restriction of generality to assume that $\mathfrak{p}$ is prime to $\mathfrak{a}$. We have $\Gamma/\Gamma[\mathfrak{c}] \cong \mathrm{PSL}(2, \mathbb{F}_q)$ with $q = \mathrm{N}(\mathfrak{p})$. We assume that $(q, 6) = 1$. Then the irreducible representations $R_{\dim}$ of $\mathrm{PSL}(2, \mathbb{F}_q)$ are indicated in the following character table.

	R_1	R_q	$R^{(i)}_{(q+e)/2}$	$R^{(j)}_{q+1}$	$R^{(l)}_{q-1}$
E	1	q	$(q+e)/2$	$q+1$	$q-1$
P	1	0	$(e \pm \sqrt{eq})/2$	1	-1
Q	1	0	$(e \mp \sqrt{eq})/2$	1	-1
A	1	1	$(-1)^a(1+e)/2$	$\rho^{ja} + \rho^{-ja}$	0
B	1	-1	$-(-1)^b(1-e)/2$	0	$-(\sigma^{lb} + \sigma^{-lb})$

Notations are as follows: $e = (-1)^{(q-1)/2}$; $1 \leqslant i \leqslant 2$, $1 \leqslant j \leqslant (q-4-e)/4$, $1 \leqslant l \leqslant (q-2+e)/4$; $E = \begin{pmatrix} 1 & 0 \\ 0 & 1 \end{pmatrix}$, $P = \begin{pmatrix} 1 & 1 \\ 0 & 1 \end{pmatrix}$, $Q = \begin{pmatrix} 1 & \delta \\ 0 & 1 \end{pmatrix}$ with $\delta \in \mathbb{F}_q^* - \mathbb{F}_q^{*2}$; A (resp. B) denotes an element of order $q-1$ (resp. $q+1$); $\rho = \exp(4\pi i/(q-1))$, $\sigma = \exp(4\pi i/(q+1))$ and a and b are integers depending on A and B.

We fix $R^{(1)}_{(q+e)/2}$ by requiring that $\operatorname{Tr} P = (e + \sqrt{eq})/2$ for this representation.

The multiplicities of the various representations R_d in the representation r of $\mathrm{PSL}(2, \mathbb{F}_q)$ on $S_k(\Gamma[\mathfrak{c}])$ can be computed by using the methods of section 4 to compute $\dim S_k(\Gamma')$ for certain $\Gamma' \supset \Gamma[\mathfrak{c}]$, e.g. the multiplicity of R_1 equals $\dim S_k(\Gamma)$ and that of R_q is $\dim S_k(\Gamma_0[\mathfrak{c}])$, where $\Gamma_0[\mathfrak{c}] = \left\{\begin{pmatrix}\alpha & \beta\\ \gamma & \delta\end{pmatrix} \in \mathrm{SL}(\mathfrak{o} \oplus \mathfrak{a})\colon \gamma \in \mathfrak{c}\mathfrak{a}\right\}\Big/\{\pm 1\}$.

Let y_1 (resp. y_2) be the multiplicity of $R^{(1)}_{(q+e)/2}$ (resp. $R^{(2)}_{(q+e)/2}$) in r. From the character table it follows that

$$(y_1 - y_2)\sqrt{eq} = \mathrm{Tr}\, r(P) - \mathrm{Tr}\, r(Q) \tag{2}$$

(5.1) Proposition. *Let $\mathfrak{p}$ be a prime ideal of $\mathfrak{o}$ with $(\mathfrak{p}, 6) = 1$ and $\mathfrak{p} \neq (\sqrt{5})$ for $D = 5$. Then*

$$(y_1 - y_2)\sqrt{eq} = \frac{2}{[U : U_{\mathfrak{p}}]} \sum_{i=1}^{h} \sum_{\substack{\mu \in \mathfrak{b}_i^{-2}\mathfrak{a}\\ \mu \bmod \mathfrak{p}}} \left(\frac{\mu}{\mathfrak{p}}\right) \chi(\mathfrak{b}_i^{-2}\mathfrak{a}\mathfrak{p} + \mu, U_{\mathfrak{p}}^2),$$

where $U_{\mathfrak{p}} = \{\varepsilon \in U_K\colon \varepsilon \equiv 1 \bmod \mathfrak{p}\}$, $\mathfrak{b}_i$, $i = 1, \ldots, h$ are integral ideals representing the elements of $\mathcal{Cl}(K)$, $\chi(M + \mu, V)$ is as in Section 4 and $\left(\frac{\cdot}{\mathfrak{p}}\right)$ is the quadratic residue symbol in K as defined in [67].

Proof. In order to apply the Lefschetz Theorem we must know the fixed points of the elements $\begin{pmatrix}1 & 1\\ 0 & 1\end{pmatrix}$ and $\begin{pmatrix}1 & x\\ 0 & 1\end{pmatrix}$ in $\mathrm{SL}(\mathfrak{o} \oplus \mathfrak{a})$ representing P and Q, cf. (2). The action of $\Gamma[\mathfrak{c}]$ is free if $\mathfrak{c}^2 \neq (2)$ and $\mathfrak{c}^2 \neq (3)$. The action on the cusps is described by the following lemmas.

(5.2) Lemma. *The $\Gamma[\mathfrak{c}]$-inequivalent cusps that are Γ-equivalent to the cusp corresponding to the ideal class of $\mathfrak{b}$ are in 1-1 correspondence with pairs (α, β) with $\alpha\mathfrak{o} + \beta\mathfrak{a}^{-1} = \mathfrak{b}$ modulo the equivalence relation: $(\alpha, \beta) \sim (\gamma, \delta) \Leftrightarrow$ there exists $\varepsilon \in \mathfrak{o}^*$ such that $\varepsilon\alpha \equiv \gamma$, $\varepsilon\beta \equiv \delta \pmod{\mathfrak{c}}$. Their number is $\mathrm{N}(\mathfrak{c})^2 \prod_{\mathfrak{p}|\mathfrak{c}}(1 - \mathrm{N}(\mathfrak{p})^{-2})/[U : U_{\mathfrak{c}}]$.*

Proof. In order that $\sigma(\alpha/\beta) = \gamma/\delta$ for some $\sigma = \begin{pmatrix}a & b\\ c & d\end{pmatrix}$ as in (1) (i.e. $a\alpha + b\beta = \lambda\gamma$, $c\alpha + d\beta = \lambda\delta$ and $\alpha\mathfrak{o} + \beta\mathfrak{a}^{-1} = \gamma\mathfrak{o} + \delta\mathfrak{a}^{-1}$) we must have $\lambda \in \mathfrak{o}^*$ and $\alpha \equiv \lambda\gamma$, $\beta \equiv \lambda\delta$ (mod $\mathfrak{c}$). This is also sufficient. To compute the number, note that

$$[\Gamma : \Gamma[\mathfrak{c}]] = e(\mathfrak{c})\mathrm{N}(\mathfrak{c})^3 \prod_{\mathfrak{p}|\mathfrak{c}} (1 - \mathrm{N}(\mathfrak{p})^{-2}), \tag{3}$$

where the product is over the prime ideals dividing $\mathfrak{c}$ and $e(\mathfrak{c}) = \frac{1}{2}$ if $2 \notin \mathfrak{c}$, $e(\mathfrak{c}) = 1$ if $2 \in \mathfrak{c}$.

The isotropy group of the cusp (α, β) in $\Gamma[\mathfrak{c}]$ is

$$\left\{\begin{pmatrix}\varepsilon^2 & \mu\\ 0 & 1\end{pmatrix} : \varepsilon \text{ unit}, \ \varepsilon \equiv 1 \ (\mathrm{mod}\ \mathfrak{c}), \mu \in \mathfrak{a}^{-1}\mathfrak{b}^{-2}\mathfrak{c}\right\},$$

which has index $\mathrm{N}(\mathfrak{c}) \cdot [U^2 : U_{\mathfrak{c}}^2] = e(\mathfrak{c})\mathrm{N}(\mathfrak{c})[U : U_{\mathfrak{c}}]$ in the isotropy group in Γ. □

(5.3) Lemma. *If $\mathfrak{p}$ is a prime ideal then the number of cusps of $\Gamma[\mathfrak{p}]$ fixed by P (resp. Q) equals $h(\mathsf{N}(\mathfrak{p})-1)/[U:U_{\mathfrak{p}}]$.*

Proof. Suppose the cusp represented by α/β with $\alpha\mathfrak{o}+\beta\mathfrak{a}^{-1}=\mathfrak{b}$ is fixed by $\begin{pmatrix}1 & 1\\ 0 & 1\end{pmatrix}$. Then there exists $\gamma\in\Gamma[\mathfrak{c}]$ with $\gamma\begin{pmatrix}1 & 1\\ 0 & 1\end{pmatrix}\alpha/\beta=\alpha/\beta$. Assuming that $\mathfrak{b}$ is coprime with $\mathfrak{p}$ we find $\beta\equiv 0 \pmod{\mathfrak{p}}$, $\alpha\not\equiv 0 \pmod{\mathfrak{p}}$. This also suffices. Using the lemma above one sees that out of the $\mathsf{N}(\mathfrak{p})-1$ residue classes of $\alpha \bmod \mathfrak{p}$ $(\mathsf{N}(\mathfrak{p})-1)/[U:U_{\mathfrak{p}}]$ classes represent a fixed cusp. A similar thing holds for Q. □

Proof of (5.1) *continued.* If α/β is a fixed cusp under P (resp. Q) then for $S\in \mathrm{PSL}(2,K)$ with $S(\alpha/\beta)=\infty$ the conjugate SPS^{-1} (resp. SQS^{-1}) acts by $z\mapsto z+\mu$ with $\mu\in\mathfrak{b}^{-2}\mathfrak{a}^{-1}$ satisfying $\alpha^2\mu\equiv 1 \pmod{\mathfrak{p}}$ (resp. $\alpha^2\mu\equiv\delta \pmod{\mathfrak{p}}$). It follows that the contribution of the cusp α/β to $\chi(Y_{\Gamma[\mathfrak{p}]},\Omega^2,g)$ with $g=P$ or Q is given by $\chi(\mathfrak{b}^{-2}\mathfrak{a}^{-1}\mathfrak{p}+\mu,U_{\mathfrak{p}}^2)$, cf. Section 4.

We have $S_{2k}(\Gamma[\mathfrak{p}]\simeq H^0(Y_{\Gamma[\mathfrak{p}]},\mathcal{O}(k\mathrm{K}+(k-1)S))$ with K the canonical divisor (class) and with S the resolution divisor of the cusps. Since $L=\mathcal{O}(\mathrm{K}+S)$ is trivial in a neighbourhood of the cusps the contributions of the cusps to $\chi(Y_{\Gamma[\mathfrak{p}]},\Omega^2\otimes L^{\otimes k-1};g)$ do not depend on k. The theorem now follows from Serre duality, $\mathrm{Tr}(g^*|H^0(Y_{\Gamma[\mathfrak{p}]},\mathcal{O}))=1$ and $H^1(Y_{\Gamma[\mathfrak{p}]},\Omega^2\otimes L^{\otimes k-1})=(0)$ by the Kodaira Vanishing Theorem, see Section 4. □

Now let χ_0 be a real character of the group of all principal ideals of K prime to $\mathfrak{p}$. The character χ_0 admits $h=\#\mathscr{C}\ell(K)$ extensions to a character of the group of all ideals prime to $\mathfrak{p}$. Let χ be such an extension. Define $L(s,\chi)=\sum_{\mathfrak{a}}\chi(\mathfrak{a})\mathsf{N}(\mathfrak{a})^{-s}$, ($\mathfrak{a}$ runs through the non-zero ideals in $\mathfrak{o}$ prime to $\mathfrak{p}$) and $L(s,\chi,A)$ for $A\in\mathscr{C}\ell(K)$ by restricting the summation in $L(s,\chi)$ to those $\mathfrak{a}$ lying in A.

Let $\chi_1,\ldots,\chi_r$ be the *real* characters of the group of all ideals prime to $\mathfrak{p}$ which extend χ_0. Moreover, let $\mathfrak{a}$ be an ideal prime to $\mathfrak{p}$ whose ideal class is A. Then we have

$$\sum_{\mathfrak{b}}\bar{\chi}(\mathfrak{a}\mathfrak{b}^{-2})L(s,\chi,AB^{-2})=\begin{cases}0 & r=0\\ \sum_{j=1}^{r}\chi_j(\mathfrak{a})L(s,\chi_j) & \text{otherwise}\,,\end{cases} \tag{4}$$

where $\mathfrak{b}$ is prime to $\mathfrak{p}$ and runs through a complete set of representatives of $\mathscr{C}\ell(K)$ and $B\in\mathscr{C}\ell(K)$ is the class of $\mathfrak{b}$. We leave the proof to the reader. Note the similarity with (3.1).

A term in the left-hand side of (4) can be written as

$$\mathsf{N}(\mathfrak{a}\mathfrak{b}^{-2})^{-s}\sum_{\mu\in\mathfrak{a}\mathfrak{b}^{-2}}\chi_0((\mu))\mathsf{N}((\mu))^{-s}\,. \tag{5}$$

Suppose in addition that χ_0 satisfies

$$\chi_0((\mu))=\left(\frac{\mu}{\mathfrak{p}}\right)\operatorname{sgn}\mathsf{N}(\mu) \qquad \mu\in\mathfrak{o}\,. \tag{6}$$

(Note that this implies $\left(\frac{-1}{\mathfrak{p}}\right) = +1$, so $\mathsf{N}(\mathfrak{p}) \equiv 1 \pmod 4$.) Then we have for any ideal $\mathfrak{c}$

$$\sum_{\mu \in \mathfrak{c}} \chi_0((\mu))\mathsf{N}((\mu))^{-s} = [U : U_{\mathfrak{p}}^2]^{-1} \sum_{\substack{\mu \in \mathfrak{c} \\ \mu \bmod U_{\mathfrak{p}}^2}} \left(\frac{\mu}{\mathfrak{p}}\right) \frac{\operatorname{sgn} \mathsf{N}(\mu)}{|\mathsf{N}(\mu)|^s}$$

$$= [U : U_{\mathfrak{p}}^2]^{-1} \sum_{\substack{\mu \in \mathfrak{c} \\ \mu \bmod U_{\mathfrak{p}}^2}} \left(\frac{\mu}{\mathfrak{p}}\right) L(\mathfrak{c}\mathfrak{p}, U_{\mathfrak{p}}^2, \mu; s),$$

where we refer to Section 4 for the definition of $L(M, V, \mu; s)$. Using this and (2) of Section 4

$$L(M, V, \mu; 1) = \frac{-4\pi^2}{\sqrt{\Delta_M}} \chi(M + \mu, V)$$

we find by substitution in (4)

$$\sum_{\mathfrak{b}} \sum_{\substack{\mu \in \mathfrak{b}^{-2}\mathfrak{a} \\ \mu \bmod \mathfrak{p}}} \left(\frac{\mu}{\mathfrak{p}}\right) \chi(\mathfrak{b}^{-2}\mathfrak{a}\mathfrak{p} + \mu; U_{\mathfrak{p}}^2) = -\frac{[U : U_{\mathfrak{p}}^2] D^{1/2} \mathsf{N}(\mathfrak{p})}{4\pi^2} \sum_{j=1}^{r} \chi_j(\mathfrak{a}) L(1, \chi_j)$$

This leads to the following analogue of (4.2) and (3.2), due to Meyer and Sczech [113].

(5.4) Theorem. *Let $\mathfrak{p}$ be a prime ideal of K with $(\mathfrak{p}, 6) = 1$ and $\mathfrak{p} \neq (\sqrt{5})$ if $D = 5$. Then*

$$\mathsf{N}(\mathfrak{p})^{-1/2} [U : U_{\mathfrak{p}}]^{-1} \sum_{\mathfrak{b}} \sum_{\substack{\mu \in \mathfrak{b}^{-2}\mathfrak{a} \\ \mu \bmod \mathfrak{p}}} \left(\frac{\mu}{\mathfrak{p}}\right) \chi(\mathfrak{b}^{-2}\mathfrak{a}\mathfrak{p} + \mu; U_{\mathfrak{p}}^2) = -\sum_{L} \chi_L(\mathfrak{a}) h'_{L/K},$$

where $\mathfrak{b}$ runs through a system of representatives of $\mathscr{Cl}(K)$ which are prime to $\mathfrak{p}$ and where L runs through all totally imaginary quadratic extensions of K with relative discriminant $\mathfrak{p}$ and χ_L denotes the corresponding character (on ideals prime to $\mathfrak{p}$). Moreover, $h'_{L/K}$ denotes the modified relative class number: $h'_{L/K} = \frac{w_K}{w_L} h_{L/K}$ with $w_F =$ # roots of 1 in $\mathfrak{o}_F$.

Proof. The characters χ_j, $j = 1, \ldots, r$ correspond 1-1 to the imaginary quadratic extensions L of K with relative discriminant $\mathfrak{p}$. For such L we have the analytic class number formula

$$\frac{h_L R_L}{h_K R_K} = \frac{w_L}{w_K} \sqrt{\frac{D_L}{D_K}} \frac{L(1, \chi)}{\pi^2}$$

with R_L (resp. R_K) the regulator of L (resp. K), cf. any standard text on algebraic number theory. In view of $(\mathfrak{p}, 6) = 1$ we have $R_L = 2R_K$ and $w_L = w_K = 2$ unless $\mathsf{N}(\mathfrak{p}) = D_K = 5$ (then $w_L = 10$). Finally, $[U : U_{\mathfrak{p}}^2] = 2[U : U_{\mathfrak{p}}]$. □

Combining (5.1) with (5.4) yields

(5.5) Theorem. *Let $\mathfrak{p}$ be a prime ideal of K which is relatively prime to (6) and $\mathfrak{a}$. Suppose $\mathfrak{p} \neq (\sqrt{5})$ if $K = \mathbb{Q}(\sqrt{5})$. Then*

$$y_1 - y_2 = -2 \sum_{L/K} \chi_L(\mathfrak{a}) h'_{L/K} ,$$

where the sum is over all totally imaginary quadratic extensions L of K with relative discriminant $\mathfrak{p}$. (The sum is zero if no such L exist.)

In the case $\mathrm{N}(\mathfrak{p}) = D_K = 5$ one finds $y_1 = y_2$ due to the fact that the quotient singularities give a contribution that cancels the one from the cusps.

6. The Vanishing of the Fundamental Group

In this section we prove that the fundamental group of the surface Y_Γ with $\Gamma = \mathrm{PSL}(\mathfrak{o} \oplus \mathfrak{a})$ vanishes, a result due to Svarčman [160].

(6.1) Theorem. *The Hilbert modular surface Y_Γ with $\Gamma = \mathrm{PSL}(\mathfrak{o} \oplus \mathfrak{a})$ is simply connected.*

Proof. Let $\pi\colon Y_\Gamma \to X_\Gamma$ be the natural map and for a singularity $x \in X_\Gamma$ let

$$R_x = \pi^{-1}(x) .$$

Choose suitable open neighbourhoods U_x of R_x in Y_Γ, e.g. for a cusp σ one takes the interior of the closure of $\Gamma_\sigma \backslash W(\sigma, r)$ for large r and for a quotient singularity x one takes $\Gamma_{x'} \backslash V_{x'}$, where $x' \in \mathfrak{H}^2$ is a representative of x and where $V_{x'}$ is a $\Gamma_{x'}$-invariant neighbourhood of x' in which the action of Γ reduces to that of the isotropy group $\Gamma_{x'}$. Choose a base point p of $Y^0 = Y_\Gamma - \sum R_x$ and choose base points p_x in $U_x - R_x$ for each singularity x. By van Kampen's theorem the fundamental group of $Y_\Gamma = Y^0 \cup \bigcup_x U_x$ is given by

$$\pi_1(Y_\Gamma, p) = \pi_1(Y^0, p)/N ,$$

where N is the smallest normal subgroup of $\pi_1(Y^0, p)$ containing the image under $i_x\colon \pi_1(U_x - R_x, p_x) \to \pi_1(Y^0, p)$ of

$$\ker\{\pi_1(U_x - R_x, p_x) \overset{d_x}{\to} \pi_1(U_x, p_x)\}$$

for all x. We can identify $\pi_1(Y^0, p)$ with Γ.

For a quotient singularity x one has $\pi_1(U_x, p_x) = 0$. We thus get

$$\bigcup_{\gamma \in \Gamma} \bigcup_x \gamma(i_x \ker d_x)\gamma^{-1} = \{\gamma \in \Gamma\colon \gamma \text{ is of finite order}\} .$$

If $x = \sigma$ is a cusp singularity of type (M_σ, V_σ), then $\pi_1(U_\sigma) = V_\sigma$ by Lemma (2.5). (Note that Y^+ is simply connected.) So we find that $i_x \ker d_x$ consists of the "translations" in Γ_σ. For the cusp at ∞ these are

$$\left\{\pm\begin{pmatrix}1 & \mu\\ 0 & 1\end{pmatrix} : \mu \in \mathfrak{a}^{-1}\right\}$$

and for the cusp at 0

$$\left\{\pm\begin{pmatrix}1 & 0\\ \mu & 1\end{pmatrix} : \mu \in \mathfrak{a}\right\}.$$

According to a deep theorem of Vaseršteĭn [166], cf. also [144], these elements generate Γ. This finishes the proof. (Vaseršteĭn proves that the groups

$$\left\{\begin{pmatrix}a & b\\ c & d\end{pmatrix} \in \mathrm{SL}(2,K) : a \equiv d \equiv 1 \bmod \mathfrak{ab}, c \in \mathfrak{a}, b \in \mathfrak{b}\right\}$$

where $\mathfrak{a}, \mathfrak{b}$ are entire ideals of K is generated by elementary matrices. It is not difficult to deduce that $\mathrm{SL}(\mathfrak{o} \oplus \mathfrak{a})$ is generated by elementary matrices.) □

(6.2) Corollary. *The Hilbert modular surfaces Y_{Γ_m} are simply connected.*

Proof. Since Γ/Γ_m is an abelian group of exponent 2, say with generators $a_1, \ldots, a_{t-1}$, we can factor the natural map $Y_\Gamma \to Y_\Gamma/\Gamma_m$ into $t-1$ maps of degree 2. We shall see later that all these maps are ramified in at least one point. It is well-known that if A is a topological space with $\pi_1 = 0$ and if a is an involution on A with at least one fixed point then $\pi_1(A/a) = 0$. Using this $t-1$ times we see that also Y_Γ/Γ_m is simply connected. By resolving the singularities of Y_Γ/Γ_m the fundamental group does not change. □

The vanishing of the fundamental group implies $b_1 = 2q = 0$. This result also follows from I (6.3) and holds for any Hilbert modular group. Indeed, suppose that ω is a holomorphic 1-form on Y. Let $f_1\, dz_1 + f_2\, dz_2$ be the pull-back of ω to $\mathfrak{H}^2$. Then f_1 (resp. f_2) is a Hilbert modular form of weight $(2,0)$ (resp. $(0,2)$), and therefore has to vanish.

However, there are Hilbert modular surfaces which are not simply connected. As an example, take for Γ the principal congruence subgroup of level $(\sqrt{2})$ in Γ_K with $K = \mathbb{Q}(\sqrt{2})$. As we shall show in Chapter VII the surface is a K-3 surface; in particular it is simply connected. The transformation $(z_1, z_2) \to (\varepsilon z_1, \varepsilon' z_2)$ with $\varepsilon = 3 + 2\sqrt{2}$ of $\mathfrak{H}^2$ defines a fixed point free involution of Y_Γ. So the quotient is not simply connected. It is an Enriques surface, see Ch. VIII.

7. Rigidity

In this section we use an argument of Faltings (see [29]) to show that Hilbert modular surfaces are defined over a number field. The idea is that a complex variety that admits no deformations is defined over a number field. Unlike compact quotients of bounded symmetric domains Hilbert modular surfaces admit in general many deformations, but—and this is the point—the whole "tower" of quotients $\Gamma\backslash\mathfrak{H}^2$, where Γ varies, does not.

If $\mathscr{X} \to B$ is a deformation of a complex manifold X with B a small (poly)disc around $0 \in \mathbb{C}^n$ then a tangent vector $\partial/\partial t$ to $0 \in \mathbb{C}^n$ defines a Kodaira-Spencer class $K(\partial/\partial t) \in H^1(X, \Theta)$, with Θ the tangent sheaf of X. The family $\mathscr{X}$ is locally trivial near 0 if and only if all Kodaira-Spencer classes $K(\partial/\partial t) \in H^1(X, \Theta)$ vanish. Similarly, if D is a divisor on X with normal crossings and $(\mathscr{X}, \mathscr{D})$ is a deformation of (X, D), then $\partial/\partial t$ defines a Kodaira-Spencer class in $H^1(X, \Omega^1(\log D)^\vee)$.

Let now X be a smooth variety over $\mathbb{C}$ and D a divisor with normal crossings. If $\partial: \mathbb{C} \to \mathbb{C}$ is a derivation (over $\mathbb{Q}$, and hence over $\bar{\mathbb{Q}}$), then ∂ defines a Kodaira-Spencer class $K(\partial)$ as follows. Let $\{U_i : i \in I\}$ be an open covering of X with coordinate neighbourhoods. On $U_i \cap U_j$ the coordinates $x_\alpha^{(i)}$ are functions $g_\alpha^{(ij)}$ of the coordinates $x_\alpha^{(j)}$. Associate to ∂ the cocycle defined on $U_i \cap U_j$ by $\sum_\alpha \partial g_\alpha^{(ij)} \dfrac{\partial}{\partial x_\alpha^{(j)}}$, and let $K(\partial)$ be the cohomology class of this cocycle. We claim:

i) $K(\partial)$ does not depend on the choices made,

ii) X is defined over $\bar{\mathbb{Q}}$ if and only if $K(\partial) = 0$ for all $\bar{\mathbb{Q}}$-derivations. Both assertions can be proved using the theory over $\mathbb{C}$: Let E be a field extension of finite type over $\mathbb{Q}$ over which X is defined. Consider $\mathscr{X} = X \otimes_{\bar{\mathbb{Q}}} \mathbb{C}$ over $E' = E \otimes_{\bar{\mathbb{Q}}} \mathbb{C}$. Then ∂ defines a $\mathbb{C}$-derivation of E' whose Kodaira-Spencer class in $H^1(X, \Theta)$ is just $K(\partial) \otimes 1$. Since this class is independent of the choices made i) follows. Moreover, ii) follows from: X is defined over $\bar{\mathbb{Q}}$ if and only if $\mathscr{X}$ has no deformations, and this is again equivalent to: $K(\partial) = 0$ for all $\mathbb{C}$-derivations ∂ of E'.

Similar things hold for pairs (X, D).

A morphism of pairs $(X_1, D_1) \to (X_2, D_2)$ induces both

$$f^*: H^1(X_2, \Theta_2) \to H^1(X_1, f^*\Theta_2)$$

and

$$df: H^1(X_1, \Theta_1) \to H^1(X_1, f^*\Theta_2)$$

A computation then shows that for any ∂

$$f^*(K_2(\partial)) = df(K_1(\partial)) . \tag{1}$$

Similarly for any morphism $f: (X_1, D_1) \to (X_2, D_2)$ of pairs there is a Kodaira-Spencer class $K(\partial, f) \in H_1(X_1, f^*\Theta_2)$. Again, f is defined over $\bar{\mathbb{Q}} \Leftrightarrow K(\partial, f) = 0$ for all derivations ∂ of $\mathbb{C}$.

We apply this to $X = Y_\Gamma$, $D =$ divisor at ∞, $\Theta = F^\vee = \Omega^1(\log D)^\vee$. For simplicity we assume that Γ acts freely. For any ∂ we find a class $K(\partial)$. It can be represented by a harmonic $F^\vee$-valued (0, 1)-form ϕ on Y_Γ. By pulling back to $\mathfrak{H}^2$ we find a harmonic form ϕ' on $\mathfrak{H}^2$ that is invariant under Γ. We claim that ϕ' is invariant under $\mathrm{SL}(2, \mathbb{R})^2$, hence zero. To see this, choose $\gamma \in \mathrm{SL}(2, K)$. Choose $\Gamma_1 \subset \mathrm{GL}^+(2, K)$ commensurable with Γ such that $\gamma\Gamma_1\gamma^{-1} \subset \Gamma$. The map $\mathfrak{H}^2 \to \mathfrak{H}^2$, $z \mapsto \gamma z$ defines a map $f_\gamma: Y'_{\Gamma_1} \to Y_\Gamma$, where Y'_{Γ_1} is a surface obtained by blowing up Y_{Γ_1} finitely many times. From (1) it follows that ϕ' is invariant under γ. Doing this for all $\gamma \in \mathrm{SL}(2, K)$ we find that ϕ' is $\mathrm{SL}(2, K)$-invariant, hence $\mathrm{SL}(2, \mathbb{R})$-invariant, hence zero.

We would like to conclude that $\phi = 0$. To see this we use Serre duality and show that for every F-valued (2, 1)-form ψ (representing an element of $H^1(Y_\Gamma, \Omega^2 \otimes F)$)

$$\int_{Y_\Gamma} \phi \wedge \psi = 0 .$$

Now $L_1 \otimes L_2 = \Omega^2(\log D)$ carries an hermitian metric (see IV.2). On $\Omega^2 \otimes F$ we thus have an hermitian metric defined by $\Omega^2 \otimes F \subset \Omega^2(\log D) \otimes F$. By the behaviour of this metric near D we have $\|\chi\| \to 0$ near D for any C^∞-section χ of $\Omega^2 \otimes F$. So the norm of ψ goes to zero near D. This shows that the integral, and hence ϕ is zero.

(7.1) Theorem. 1) *All pairs* (Y_Γ, D) *with* $D = Y_\Gamma - \Gamma\backslash\mathfrak{H}^2$, *and hence all* X_Γ *are defined over* $\bar{\mathbb{Q}}$.

2) *The morphisms* $f_\gamma\colon X_{\Gamma_1} \to X_\Gamma$ (*with* $\gamma\Gamma_1\gamma^{-1} \subset \Gamma$) *are all defined over* $\bar{\mathbb{Q}}$.

3) *The cusps of* X_Γ *are defined over* $\bar{\mathbb{Q}}$.

Proof. 1) If Γ acts fixed point freely then we find $K(\partial) = 0$ for all ∂ as shown above. If Γ has elliptic fixed points then a modification of our argument gives the result.

2) Similarly one shows $K(\partial, f_\gamma) = 0$ for all ∂.

3) $L = \Omega^2(\log D)$ is defined over $\bar{\mathbb{Q}}$, hence $X_\Gamma = \operatorname{Proj}(\bigoplus_m \Gamma(Y_\Gamma, L^{\otimes m}))$ is defined over $\bar{\mathbb{Q}}$, hence $\operatorname{Sing}(X_\Gamma)$, hence each of its points. □

(7.2) Corollary. *The* $\bar{\mathbb{Q}}$-*structure of* X_Γ *is the unique one for which all* f_γ *are defined over* $\bar{\mathbb{Q}}$.

(7.3) Theorem. *Let* $f \in S_{2m}(\Gamma)$ *be a modular form of weight* $2m$ *with Fourier expansion* $\sum a_\nu e^{2\pi i \operatorname{Tr} \nu z}$ *at* ∞. *Then* f *is defined over* $\bar{\mathbb{Q}}$ *if and only if all* $a_\nu \in \bar{\mathbb{Q}}$.

Proof. Let p be a singular point of the resolution cycle of $\sigma = \infty$ on Y_Γ. Suppose $uv = 0$ defines locally the resolution cycle near p. The completion R' of the local ring of p on Y_Γ over $\mathbb{C}$ is $\mathbb{C}[[u, v]]$. We let R be the completion of the local ring of p over $\bar{\mathbb{Q}}$. We are going to define an action of $K = M_\sigma \otimes \mathbb{Q} \subset \mathrm{SL}(2, K)$ on R' which preserves R. (Here M_σ denotes the module of translations, i.e. the unipotent part of Γ_σ, see II.1.) On R' this action is induced by $z \mapsto z + \mu$ $(\mu \in K)$ on $\mathfrak{H}^2$, cf. (2) of IV.4. To see that this preserves R choose $\Gamma_1 \subseteq \Gamma$ of finite index such that $\mu\Gamma_1\mu^{-1} \subseteq \Gamma$ and $\mu M_{\sigma_1}\mu^{-1} = M_\sigma$ with σ_1 a cusp of Y_{Γ_1} lying over σ. Choose a singular point p_1 of the resolution cycle of σ_1 such that $f_{\mathrm{id}}, f_\mu\colon Y_{\Gamma_1} \to Y_\Gamma$ both map p_1 to p. (This is possible.) Now f_{id} induces isomorphisms $R \cong R_1$, $R' \cong R_1' \cong \mathbb{C}[[u, v]]$, where R_1 (resp. R_1') is the completion of the local ring of p_1 over $\bar{\mathbb{Q}}$ (resp. over $\mathbb{C}$). Moreover, $f_\mu^* R \subset R_1$ since both f_{id} and f_μ are defined over $\bar{\mathbb{Q}}$.

Let $\mathfrak{m}$ (resp. $\mathfrak{m}'$) be the maximal ideal of R (resp. R'). The action of K/M_σ on the vector spaces $R/\mathfrak{m}^r$ and $R'/(\mathfrak{m}')^r$, $r \in \mathbb{Z}_{>0}$, can be diagonalized using the fact that K/M_σ is an inductive limit of finite abelian groups. The eigenspaces are spanned by monomials in u and v. Therefore, $\bar{\mathbb{Q}} \subset R$ are the constants (invariant elements), and there are $c_u, c_v \in \mathbb{C}$ such that $c_u u, c_v v \in R$ and $R = \bar{\mathbb{Q}}[[c_u u, c_v v]]$.

So we must show $c_u, c_v \in \bar{\mathbb{Q}}$. Let V be the group of (totally positive) units defined by the cusp at ∞. For $\varepsilon \in V \otimes \mathbb{Q} = K^* \subset \mathrm{GL}^+(2, K)$ we can choose $\Gamma_1 \subseteq \Gamma$ of finite index such that $\varepsilon \Gamma_1 \varepsilon^{-1} \subseteq \Gamma$. Take p_1 on Y_{Γ_1} as above. We assume that the action of ε on $\mathbb{R}^2_+/V_1$ maps the cone corresponding to p_1 (cf. Ch. II.) to itself. Using the definition of f_ε: $(z \mapsto \varepsilon z$ on $\mathfrak{H}^2)$ one checks that for all a, $b \in \mathbb{Z}_{>0}$ there exists an $N \in \mathbb{Z}_{>0}$ such that in R'_1 the equality $f_\varepsilon^*(u^a v^b)^N = (\mathrm{id})^*(u^c v^d)$ for some $c, d \in \mathbb{Z}$ holds. Since f_ε, f_{id} are defined over $\bar{\mathbb{Q}}$ this implies $c_u^{Na-c} c_v^{Nb-d} \in \bar{\mathbb{Q}}$. Choosing two such ε's such that the corresponding vectors $(Na - c, Nb - d)$ are linearly independent over $\mathbb{Z}$ gives that $c_u, c_v \in \bar{\mathbb{Q}}$. □

Since the field of definition is finitely generated over the prime field it follows that the field of definition of X_Γ is a number field. For more information about this number field see Chapter X.

V. Modular Curves on Modular Surfaces

In this chapter we introduce the modular curves on Hilbert modular surfaces. It is the presence of these curves that makes the geometry and arithmetic of these surfaces so rich. If one considers the surface $(\mathrm{PSL}(2,\mathbb{Z})\backslash\mathfrak{H})^2$ as a degenerate Hilbert modular surface, then on this surface the modular curves are just the Hecke correspondences T_N. If not by this analogy, one arrives at these modular curves by viewing Hilbert modular surfaces as moduli spaces of abelian surfaces with real multiplication: the points of the modular curves correspond to abelian surfaces whose endomorphism ring contains an order of a quaternion algebra, see Ch. IX. In spite of these heuristic indications and of the fact that the equations for these curves already occur in Hecke's thesis, nobody looked at them seriously before the seventies. Hirzebruch then studied them in connection with the classification of Hilbert modular surfaces. The development was then triggered by a letter of Serre to Hirzebruch (8 dec. 1971), where Serre noticed a remarkable coincidence of the dimension of a space of modular forms in one variable and the dimension of a certain part of the cohomology group $H^2(Y_{\Gamma_K}, \mathbb{C})$ containing the classes of these curves. The outcome of this will be the topic of the next chapter.

Here we discuss basic properties of the modular curves on our surfaces, such as how they intersect the resolution of the singularities, the number of their irreducible components, their volumes and how they intersect each other. Throughout this chapter we have $\Gamma = \Gamma(\mathfrak{o} \oplus \mathfrak{a})$ for some integral ideal $\mathfrak{a}$, unless stated otherwise. Some of the sections of this chapter are quite technical; the reader with no interest in these technicalities is advised to read only Sections 1, 2, 4, 5 and 8, which contain, respectively, the definition of the curves T_N, the description of their intersections with the cusp resolutions, an interpretation in terms of Grassmannians, a formula for the volume of T_N and the computation of the intersection number $T_1 \cdot T_N$.

1. The Curves F_N and T_N

In this chapter K will be a fixed real quadratic field with discriminant D and ring of integers $\mathfrak{o}$. The action of $\mathrm{Gal}(K/\mathbb{Q})$ is denoted by $x \mapsto x'$.

In order to construct curves on Hilbert modular surfaces other than those coming from the resolution of singularities one may start with the obvious candidate: the curve coming from the diagonal in $\mathfrak{H} \times \mathfrak{H}$. To see what we get, let $\mathfrak{b}$ be an integral ideal of $\mathfrak{o}$ which is not divisible by a natural number >1 and take $\Gamma = \Gamma(\mathfrak{o} \oplus \mathfrak{b})$.

By $\Gamma_0(N)$ for any natural number N we mean as usual the following subgroup of $\mathrm{SL}(2, \mathbb{Z})$:

$$\left\{\begin{pmatrix} a & b \\ c & d \end{pmatrix} \in \mathrm{SL}(2, \mathbb{Z}) \colon c \equiv 0 \pmod{N}\right\}. \tag{1}$$

(1.1) Proposition. *Suppose that $\mathfrak{b}$ is an integral ideal not divisible by a natural number >1. Then the subgroup of elements of $\mathrm{SL}(\mathfrak{o} \oplus \mathfrak{b})$ which carry the diagonal to itself equals $\Gamma_0(\mathrm{N}(\mathfrak{b}))$ if $\mathrm{N}(\mathfrak{b})$ is not divisible by the square-free part d of D. Otherwise it is an extension of index 2 of $\Gamma_0 \mathrm{N}(\mathfrak{b})$.*

Proof. A matrix B of $\mathrm{SL}(\mathfrak{o} \oplus \mathfrak{b})$ transforms the diagonal into itself if and only if

$$B = B' \quad \text{or} \quad B = -B',$$

where the prime means conjugation of all entries. If $B = B'$ then $B \in \mathrm{SL}(2, \mathbb{Q})$ and one sees that $\Gamma_0(\mathrm{N}(\mathfrak{b}))$ is the intersection of $\mathrm{SL}(2, \mathbb{Q})$ and $\mathrm{SL}(\mathfrak{o} \oplus \mathfrak{b})$ in $\mathrm{SL}(2, K)$ (use that for a primitive ideal $\mathfrak{b}$ one has $\mathfrak{b} \cap \mathbb{Q} = \mathrm{N}(\mathfrak{b})\mathbb{Z}$, $\mathfrak{b}^{-1} \cap \mathbb{Q} = \mathbb{Z}$). If $B = -B'$, we write $B = \sqrt{d}\begin{pmatrix} e & f \\ g & h \end{pmatrix}$ with $e, f, g, h \in \mathbb{Q}$ and d the square free part of D. Then $e, g, h \in \mathbb{Z}$. Suppose that $\mathfrak{b}$ is not divisible by $(\sqrt{d})$. Then a prime ideal dividing $(\sqrt{d})$ occurs with positive multiplicity in $(f\sqrt{d})$, hence in the determinant $d(eh - fg) = 1$, a contradiction. So $(\sqrt{d})$ divides $\mathfrak{b}$ and the elements B with $B = -B'$ are exactly the matrices of the form

$$\sqrt{d}\begin{pmatrix} e_0 & f_0/d \\ g_0 & h_0 \end{pmatrix}$$

of determinant 1 and with $e_0, f_0, g_0, h_0 \in \mathbb{Z}$, $g_0 \equiv 0 \pmod{\mathrm{N}(\mathfrak{b})/d}$. Such matrices exist since $(\mathrm{N}(\mathfrak{b})/d, d) = 1$. □

Note that in the particular case that $\mathfrak{b} = (\sqrt{d})$ we get as stabilizer of the diagonal the group $\Gamma^*(d)$: the extension of $\Gamma_0(d)$ of index 2 defined by $\Gamma^*(d) = \Gamma_0(d) \cup \Gamma_0(d) W_d$ with W_d the transformation given by the matrix

$$\begin{pmatrix} 0 & -1/\sqrt{d} \\ \sqrt{d} & 0 \end{pmatrix}.$$

This group was studied by Fricke.

Now suppose that $\mathfrak{a}$ and $\mathfrak{b}$ are ideals of $\mathfrak{o}$ lying in the same genus. Then there exists a matrix $M \in \mathrm{GL}^+(2, K)$ (cf. I.4) such that

$$M^{-1}\mathrm{SL}(\mathfrak{o} \oplus \mathfrak{a})M = \mathrm{SL}(\mathfrak{o} \oplus \mathfrak{b}).$$

By pulling back the image of the diagonal in $\Gamma(\mathfrak{o} \oplus \mathfrak{b})\backslash\mathfrak{H}^2$ under the isomorphism

$$\Gamma(\mathfrak{o} \oplus \mathfrak{a})\backslash\mathfrak{H}^2 \to \Gamma(\mathfrak{o} \oplus \mathfrak{b})\backslash\mathfrak{H}^2$$

induced by M we get a curve in $\Gamma(\mathfrak{o} \oplus \mathfrak{a})\backslash\mathfrak{H}^2$. It arises from the curve

$$\{(M\zeta, M'\zeta): \zeta \in \mathfrak{H}\}$$

in $\mathfrak{H}^2$. It can be considered as a "*twisted*" *form of the diagonal.* The curve in $\Gamma(\mathfrak{o} \oplus \mathfrak{a})\backslash\mathfrak{H}^2$ arising from such a twisted diagonal is the image under a map

$$\Delta\backslash\mathfrak{H} \to \Gamma(\mathfrak{o} \oplus \mathfrak{a})\backslash\mathfrak{H}^2 \tag{2}$$

where Δ is $\Gamma_0(N)$ or an extension of index 2 of a $\Gamma_0(N)$. The map (2) is *generically of degree* 1 *but not necessarily injective.* We shall later explain which image points are points of self-intersection of the image curve. All the curves we get in this way are "non-compact", that is, they are the image of a non-compact quotient $\Gamma\backslash\mathfrak{H}$ and are non-compact in $X_{\Gamma(\mathfrak{o}\oplus\mathfrak{a})}$.

To motivate the systematic treatment, notice that a twisted diagonal $\{(M\zeta, M'\zeta): \zeta \in \mathfrak{H}\}$ in $\mathfrak{H} \times \mathfrak{H}$ is the graph of a linear fractional transformation

$$z_2 = M'M^*z_1 = \begin{pmatrix} \lambda' & -b\sqrt{D} \\ a\sqrt{D} & \lambda \end{pmatrix} z_1 = Bz_1$$

with $a, b \in \mathbb{Q}$, $\lambda \in K$ and $M^* = (\det M)M^{-1}$. Equivalently, we may write it as

$$a\sqrt{D}\, z_1 z_2 - \lambda' z_1 + \lambda z_2 + b\sqrt{D} = 0 \tag{3}$$

or

$$(z_2 \quad 1)\begin{pmatrix} a\sqrt{D} & \lambda \\ -\lambda' & b\sqrt{D} \end{pmatrix}\begin{pmatrix} z_1 \\ 1 \end{pmatrix} = 0 \tag{4}$$

To fix notations, we choose an ideal $\mathfrak{a} \subset \mathfrak{o}$ with norm $\mathrm{N}(\mathfrak{a}) = A$. Since $X_{\Gamma(\mathfrak{o}\oplus\mathfrak{a})}$ only depends on the genus of $\mathfrak{a}$ we can choose $\mathfrak{a}$ appropriately. Set $\Gamma = \Gamma(\mathfrak{o} \oplus \mathfrak{a})$.

(1.2) Definition. A matrix $B \in M_2(K)$ is called *skew-hermitian* if ${}^tB' = -B$. A skew-hermitian matrix is called *integral* w. r. t. $\mathfrak{a}$ if it is of the form

$$\begin{pmatrix} a\sqrt{D} & \lambda \\ -\lambda' & \dfrac{b}{A}\sqrt{D} \end{pmatrix}$$

with $a, b \in \mathbb{Z}$, $\lambda \in \mathfrak{a}^{-1}$. It is called *primitive* if it is not divisible by a natural number $m > 1$.

(1.3) Definition. Let B be an integral skew-hermitian matrix. The curve F_B in X_Γ is the image in X_Γ of the set

$$\left\{(z_1, z_2) \in \mathfrak{H}^2 \cup \mathbb{P}^1(K): (z_2 \quad 1)B\begin{pmatrix} z_1 \\ 1 \end{pmatrix} = 0\right\}.$$

For $N \in \mathbb{Z}_{>0}$ the curve F_N is the union of all F_B with B skew-hermitian, integral, primitive and of determinant N/A.

Using Chow's lemma one can easily check that F_B is an algebraic curve on X_Γ.

If we omit the primitivity condition in (1.3) we get the *definition of* T_N; clearly

$$T_N = \bigcup_{\substack{t \geqslant 1 \\ t^2 | N}} F_{N/t^2} .$$

Note that if F_N contains one of our "twisted diagonals" then N/A (or NA) must be the norm of an element of K^* since in this case B has the form $M'M^*$ for some $M \in \mathrm{GL}^+(2, K)$.

(1.4) Lemma. *The curve F_N on X_Γ is non-empty if and only if for each prime q dividing D and not dividing N we have $\chi_{D(q)}(N) = (A, D)_q$. (For notations see p. 2.)*

Proof. We have $F_N \neq \emptyset \Leftrightarrow$ there exist $a, b \in \mathbb{Z}$, $\lambda \in \mathfrak{a}^{-1}$ such that $N = abD + A\lambda\lambda'$, that is, N is congruent modulo D to the norm of an ideal $\mathfrak{a}\lambda$ in the class of $\mathfrak{a}$. But this is equivalent to the condition given in the lemma. □

The group Γ_m/Γ acts on F_N and T_N and on the sets of their irreducible components. Under the natural map $X_\Gamma \to X_{\Gamma_m}$, F_N (resp. T_N) projects to a (reduced) curve in X_{Γ_m}, denoted by F_N^m (resp. T_N^m) or simply by F_N (resp. T_N) if no confusion is likely to arise.

The curves F_N and T_N on X_Γ give rise to curves on the non-singular model Y_Γ of X_Γ, namely their *strict transforms* under $Y_\Gamma \to X_\Gamma$, again denoted by F_N and T_N. All these curves F_N and T_N will be called loosely the *modular curves*. The T_N's are also called *arithmetic cycles*.

Now let us describe the components of F_N more closely. For an integral primitive skew-hermitian matrix B we consider

$$Q_B = \{M \in M_2(K)\colon {}^t M' B = BM^*\} , \tag{5}$$

where for $M = \begin{pmatrix} a & b \\ c & d \end{pmatrix}$ we put $M^* = \begin{pmatrix} d & -b \\ -c & a \end{pmatrix}$. Moreover we let

$$\mathfrak{o}_B = Q_B \cap \begin{pmatrix} \mathfrak{o} & \mathfrak{a}^{-1} \\ \mathfrak{a} & \mathfrak{o} \end{pmatrix}$$

and

$$E_B = \{M \in \mathrm{SL}(\mathfrak{o} \oplus \mathfrak{a})\colon {}^t M' B M = B\} = \{M \in \mathfrak{o}_B^*\colon \det M = 1\}$$

Then the stabilizer

$$\{M \in \mathrm{SL}(\mathfrak{o} \oplus \mathfrak{a})\colon {}^t M' B M = \pm B\}$$

of $\mathfrak{H}_B = \{(z_1, z_2) \in \mathfrak{H}^2 \cup \mathbb{P}^1(K)\colon (z_2 \quad 1) B^t (z_1 \quad 1) = 0\}$ is an extension of degree 1 or 2 of E_B. Now $\mathfrak{H}_B \cap \mathfrak{H}^2$ is the graph of a linear fractional transformation, so it is a copy of $\mathfrak{H}$ embedded in $\mathfrak{H}^2$. The quotient $E_B \backslash \mathfrak{H}_B$ is a compact non-singular model of F_B and we have a commutative diagram

$$\begin{array}{ccc} \mathfrak{H}_B & \hookrightarrow & \mathfrak{H}^2 \cup \mathbb{P}^1(K) \\ \downarrow & & \downarrow \\ E_B\backslash\mathfrak{H}_B & \longrightarrow & X_\Gamma \end{array}$$

The map $E_B\backslash\mathfrak{H}_B \to F_B$ induced by the lower vertical map is the normalization of the (possibly singular) curve F_B. Singularities of F_B will be discussed in Section 6.

In order to describe Q_B and E_B more precisely we introduce some notation.

For $a, b \in \mathbb{Q}^*$ we denote by $\left(\frac{a, b}{\mathbb{Q}}\right)$ the quaternion algebra over $\mathbb{Q}$ having a basis 1, i, j, k such that 1 is the unit element, $i^2 = a, j^2 = b, k = ij = -ji$. There is an injective algebra homomorphism

$$\left(\frac{a, b}{\mathbb{Q}}\right) \hookrightarrow M_2(\mathbb{Q}(\sqrt{a}))$$

with $i \mapsto \sqrt{a}\begin{pmatrix} 1 & 0 \\ 0 & -1 \end{pmatrix}$, $j \mapsto \begin{pmatrix} 0 & b \\ 1 & 0 \end{pmatrix}$. If a is not a square in $\mathbb{Q}$ then we also have the description

$$\left(\frac{a, b}{\mathbb{Q}}\right) \cong \mathbb{Q}(\sqrt{a}) \oplus \mathbb{Q}(\sqrt{a})j$$

with $j^2 = b, j\alpha = \alpha' j$ for all $\alpha \in \mathbb{Q}(\sqrt{a})$. The discriminant of a quaternion algebra Q is the product of the ramifying primes, i.e. the primes p such that $Q \otimes \mathbb{Q}_p$ is a skew field.

(1.5) Proposition. i) *Q_B is an indefinite quaternion algebra isomorphic to* $\left(\frac{D, -N/AD}{\mathbb{Q}}\right)$.

ii) *$\mathfrak{o}_B$ is an order of Q_B of discriminant N^2; more precisely,* $\mathfrak{o}_B = \mathfrak{o}_\mathfrak{a} + \left(\frac{\lambda}{\sqrt{D}} + j\right) M_\mathfrak{a} \subset K \oplus Kj$ *with $\mathfrak{o}_\mathfrak{a}$ the order of K of conductor* $\mathfrak{a}$ *and with $M_\mathfrak{a}$ the* $\mathfrak{o}_\mathfrak{a}$*-module* $\left\{x \in \frac{1}{\mathfrak{a}}\mathfrak{a} \colon \lambda x - \lambda' x' \in \mathbb{Z}\right\}$.

Proof. i) Replacing B by ${}^tR'BR$ where $R = \sqrt{a}\begin{pmatrix} 1/a & -\lambda/a\sqrt{D} \\ 0 & 1 \end{pmatrix}$ puts Q_B in the standard form for $\left(\frac{D, -N/AD}{\mathbb{Q}}\right)$ given above and makes ii) evident.

(1.6) Corollary. *The discriminant g_B of Q_B equals $g_B = \prod_q q$, where the product is over those primes q dividing N with either*

$$\left(\frac{D}{q}\right) = -1 \quad \textit{and} \quad \operatorname{ord}_q(N) \equiv 1 \pmod 2$$

or

$$\left(\frac{D}{q}\right) = 0 \quad and \quad (NA, D)_q = -1 \ .$$

By $Q_B \otimes \mathbb{R} \cong M_2(\mathbb{R})$ we have an action of the group of units of norm 1 of an (maximal) order of Q_B on $\mathfrak{H}$. It is well-known that a quotient of such an action is compact if and only if $g_B > 1$. Equivalently, using (1.6) we find

(1.7) Corollary. *The curve F_B on Y_Γ intersects the cusp resolutions if and only if $N = (\det B)A$ is the norm of an ideal from the genus of $\mathfrak{a}$.*

The question how it intersects the cusp resolutions will be answered in the next section.

2. Intersections with the Cusp Resolutions

Now we describe how the modular curves on the surface Y_Γ intersect the resolution cycles of the cusps. Instead of giving the intersections of a fixed F_N with the cusp resolutions we shall tell which curves F_N pass through a given point of the cusp resolutions.

Recall that the cusps of X_Γ with $\Gamma = \Gamma(\mathfrak{o} \oplus \mathfrak{a})$ are in $1-1$ correspondence with the ideal classes of K. Suppose that σ is a cusp corresponding to the ideal class of $\mathfrak{c}$. After transforming σ to ∞ as in I.1 we may assume that its isotropy group is of the form $G(M, V)$ with V an infinite cyclic group of totally positive units of $\mathfrak{o}$ preserving the module $M = \mathfrak{a}^{-1}\mathfrak{c}^{-2}$.

Consider as in II.2 the embedding of M_+ in $\mathbb{R}_+^2$. The boundary points of the convex hull of M_+ in $\mathbb{R}_+^2$ are denoted by $A_k, k \in \mathbb{Z}$. Two consecutive boundary points A_{k-1}, A_k form a basis of M and such a pair determines an embedding

$$M\backslash\mathfrak{H}^2 \to (\mathbb{C}^2)_k \subset Y^+$$

given in coordinates by

$$2\pi i z_j = A_{k-1}^{(j)} \log u_k + A_k^{(j)} \log v_k \qquad j = 1, 2 \, .$$

Let $\lambda \in M$ and consider the curves in $\mathfrak{H}^2/M$ defined by the equations

$$-\lambda' z_1 + \lambda z_2 + b\sqrt{\Delta_M} = 0 \, , \tag{1}$$

where b runs through $\mathbb{Z}$ and where $\sqrt{\Delta_M} = A_{k-1}A_k' - A_{k-1}'A_k$. Since we work on $\mathfrak{H}^2/M$ we may take b modulo the ideal $n\mathbb{Z} = \sqrt{\Delta_M^{-1}}\,\{-\lambda'\alpha + \lambda\alpha' \colon \alpha \in M\}$ of $\mathbb{Z}$. If $\lambda = pA_{k-1} + qA_k$ then $n = \text{g.c.d.}(p, q)$ and in the k-th coordinate system u_k, v_k (1) yields a relation

$$u_k^p = v_k^q \, . \tag{2}$$

Conversely, any branch $u_k^{p/n} = \zeta v_k^{q/n}$ with ζ a n-th root of 1 comes from a curve defined by an Equation (1).

The basis (A_{k-1}, A_k) of M determines a primitive binary integral quadratic form $Q_k(p,q)$, namely the norm form on M expressed in this basis after normalization such that it is primitive and integral, see II.5.

(2.1) Proposition. *The intersection of T_N with a sufficiently small neighbourhood of the coordinate axes in the k-th coordinate system (u_k, v_k) is given by all equations $u_k^p = v_k^q$ with $p, q \geqslant 0$ such that $Q_k(p,q) = N$. The primitive branches $u_k^{p/n} = \zeta v_k^{q/n}$ with $n = (p,q)$ and ζ a primitive n-th root of 1 belong to F_N.*

Proof. A curve defined by $a\sqrt{D}\,z_1 z_2 - \lambda' z_1 + \lambda z_2 + b\sqrt{D} = 0$ has a non-empty intersection with $W(\infty, r)$ for all large r if and only if $a = 0$. Take a cusp σ and transform it to ∞ by the matrix M_σ as in I.1.

The equations we now have to consider are given by skew-hermitian matrices ${}^tM'_\sigma B M_\sigma$ which we multiply by $\mathsf{N}(\mathfrak{c})^{-1}$ to make them integral with respect to $\mathfrak{a}^{-1}\mathfrak{c}^{-2}$. The first half of the proposition follows from the relation between (1) and (2) and the observation that $-\lambda' z_1 + \lambda z_2 + \beta = 0$ belongs to T_N (with $\lambda = pA_{k-1} + qA_k$, $p, q \in \mathbb{Z}_{\geqslant 0}$) if and only if $N \cdot \mathsf{N}(\mathfrak{a}^{-1}\mathfrak{c}^{-2}) = \mathsf{N}(\lambda) = \mathsf{N}(M) Q_k(p,q)$, i.e. $N = Q_k(p,q)$. □

We find again the result of the preceding section: F_N intersects the cusps if and only if N is the norm of an ideal in the genus of $\mathfrak{a}$. Let

$$Y_m(D) = \bigcup_{\mathfrak{a}} Y_{\Gamma_m}(\mathfrak{o} \oplus \mathfrak{a}),$$

where $\mathfrak{a}$ runs through a set of 2^{t-1} representatives of the genera. The h^+ cusps of $Y_m(D)$ correspond to the h^+ strict ideal classes of $\mathscr{C}\ell^+$. Take one of these cusps, say σ. One can now associate to a triple $(k; p, q)$ with $k \in \mathbb{Z}/r\mathbb{Z}$, p, $q \in \mathbb{Z}_{\geqslant 0}$ the integral ideal λM_σ^{-1}, where $\lambda = pA_{k-1} + qA_k$. In this way one gets all integral ideals strictly equivalent to M_σ^{-1}. (We have to identify $(k; 0, 1)$ with $(k+1; 1, 0)$ to get all ideals once.) In this way the h^+ cusps nicely display all integral ideals of K.

Examples. 1) Let $D = 13$. Then Y_{Γ_K} has only one cusp with cycle $(5, 2, 2)$. The quadratic forms are $Q_1 = x^2 + 5xy + 3y^2$, $Q_2 = 3x^2 + 7xy + 3y^2$ and $Q_3 = 3x^2 + 5xy + y^2$. We indicate some of the curves F_N in the following picture.

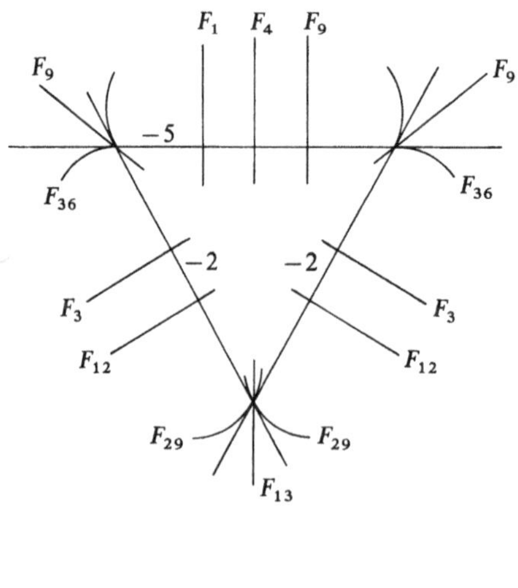

2) Let $D = 60$. There are four genera. Each surface $Y_{\Gamma_m(\mathfrak{o}\oplus\mathfrak{a})}$ has one cusp. In the following pictures some curves F_N which intersect the local branches of the cusp resolution transversally are indicated. The notation for the genera is as defined on p. 3.

genus $+++$

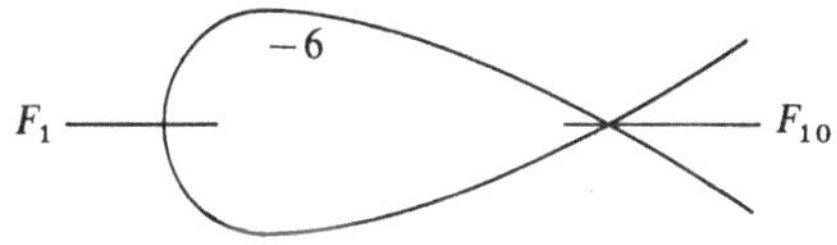

genus $+--$

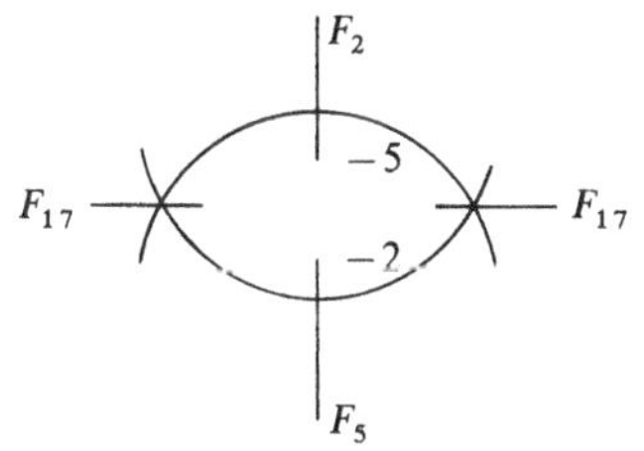

genus $-+-$

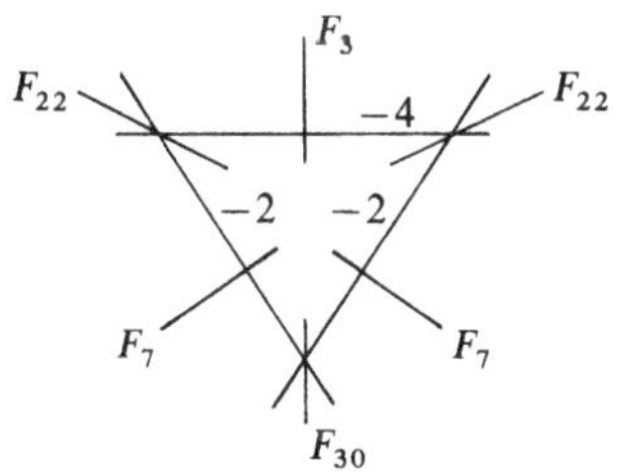

genus $--+$

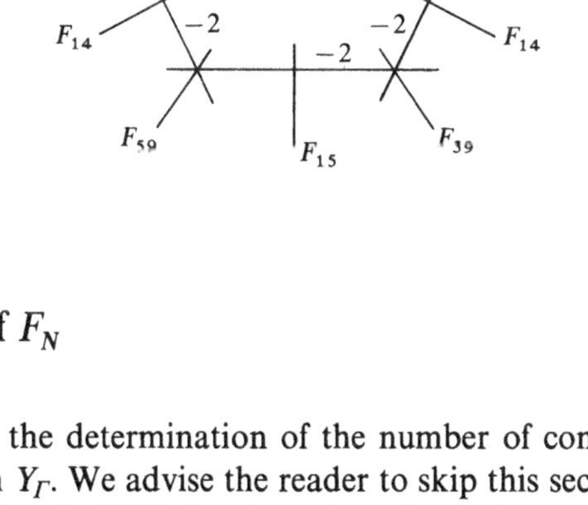

3. The Components of F_N

This section is devoted to the determination of the number of components of a modular curve F_N or T_N on Y_Γ. We advise the reader to skip this section on a first reading. It was shown by Franke [32] that for prime discriminants F_N has one or two components, the latter if and only if N is divisible by the square of the discriminant. Franke's results were extended to the general case by Hausmann [62]. Both of them use the theory of hermitian lattices. *We shall assume that* $\mathfrak{a}$ *is an ideal prime to* D.

Let $F(N, \mathfrak{a})$ be the set of primitive skew-hermitian matrices which are integral with respect to $\mathfrak{a}$ and of determinant N/A with $A = \mathrm{N}(\mathfrak{a})$. The groups $\mathrm{GL}(\mathfrak{o} \oplus \mathfrak{a})$ and $\mathrm{SL}(\mathfrak{o} \oplus \mathfrak{a})$ act on $F(N, \mathfrak{a})$ by conjugation: $B \to {}^tT'BT$. It is obvious that B_1 and B_2 define the same component of F_N if and only if

$$B_1 = \pm {}^tT'B_2T \quad \text{for some } T \in \mathrm{SL}(\mathfrak{o} \oplus \mathfrak{a}) .$$

We first study the number of $\mathrm{SL}(\mathfrak{o} \oplus \mathfrak{a})$-orbits of $F(N, \mathfrak{a})$ and deal with the $\pm$ sign later.

Let

$$R_1 = \{p\colon p \text{ prime}, p|D, p \nmid N\}$$

$$R_2 = \{p\colon p \text{ prime}, p|D, \mathrm{ord}_p(N) \geqslant 2\, \mathrm{ord}_p(D)\}$$

and let

$$r_i = \# R_i .$$

Now suppose that F_N is not empty. In order to distinguish the different orbits we introduce $r_1 + r_2$ invariants, each of which can assume two values.

First, let $p \in R_1$ and let

$$B = \begin{pmatrix} a\sqrt{D} & \lambda \\ -\lambda' & \frac{b}{A}\sqrt{D} \end{pmatrix}$$

be an element of $F(N, \mathfrak{a})$. Then A lies in $\mathfrak{o}$ and satisfies

$$A\lambda A\lambda' \equiv NA \pmod{p^{\mathrm{ord}_p(D)}} \tag{1}$$

Let $\mathfrak{p}$ be the ideal of $\mathfrak{o}$ satisfying $\mathfrak{p}^2 = p$. Note that A is co-prime with $\mathfrak{p}$. By (1) there are two residue classes modulo $\mathfrak{p}^{\mathrm{ord}_p(D)}$ to which $A\lambda$ (or what is the same, $A\lambda'$) can belong. If x is one residue class, then $-x$ is the other one except when $p = 2$ and $\mathrm{ord}_2(D) = 2$, but then 1 and $\sqrt{d}$ represent the two classes. One easily checks that the residue class of $A\lambda \pmod{\mathfrak{p}^{\mathrm{ord}_p(D)}}$ is a $\mathrm{SL}(\mathfrak{o} \oplus \mathfrak{a})$-invariant of B. We denote it by $\theta_p(B)$.

Secondly, let $p \in R_2$. We need a little lemma.

(3.1) Lemma. *Let* $B \in F(N, \mathfrak{a})$. *The* $\mathrm{SL}(\mathfrak{o} \oplus \mathfrak{a})$-*orbit of* B *contains an element*

$$\begin{pmatrix} a\sqrt{D} & \lambda \\ -\lambda' & \frac{b}{A}\sqrt{D} \end{pmatrix}$$

satisfying i) $(a, p) = 1$, ii) $A\lambda$ *and* b *belong to* $(\mathfrak{p}^{\mathrm{ord}_p(D)})$.

Choose such an element verifying i) and ii) and define

$$\eta_p(B) := (a, D)_p .$$

Of course we have to check that this does not depend on the particular choice that we have made. We leave the verification of this and the proof of (3.1) to the reader. In all we have $r_1 + r_2$ invariants of the orbits.

(3.2) Theorem. *Two elements B_1 and B_2 in $F(N, \mathfrak{a})$ belong to the same* $\mathrm{SL}(\mathfrak{o} \oplus \mathfrak{a})$*-orbit if and only if the r_1 invariants θ_p and the r_2 invariants η_p assume the same values for both of them. Moreover, if for none of the primes q dividing D one has $\chi_{D(q)}(N)(A,D)_q = -1$ then there are $2^{r_1+r_2}$ orbits.*

A sketch of the proof of this theorem will be given at the end of this section. We set

$$G_B = \{T \in \mathrm{SL}(\mathfrak{o} \oplus \mathfrak{a}) \colon {}^tT'BT = \pm T\}$$

$$E_B = \{T \in G_B \colon {}^tT'BT = T\} .$$

Of course, $[G_B : E_B]$ is either 1 or 2. The main result of this section is

(3.3) Theorem (cf. [62]). *The curve F_N on Y_Γ, if non-empty, has $2^{r_1+r_2-1}[G_B : E_B]$ components. Moreover, if d is the square-free part of D, then*

$$[G_B : E_B] = 2 \Leftrightarrow \begin{cases} (-1, D)_p = +1 & \text{for all } p \in R_2, \text{ and} \\ R_1 \text{ contains no prime dividing } d . \end{cases}$$

Proof. By (3.2) F_N has $2^{r_1+r_2}$ or $2^{r_1+r_2-1}$ components depending on whether B and $-B$ in $F(N, \mathfrak{a})$ belong to the same orbit, or not. They do if and only if $[G_B : E_B] = 2$. To check whether B and $-B$ belong to the same orbit one has to look at their θ_p and η_p. For the η_p it depends on $(-1, D)_p$ and for the θ_p one observes: $\theta_p(B) = \theta_p(-B) \Leftrightarrow p = 2$ and $\mathrm{ord}_2(D) = 2$. □

Now we look at the action of Γ_m/Γ on the set of components of F_N. By (3.2) it suffices to look at the action on their invariants. Recall that for a square-free divisor w of D we have an extension Γ_w of Γ. The index of this extension is 2 unless $w = 1$ or $w =$ square-free part of D. Moreover, $\Gamma_w = \Gamma_{\tilde{w}}$ if $\tilde{w} =$ square-free part of D/w, see I.4.

(3.4) Lemma. *Let $B \in F(N, \mathfrak{a})$ and let $B_w = \frac{1}{w}\,{}^tT'BT$ with $T = \begin{pmatrix} \alpha & \beta \\ \gamma & \delta \end{pmatrix} \in \Gamma_w - \Gamma$. Then*

i) *For $p \in R_1$ one has $\theta_p(B_w) = \theta_p(B) \Leftrightarrow p \nmid w$*

ii) *For $p \in R_2$ one has $\eta_p(B_w) = \eta_p(B)(w, D)_p$.*

Proof. One computes

$$B_w = \frac{1}{w}\begin{pmatrix} \alpha\alpha' a\sqrt{D} + \lambda\alpha'\gamma - \lambda'\alpha\gamma' + \gamma\gamma'\frac{b}{A}\sqrt{D} & * \\ -\alpha'\beta a\sqrt{D} - \lambda\alpha'\delta + \lambda'\gamma'\beta - \gamma'\delta\frac{b}{A}\sqrt{D} & * \end{pmatrix}$$

$$= \begin{pmatrix} c\sqrt{D} & \mu \\ -\mu' & * \end{pmatrix} .$$

For i) one notes $A\mu \equiv A\lambda(\alpha'\delta - \beta\gamma')/w \pmod{\mathfrak{p}^{\mathrm{ord}_p(D)}}$ with the notation as above. Assume that not both $p = 2$ and $\mathrm{ord}_2(D) = 2$ hold. Then $\frac{1}{w}(\alpha'\delta - \beta\gamma') \equiv \pm\frac{1}{w}(\alpha\delta - \beta\gamma) \equiv \pm 1 \pmod{\mathfrak{p}^{\mathrm{ord}_p(D)}}$, the sign depending on whether $p|w$ or not. This gives $\theta_p(B_w) = \theta_p(-B) \neq \theta_p(B)$ if $p|w$. If $p = 2$ and $\mathrm{ord}_2(D) = 2$ then $\alpha\delta - \beta\gamma = 2$ implies that either $\frac{\alpha\delta}{2}$ or $\frac{\beta\gamma}{2}$ lies in $(\sqrt{2})$ so $\frac{\alpha'\delta - \beta\gamma'}{2} \equiv \sqrt{d} \pmod 2$.

For ii) we point out that we may choose α such that $\left(\frac{\alpha\alpha'}{w}, D\right) = 1$. In fact choose α, γ with $0 \neq \gamma \in D\cdot\mathfrak{a}$ such that $(\sqrt{w}) = (\alpha, \gamma)$, cf. [67], Satz 74. If $w = x\alpha^2 + y\alpha\gamma + z\gamma^2$ for some $x, y, z \in \mathbb{Z}$ let $\beta = -y\alpha - z\gamma$ and $\delta = x\alpha$. Suppose that A satisfies i) and ii) of lemma (3.1). Then $\eta_p(B_w) = (c, D)_p = \left(\frac{a\alpha\alpha'}{w}, D\right)_p = \left(\frac{a}{w}, D\right)_p = (a, D)_p \cdot (w, D)_p$. □

Note that the condition $(p|w)$ in i) is not symmetric with respect to w and $\tilde{w}$, the square-free part of D/w. So in general Γ_w does not act on the $\mathrm{SL}(\mathfrak{o} \oplus \mathfrak{a})$ — orbits of $F(N, \mathfrak{a})$. Of course it does act on the equivalence classes of the relation $B \sim \pm {}^tT'BT$, $T \in \Gamma$.

(3.5) Theorem ([62]). *The group Γ_w/Γ either acts freely or trivially on the set of components of F_N. It acts trivially if and only if the following holds:*
either i) $p \in R_1 \Rightarrow p \nmid w$, $p \in R_2 \Rightarrow (w, D)_p = 1$
or ii) $p \in R_1 \Rightarrow p \nmid \tilde{w}$, $p \in R_2 \Rightarrow (\tilde{w}, D)_p = 1$.

Proof. This follows from Lemma (3.4) since the action of Γ_w/Γ on the curve corresponding to it is given by either $B \to B_w$ or $B \to B_{\tilde{w}}$ and the conditions depend only on N. □

Define $R_3 = \{p\colon p \text{ prime}, p|D, p|N\}$.

(3.6) Theorem. *Suppose that F_N is not empty on X_{Γ_m}. Then the number of irreducible components of F_N on X_{Γ_m} is $2^{r_2 - v}$, where v is the rank over $\mathbb{F}_2$ of the matrix $(m_{pq})_{p \in R_2, q \in R_3}$ with $(-1)^{m_{pq}} = (p, D)_q$.*

Proof. By Theorem (3.5) and the fact that $\#\Gamma_m/\Gamma = 2^{t-1}$ with t the number of primes dividing D one has almost immediately that the number in question is (with $P_3 = \prod_{p \in R_3} p$)

$$\frac{1}{2^{t-r_1-r_2}}\#\{w\colon w|P_3,(w,D)_q=1\text{ for all } q\in R_2\}$$

$$=\frac{1}{2^{t-r_1}}\sum_{w|P_3}\prod_{q\in R_2}(1+(w,D)_q)$$

$$=2^{-r_3}\sum_{w|P_3}\prod_{f\in P_2}(w,D)_f$$

$$=\sum_{f|P_2}\prod_{p\in P_3}\frac{1+(p,D)_f}{2},$$

where $P_2=\prod_{q\in R_2}q$ and $(x,D)_f=\prod_{q|f}(x,D)_q$. Now this last expression equals $\#\{f\colon f|P_2,(p,D)_f=+1\text{ for all } p\in R_3\}$. Each f from this set corresponds to a vector $(f_q)_{q\in R_2}\in\mathbb{F}_2^{r_2}$ (with $f_q=1$ if $q|f$, $f_q=0$ otherwise) with the property that

$$\sum_q m_{pq}f_q=0\quad\text{for all } p\in R_3\,.$$

So the desired number equals $\#\ker M$ with M the linear map $\mathbb{F}_2^{r_2}\to\mathbb{F}_2^{r_3}$ given by the matrix (m_{pq}).

(3.7) Corollary. *The action of Γ_m/Γ on the set of irreducible components of F_N on X_Γ is transitive if $r_2=0$.*

We end this section with a sketch of proof of Theorem (3.2).

Let $H(N,\mathfrak{a})=\{\sqrt{D}B\colon B\in F(N,\mathfrak{a})\}$ be the set of $\mathfrak{a}$-integral primitive hermitian matrices of determinant ND/A. It suffices to classify the $\mathrm{SL}(\mathfrak{o}\oplus\mathfrak{a})$-orbits of $H(N,\mathfrak{a})$ (action: $C\to {}^tT'CT$). If V is a 2-dimensional K-vector space then a hermitian form is a $\mathbb{Q}$-bilinear map $\phi\colon V\times V\to K$ which satisfies $\phi(x,y)=\phi(y,x)'$ and $\phi(\alpha x,\beta y)=\alpha\beta'\phi(x,y)$ for all $\alpha,\beta\in K$ and $x,y\in V$. Let (V,ϕ) be a two-dimensional hermitian vector space over K and let L be a lattice ($\mathfrak{o}$-module of rank 2) in V. Then $L\cong\mathfrak{o}\oplus\mathfrak{a}$ for some ideal $\mathfrak{a}$ as $\mathfrak{o}$-modules. Choose a basis v_1, v_2 of V such that $L=\mathfrak{o}v_1\oplus\mathfrak{a}v_2$. Then ϕ restricted to L corresponds to a hermitian matrix C and C is $\mathfrak{a}$-integral and primitive if and only if $\phi|L$ is integral valued and primitive. We call L *admissible* if $C\in H(N,\mathfrak{a})$.

Given lattices L_1, L_2 with $\mathfrak{o}$-module isomorphisms $\varepsilon_i\colon L_i\simeq\mathfrak{o}\oplus\mathfrak{a}$ such that under ε_i the lattice L_i corresponds to $C_i\in H(N,\mathfrak{a})$, then C_1, C_2 belong to the same $\mathrm{GL}(\mathfrak{o}\oplus\mathfrak{a})$-orbit if and only if there exists $g\in\mathrm{U}(V,\phi)$ (= unitary group of ϕ) with $gL_1=L_2$.

We may and do assume for the rest of this section that $\mathfrak{a}$ is a prime ideal of $\mathfrak{o}$ whose norm A is a prime that does not divide ND. The proof now proceeds in two steps. One first does a local classification of hermitian lattices and then makes the step from local to global.

To begin with local classification, let q be a prime of $\mathbb{Z}$ and localize everything at q: $K_q=K\otimes_{\mathbb{Q}}\mathbb{Q}_q$, $V_q=V\otimes_{\mathbb{Q}}\mathbb{Q}_q$, $\mathfrak{o}_q=\mathfrak{o}\mathbb{Z}_q$ and $L_q=\mathfrak{o}_qL\subset V_q$.

(3.8) Definition. Two admissible lattices L, $M\subset V$ are said to belong to the same genus (resp. strict genus) if L_q and M_q are $\mathrm{U}(V_q,\phi)$-equivalent (resp. $\mathrm{SU}(V_q,\phi)$-equivalent) for all q.

The introduction of genera for lattices induces the notion of a genus for $H(N, \mathfrak{a})$. We write $L \approx_q M$ for $\mathrm{SU}(V_q, \phi)$-equivalence and $L \sim_q M$ for $\mathrm{U}(V_q, \phi)$-equivalence. Our first task is to classify the genera of admissible lattices.

(3.9) Proposition. *If $q \nmid D$ then $L_q \approx_q M_q$.*

One shows that L_q and M_q have orthogonal bases with equal hermitian matrices.

(3.10) Lemma. *If L is an admissible lattice and if $\mathrm{ord}_q(N) \geqslant \mathrm{ord}_q(D) > 0$ then L_q has an orthogonal basis v, w such that $\phi(v, v)/D$ is a unit in $\mathbb{Z}_q$.*

Proof. Let $L = \mathfrak{o}x + \mathfrak{a}y$. Then $L_q = \mathfrak{o}_q x + \mathfrak{o}_q y$ since $q|D$ implies that $q \neq A$. If $\phi(x, x)/D$ is a unit then $v = x$, $w = y - (\phi(x, y)/\phi(x, x))x$ will do. If it is not, then interchange x and y or replace x by $x + y$. This suffices.

(3.11) Proposition. *Let L and M be admissible lattices and $q|D$.*

i) *If $\mathrm{ord}_q(N) \leqslant \mathrm{ord}_q(D)$ then $L_q \sim_q M_q$.*

ii) *If $2\,\mathrm{ord}_q(D) > \mathrm{ord}_q(N) > \mathrm{ord}_q(D)$ (hence $q = 2$) then $L_q \sim_q M_q$.*

iii) *If $\mathrm{ord}_q(N) \geqslant 2\mathrm{ord}_q(D)$ and if (v_1, w_1) and (v_2, w_2) are bases of L_q and M_q as in* (3.10) *then:*

$$L_q \sim_q M_q \text{ if and only if } (\phi(v_1, v_1)/D, D)_q = (\phi(v_2, v_2)/D, D)_q\,.$$

This proposition can be deduced from the literature on hermitian forms over local fields, cf. [91], esp. Prop. (5.2) and [92], esp. (8.1), (10.4) and Thm. (8.2).

Part iii) of Proposition (3.11) allows the introduction of a genus invariant. If L is admissible and (v, w) is a basis as in (3.10) then we take for $q \in R_2$ the invariant $(\phi(v, v)/D, D)_q$. It is a genus invariant, hence a $\mathrm{GL}(\mathfrak{o} \oplus \mathfrak{a})$-invariant of the corresponding matrix $C \in H(N, \mathfrak{a})$. Obviously, this is the invariant $\eta_q(C/\sqrt{D})$ defined after (3.1).

(3.12) Corollary. *$H(N, \mathfrak{a})$ consists of at most 2^{r_2} genera (distinguished by their η_q-invariants).*

We now go from local to global. Let p be a prime with $p|D$ and let L be an admissible lattice in (V, ϕ). We define

$$E_{\mathfrak{o}_p} = \{e \in \mathfrak{o}_p \colon ee' = 1\}$$

$$E_p(L) = \{\det g \colon g \in U(V, \phi), g(L_p) = L_p\}$$

$$G_p(L) = E_{\mathfrak{o}_p}/E_p(L)\,.$$

Note that $E_p(L)$ and hence $G_p(L)$ only depend on the genus of L.

(3.13) Lemma. *We have $\# G_p(L) = 1$ or 2. Moreover*

i) *If L_p has an orthogonal basis then $\# G_p(L) = 1$.*

ii) *If $\mathrm{ord}_p(A \det \phi) > \mathrm{ord}_p(D)$ then $\# G_p(L) = 1$.*

iii) *$\# G_p(L) = 2$ if and only if $E_p(L) = \{e^{-1}e' \colon e \in \mathfrak{o}_p^*\}$.*

This lemma can be deduced from [146] and [92]. The basic step now comes from the following theorem of Shimura (see [146] (5.19) and (5.27) ii)).

(3.14) Theorem. *Let L, M be admissible lattices belonging to the same* $\mathrm{GL}(\mathfrak{o} \oplus \mathfrak{a})$-*genus. Then L and M are $\mathrm{U}(V, \phi)$-equivalent if and only if the following holds: there is $x \in \mathbb{Q}(\sqrt{D})$ with $xx' = 1$ and for each prime $p|D$ an element $s_p \in U(V, \phi)$ with $s_p(L_p) = M_p$ such that*

i) $\mathfrak{o}x =$ *$\mathfrak{o}$-module generated by* $\{\det g : g \in \mathrm{End}(V), g(L) \subset M\}$

ii) $[x]_p = [\det s_p]_p$ *for every $p|D$, where $[\ \ ]_p$ means the class in $G_p(L) = G_p(M)$.*

Moreover, L and M belong to the same strict genus if and only if they are $\mathrm{SU}(V, \phi)$-equivalent.

(3.15) Corollary. *Every $\mathrm{GL}(\mathfrak{o} \oplus \mathfrak{a})$-genus of $H(N, \mathfrak{a})$ contains at most $2^{r_1}/\delta(L)$ $\mathrm{GL}(\mathfrak{o} \oplus \mathfrak{a})$-equivalence classes, where $\delta(L)$ is the order of the subgroup of $\prod_{p|D} G_p(L)$ generated by units of norm* 1.

Proof. Let L and M be admissible lattices with bases (v_1, w_1) and (v_2, w_2) whose associated matrices C_L, C_M belong to the same genus. If $g \in \mathrm{End}(V)$ sends (v_1, w_1) to (v_2, w_2) then $(\det g)(\det g') = 1$ since $\det C_L = \det C_M$. Moreover, $\det g$ generates the $\mathfrak{o}$-module of K generated by $\{\det g \colon g \in \mathrm{End}(V), gL \subset M\}$. According to (3.13) we find conditions exactly for those primes $p|D$ for which $E_p(L) = \{e^{-1}e' \colon e \in \mathfrak{o}_p^*\}$ with $\mathrm{ord}_p(ND) \leqslant \mathrm{ord}_p(D)$; i.e. $p|D$, $p \nmid N$, giving 2^{r_1} possibilities. However, since $(\det g)\mathfrak{o} = (\varepsilon \det g)\mathfrak{o}$ for all $\varepsilon \in \mathfrak{o}^*$ we may divide by $\delta(L)$. □

(3.16) Corollary. *Let L be a $\mathrm{GL}(\mathfrak{o} \oplus \mathfrak{a})$-genus in $H(N, \mathfrak{a})$. Then the number of $\mathrm{SL}(\mathfrak{o} \oplus \mathfrak{a})$-orbits in $H(N, \mathfrak{a})$-orbits is at most 2^{r_1}.*

Proof. By (3.15) we must show that the number of $\mathrm{SL}(\mathfrak{o} \oplus \mathfrak{a})$-orbits in a $\mathrm{GL}(\mathfrak{o} \oplus \mathfrak{a})$-orbit is at most $\delta(L)$. If $C \in H(N, \mathfrak{a})$, let for $\varepsilon \in \mathfrak{o}$ with norm 1

$$C_\varepsilon = \begin{pmatrix} \varepsilon' & 0 \\ 0 & 1 \end{pmatrix} C \begin{pmatrix} \varepsilon & 0 \\ 0 & 1 \end{pmatrix}.$$

It suffices to show: C_ε belongs to the $\mathrm{SL}(\mathfrak{o} \oplus \mathfrak{a})$-orbit of C if $[\varepsilon] = 1 \in \prod_{p|D} G_p(L)$. If $L = \mathfrak{o}v_1 + \mathfrak{o}w_1$ has matrix C then $M = \mathfrak{o}v_1 + \mathfrak{o}w_2$ with $w_2 = w_1 + (\varepsilon' - 1)(\phi(w_1, v_1)/\phi(v_1, v_1))v_1$ has matrix C. (We may assume $\phi(v_1, v_1) \neq 0$.) Now we show: there exists $s \in \mathrm{SU}(V, \phi)$ with $s(L) = M$. According to (3.14) it suffices to show that for every prime p there exists a $s_p \in \mathrm{SU}(V, \phi)$ with $s_p(L_p) = M_p$. If $p \nmid D$ then (3.9) assures the existence of such a s_p. If $p|D$ we have $s \in U(V, \phi)$ with $s(L) = M$ and we know $\varepsilon\varepsilon' = 1$ if $\varepsilon = \det s$. Compose s with $t \in \mathrm{End}(V, \phi)$, $t(L) = L$ with $t(v_1) = \varepsilon' v_1$, $t(w_1) = w_1$. □

It follows from (3.15) and (3.16) that there are at most $2^{r_1+r_2}$ $\mathrm{SL}(\mathfrak{o} \oplus \mathfrak{a})$-orbits in $F(N, \mathfrak{a})$. For (3.2) we still have to show that for each $r_1 + r_2$-tuple of invariants there is an element $B \in F(N, \mathfrak{a})$ with these invariants. Look at the quadratic form ϕ: $R = \mathbb{Z}^2 \oplus \mathfrak{a}^{-1} \to \mathbb{Z}$ of rank 4 given by $(a, b, \lambda) \to abD + A\lambda\lambda'$. The θ_p and η_q-invariants of

(a, b, λ) only depend on its class in R/DR. The question is whether ϕ restricted to this class represents N. By [181] it suffices to check that the residue class of N modulo an arbitrary integer is represented by ϕ restricted to this class. The theory of norm residues in K (cf. [67]) tells us that there are no problems with the θ_p. Moreover, choose a prime r which does not divide ADN and such that $\eta_q = \left(\frac{D(q)}{r}\right)$ for all $q \in R_2$. Fixing the residue class of $a \pmod{D}$ by $a \equiv r \pmod{D}$ puts no restrictions on the residue class of $ab \pmod{D}$. Also ii) in (3.1) can be satisfied then. This finishes the proof of (3.2).

4. The Geometry of SO(2, 2)

In this section we look at the modular curves from a somewhat different point of view.

Let V be the 4-dimensional vector space over $\mathbb{Q}$ of all skew-hermitian matrices over K. The determinant induces a quadratic form φ on V which we can extend to $V_{\mathbb{R}} = V \otimes \mathbb{R}$. On $V_{\mathbb{R}}$ the extension of φ (again denoted φ) has signature (2, 2). The vector space V contains the lattice of integral skew-hermitian matrices (with respect to a given $\mathfrak{b}$), where integral has the meaning defined in Section 1.

Choose a vector $v \in V_{\mathbb{R}}$. The set

$$\left\{z \in \mathfrak{H}^2 : (z_2 \quad 1) v \begin{pmatrix} z_1 \\ 1 \end{pmatrix} = 0\right\}$$

is non-empty if and only if $\varphi(v) > 0$. Conversely, to $z \in \mathfrak{H}^2$ we can associate

$$V_z = \left\{v \in V_{\mathbb{R}} : (z_2 \quad 1) v \begin{pmatrix} z_1 \\ 1 \end{pmatrix} = 0\right\},$$

which is a 2-dimensional linear subspace of $V_{\mathbb{R}}$ on which φ is positive definite. This construction provides us with a bijective map

$$\begin{aligned} \mathfrak{H}^2 &\to \mathrm{Gr}^+ \\ z &\mapsto V_z , \end{aligned} \tag{1}$$

where Gr^+ is the Grassmannian of all 2-dimensional linear subspaces of $V_{\mathbb{R}}$ on which φ is positive definite. Let SO(2, 2) be the special orthogonal group of $(V_{\mathbb{R}}, \varphi)$ and let $\mathrm{SO}_\circ(2, 2)$ be its identity component. The Grassmannian Gr^+ can be identified with the symmetric space

$$\mathrm{SO}_\circ(2,2)/\mathrm{SO}(2) \times \mathrm{SO}(2) ,$$

which combined with (1) yields an isomorphism of symmetric spaces $\mathfrak{H}^2 \approx \mathrm{SO}_\circ(2,2)/\mathrm{SO}(2) \times \mathrm{SO}(2)$.

In this terminology the skew-hermitian curve defined by $v \in V$ with $\varphi(v) > 0$ is given as

$$\{W \in \mathrm{Gr}^+(2, V_{\mathbb{R}}, \varphi) \colon W \ni v\},$$

which as a symmetric space is $SO_\circ(2,1)/SO(2) \times SO(1) \approx \mathfrak{H}$.

The natural boundary of $\mathrm{Gr}^+(2, V_{\mathbb{R}}, \varphi)$ inside the Grassmannian of all 2-planes in $V_{\mathbb{R}}$ is the set of all two-planes on which φ is only semi-definite ($\geqslant 0$) and not zero. A boundary point is called rational if the corresponding W is defined over $\mathbb{Q}$. If Γ is a Hilbert modular group then it corresponds to a discrete (arithmetic) subgroup of $SO_\circ(2,2)$. The orbits of the rational boundary points are exactly the cusps. The cusps lying on the skew-hermitian curve defined by v correspond to the 2-planes in V containing v on which φ is positive semi-definite, and not definite.

Now define for $v \in V_{\mathbb{R}}$

$$\mathfrak{H}_v = \left\{ z \in \mathfrak{H}^2 \cup \mathbb{P}^1(K) : (z_2 \quad 1) v \begin{pmatrix} z_1 \\ 1 \end{pmatrix} = 0 \right\}.$$

For $\varphi(v) > 0$ $\mathfrak{H}_v$ is the union of a copy of $\mathfrak{H}$ embedded in $\mathfrak{H}^2$ by a linear fractional transformation and a set of "boundary points" from $\mathbb{P}^1(K)$. Its image in $\overline{\Gamma\backslash\mathfrak{H}^2}$ is denoted by F_v if v is primitive.

We fix $\Gamma = \Gamma(\mathfrak{o} \oplus \mathfrak{a})$. Let $L(\mathfrak{a}) \subset V_{\mathbb{R}}$ be the lattice of all skew-hermitian matrices that are integral w.r.t. $\mathfrak{a}$. The curve T_N can be defined as

$$T_N = \Gamma \Bigg\backslash \left\{ \bigcup_{\substack{v \in L(\mathfrak{a}) \\ \varphi(v) = N/A}} \mathfrak{H}_v \right\}$$

The irreducible components of T_N correspond to the orbits of Γ on $\{v \in L(\mathfrak{a}) \colon \varphi(v) = N/A\}$. The intersection $V_z \cap V$ can be a 0-, 1- or a 2-dimensional vector space over $\mathbb{Q}$. The meaning of this is:

$$\dim V_z \cap V \geqslant 1 \Leftrightarrow z \in F_N \text{ for some } N$$

$$\dim V_z \cap V = 2 \Leftrightarrow z \in F_N \text{ for infinitely many } N.$$

To be more precise we have

(4.1) Lemma. *Let v_1, v_2 be linearly independent vectors in $V_{\mathbb{R}}$ and let W be the 2-dimensional subspace of $V_{\mathbb{R}}$ spanned by v_1 and v_2. Then*

i) $\varphi | W > 0 \Leftrightarrow \mathfrak{H}_{v_1} \cap \mathfrak{H}_{v_2}$ *is a point of* $\mathfrak{H}^2$

ii) $\varphi | W \geqslant 0$, $\phi | W \not> 0 \Leftrightarrow \mathfrak{H}_{v_1} \cap \mathfrak{H}_{v_2}$ *is a point of* $\mathbb{P}^1(K)$

iii) $\varphi | W \not\geqslant 0 \Leftrightarrow \mathfrak{H}_{v_1} \cap \mathfrak{H}_{v_2} = \emptyset$.

Proof. This can be easily seen using the bijection $\mathfrak{H}^2 \leftrightarrow \mathrm{Gr}^+$. □

5. The Volume of the Modular Curves

Let F_B be a skew-hermitian curve on $\Gamma\backslash\mathfrak{H}^2$. Its normalization is $\Gamma_B\backslash\mathfrak{H}$. We can measure its volume

$$\mathrm{vol}(F_B) := \int_{\Gamma_B\backslash\mathfrak{H}} \omega,$$

where $\omega = -\frac{1}{2\pi}\frac{dx \wedge dy}{y^2}$ is the invariant volume element on $\mathfrak{H}$. We write $\mathrm{vol}(T_N)$ for the sum of the volumes of the components F_B of T_N. For instance, if $\Gamma = \Gamma_K$ $\mathrm{vol}(T_1) = -1/6$ because T_1 has one component isomorphic to $\mathrm{SL}(2,\mathbb{Z})\backslash\mathfrak{H}$ and the integral of ω over a fundamental domain for $\mathrm{SL}(2,\mathbb{Z})$ equals $-1/6$, cf. IV.1. By (2.3) of Ch. IV we can view this number as $\frac{1}{2}$ times the degree of the line bundle L on X_Γ when restricted to the completion of F_A in X_Γ if Γ acts freely.

(5.1) Theorem. *Let $\Gamma = \Gamma(\mathfrak{o} \oplus \mathfrak{a})$. Then the generating series*

$$\frac{1}{4}\mathrm{vol}(\Gamma\backslash\mathfrak{H}^2) + \sum_{N=1}^{\infty} \mathrm{vol}(T_N)q^N$$

equals the Eisentein series of weight 2 in $M_2^\gamma(\Gamma_0(D), \chi_D)$ with γ the genus of $\mathfrak{a}$.[1]) *In particular, if $\mathfrak{a} \subset \mathfrak{o}$ has norm A with $(A, D) = 1$, then*

$$\mathrm{vol}(T_N) = -\frac{1}{12}\sum_{d|N} d \prod_{p|D} (\chi_{D(p)}(d) + \chi_{D(p)}(NA/d)) .$$

We do not give a proof here. The theorem is a consequence of Siegel's Mass-formula ([158]). One has to use the description of $\Gamma\backslash\mathfrak{H}^2$ in terms of $\mathrm{SO}(2,2)$ as given in Section 4. One can also give a direct proof by computing the index of E_B in the group of units of norm 1 in a maximal order of Q_B containing $\mathfrak{o}_B$ and by using a theorem of Eichler which says the following: If Δ is the group of units of norm 1 of a maximal order of an indefinite quaternion algebra over $\mathbb{Q}$ with discriminant g_0 then $\mathrm{vol}(\Delta\backslash\mathfrak{H}^2) = -\frac{1}{6}\varphi(g_0)$ with φ Euler's phi-function. This was carried out by Franke for prime discriminants and by Hausmann in the general case, [32, 62]. It follows from both proofs that all components of F_N on $\Gamma\backslash\mathfrak{H}^2$ have the same volume.

6. The Intersection Points of the Modular Curves

The present section deals with the position and the properties of the intersection points of the modular curves. The reader is advised to skip this section on a first reading. In Section 4 we saw that a point $z \in \Gamma\backslash\mathfrak{H}^2$ is the intersection point of two modular curves if and only if $\mathrm{rk}(V_z \cap L(\mathfrak{a})) = 2$. We set $L = L(\mathfrak{a})$.

(6.1) Definition. A point $z \in \Gamma\backslash\mathfrak{H}^2$ is called *special* if $\mathrm{rk}(V_z \cap L) = 2$. The quadratic form of a special point z is $\varphi_z = \varphi | L_z$ with $L_z = V_z \cap L$.

Thus ϕ_z is a positive definite integral binary quadratic form and it tells us which curves T_N pass through z. In a special point two different branches of modular curves always meet transversally.

---

[1]) See the appendix at the end of this chapter.

The first question is which quadratic forms occur. We need some notation. If a binary quadratic form is given w.r.t. a basis by $aX^2 + bXY + cY^2$ then $m =$ g.c.d.(a, b, c) is called the *content* and $\Delta = b^2 - 4ac$ is called the *discriminant* of the form. The form is denoted by $[a, b, c]$.

(6.2) Proposition. *Let z be a special point with form ϕ_z. Then:*

i) *D divides the discriminant of ϕ_z and $\Delta/D \equiv 0$ or $1 \pmod 4$.*

ii) *The content of ϕ_z is divisible only by primes q with $\left(\frac{D}{q}\right) \neq -1$.*

Moreover, if p is a prime dividing D then:

iii) $\operatorname{ord}_p(m) \leqslant \operatorname{ord}_p(D)$,

iv) *if ϕ_z represents n then* $\left(\frac{D(p)}{nA}\right) \neq -1$,

v) *if* $\operatorname{ord}_p(m) = 1$ *and $\Delta/p^2 D(p)$ is a discriminant then $\frac{1}{p}\phi_z$ represents only numbers n with* $\left(\frac{D(p)}{nA}\right) \neq -(p, D)_p$.

vi) *If $p = 2$ one has the additional properties:*

$$\operatorname{ord}_2(m) = 1 \Rightarrow \Delta/4 \equiv 0 \pmod{2^{\operatorname{ord}_2(D)}}$$

$$\operatorname{ord}_2(m) = 2 \Rightarrow \Delta/16 \equiv \begin{cases} 1 \pmod 4 & \text{if } \operatorname{ord}_2(D) = 2 \\ 4 \pmod 8 & \text{if } \operatorname{ord}_2(D) = 3 \end{cases}$$

$$\operatorname{ord}_2(m) = 3 \Rightarrow \Delta/64 \equiv 1 \pmod 4.$$

Proof. The discriminant of ϕ on L is D^3. Since L/L_z is torsion-free it follows that for every prime power q^r we have a direct summand $L_z \otimes \mathbb{Z}/q^r\mathbb{Z}$ of $L \otimes \mathbb{Z}/q^r\mathbb{Z}$. These are the main ingredients for the proof. We show ii), iv) and vi) leaving the other cases to the reader.

ii) Suppose first that an odd prime q divides m. If we choose a basis of $L \otimes \mathbb{Z}/q\mathbb{Z}$ whose first two elements are in $L_z \otimes \mathbb{Z}/q\mathbb{Z}$, then $2A$ det is given by a matrix $\begin{pmatrix} 0 & C \\ {}^tC & B \end{pmatrix}$. So we see $\det \phi = D^3 \equiv (\det C)^2 \pmod q$, hence $\left(\frac{D}{q}\right) \neq -1$.

Now suppose that 2 divides m, and also that $A = 1$ (the general case is similar). We want to show $\left(\frac{D}{2}\right) \neq -1$, i.e. $D \not\equiv 5 \pmod 8$. Write ϕ with respect to a basis

$$\begin{pmatrix} a_i\sqrt{D} & \lambda_i \\ -\lambda_i' & b_i\sqrt{D} \end{pmatrix} \qquad (i = 1, 2)$$

as $aX^2 + bXY + cY^2$, i.e.

$$a = a_1 b_1 D + \lambda_1 \lambda_1'$$

$$c = a_2 b_2 D + \lambda_2 \lambda_2'$$

$$b = (a_1 b_2 + a_2 b_1)D + (\lambda_1 \lambda_2' + \lambda_2 \lambda_1')$$

If $\lambda_i = c_i + d_i \frac{1+\sqrt{D}}{2}$ and $D \equiv 5 \pmod 8$ then we must solve

$$a_1 b_1 + c_1 + d_1 + c_1 d_1 \equiv 0 \pmod 2$$
$$a_2 b_2 + c_2 + d_2 + c_2 d_2 \equiv 0 \pmod 2$$
$$a_1 b_2 + a_2 b_1 + c_1 d_2 + c_2 d_1 \equiv 0 \pmod 2$$
$$(a_1, b_1, c_1, d_1) \not\equiv (a_2, b_2, c_2, d_2) \pmod 2$$
$$(a_i, b_i, c_i, d_i) \not\equiv 0 \pmod 2$$

(the last two conditions since $\dim L_z \otimes \mathbb{Z}/2\mathbb{Z} = 2$). But this system has no solutions.

iv) From $n = abD + \lambda\lambda' A$ it follows that nA is congruent to the norm of an ideal $(\bmod D)$, i.e. $\left(\frac{D(p)}{nA}\right) \neq -1$. (Since ϕ_z represents no norm non-residues $(\bmod D)$ it follows that $D(p)$ divides Δ as a discriminant, hence i).

vi) If $\operatorname{ord}_2(m) = 1$ we may assume that λ_1 and λ_2 both belong to $\sqrt{(2)}\,\mathfrak{a}^{-1}$ but not to $2\mathfrak{a}^{-1}$. Then $\lambda_1 \lambda_2' \in (2/A)$, so $(\lambda_1 \lambda_2' + \lambda_2 \lambda_1')A = (\operatorname{Tr}(\lambda_1 \lambda_2'))A \equiv 0 \pmod 4$. If $\operatorname{ord}_2(D) = 2$ one finds $\Delta \equiv 0 \pmod{16}$. If $\operatorname{ord}_2(D) = 3$ then $(\lambda_1 \lambda_2' + \lambda_2 \lambda_1')A \equiv 4 \pmod 8$ and $ac \equiv 4 \pmod 8$ which suffices for the first case of vi).

If $\operatorname{ord}_2(m) = 2$ then $\lambda_i \in 2\mathfrak{a}^{-1}$, hence $(\lambda_1 \lambda_2' + \lambda_2 \lambda_1')A \equiv 0 \pmod 8$. Since the two basis vectors v_1, v_2 are linearly independent $(\bmod 2)$ we have $a_1 b_2 + a_2 b_1 \equiv 1 \pmod 2$. This implies the second case (and moreover that $\operatorname{ord}_2(m) \leqslant 2$ if $D \equiv 4 \pmod 8$).

If $\operatorname{ord}_2(m) = 3$ and hence $\operatorname{ord}_2(D) = 3$ then $(\lambda_1 \lambda_2' + \lambda_2 \lambda_1')A \equiv 0 \pmod{16}$ and vi) follows (and hence also that $\operatorname{ord}_2(m) < 4$). □

The preceding proposition shows that for a special point z the primitive form $\frac{1}{m}\phi_z$ represents numbers r with

$$\left(\frac{D(p)}{rA}\right) = (m, D)_p \quad \text{for all primes } p \mid D\,. \tag{1}$$

By Dirichlet's theorem we can choose among such r a prime which does not divide mDA. Then r is the norm of an ideal because A and m (by (6.2) ii)) are norms of ideals and

$$\left(\frac{D}{r}\right) = \left(\frac{D}{rA}\right) = \prod_{p|D} \left(\frac{D(p)}{rA}\right) = \prod_{p|D} (m, D)_p = 1$$

by (1). We thus find

(6.3) Lemma. *Let z be a special point whose quadratic form ϕ_z has content m. Then z lies on F_N with N the norm of an ideal $\mathfrak{b}$ from the genus of $\mathfrak{a}$ and $N = rm$, where r is a prime that does not divide mDA. Moreover, if D is odd then one may take $\mathfrak{b}$ to be a primitive ideal, whereas if D is even either N or $N/4$ is the norm of a primitive ideal.*

Now we distinguish two cases, depending on whether $\mathfrak{b}$ is primitive or not. First, let $\mathfrak{b}$ be primitive. It is easy to see that we can choose λ and $\mathfrak{c}$ such that $\mathfrak{b} = \lambda\mathfrak{c}^2\mathfrak{a}$ and such that under the isomorphism

$$\Gamma(\mathfrak{o} \oplus \mathfrak{b})\backslash\mathfrak{H}^2 \xrightarrow{\sim} \Gamma(\mathfrak{o} \oplus \mathfrak{a})\backslash\mathfrak{H}^2 \tag{2}$$

a chosen component of F_N on $\Gamma(\mathfrak{o} \oplus \mathfrak{a}) \backslash \mathfrak{H}^2$ corresponds to the image of the diagonal of $\Gamma(\mathfrak{o} \oplus \mathfrak{b})\backslash\mathfrak{H}^2$. Under (2) the lattice $L(\mathfrak{a})$ corresponds to

$$L(\mathfrak{b}) = \left\{\begin{pmatrix} a\sqrt{D} & \lambda \\ -\lambda' & c\sqrt{D}/N \end{pmatrix} : a, c \in \mathbb{Z}, \lambda \in \mathfrak{b}^{-1}\right\}$$

and the quadratic form ϕ goes over into N times the determinant.

(6.4) Proposition. *Let ϕ be a binary quadratic form of discriminant $\Delta < 0$ and content m such that $4 \nmid m$ and ϕ satisfies the properties listed in (6.2) Then the special points ζ on $\Gamma(\mathfrak{o} \oplus \mathfrak{a})\backslash\mathfrak{H}^2$ with $\phi_\zeta \cong \phi$ are exactly those which correspond under an isomorphism (2) for some $\lambda \gg 0$ and $\mathfrak{c} \subset \mathfrak{o}$ to a point (t, t) on the diagonal of $\Gamma(\mathfrak{o} \oplus \mathfrak{b})\backslash\mathfrak{H}^2$, where $t \in \mathfrak{H}$ satisfies a quadratic equation $a_0Nt^2 + b_0t + c_0 = 0$ with $a_0, b_0, c_0 \in \mathbb{Z}$ such that*

i) $a_0 > 0, (a_0, b_0, c_0) = 1,$

ii) $\Delta_1 := b_0^2 - 4a_0c_0 = \Delta/D,$

iii) $(a_0c_0, b_0, N) = m_1 := \dfrac{m}{\prod p},$

the product being taken over those primes p dividing (m, D) for which $D(p)$ does not divide Δ/p^2 as discriminant.

Proof. The special points (t, t) are those for which t satisfies a quadratic equation $a_0Nt^2 + b_0t + c_0 = 0$ with a_0, b_0, c_0 coprime integers with $a_0 > 0$. We write $\mathfrak{a}^{-1}$ as $\mathbb{Z}1 + \mathbb{Z}\dfrac{M + \sqrt{D}}{2N}$ and we put $\mu = b_0\left(\dfrac{M + \sqrt{D}}{2N}\right)$. Then for a special point (t, t) the lattice $L(\mathfrak{b})_{(t,t)} = \left\{C \in L(\mathfrak{b}) \colon (t, 1)C\begin{pmatrix} t \\ 1 \end{pmatrix} = 0\right\}$ equals $\mathbb{Z}C_t + \mathbb{Z}E$ with

$$E = \begin{pmatrix} 0 & 1 \\ -1 & 0 \end{pmatrix} \text{ and } C_t = \begin{pmatrix} a_0\sqrt{D} & \mu \\ -\mu' & c_0\sqrt{D}/N \end{pmatrix}.$$

For a fixed map (2) the bases (C_t, E) either all have the right or all have the wrong orientation. (To see this use a homotopy $(sC_t + (1 - s)C_{t'}, E)$, $0 \leqslant s \leqslant 1$.) Assuming that they have the right orientation we find as quadratic form ϕ_t of (t, t):

$$\left[a_0c_0D + b_0^2\frac{M^2 - D}{4N}, b_0M, N\right]. \tag{3}$$

If q is a prime which divides m but not D then q does not divide M since $M^2 \equiv D \pmod{4N}$ and hence $\operatorname{ord}_q(b_0) \geqslant \operatorname{ord}_q(m)$ and $\operatorname{ord}_q(m) = \operatorname{ord}_q(a_0c_0, b_0, N)$. A prime p with $p|D$ divides the content m of ϕ_t if and only if p divides b_0. By the assumption $4 \nmid m$ and (6.2) iii) we have $\operatorname{ord}_p(m) \leqslant 1$. Note that for $p|D$ one has

$\operatorname{ord}_p(a_0c_0, b_0, N) = 1 \Leftrightarrow p|b_0$ and $p|a_0c_0$. The formula in iii) now follows from: if $p|b_0$ then $p|a_0c_0 \Leftrightarrow D(p)$ divides the discriminant of $\frac{1}{p}\phi_t$ (as discriminant). □

In case $\mathfrak{b}$ is not primitive the idea is essentially the same; we only sketch the situation. So assume that D and m are divisible by 4. Instead of (2) we now consider the maps

$$S^{-1}\Gamma(\mathfrak{o} \oplus \mathfrak{b})S\backslash\mathfrak{H}^2 \to \Gamma(\mathfrak{o} \oplus \mathfrak{b})\backslash\mathfrak{H}^2 \to \Gamma(\mathfrak{o} \oplus \mathfrak{a})\backslash\mathfrak{H}^2\,, \tag{4}$$

where

$$S = \begin{pmatrix} 1 & \sqrt{D}/(N,8) \\ 0 & 2 \end{pmatrix}$$

and the last map is as in I.4. A component of F_N now comes from the diagonal on $S^{-1}\Gamma(\mathfrak{o} \oplus \mathfrak{b})S\backslash\mathfrak{H}^2$ under such a composition of maps. One finds that the stabilizer of this diagonal is an extension of index 2 or 4 of $\Gamma_0(N)$ depending on whether $d \nmid N$ or $d|N$. The analogue of (6.4) is:

(6.5) Proposition. *Let ϕ be a binary quadratic form of discriminant $\Delta < 0$ and content m such that $4|m$ and ϕ satisfies the properties listed in (6.2). Then the special points ζ on $\Gamma(\mathfrak{o} \oplus \mathfrak{a})\backslash\mathfrak{H}^2$ with $\phi_\zeta \simeq \phi$ are exactly those which correspond under an isomorphism (4) to a point (t,t) on the diagonal of $S^{-1}\Gamma(\mathfrak{o} \oplus \mathfrak{b})S\backslash\mathfrak{H}^2$ satisfying the following equations (where $a_0, b_0, c_0 \in \mathbb{Z}, a_0 > 0$):*

i) *If* $N \equiv 4 \pmod 8$, $D \equiv 0 \pmod 8$: $a_0\frac{N}{2}t^2 + 4b_0t + c_0 = 0$ *with* $(a_0, b_0, c_0) = 1$ *and with discriminant* Δ/D.

ii) *If* $N \equiv D \equiv 0 \pmod 8$: $a_0Nt^2 + 16b_0t + c_0 = 0$ *with* $(a_0, 16b_0, c_0) = 1$ *and with discriminant* $4\Delta/D$.

iii) *If* $N \equiv D \equiv 4 \pmod 8$: $a_0Nt^2 + 8b_0t + c_0 = 0$ *with* $(a_0, 8b_0, c_0) = 1$ *and with discriminant* $4\Delta/D$.

One can also give the content in these cases:

i) $m_1 = (a_0c_0, 4b_0, N/2) = m/4\prod_{p|(D,N/4)} p$

ii) $m_1 = (a_0c_0, 16b_0, N) = m/8\prod_{p|(D,N/8)} p$

iii) $m_1 = (a_0c_0, 8b_0, N) = m/4\prod_{p|(D,N/4)} p$,

where the product is restricted to those primes p for which $D(p)$ does not divide Δ/p^2 as discriminant.

Using the explicit description of (6.4) and (6.5) and the $\Gamma_0(N)$ classification of quadratic forms one can derive formulae for the number of special points of $\Gamma\backslash\mathfrak{H}^2$ with quadratic form ϕ, see [85, 62].

7. Classification of Elliptic Fixed Points

As spin-off of the somewhat tedious computations of Section 6 we have a very explicit description of the orbits of elliptic fixed points of $\Gamma_m\backslash\mathfrak{H}^2$, $\Gamma_m = \Gamma_m(\mathfrak{o} \oplus \mathfrak{a})$

following Hausmann [62]. We advise the reader to skip this section on a first reading.

Recall from Chapter I.5 that an elliptic fixed point of $\Gamma = \Gamma(\mathfrak{o} \oplus \mathfrak{a})$ has order 2 or 3 if $D \neq 5, 8, 12$. (Here the order of ζ is $\# \Gamma_\zeta$ of course.) For $D = 5$ (resp. 8, resp. 12) we also have points of order 5 (resp. 4, 6). The fact that Γ_m/Γ is of the form $(\mathbb{Z}/2\mathbb{Z})^{t-1}$ implies that for $D \neq 5$ the order of an elliptic fixed point is divisible only by the primes 2 or 3.

For $D = 5$ there are two points of order 5, see [129]. These points are not special points as can easily be checked.

(7.1) Proposition. *An elliptic fixed point of $\Gamma_m(\mathfrak{o} \oplus \mathfrak{a})$ which is not one of the two order 5 elliptic fixed points of $\Gamma_{\mathbb{Q}(\sqrt{5})}$ is a special point.*

Proof. The isotropy group of such a point ζ contains an element of order 2 or 3. If it is of order 3 we may even assume that it belongs to $\mathrm{PSL}(\mathfrak{o} \oplus \mathfrak{a})$. So let $S = \begin{pmatrix} \alpha & \beta \\ \gamma & \delta \end{pmatrix}$ be a matrix representing this element with $(\mathrm{tr}\, S)^2 - 4 \det S \ll 0$ and $\det S = w$, a square-free divisor of D. We then have $(\mathrm{tr}\, S)^2 - 4 \det S = a \in \mathbb{Z}_{<0}$, since $\mathrm{tr}\, S = 0$ or ± 1. The corresponding fixed point $\zeta = (z_1, z_2)$ is

$$z_i = \frac{\alpha^{(i)} - \delta^{(i)}}{2\gamma^{(i)}} + \frac{1}{2|\gamma^{(i)}|}\sqrt{a} \qquad i = 1, 2\,.$$

The equation $c\sqrt{D}\, z_1 z_2 + \lambda z_2 - \lambda' z_1 + \frac{b}{A}\sqrt{D} = 0$ is of the form $\mu c\sqrt{D} + \nu\sqrt{D} + \frac{b}{A}\sqrt{D} = 0$ with $\mu, \nu \in \mathbb{Q}(\sqrt{a})$. This shows that $(L(\mathfrak{a}) \otimes \mathbb{Q})_\zeta$ has $\mathbb{Q}$-dimension 2. $\square$

For elliptic fixed points of order $r \neq 5$ on $\Gamma \backslash \mathfrak{H}^2$ the action of a generator of the isotropy group is locally given by $(u, v) \to (\zeta u, \zeta v)$ or by $(u, v) \to (\zeta u, \zeta^{-1} v)$ with ζ a primitive r-th root of 1. We call these fixed points of type $(r; 1, 1)$ and $(r; 1, -1)$ respectively. The possible types here (excluding $r = 5$) are $r = 2, 3, 4, 6$. The theorems in this section are due to Hausmann [62].

(7.2) Theorem. *Let $D \neq 8$, 12 and let $\zeta \in \Gamma(\mathfrak{o} \oplus \mathfrak{a}) \backslash \mathfrak{H}^2$ be a special point whose quadratic form has discriminant Δ and content m. Then*

i) *ζ is an elliptic fixed point of order* $2 \Leftrightarrow \Delta = -4D$, *or* $\Delta = -4dD$, $d|m$, *or* $\Delta = -dD$, $d|m$.

ii) *ζ is an elliptic fixed point of type* $(3, 1, 1) \Leftrightarrow \Delta = -3D$

iii) *ζ is an elliptic fixed point of type* $(3, 1, -1) \Leftrightarrow \Delta = -3D^2$, $m = D$, *or* $\Delta = -D^2/3$, $m = D/3$.

Remark. Note that

$$\Delta = -3D^2, m = D \Leftrightarrow \varphi \cong [D, D, D]$$

$$\Delta = -D^2/3, m = D/3 \Leftrightarrow \varphi \cong [D/3, D/3, D/3]\,.$$

The elliptic fixed points of $\Gamma_m(\mathfrak{o} \oplus \mathfrak{a})\backslash\mathfrak{H}^2$ either come from elliptic fixed points of $\Gamma(\mathfrak{o} \oplus \mathfrak{a})\backslash\mathfrak{H}^2$ or are "new". We now look first at the action of Γ_m/Γ on the elliptic fixed points of $\Gamma\backslash\mathfrak{H}^2$. It suffices (see Ch. I.4) to look at the action of Γ_w/Γ with w a square-free divisor of D. Recall that $\Gamma_w = \Gamma_{\tilde{w}}$ if $\tilde{w}$ = square-free part of D/w.

(7.3) Theorem. *Let $D \neq 8, 12$. Then*

a) *Γ_w/Γ acts freely on the set of elliptic fixed points of $\Gamma\backslash\mathfrak{H}^2$ of order* 2 *(resp.* 3) *if $w \neq 2, \tilde{w} \neq 2$ (resp. $w \neq 3, \tilde{w} \neq 3$.).*

b) *If $w = 2$ then $\Gamma_w\backslash\mathfrak{H}^2$ has elliptic fixed points of order* 4. *If we denote by ζ a special point whose quadratic form has discriminant Δ and content m then*

i) *ζ is an elliptic fixed point of type* $(4, 1, 1) \Leftrightarrow \Delta = -4D, 2|m$

ii) *ζ is an elliptic fixed point of type* $(4, 1, -1) \Leftrightarrow \Delta = -4dD, 2d|m$, *or* $\Delta = -dD$, $d = m$.

Γ_2/Γ acts freely on all other elliptic fixed points of Γ.

c) *If $w = 3$ then $\Gamma_w\backslash\mathfrak{H}^2$ has elliptic fixed points of order* 6.

i) *ζ is an elliptic fixed point of type* $(6, 1, 1) \Leftrightarrow \Delta = -3D, 3|m$

ii) *ζ is an elliptic fixed point of type* $(6, 1, -1) \Leftrightarrow \Delta = -D^2/3, m = D/3$

Γ_3/Γ acts freely on all other elliptic fixed points.

The elliptic fixed points that are "new" for Γ_w (not an image of an elliptic fixed point of $\Gamma\backslash\mathfrak{H}^2$ under $\Gamma\backslash\mathfrak{H}^2 \to \Gamma_w\backslash\mathfrak{H}^2$) are necessarily of order 2. They are as follows

(7.4) Theorem. *Let $D \neq 8, 12$. Notation is as in* (7.3).

a) $w \neq 3, \tilde{w} \neq 3$.

ζ is a new elliptic fixed point of order 2 *of* $\Gamma_w \Leftrightarrow$

i) *$w|m$ and ($\Delta = -4wD$ or $\Delta = -wD$), or*

ii) *$\tilde{w}|m$ and ($\Delta = -4\tilde{w}D$ or $\Delta = -\tilde{w}D$)*

b) $w = 3$.

The new elliptic fixed points of order 2 *are those ζ that satisfy* i) *or* ii) *of* a) *but are not of order* 6 *(described in* (7.3).

Proof of (7.2), (7.3) *and* (7.4). We first prove the implications $\Rightarrow$ of these theorems.

Let ζ be an elliptic fixed point of $\Gamma_m\backslash\mathfrak{H}^2$ whose order is divisible by 2. By (6.4) and (6.5) ζ corresponds to a point t on the diagonal of $\Gamma'\backslash\mathfrak{H}^2$, where $\Gamma' = \Gamma(\mathfrak{o} \oplus \mathfrak{b})$ or $\Gamma' = S^{-1}\Gamma(\mathfrak{o} \oplus \mathfrak{b})S$. Let T be a matrix and w a square-free divisor of D such that

$$\operatorname{tr} T = 0\,, \qquad \det T = w\,, \qquad Tt = t\,,$$

which represents an element of order 2 of $(\Gamma')_\zeta$.

If $w = 1$ or $w = d$ then ζ is already an elliptic fixed point of $\Gamma(\mathfrak{o} \oplus \mathfrak{b})$, otherwise it is a fixed point of Γ_w/Γ.

Now $Tt = t$ implies a quadratic equation

$$\gamma t^2 + (\delta - \alpha)t + \beta = 0\,, \tag{1}$$

But by (6.4), (6.5) t satisfies also a quadratic equation

$$\begin{aligned} a_0Nt^2 + b_0t + c_0 = 0 \quad &\text{with } b_0^2 - 4a_0c_0N = \Delta/D \text{ if } 4 \nmid m \\ a_0Nt^2 + 4b_0t + c_0 = 0 \quad &\text{with } b_0^2 - 4a_0c_0N = 4\Delta/D \text{ if } 4|m\,. \end{aligned} \tag{2}$$

Suppose $4 \nmid m$. Combination of (1) and (2) yields

$$a_0 N = \gamma k\,, \qquad b_0 = (\delta - \alpha)k\,, \qquad c_0 = \beta k\,, \qquad \Delta = -4wk^2 D$$

for some $k \in \mathbb{Q}(\sqrt{D})$. If $k \notin \mathbb{Q}$ then we replace T by $\frac{(w,\tilde{w})\sqrt{d}}{w} T$. Then $k \in \mathbb{Q}$. Clearly $k \in \frac{1}{2}\mathbb{Z}$. By (1.1) $\gamma \equiv 0 \pmod N$, hence $(a_0, b_0, c_0) = 1$ implies $k = \frac{1}{2}$ or $k = 1$ (assuming $k > 0$), and $k = \frac{1}{2}$ only if $w = 3 \pmod 4$. Moreover the choice of T implies $w|(c_0^2 N, a_0^2 N, b_0^2/4)/k^2$. From $b_0^2 - 4a_0 c_0 N = -4wk^2$ we deduce $w|N$ hence $w|m$.

If $4|m$ one finds using (6.5) that $16b_0^2 - 4a_0 c_0 N = -16w$ and $w|N$. This proves $\Rightarrow$ of (7.2) i).

The other direction ($\Leftarrow$) of (7.2) i) as well as all $\Leftarrow$ of (7.2), (7.3) and (7.4) are shown by listing elliptic matrices which have ζ as fixed point. A list of these elliptic matrices can be found in [62].

If ζ is an elliptic fixed point of order 4 then we can find a matrix T with

$$\operatorname{tr} T = \pm\sqrt{2}\sqrt{\det T} = \pm\sqrt{2w}\,, \qquad Tt = t\,.$$

This implies that $w = 2$ or $\tilde{w} = 2$. If e.g. $w = 2$ and $4 \nmid m$ then as above we find $k = 1$ and $w = 2|m$. The other cases are dealt with similarly. A list of elliptic matrices then finishes the proof of (7.3) b).

The elliptic fixed points of order 3 and order 6 can be left to the reader. □

(7.5) *Remark.* Let $\mathfrak{H}_- = \{z \in \mathbb{C}\colon \operatorname{Im} z < 0\}$. The map

$$\mathfrak{H} \times \mathfrak{H} \to \mathfrak{H} \times \mathfrak{H}_- \qquad (z_1, z_2) \to (z_1/\sqrt{D}, -z_2/\sqrt{D})$$

establishes an isomorphism $\Gamma(\mathfrak{o} \oplus \mathfrak{a})\backslash\mathfrak{H} \times \mathfrak{H}_- \to \Gamma(\mathfrak{o} \oplus \mathfrak{a}\cdot(\sqrt{D}))\backslash\mathfrak{H}^2$. But under $(z_1, z_2) \mapsto (z_1, \bar{z}_2)$ a fixed point of $\Gamma(\mathfrak{o} \oplus \mathfrak{a})\backslash\mathfrak{H}^2$ of type $(r; a, b)$ goes to one of $\Gamma(\mathfrak{o} \oplus \mathfrak{a})\backslash \mathfrak{H} \times \mathfrak{H}_-$ of type $(r; a, -b)$. This yields a bijection between the points of type $(r; a, b)$ for $\Gamma(\mathfrak{o} \oplus \mathfrak{a})$ and those of type $(r; a, -b)$ for $\Gamma(\mathfrak{o} \oplus \mathfrak{a}(\sqrt{D}))$.

Finally we describe the elliptic fixed points of the principal congruence subgroups.

Let $\mathfrak{c} \subset \mathfrak{o}$ be an ideal and let $\Gamma[\mathfrak{c}]$ be the image in $\Gamma(\mathfrak{o} \oplus \mathfrak{a})$ of the group

$$\left\{\begin{pmatrix} \alpha & \beta \\ \gamma & \delta \end{pmatrix} \in \mathrm{SL}(\mathfrak{o} \oplus \mathfrak{a}) : \alpha \equiv \delta \equiv 1 \ (\mathrm{mod}\ \mathfrak{c}),\ \gamma \in \mathfrak{c}\mathfrak{a},\ \beta \in \mathfrak{c}\mathfrak{a}^{-1}\right\}.$$

$\Gamma[\mathfrak{c}]$ acts freely unless $\mathfrak{c}^2 = (2)$ or $\mathfrak{c}^2 = (3)$. If $\mathfrak{c}^2 = (2)$ then $\Gamma[\mathfrak{c}]$ has fixed points of order 2, if $\mathfrak{c}^2 = (3)$, then $\Gamma[\mathfrak{c}]$ has fixed points of order 3.

(7.6) Theorem. *Let* $D \neq 8, 12$.

i) *If* $\mathfrak{c}^2 = 2$ *then*: ζ *is an elliptic fixed point of order* 2 *for* $\Gamma[\mathfrak{c}]$ *if and only if* ζ *is an elliptic fixed point of order* 4 *for* $\Gamma_2(\mathfrak{o} \oplus \mathfrak{a})$.

ii) *If* $\mathfrak{c}^2 = 3$ *then*: ζ *is an elliptic fixed point of type* $(3, 1, 1)$ (*resp.* $(3, 1, -1)$) *for* $\Gamma[\mathfrak{c}]$ *if and only if it is an elliptic fixed point of type* $(6, 1, 1)$ (*resp. type* $(6, 1, -1)$) *for* $\Gamma_3(\mathfrak{o} \oplus \mathfrak{a})$.

Proof. We only prove i) since ii) is similar. Let $T = \begin{pmatrix} \alpha & \beta \\ \gamma & \delta \end{pmatrix} \in \Gamma[\mathfrak{c}]$ with $T^2 = -\mathrm{id}$ then $\operatorname{tr} T = 0$. Therefore, $\begin{pmatrix} \alpha+1 & \beta \\ \gamma & \delta+1 \end{pmatrix} \in \mathfrak{c}\begin{pmatrix} \mathfrak{o} & \mathfrak{a}^{-1} \\ \mathfrak{a} & \mathfrak{o} \end{pmatrix}$ and its square equals $2T$. Conversely, let a matrix $T \in \mathfrak{c}\begin{pmatrix} \mathfrak{o} & \mathfrak{a}^{-1} \\ \mathfrak{a} & \mathfrak{o} \end{pmatrix}$ with $\det T = 2$, $\operatorname{tr} T = 2$ belong to an (arbitrary) elliptic fixed point of order 2 of $\Gamma_2(\mathfrak{o} \oplus \mathfrak{a})$. Then $\frac{1}{2}T^2 \in \Gamma[\mathfrak{c}]$. This proves i). □

The number of fixed points can be read off from the tables on p. 267. If $\mathfrak{c}^2 = (2)$ then (for $(A, D) = 1$)

$$a_2 = \begin{cases} 6h(-D) & D \equiv 0 \pmod 8 \\ 12h(-D/4) & D/4 \equiv 7 \pmod 8 \\ 24h(-D/4) & D/4 \equiv 3 \pmod 8\,. \end{cases}$$

If $\mathfrak{c}^2 = (3)$ then

$$a_3^+ = \begin{cases} 12h(-D/3) & AD \equiv 6 \pmod 9 \\ 0 & \text{otherwise} \end{cases}$$

$$a_3^- = \begin{cases} 12h(-D/3) & AD \equiv 3 \pmod 9 \\ 0 & \text{otherwise}\,. \end{cases}$$

For the special cases $D = 5, 8, 12$ we refer to the tables on p. 268.

8. The Intersection Number of T_1 and T_N

As a second application of V.6 we compute the intersection number of T_N and T_1 on X_Γ (in a suitable sense). The general case (intersection number of T_N and T_M) will be deduced from this in the next chapter. These intersection numbers were calculated by Hirzebruch and Zagier in [85] (for prime discriminants).

Let Z_Γ be the surface obtained from X_Γ by resolving only the cusp singularities. Let $t_N \in H^2(Z_\Gamma, \mathbb{Q})$ be the cohomology class which is the Poincaré dual of the pullback of the class of T_N under the natural map $Z_\Gamma \to X_\Gamma$, i.e.

$$t_N = \text{class of } T_N + \sum \alpha(N,k) S_k\,,$$

where the $\alpha(N,k)$ are uniquely defined rational numbers determined by $t_N \cdot S_k = 0$ for all resolution curves S_k. Now $\Gamma \backslash \mathfrak{H}^2 \subset Z_\Gamma$ is a rational homology manifold, so there is an intersection pairing $H^2(Z_\Gamma, \mathbb{Q}) \times H^2(Z_\Gamma, \mathbb{Q}) \to \mathbb{Q}$, and intersection numbers are rational numbers. In fact, if MN is not a square, so that T_M and T_N have no common components then the local intersection number $(T_M \cdot T_N)_z$ in a special point z is $\dfrac{1}{4\#\Gamma_z}$ times the number of ordered pairs $v, w \in L_z$ (see 6.1) with v and w

linearly independent such that $\varphi_z(v) = M$, $\varphi_z(w) = N$. Here the 4 comes from the fact (v, w), $(-v, w)$, $(v, -w)$ and $(-v, -w)$ determine the same pair of branches. (To avoid the problems created by the quotient singularities of Z_Γ one could replace Γ by a normal subgroup Γ' of finite index which acts freely on $\mathfrak{H}^2$, compute intersection numbers of $\Gamma'\backslash\mathfrak{H}^2$ and divide by the degree of $\Gamma'\backslash\mathfrak{H}^2 \to \Gamma\backslash\mathfrak{H}^2$.)

Suppose first that N is not a square, so T_N and T_1 have no common components. We work on Z_{Γ_m} instead of Z_Γ because T_1 has only one component ($\cong \overline{\mathrm{SL}(2, \mathbb{Z})\backslash\mathfrak{H}}$) there. (To get the intersection number on Z_Γ multiply by 2^{t-1}.)

(8.1) Lemma. *There is a bijective correspondence between* SL(2, ℤ)*-equivalence classes of positive definite integral binary quadratic forms with a discriminant of the form* $(s^2 - 4N)/D$ *for some* $s \in \mathbb{Z}$ *with* $s^2 < 4N$ *and local branches of* T_N *through the intersection points of* T_N *and* T_1 *on* Z_{Γ_m}.

Proof. An intersection point $z \in T_1 \cap T_N$ has quadratic form φ_z which represents 1, hence φ_z is the norm form of an order R of discriminant Δ in $\mathbb{Q}(\sqrt{\Delta})$. A local branch of T_N through z corresponds to $r \in R$ with $N = \mathrm{N}(r)$ and $R_f = \mathbb{Z}1 + \mathbb{Z}r$ is an order of index f in R. By (6.4) we know that $z = (t, t)$, where t satisfies $a_0 t^2 + b_0 t + c_0 = 0$ with $a_0, b_0, c_0 \in \mathbb{Z}$, $a_0 > 0$, $(a_0, b_0, c_0) = 1$ and $b_0^2 - 4a_0c_0 = \Delta/D$. We associate to r the quadratic form $[fa_0, fb_0, fc_0]$.

Conversely, every quadratic form $[fa_0, fb_0, fc_0]$ with a_0, b_0, $c_0 \in \mathbb{Z}$, $a_0 > 0$, $(a_0, b_0, c_0) = 1$ and $(b_0^2 - 4a_0c_0)f^2 = (s^2 - 4N)/D$ determines an order R_f of index f in the order of discriminant $(b_0^2 - 4a_0c_0)$ in $\mathbb{Q}(\sqrt{\Delta})$. By (6.4) this last order corresponds to a special point (t, t) whose form represents N. It is now easy to check that this defines the desired bijection. □

We note that the contribution to the rational intersection number from a local branch of T_N through z is $\frac{1}{w}$, where $2w$ = number of SL(2, ℤ)-automorphisms of φ_z.

Let $h'(n)$ be the number of SL(2, ℤ)-equivalence classes of positive definite primitive binary integral quadratic forms with discriminant n, each form φ being counted by $2/\#\,\mathrm{Aut}\,\varphi$ with $\mathrm{Aut}\,\varphi$ = group of orientation preserving automorphisms of φ (i.e. $h'(-3) = 1/3$, $h'(-4) = 1/2$, $h'(n) = h(n)$ otherwise). We let $H(n)$ (for $n \in \mathbb{Z}_{>0}$) be the number of SL(2, ℤ)-equivalence classes of positive definite quadratic forms of discriminant $-n$, again counted with multiplicities as above. We have

$$H(n) = \sum_{d^2 | n} h'(-n/d^2)\,.$$

We define also $H(0) = -1/12$.

We have proved for N not a square:

(8.2) Proposition. $(T_1 \cdot T_N)_{\Gamma_m\backslash\mathfrak{H}^2} = \sum\limits_{\substack{s \in \mathbb{Z} \\ s^2 < 4N \\ s^2 \equiv 4N \ (\mathrm{mod}\ D)}} H\left(\frac{4N - s^2}{D}\right).$

So far we have been working on $\Gamma_m\backslash\mathfrak{H}^2$ and we still have to take the intersections in the cusps into account. One has to invert the intersection matrix (S_k, S_l) to determine the $\alpha(N,k)$ defined above. To do this we put for an ideal $\mathfrak{c} \subset \mathfrak{o}$

$$f(\mathfrak{c}) = \frac{1}{\sqrt{D}} \sum_{\substack{\alpha\in\mathfrak{o}\\ (\alpha)=\mathfrak{c},\,\alpha\gg 0}} \min(\alpha,\alpha') \tag{1}$$

Clearly $f(\mathfrak{c}) = 0$ if $\mathfrak{c}$ is not a principal ideal in the narrow sense. Let ε be a generator of the infinite cyclic group U^+ of totally positive units of $\mathfrak{o}$ which satisfies $\varepsilon < 1$, $\varepsilon' > 1$. Define for a totally positive $\mu \in \mathfrak{o}$

$$g(\mu) = \frac{1}{\sqrt{D}}\left(\frac{\mu}{1-\varepsilon} + \frac{\mu'\varepsilon}{1-\varepsilon}\right).$$

Then if $\varepsilon^2 \leqslant \mu/\mu' \leqslant 1$ we have

$$f((\mu)) = g(\mu)\,, \tag{2}$$

since the possible α in the definition of $f((\mu))$ are $\mu\varepsilon^n (n \geqslant 0)$ and $\mu'\varepsilon^n (n \geqslant 1)$. Note that $g(\mu)$, and hence $f(\mathfrak{c})$ for any $\mathfrak{c}$, is a rational number.

The intersection matrix $(S_k \cdot S_l)$ decomposes into blocks corresponding to the different cusps. Given a cycle of curves S_k ($k \bmod r$) corresponding to the boundary points A_k of the convex hull of the totally positive elements of $M = \mathbb{Z}1 + \mathbb{Z}w_0$ in $\mathbb{R}_+^2$ (cf. Ch. II.2) we can associate an integral ideal $\mathfrak{a}_k$ to A_k by $\mathfrak{a}_k = A_k M^{-1}$. (We have $A_{-1} = w_0, A_0 = 1, -A_{k-1} + b_k A_k - A_{k+1} = 0$, see Ch. II).

(8.3) Proposition. *The inverse matrix of* $(S_k \cdot S_l)$ *is the matrix* $-f(\mathfrak{a}_k\mathfrak{a}_l')$.

Proof. We may consider just one cycle since $f(\mathfrak{a}_k\mathfrak{a}_l') = 0$ if $\mathfrak{a}_k$ and $\mathfrak{a}_l$ do not belong to the same ideal class. We can choose w_0 such that a given ideal $\mathfrak{a}_l$ becomes $\mathfrak{a}_0$. We then have to check

$$-f(\mathfrak{a}_{-1}\mathfrak{a}_0') + b_0 f(\mathfrak{a}_0\mathfrak{a}_0') - f(\mathfrak{a}_1\mathfrak{a}_0') = 1$$
$$-f(\mathfrak{a}_{k-1}\mathfrak{a}_0') + b_k f(\mathfrak{a}_k\mathfrak{a}_0') - f(\mathfrak{a}_{k+1}\mathfrak{a}_0') = 0 \quad \text{for } k \neq 0\,.$$

(even if the cycle consists of one or two curves only). Now $\mathfrak{a}_k\mathfrak{a}_l' = (A_k A_l' \mathrm{N}(\mathfrak{a}_0)) = (A_k A_l' N_0)$, $N_0 = \mathrm{Tr}(w_0/\sqrt{D})^{-1}$. So we have to prove (writing $f(\lambda)$ for $f((\lambda))$)

$$-f(A_{-1}) + b_0 f(A_0) - f(A_1) = \mathrm{Tr}(w_0/\sqrt{D})$$
$$-f(A_{k-1}) + b_k f(A_k) - f(A_{k+1}) = 0 \quad \text{for } k = 1, \dots, r-1\,.$$

Since g is linear it satisfies the second equation. The A_k with $0 \leqslant k \leqslant r$ satisfy $\varepsilon^2 \leqslant A_k/A_k' \leqslant 1$, hence by (2) f satisfies the second equation. If $1 \leqslant \mu/\mu' \leqslant \varepsilon^{-2}$ then $f((\mu)) = g(\mu) - \mathrm{Tr}(\mu/\sqrt{D})$. The linearity of g now also implies the first equation for f. □

(8.4) Corollary. $\alpha(N,k) = \sum_{\mathfrak{c}} f(\mathfrak{a}_k\mathfrak{c})$, *where the sum is over all ideals of norm* N *in* $\mathfrak{o}$.

Proof. A branch of T_N in a neighbourhood of the cycle determined by M is given by an element $\gamma \in M_+$ modulo totally positive units (preserving M). If $\gamma = pA_{l-1} + qA_l$ then for the branch T_γ we have $T_\gamma \cdot S_{l-1} = p$, $T_\gamma \cdot S_l = q$. The ideal $\mathfrak{c} = \gamma M^{-1}$ is an integral ideal of norm N. The map $\gamma \mapsto \mathfrak{c}$ sets up a bijection between the non-zero integral ideals equivalent to M^{-1} and the elements $\gamma \in M_+$ modulo totally positive units. Thus the corollary follows from

$$\alpha(N, k) = \sum_l f(\mathfrak{a}_k \mathfrak{a}_l')(T_N \cdot S_l)$$

and (2.1). □

We now define for MN not a square

$$(T_M \cdot T_N)_\infty := t_M \cdot t_N - (T_M \cdot T_N)_{\Gamma_m \backslash \mathfrak{H}^2}.$$

We have

$$\begin{aligned}(T_M \cdot T_N)_{\Gamma_m \backslash \mathfrak{H}^2} + (T_M \cdot T_N)_{\cup S_k} &= T_M \cdot T_N \\ &= t_M \cdot t_N + \sum_{k,l} \alpha(M,k)\alpha(N,l) S_k \cdot S_l,\end{aligned}$$

so

$$(T_M \cdot T_N)_\infty = (T_M \cdot T_N)_{\cup S_k} + \sum_{k,l} f(\mathfrak{a}_k \mathfrak{a}_l')(T_M \cdot S_k)(T_N \cdot S_l). \tag{3}$$

Using (2.1) one sees $(T_1 \cdot T_N)_{\cup S_k} = 0$. So we find

(8.5) Corollary. *If N is not a square then*

$$(T_1 \cdot T_N)_\infty = \sum_{N(\mathfrak{n})=N} f(\mathfrak{n}).$$

Essentially the same argument gives if MN is not a square

$$(T_M \cdot T_N)_\infty = \sum_{\substack{\mathfrak{m},\mathfrak{n} \in G \\ N(\mathfrak{m})=M \\ N(\mathfrak{n})=N}} f(\mathfrak{m}\mathfrak{n}) \quad \text{on } Z_{\Gamma_m(\mathfrak{o} \oplus \mathfrak{a})},$$

where G is the genus of $\mathfrak{a}$.

We define

$$I_D(N) = \frac{1}{\sqrt{D}} \sum_{\substack{\lambda \in \mathfrak{o} \\ \lambda \gg 0 \\ N(\lambda)=N}} \min(\lambda, \lambda')$$

and

$$H_D(N) = \sum_{\substack{s \in \mathbb{Z} \\ s^2 \leqslant 4N \\ s^2 \equiv 4N \ (\mathrm{mod}\ D)}} H\left(\frac{4N - s^2}{D}\right).$$

(8.6) Theorem. *Let $\Gamma_m = \Gamma_m(\mathfrak{o} \oplus \mathfrak{a})$ with $\mathfrak{a}$ in the principal genus. The intersection number of the cohomology classes t_1 and t_N on Z_{Γ_m} is given by*

$$t_1 \cdot t_N = H_D(N) + I_D(N).$$

Proof. If N is not a square the result follows from (8.2) and (8.5). If N is a square we still have

$$t_1 \cdot t_N = T_1 \cdot T_N + I_D(N).$$

Here T_1 and T_N have a common component, namely T_1. Now $T_1 \cdot T_N = T_1 \cdot T_1 + (T_1 \cdot T_N)_{\text{trans}}$, where the latter term comes from transversal intersections and is still given by (8.2). The extra terms $s = \pm 2\sqrt{N}$ in the definition of $H_D(N)$ give a contribution $2H(0) = -\frac{1}{6}$ and this is the value of $T_1 \cdot T_1$ as we shall see in Chapter VII. □

Let $Z_m(D)$ be the disjoint union of 2^{t-1} surfaces $Z_{\Gamma_m(\mathfrak{o}\oplus\mathfrak{a})}$, where $\mathfrak{a}$ runs through the 2^{t-1} genera. Since $T_1 = \emptyset$ on $Z_{\Gamma_m(\mathfrak{o}\oplus\mathfrak{a})}$ if $\mathfrak{a}$ is not in the principal genus we can reformulate our result as

(8.7) Corollary. *On* $Z_m(D)$ we have

$$t_1 \cdot t_N = H_D(N) + I_D(N).$$

One can, of course, also ask what the intersection number t_M and t_N is in general. A formula for $t_M \cdot t_N$ (for D prime) was worked out by Hirzebruch and Zagier in [85] by geometric arguments like those in this section, but requiring more work especially when M and N have a common factor. We shall deduce the formula for $t_M \cdot t_N$ by an abstract argument from the formula for $t_1 \cdot t_N$ in the next chapter.

The main result of [85] was that the generating series

$$\frac{1}{2}\operatorname{vol}(T_M) + \sum_{N=1}^{\infty} t_M \cdot t_N e^{2\pi i N z} \qquad (z \in \mathfrak{H})$$

is a modular form (in one variable). This will be proved in the next chapter. Here we discuss briefly a direct proof for this fact for the case $M = 1$, i.e. for the fact that

$$g = \sum_{N=0}^{\infty} (H_D(N) + I_D(N)) e^{2\pi i N z}$$

is a modular form of weight 2 in $M_2(\Gamma_0(D), \chi_D)$.

Let

$$\mathscr{H} = \sum_{m=0}^{\infty} H(m) q^m, \qquad \theta = \sum_{s=-\infty}^{\infty} q^{s^2} \qquad (q = e^{2\pi i z}).$$

Clearly

$$\sum_{N=0}^{\infty} H_D(N) q^N = \mathscr{H}(Dz)\theta(z) | U_4,$$

where the operator U_4 acts on Fourier series by $\sum_r a_r q^r \mapsto \sum_r a_{4r} q^r$. Since θ is a modular form of weight 1/2 and U_4 takes modular forms to modular forms, this is a modular form if $\mathscr{H}(z)$ is a modular form of weight 3/2, i.e. if

$$\mathscr{H}\left(\frac{az+b}{cz+d}\right) = \left(\frac{-1}{d}\right)^{1/2} \left(\frac{c}{d}\right) (cz+d)^{3/2} \mathscr{H}(z) \quad \text{for all} \begin{pmatrix} a & b \\ c & d \end{pmatrix} \in \Gamma_0(4).$$

This is not true, but nearly: the non-holomorphic function

$$\mathscr{H}^*(z) = \mathscr{H}(z) + \frac{1}{\sqrt{y}} \sum_{t=-\infty}^{\infty} \varphi(t^2 y) e^{-2\pi i t^2 z} \qquad z = x + iy$$

with

$$\varphi(y) = \frac{1}{16\pi} \int_1^{\infty} u^{-3/2} e^{-4\pi y u}\, du \qquad (y \geqslant 0)$$

does transform this way. $\mathscr{H}^*$ is the Eisentein series of weight 3/2 for $\Gamma_0(4)$, computed using Hecke's trick, see [85, p. 92]. Hence the non-holomorphic function

$$g^*(z) = \mathscr{H}^*(Dz)\theta(z) | U_4 ,$$

where U_4 is now defined by $f | U_4 = \frac{1}{4}\sum_{m=1}^{4} f\left(\frac{z+m}{4}\right)$, transforms like a modular form of weight 2 and level D. It has the Fourier series

$$\begin{aligned} g^*(z) &= \sum_{N=0}^{\infty} H_D(N) q^N + \frac{2}{\sqrt{Dy}} \sum_{\substack{s,t \in \mathbb{Z} \\ s^2 \equiv t^2 D \,(\mathrm{mod}\, 4)}} \varphi\left(\frac{Dt^2 y}{4}\right) q^{(s^2 - t^2 D)/4} \\ &= \sum_{N=0}^{\infty} H_D(N) q^N + \frac{2}{\sqrt{Dy}} \sum_{\lambda \in \mathfrak{o}} \varphi\left(\left(\frac{\lambda - \lambda'}{2}\right)^2 y\right) q^{\mathrm{N}(\lambda)} . \end{aligned}$$

We now use the holomorphic projection principle.

(8.8) Proposition. *Let $g^*(z) = \sum_{N=-\infty}^{\infty} a_N(y) q^N$ be a non-holomorphic modular form of weight 2 and some level D and non-trivial character χ (i.e. $g^*((az+b)/(cz+d)) = \chi(d)(cz+d)^2 g^*(z)$ for all $\left(\begin{smallmatrix} a & b \\ c & d \end{smallmatrix}\right) \in \Gamma_0(D)$) and suppose that*

$$g^*(z) = a + O(y^{-\lambda})$$

for some constant $a \in \mathbb{C}$ and $\lambda \in \mathbb{R}_{>0}$. Define a_N for $N > 0$ by

$$a_N = 4\pi N \int_0^{\infty} a_N(y) e^{-4\pi N y}\, dy .$$

Then the function $g(z) = a + \sum_{N=1}^{\infty} a_N q^N$ is a holomorphic modular form of weight 2, level D and character χ.

This principle was defined by Sturm for modular forms of arbitrary weight, but under somewhat restricted hypotheses. A principle similar to our principle can be found in [49] (There the case of trivial character is considered; the presence of a non-holomorphic Eisenstein series leads to considerable complications; these difficulties disappear for χ non-trivial.)

In our situation

$$a_N(y) = H_D(N) + \frac{2}{\sqrt{Dy}} \sum_{\lambda\lambda' = N} \varphi\left(\left(\frac{\lambda - \lambda'}{2}\right)^2 y\right)$$

and the hypotheses of (8.8) are fulfilled with $a = H_D(0)$ and $\lambda = \frac{1}{2}$. The formula for the N-th coefficient of g then gives

$$a_N = H_D(N) + \frac{16\pi N}{\sqrt{D}} \sum_{\substack{\lambda \gg 0 \\ \lambda\lambda' = N}} \int_0^\infty y^{-1/2} \varphi\left(\left(\frac{\lambda - \lambda'}{2}\right)^2 y\right) e^{-4\pi N y}\, dy\,.$$

This integral is easy to compute. Using $\int_0^\infty y^{-1/2} e^{-ty}\, dy = \sqrt{\pi/t}$, we find

$$\begin{aligned}\int_0^\infty y^{-1/2} \varphi\left(\left(\frac{\lambda - \lambda'}{2}\right)^2 y\right) e^{-4\pi\lambda\lambda' y}\, dy &= \frac{1}{16\pi} \int_1^\infty \frac{u^{-3/2}\, du}{\sqrt{(\lambda - \lambda')^2 u + 4\lambda\lambda'}} \\ &= \frac{-1}{32\pi\lambda\lambda'} \sqrt{(\lambda - \lambda')^2 + \frac{4\lambda\lambda'}{u}}\,\Bigg]_1^\infty \\ &= \frac{1}{32\pi\lambda\lambda'}(\lambda + \lambda' - |\lambda - \lambda'|) = \frac{\min(\lambda, \lambda')}{16\pi\lambda\lambda'}\,.\end{aligned}$$

The formula follows. This proof is considerably simpler than the one in [85] and replaces their § 2.3.

9. The Fixed Points of the Galois Involution

The involution $(z_1, z_2) \mapsto (-1/Az_2, -1/Az_1)$ of $\mathfrak{H}^2$ defines an involution τ of X_Γ ($\Gamma = \Gamma(\mathfrak{o} \oplus \mathfrak{a})$ and $\mathsf{N}(\mathfrak{a}) = A$) and since it commutes with all elements of Γ_m/Γ also on X_{Γ_m}. It can be lifted to an involution (again denoted by τ) of Y_Γ (or Y_{Γ_m}).

For a square-free divisor w of D we let α_w be the automorphism of Y_Γ defined by a generator of Γ_w/Γ. In particular, $\alpha_w = \alpha_{\tilde{w}}$ if $\tilde{w}$ is the square-free part of D/w. The fixed points of τ were determined by Prestel [130] for the case where D is prime and $\mathfrak{a}$ belongs to the principal genus. His results were extended by Hausmann [63].

Let $(z_1, z_2) \bmod \Gamma$ be a fixed point of $\alpha_w\tau$. Then there exists a matrix M of determinant w representing α_w such that (z_1, z_2) is a fixed point of $MW_AM'W_A'$ with $W_A = \begin{pmatrix} 0 & 1 \\ -A & 0 \end{pmatrix}$. So either (z_1, z_2) is an elliptic fixed point of $C = (wA)^{-1} MW_AM'W_A' \in \mathrm{SL}(\mathfrak{o} \oplus \mathfrak{a})$ or $C = \pm id$. In the latter case we may assume $C = id$ (if $C = -\text{id}$ replace M by $\sqrt{d}M/(w, d)$.). Then the equation $MW_A(z_1, z_2) = (z_2, z_1)$ can be written as

$$(z_2 \quad 1)\begin{pmatrix} a\sqrt{d} & -\lambda'/A \\ \lambda/A & b\sqrt{d}/A \end{pmatrix}\begin{pmatrix} z_1 \\ 1 \end{pmatrix} = 0$$

with $\lambda \in \mathfrak{a}$, $a, b \in \mathbb{Z}$ and $(abd + \lambda\lambda')/A = w$. So (z_1, z_2) lies on F_w or on F_{4w}. If (z_1, z_2) lies on F_{4w} but not on F_w then $D \equiv 0 \pmod 2$ and $2 \nmid w$ if $D \equiv 4 \pmod 8$. One easily checks that F_w and F_{4w} (if $4w|D$) are pointwise fixed by $\alpha_w\tau$. We have proved

(9.1) Proposition. *Let w be a square-free divisor of D. All isolated fixed points of $\alpha_w\tau$ in $\Gamma\backslash\mathfrak{H}^2$ are elliptic fixed points. The 1-dimensional fixed point set of $\alpha_w\tau$ on X_Γ is*

$$\bigcup_{\substack{N \\ \alpha_N = \alpha_w}} F_N = \begin{cases} F_w \cup F_{D/w} & D \text{ odd or } D \equiv 4 \pmod 8 \text{ and } 2|w \\ F_w \cup F_{4w} \cup F_{D/4w} \cup F_{D/w} & \text{otherwise}\,. \end{cases}$$

(9.2) Corollary. *The 1-dimensional fixed point set of τ on X_{Γ_m} is $\bigcup_{w|D} F_w$.*

Now assume $D > 12$. Then all elliptic fixed points of Γ have order 2 and 3 and moreover all elliptic fixed points are special points. These were determined in Section 7, see table on p. 267.

If a special point z with form ϕ_z is fixed by $\alpha_w\tau$ then ϕ_z is ambiguous, i.e. admits an orientation reversing automorphism. Using the results of Section 7 one can determine which curves $F_u\ (u|D)$ pass through a fixed point of $\alpha_w\tau$ and with how many local branches. This was carried at by Hausmann in [62, 63]. Similarly, one can determine the τ-invariant elliptic fixed points of X_{Γ_m}. We have (cf. [11]) for $D > 12$:

(9.3) Proposition. i) *An elliptic fixed point of order* 2 *on* X_{Γ_m} *is a fixed point of* τ *if and only if* z *is the intersection point of two branches of the* 1*-dimensional fixed point set* $\bigcup_{w|D} F_w$ *or of two curves* $F_{2w}, F_{2D/w}$ *with* $w|D$, w *square-free and* $D \equiv 1 \pmod 4$.

ii) *All elliptic fixed points on* X_{Γ_m} *of type* $(r, 1, -1)$ *with* $r \geqslant 3$ *are fixed points of* τ.

iii) *An elliptic fixed point on* X_{Γ_m} *of type* $(r, 1, 1)$ *with* $r \geqslant 3$ *is a fixed point of* τ *if and only if it belongs to the* 1*-dimensional fixed point set* $\bigcup_{w|D} F_w$.

The involution τ also acts on the cusps of X_Γ and X_{Γ_m}. If we identify the set of cusps with the class group of K then the action is given by sending the class of $\mathfrak{b}$ in $\mathscr{C}\ell$ to that of $\mathfrak{a}^{-1}\mathfrak{b}^{-1}$. The action of α_w on the cusps corresponds to sending the class of $\mathfrak{b}$ to that of $\sqrt{(w)}\mathfrak{b}$. A cusp corresponding to $C \in \mathscr{C}\ell$ is fixed by $\alpha_w\tau$ if and only if $\sqrt{(w)}\mathfrak{a}^{-1}$ belongs to C^2. The isotropy group of a cusp of X_Γ has an isotropy group of order 1, 2 or 4 in $\Gamma_m \cup \Gamma_m\tau$. The cusps of X_{Γ_m} correspond to the strict ideal classes in the genus of $\mathfrak{a}$ (the type of such a cusp is (C, U^+) with $C \in \mathscr{C}\ell^+$). A cusp is fixed by τ if and only if the class of C is ambiguous, i.e. it is the class of $\sqrt{(w)}$ for some $w|D$. The action of $\alpha_w\tau$ on the resolution cycles was computed by Hausmann [63].

(9.4) Theorem. *The cusps of* X_{Γ_m} *which are fixed by* τ *correspond to the strict ideal classes of the form* $\sqrt{(w)}\ (w|D)$ *in the genus of* $\mathfrak{a}$. *The fixed curve* F_w *on* Y_{Γ_m} *intersects the resolution cycle of a cusp corresponding to* $\sqrt{(w)}$ *in one point. The following diagrams illustrate the action of* τ *near this cusp.*

$D \equiv 1 \pmod 4$ or
$D \equiv 4 \pmod 8$, $w \equiv 2 \pmod 4$

if $w < D/w$

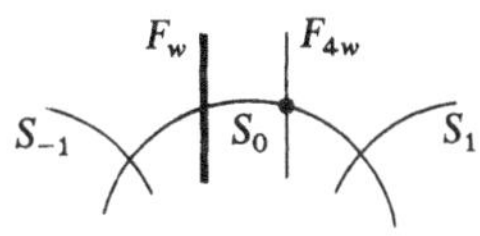

if $w > D/w$

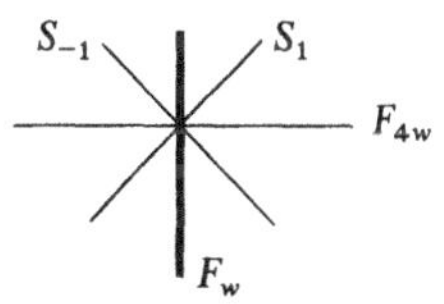

$D \equiv 0 \pmod 4$ or
$D \equiv 4 \pmod 8$, $w \equiv 1 \pmod 2$.

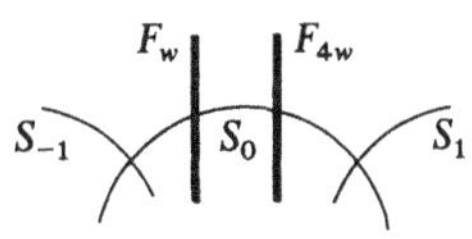

Here τ acts by $\tau(S_k) = S_{-k}$. Fixed point sets are drawn heavily

Finally we describe the action of τ on the set of components of F_N.

(9.5) Proposition ([11]). *The element $\alpha_w\tau$ acts trivially on the set of components of F_N on X_Γ if and only if for all primes p dividing $(D, N/(D,N))$ one has*

$$w|D \quad \text{and} \quad (w,D)_p = \left(\frac{-1}{p}\right)(A,D)_p$$

or

$$D/w|D \quad \text{and} \quad (w,D)_p = (A,D)_p\,.$$

If the action is non-trivial, then it is free.

Appendix: Modular Forms on $\Gamma_0(D)$

In this Appendix we collect some results which are used in Chapter V and VI. Let D be the discriminant of a real quadratic field K. The group $\Gamma_0(D) = \{(\begin{smallmatrix} a & b \\ c & d \end{smallmatrix}) \in \mathrm{SL}(2,\mathbb{Z})\colon c \equiv 0 \pmod D\}$ acts on $\mathfrak{H}$ and the modular curve $\Gamma_0(D)\backslash(\mathfrak{H} \cup \mathbb{P}^1(\mathbb{Q}))$ is a non-singular algebraic curve defined over $\mathbb{Q}$. The cusps of this curve $X_0(D)$ correspond $1-1$ to the positive divisors of D via $\dfrac{m}{n} \mapsto (n, D)$.

Often we shall identify $\Gamma_0(D)$ with its image in $\mathrm{PSL}(2,\mathbb{R})$. Let $\Gamma^*(D)$ be the following subgroup of $\mathrm{PSL}(2,\mathbb{R})$:

$$\Gamma^*(D) = \left\{\frac{1}{\sqrt{m}}\begin{pmatrix} a & b \\ c & d\end{pmatrix} \in \mathrm{PSL}(2,\mathbb{R});\, m \in \mathbb{Z}_{>0},\, m|D,\, a,b,c,d \in \mathbb{Z},\, m|a,\, m|d,\, D|c\right\}. \tag{1}$$

One can show that $\Gamma^*(D)$ is the abelian normalizer of $\Gamma_0(D)$ in $\mathrm{PSL}(2,\mathbb{R})$ (even the full normalizer if $4 \nmid D$). The quotient group is of type $(2,\ldots,2)$; its order is 2^t with t the number of primes dividing D. The elements of this quotient group correspond to the (ordered) decompositions

$$D = D_1 D_2$$

of D as a product of fundamental discriminants. (Note that $(m, D/m) = 1$ above.)

The elements of $\Gamma^*(D)/\Gamma_0(D)$ act on $M_k = M_k(\Gamma_0(D), \chi_D)$ and on $S_k = S_k(\Gamma_0(D), \chi_D)$. The spaces M_k and S_k decompose in 2^t eigenspaces.

Let $\gamma = (\ldots, \gamma_p, \ldots)$ be a t-tuple of signs $\pm$, one sign for each prime dividing D. (The primes are listed in increasing order.) We let

$$M_k^\gamma = \{f = \textstyle\sum a(n)q^n \in M_k\colon a(n) = 0 \text{ whenever } \chi_{D(q)}(n) = -\gamma_q\}\,.$$

If γ has the property that the product of all γ_p is $+$ then γ occurs as the t-tuple $(\ldots, \mathrm{sgn}(\mathsf{N}(\mathfrak{a}), D)_p, \ldots)_{p|D}$ for some ideal $\mathfrak{a}$ in K. So we define

$$M_k^+ = \bigoplus_{\Pi\gamma_p=+} M_k^\gamma \qquad M_k^- = \bigoplus_{\Pi\gamma_p=-} M_k^\gamma$$

We have

$$M_k^+ = \{f = \sum a(n)q^n\ M_k\colon a(n) \neq 0 \text{ implies } \chi_D(n) \neq -1\}\,.$$

The codimension of S_2 in M_2 equals 2^t. We have the following 2^t Eisenstein series in M_2

$$E_{D,1} = \frac{1}{2}L(-1,\chi_D) + \sum_{N=1}^{\infty}\left(\sum_{d|N}\chi_D(d)d\right)q^N$$

$$E_{D_1,D_2} = \sum_{N=1}^{\infty}\left(\sum_{d|D}\chi_{D_1}(d)\chi_{D_2}(N/d)d\right)q^N \qquad (D_2 \neq 1)\,.$$

By combining these we can construct Eisenstein series in M_2^γ. In particular, if $\mathfrak{a}$ is an ideal of K whose norm is A (prime to D), then

$$\sum_{D=D_1D_2}\left(\prod_{p|D_2}(A,D)_p\right)E_{D_1,D_2}$$
$$= \frac{1}{2}L(-1,\chi_D) + \sum_{N=1}^{\infty}\sum_{d|D} d\left(\prod_{p|D}(\chi_{D(p)}(d) + \chi_{D(p)}(NA/d))\right)q^N$$

belongs to M_2^γ with $\gamma_p = (A,D)_p$. Summarizing we have

(1.1) Proposition. *M_2^+ can be decomposed into 2^{t-1} linear subspaces M^γ, one for each genus of K. Each of these contains an Eisenstein series and the codimension of $M_2^\gamma \cap S_2$ in M_2^γ is 1.*

The Hecke operator $T(p)$ (p prime) acts on M_k by

$$T(p)f = p^{k/2-1}\left\{\chi_D(p)f\left|\begin{pmatrix}p & 0\\ 0 & 1\end{pmatrix}\right. + \sum_{r=0}^{p-1} f\left|\begin{pmatrix}1 & r\\ 0 & p\end{pmatrix}\right.\right\}.$$

Let $f = \sum_{n\geq 1} a_n q^n$ be a normalized eigenform ($a(1) = 1$) of the Hecke algebra. We define

$$f^{(D_1)} = \sum_{n\geq 1} a(n,D_1)q^n$$

where

$$a(n,D_1) = \begin{cases}\chi_{D_1}(n)a(n) & \text{if } (n,D_1) = 1\\ \chi_{D_2}(n)\overline{a(n)} & \text{if } (n,D_2) = 1\end{cases} \tag{2}$$

and

$$a(n,D_1)a(m,D_1) = a(nm,D_1) \quad \text{if } (n,m) = 1\,.$$

Note that we have

$$a(n)\langle f,f\rangle = \langle T(n)f,f\rangle = \chi_D(n)\langle f,T(n)f\rangle = \chi_D(n)\overline{a(n)}\langle f,f\rangle\,,$$

so that (2) makes sense. Then $f^{(D_1)}$ is again a normalized eigenform and its Dirichlet series has an Euler product whose p-factor is

$$(1 - a(p, D_1)p^{-s} + \chi_D(p)p^{k-1-2s})^{-1}.$$

If $W \in \Gamma^*(D)$ is a matrix with m in (1) equal to $|D_1|$, then

$$f|W = \lambda_{D_1} f^{(D_1)}$$

for some constant λ_{D_1} (essentially a Gauss sum).

References for this appendix are [5], [65].

VI. The Cohomology of Hilbert Modular Surfaces

All interesting cohomology of Hilbert modular surfaces lies in the middle dimension, i.e. in H^2. The main part of this is the cuspidal cohomology; it carries a Hodge structure and its spaces of forms of type (p, q), $p + q = 2$, are described by spaces of Hilbert modular forms of weight 2. The cohomology group $\mathbb{H}^2(X_\Gamma)$ is the cohomology generated by the cuspidal cohomology $\mathbb{H}^2_{\mathrm{sp}}(X_\Gamma)$ and the classes of the Chern forms w_i associated to the automorphy factors $(cz_i + d)^2$ $(i = 1, 2)$. It can also be described as intersection homology of X_Γ.

Since Hilbert modular surfaces are moduli spaces they are accompanied by extra pieces of information. On the one hand we have the modular curves T_N and on the other hand we have the Hecke correspondences. To begin with the first, each curve T_N on X determines a class t_N in $\mathbb{H}^2(X_\Gamma, \mathbb{Q})$ and we can form the generating series

$$\sum_{N \geqslant 0} t_N q^N \in \mathbb{H}^2(X_\Gamma, \mathbb{Q}) \otimes \mathbb{Q}[[q]],$$

where t_0 is the class of $\frac{1}{4}(w_1 + w_2)$. It turns out that for $q = e^{2\pi i z}$ this is the Fourier series of a modular form (in one variable) on $\Gamma_0(D)$ with character χ_D and with coefficients in $\mathbb{H}^2(X_\Gamma, \mathbb{Q})$, i.e. for every linear form λ on $\mathbb{H}^2(X_\Gamma, \mathbb{Q})$ the series $\sum \lambda(t_N) q^N$ is a modular form on $\Gamma_0(D)$ with character χ_D. In particular, if $\langle\,,\rangle$ is the intersection pairing on $\mathbb{H}^2(X_\Gamma, \mathbb{Q})$ the series

$$\sum_{N \geqslant 0} \langle t_N, t_M \rangle q^N$$

is a modular form of weight 2 on $\Gamma_0(D)$ with character χ_D. This is the main result of Hirzebruch and Zagier in their paper [85], where they proved this by a direct computation of the intersection numbers $\langle t_N, t_M \rangle$ (for prime discriminants). We follow a different path: we construct the dual form of t_N and consider the generating series of these dual forms. This series can be viewed as a theta function of an indefinite quadratic form of signature (2, 2) from which the modularity follows. The intersection numbers t_N, t_M can now be deduced from $\langle t_N, t_1 \rangle$ (which is easy to compute, see V.8) and the action of the Hecke operators.

One thus has a duality between the space $\mathbb{T} \subset \mathbb{H}^2(X_m)$, where X_m is a Hilbert modular surface with 2^{t-1} connected components (one for each genus), spanned by the t_N and a space $M_2^+(\Gamma_0(D), \chi_D)$ of modular forms in one variable, given by $t_N \leftrightarrow a(N) = N$-th Fourier coefficient (a linear form on modular forms). Using Poincaré duality and the description of $\mathbb{H}^2_{\mathrm{sp}}(X_m, \mathbb{C})$ by Hilbert modular cusp forms one finds a map from Hilbert modular forms to modular forms in one variable. The adjoint of this map (w.r.t. Petersson product) can be identified with the Doi-Naganuma lifting. This lifting associates to a normalized Hecke eigenform $f = \sum a(n)q^n \in M_2(\Gamma_0(D), \chi_D)$ a normalized Hilbert modular Hecke eigenform $\hat{f}$ specified by its L-series: $L(\hat{f}, s) = L(f, s)L(f^\rho, s)$ with $f^\rho = \sum \overline{a(n)} q^n$. In this way the curves T_N give a geometric interpretation of the Doi-Naganuma lifting.

This brings us to the Hecke correspondences for Hilbert modular varieties. These correspondences give rise to an algebra of operators on the cohomology and on spaces of Hilbert modular forms. The cohomology of Hilbert modular varieties of dimension one is accessible by the use of the Jacobian of an algebraic curve. In particular, a Hecke eigenform of weight 2 on $\Gamma_0(N)$ determines an isogeny factor of the Jacobian $J_0(N)$ of $X_0(N) = \overline{\Gamma_0(N)\backslash\mathfrak{H}}$. This is an abelian variety defined over $\mathbb{Q}$. The case $[K:\mathbb{Q}] > 1$ is more difficult due to the lacking of a good substitute for the Jacobian. However, Oda conjectured the existence of abelian varieties defined over K which are the analogues of the factors of $J_0(N)$. Let $\mathfrak{n}$ be an integral ideal of K and let $V_\mathfrak{n}$ be the analogue of $X_0(N)$. This is a Hilbert modular variety of dimension $n = [K:\mathbb{Q}]$. As we shall explain (for $n = 2$), the cuspidal cohomology of $V_\mathfrak{n}$ splits into a direct sum of pieces $H^n(M_f, \mathbb{Q})$ of dimension 2^n over $\mathscr{k}_f$ corresponding to Hecke eigenforms f of weight 2. Here $\mathscr{k}_f$ is the field of Fourier coefficients of f. Each piece $H^n(M_f, \mathbb{Q})$ carries a homogeneous Hodge structure. Oda's conjecture is (cf. [124]):

Conjecture. *For each primitive Hecke eigenform f which determines a piece $H^n(M_f, \mathbb{Q})$ of $\mathbb{H}^n_{\mathrm{sp}}(V_\mathfrak{n}, \mathbb{Q})$ there exists an abelian variety A_f of dimension $[\mathscr{k}_f:\mathbb{Q}]$ defined over K (and with good reduction outside $\mathfrak{n}$) and a homomorphism of $\mathscr{k}_f$ into* $\mathrm{End}(A_f) \otimes_\mathbb{Z} \mathbb{Q}$ *such that*

$$\bigotimes_{\sigma: K \to \mathbb{C}} H^1(A_f^\sigma, \mathbb{C}) \to H^n(M_f, \mathbb{C})$$

is an isomorphism of $\mathscr{k}_f$-Hodge structures. (Here A_f^σ is the complex abelian variety obtained from A_f by $\sigma: K \to \mathbb{C}$.)

Oda could prove for real quadratic fields K that $H^2(M_f, \mathbb{Q})$ is the tensor product of two rational $\mathscr{k}_f$-Hodge structures (of weight 1) coming from the isogeny classes of abelian varieties $A_f^{(1)}$ and $A_f^{(2)}$. But unfortunately not much is known about the field of definition of $A_f^{(i)}$. If f is the Doi-Naganuma lift $\hat{g}$ of g then $A_f^{(i)}$ can be identified with a factor of the Jacobian of a modular curve. This enabled Oda to prove his conjecture in this case. A consequence of the existence of the abelian varieties $A_f^{(i)}$ is that the contribution to the Picard number of $H^2(M_f, \mathbb{Q})$ equals $\mathrm{rank}_\mathbb{Q} \mathrm{Hom}(A_f^{(1)}, A_f^{(2)}) \otimes_\mathbb{Z} \mathbb{Q}$. For example, if D is prime and if $f = \hat{g}$, the lift under Doi-Naganuma of g then $\mathrm{rank}_\mathbb{Q} \mathrm{Hom}(A_f^{(1)}, A_f^{(2)}) \otimes \mathbb{Q} = [\mathscr{k}_f:\mathbb{Q}]$. This contribution comes from the curves T_N.

We start this chapter by describing the cohomology of Hilbert modular varieties and the action of the Hecke operators. In Section 2 and 3 we set up the correspondence between $\mathbb{T}$ and $M_2^+(\Gamma_0(D), \chi_D)$ and explain the relation with the Doi-Naganuma map in Section 4. In Section 5 we deduce the formula for the intersection numbers of t_N and t_M. The remaining three sections are used to give Oda's results on the conjecture above.

In Chapter XI we shall give a more detailed treatment of the arithmetic aspects of Hilbert modular surfaces that become apparent in this chapter; in particular we shall show how Oda's conjecture ties in with other conjectures.

1. Cohomology and Hilbert Modular Forms

We shall describe here the cohomology of Hilbert modular surfaces in terms of Hilbert modular forms and the action of the Hecke algebra on both cohomology and Hilbert modular forms. We try to give a down-to-earth description. A more sophisticated one is given in Chapter XI.

Recall $\mathsf{G} = \mathsf{R}_{K/\mathbb{Q}}\mathrm{GL}(2)$ and let K_f be a compact open subgroup of $\mathsf{G}(\mathbb{A}_{\mathbb{Q}}^f)$, see Ch.I.7. We put $X'_{\mathsf{K}_f} = \mathsf{G}(\mathbb{Q})\backslash\mathsf{G}(\mathbb{A}_{\mathbb{Q}})/\mathsf{K}_f\mathsf{K}_\infty$ and let X_{K_f} be the Baily-Borel compactification.

In this chapter we deal mainly with $\mathsf{K}_f = \mathsf{K}_0$, the maximal compact open subgroup. Then X_{K_0} consists of h^+ connected components of the form $\overline{\Gamma\backslash\mathfrak{H}^2}$ with $\Gamma = \mathrm{PGL}^+(\mathfrak{o} \oplus \mathfrak{a})$.

We let Z_{K_f} (resp. Y_{K_f}) be the minimal resolution of the cusp (resp. all) singularities. Then Y_{K_f} is a (possibly non-connected) manifold whereas X'_{K_f} and Z_{K_f} are only orbifolds. They are, however, rational homology manifolds.

We fix K_f and let X_Γ $(X'_\Gamma, Y_\Gamma, Z_\Gamma)$ be a connected component of X_{K_f} $(X'_{\mathsf{K}_f}$, etc.). Then Poincaré duality holds for each Z_Γ so that we have a non-degenerate pairing

$$H_2(Z_\Gamma) \times H_2(Z_\Gamma) \to \mathbb{Q}\,.$$

For each cusp σ of X_Γ we let E_σ be the subspace of $H_2(Z_\Gamma, \mathbb{Q})$ generated by the classes of the curves S_k in the resolution of σ. In Chapter II.3 we saw that the intersection pairing is non-degenerate on each E_σ. Hence, in the direct sum decomposition

$$H_2(Z_\Gamma, \mathbb{Q}) = \left(\bigoplus_\sigma E_\sigma\right) \oplus \mathrm{Im}\{j_*\colon H_2(X'_\Gamma, \mathbb{Q}) \to H_2(Z_\Gamma, \mathbb{Q})\}$$

Poincaré duality induces a non-degenerate pairing on each summand. In cohomology we have a dual decomposition

$$H^2(Z_\Gamma, \mathbb{Q}) = \left(\bigoplus_\sigma E_\sigma^\vee\right) \oplus \pi^* H^2(X_\Gamma, \mathbb{Q})\,,$$

where $\pi\colon Z_\Gamma \to X_\Gamma$ is the natural map. The exact sequence

$$0 = H^1(X_\Gamma - X'_\Gamma, \mathbb{Q}) \to H_c^2(X'_\Gamma, \mathbb{Q}) \to H^2(X_\Gamma, \mathbb{Q}) \to H^2(X_\Gamma - X', \mathbb{Q}) = 0$$

implies

$$H^2(X_\Gamma, \mathbb{Q}) \cong H_c^2(X'_\Gamma, \mathbb{Q})\,.$$

The (Betti-) cohomology groups $H^i(X'_\Gamma, \mathbb{Q})$ have a mixed Hodge structure, i.e. there are two filtrations: a decreasing filtration $\{F_p\}_{p\in\mathbb{Z}}$ on $H^i(X'_\Gamma, \mathbb{Q}) \otimes \mathbb{C}$ (the

Hodge filtration) and an increasing filtration $\{W_k\}_{k\in\mathbb{Z}}$ (the *weight filtration*). Without going into the general theory we just give the weight filtration (cf. [22]) on $H^2(X'_\Gamma, \mathbb{Q})$:

$$W_k H^2(X'_\Gamma, \mathbb{Q}) = 0 \qquad k \leqslant 1$$

$$W_2 H^2(X'_\Gamma, \mathbb{Q}) = W_3 H^2(X'_\Gamma, \mathbb{Q}) = j^* H^2(Z_\Gamma, \mathbb{Q})$$

$$W_k H^2(X'_\Gamma, \mathbb{Q}) = H^2(X'_\Gamma, \mathbb{Q}) \qquad k \geqslant 4$$

with $j\colon X'_\Gamma \hookrightarrow Z_\Gamma$ the natural inclusion. Since $j^* E^\vee_\sigma = 0$ we have

$$j^* H^2(Z_\Gamma, \mathbb{Q}) = j^* \pi^* H^2(X_\Gamma, \mathbb{Q})$$

and it is this part of the cohomology—i.e. the part that comes from the Baily-Borel compactification X_Γ—that we are really interested in. We shall use the notation:

$$\mathbb{H}^2(X_\Gamma, \mathbb{Q}) = j^* \pi^* H^2(X_\Gamma, \mathbb{Q})\,.$$

The cohomology group $\mathbb{H}^2(X_\Gamma, \mathbb{Q})$ can be interpreted as the second intersection homology group of X, see [16].

Next we describe the Hodge filtration on $\mathbb{H}^2(X, \mathbb{Q})$ with $X = X_{\mathsf{K}_f}$ or often $X = X_{\mathsf{K}_0}$. We know from Ch. I what modular forms on each connected component of X are, hence it is clear what modular forms on X are: they are given by m-tuples ($m = \#\pi_0(X)$) of modular forms on the connected components. The vector space of modular forms of weight k is denoted by M_k, the subspace of cusp forms (forms vanishing in all cusps) by S_k.

A Hilbert modular form $f = (\ldots, f_j, \ldots)$ of weight 2 defines a holomorphic 2-form ω_f on Y by

$$\omega_f = (2\pi i)^2 f_j \, dz_1 \wedge dz_2 \quad \text{on } \Gamma_j \backslash \mathfrak{H}^2$$

(see Ch. III.3). This gives a linear map $S_2 \to \mathbb{H}^2(X, \mathbb{C})$. The image is the (2,0)-component:

(1.1) Lemma. $F^2\mathbb{H}^2(X, \mathbb{C}) = \{\omega_f\colon f \in S_2\}$.

Proof. We saw in (3.6) of Chapter III that $H^{2,0}(Y) = \{\omega_f\colon f \in S_2(\Gamma)\}$. In the diagram

$$\begin{array}{ccc} H^2(Z, \mathbb{C}) & \to & \mathbb{H}^2(X, \mathbb{C}) \\ \downarrow & & \downarrow \\ H^2(Y, \mathbb{C}) & \to & H^2(X' - \{\text{quot. sing.}\}, \mathbb{C}) \end{array}$$

the kernels of the horizontal maps and the cokernels of the vertical ones consist of elements of type (1, 1). □

In order to get (1, 1)-forms we use the action of $\pi_0(\mathsf{G}(\mathbb{R})) \cong (\mathbb{Z}/2\mathbb{Z})^2$. This group $\pi_0(\mathsf{G}(\mathbb{R}))$ acts on $\mathsf{G}(\mathbb{R})/\mathsf{K}_\infty \sim (\mathbb{C} - \mathbb{R})^2$: if $\varepsilon_1 = \left(\begin{pmatrix} -1 & 0 \\ 0 & 1 \end{pmatrix}, 1\right)$, $\varepsilon_2 = \left(1, \begin{pmatrix} -1 & 0 \\ 0 & 1 \end{pmatrix}\right)$ then $\varepsilon_1, \varepsilon_2$ act on $(\mathbb{C} - \mathbb{R})^2$ by

$$\varepsilon_1 : (z_1, z_2) \mapsto (\bar{z}_1, z_2), \qquad \varepsilon_2 : (z_1, z_2) \to (z_1, \bar{z}_2).$$

We define

$$\eta_f = \varepsilon_2^* \omega_f$$

$$\eta'_f = \varepsilon_1^* \omega_f$$

These are (1, 1)-forms on X and classes in $\mathbb{H}^2(X, \mathbb{C})$. E.g. if $h^+(K) = 1$, $X_{\mathsf{K}_0} = \mathrm{PGL}^+(2, \mathfrak{o}) \backslash \mathfrak{H}^2$, f defines the differential forms

$$(2\pi i)^2 f(z_1, z_2)\, dz_1 \wedge dz_2$$

$$(2\pi i)^2 f(\varepsilon' \bar{z}_1, \varepsilon z_2)\, dz_1 \wedge dz_2$$

$$(2\pi i)^2 f(\varepsilon z_1, \varepsilon' \bar{z}_2)\, dz_1 \wedge dz_2$$

$$(2\pi i)^2 f(-\bar{z}_1, -\bar{z}_2)\, d\bar{z}_1 \wedge d\bar{z}_2 . \quad \text{(Here } \varepsilon \text{ is a unit with } \varepsilon > 0,\ \varepsilon' < 0.)$$

Moreover, on each component of X we have the forms

$$w_1 = -\frac{1}{2\pi} \frac{dx_1 \wedge dy_1}{y_1^2} \qquad w_2 = -\frac{1}{2\pi} \frac{dx_2 \wedge dy_2}{y_2^2}$$

giving each a class in $F^1 \mathbb{H}^2(X, \mathbb{C})$, see Ch. IV.2.

(1.2) Proposition. *$\mathbb{H}^2(X, \mathbb{C})$ is the direct sum of*
- i) *its (2, 0)-component $\{\omega_f : f \in S_2\}$*
- ii) *its (0, 2)-component $\{\bar{\omega}_f : f \in S_2\}$*
- iii) *its (1, 1)-component $\{\eta_f + \eta'_g : f, g \in S_2\} \oplus W$,*

where W is the $2(\#\pi_0(X))$-dimensional space spanned by the forms w_1, w_2 on all components of X.

Proof. We have $\int w_1 \wedge w_2 \neq 0$ on each component, hence $\dim W = 2\#\pi_0(X)$. Moreover, $\varepsilon_j^* W = W$ $(j = 1, 2)$. We already know the (2, 0)-component by (1, 1), hence also the (0, 2)-component. Using ε_1^*, ε_2^* it follows that the (1, 1)-component of $\mathbb{H}^2(X, \mathbb{C})$ contains $\{\eta_f + \eta'_g : f, g \in S_2\}$ and this is orthogonal to W. But we know

$$e(X') = \dim \mathbb{H}^2(X, \mathbb{C}) + 2\#\pi_0(X) = 4 \dim S_2 + 4\, \#\, \pi_0(X)$$

by Chapter IV (2.7) and (1.1). So we have found all of $\mathbb{H}^2(X, \mathbb{C})$. □

(1.3) Definition. $\mathbb{H}^2_{\mathrm{sp}}(X, \mathbb{Q})$ = orthogonal complement of W in $\mathbb{H}^2(X, \mathbb{C})$. We call it the *cuspidal* cohomology. (sp from Dutch "spits")

To decompose the cohomology we use the action of the Hecke operators. Let $g \in \mathsf{G}(\mathbb{A}^f)$ and consider $\mathsf{L}_f = g \mathsf{K}_f g^{-1} \cap \mathsf{K}_f$. We have a diagram

$$\begin{array}{ccccc} & & X'_{\mathsf{L}_f} & & \\ & \overset{I(1)}{\swarrow} & & \overset{I(g)}{\searrow} & \\ X'_{\mathsf{K}_f} & & & & X'_{\mathsf{K}_f}, \end{array}$$

where $I(1)$, $I(g)$ are induced by right multiplication by 1 and g, see Ch. I.7. Now $I(1)$ and $I(g)$ define algebraic cycles in $X'_{\mathsf{L}_f} \times X'_{\mathsf{K}_f}$. The Hecke correspondence T_g associated to g is the cycle $I(g) \circ {}^tI(1)$ on $X'_{\mathsf{K}_f} \times X'_{\mathsf{K}_f}$. It is not difficult to see that it extends to a correspondence on $X_{\mathsf{K}_f} \times X_{\mathsf{K}_f}$.

Since we deal with a rather explicit case here—mostly $\mathsf{K}_f = \mathsf{K}_0$—we prefer to give a classical description of the Hecke algebra and its action on cohomology and modular forms. For the language of automorphic representations we refer to Chapter XI. We assume that the reader is acquainted with the classical description of the Hecke algebra for the 1-dimensional case, see e.g. Shimura's book [150]. Reference: [152].

We put

$$B = \mathsf{G}(\mathbb{A}_{\mathbb{Q}}) \cap \left(\mathsf{G}(\mathbb{R})^0 \times \prod_{v \in S_f} B_v\right)$$

with

$$B_v = \left\{\begin{pmatrix} a & b \\ c & d \end{pmatrix} \in \mathrm{GL}(2, K_v) \colon a, b, c, d \in \mathfrak{o}_v\right\}.$$

The formal finite sums

$$\sum c_b RbR \quad \text{with } b \in B\,,\ c_b \in \mathbb{C}\,,\ R = \mathsf{G}(\mathbb{R})^0 \times \mathsf{K}_0$$

with the obvious addition and with multiplication as defined in Shimura's book [150] define an algebra A.

Let $\mathfrak{m}$ be an integral ideal in $\mathfrak{o}$ and define $T(\mathfrak{m})$ in A to be the sum $\sum_b RbR$ over all $b \in B$ with $(\det b)\mathfrak{o} = \mathfrak{m}$. We define $S(\mathfrak{m})$ as RaR with $a \in \mathbb{A}_K^*$ such that $a\mathfrak{o} = \mathfrak{m}$. The algebra A is generated by the elements $T(\mathfrak{p})$ and $S(\mathfrak{p})$ with $\mathfrak{p}$ running through the prime ideals. We have in A the relation

$$T(\mathfrak{m})T(\mathfrak{n}) = \sum_{\mathfrak{m}+\mathfrak{n} \subset \mathfrak{a}} \mathsf{N}(\mathfrak{a})S(\mathfrak{a})T(\mathfrak{m}\mathfrak{n}\mathfrak{a}^{-2})\,.$$

One deduces the formal Euler product formula:

$$\sum_{\mathfrak{m}} T(\mathfrak{m})\mathsf{N}(\mathfrak{m})^{-s} = \prod_{\mathfrak{p}} [1 - T(\mathfrak{p})\mathsf{N}(\mathfrak{p})^{-s} + S(\mathfrak{p})\mathsf{N}(\mathfrak{p})^{1-2s}]^{-1}\,. \tag{1}$$

A double coset RbR defines an algebraic correspondence R_{RbR} on $X_{\mathsf{K}_0} \times X_{\mathsf{K}_0}$: the closure of the image of $\mathsf{G}(\mathbb{A}_{\mathbb{Q}}) \to \mathsf{G}(\mathbb{A}_{\mathbb{Q}}) \times \mathsf{G}(\mathbb{A}_{\mathbb{Q}}) \to X_{\mathsf{K}_0} \times X_{\mathsf{K}_0}$, $a \mapsto (a, ab)$. Consider the two projections $X_{\mathsf{K}_0} \times X_{\mathsf{K}_0} \to X_{\mathsf{K}_0}$. When restricted to the correspondence both are finite morphism (p_i, $i = 1, 2$) which are unramified outside finitely many points. For $\mathfrak{m}$ an integral ideal the correspondence $T(\mathfrak{m})$ is defined by $\sum RbR$ with $(\det b)\mathfrak{o} = \mathfrak{m}$.

We have an action of T_{RbR} on $H_2(X'_{\mathsf{K}_0}, \mathbb{Q})$ by

$$(T_{RbR})_* c = p_2(p_1^{-1}c) \qquad c \in H_2(X'_{\mathsf{K}_0}, \mathbb{Q})$$

An action on $H^2_{DR}(X'_{\mathsf{K}_0}, \mathbb{Q})$ is defined by

$$(T_{RbR})^*\omega = p_{1*}(p_2^*\omega)\,, \qquad \omega \in H^2_{DR}(X'_{\mathsf{K}_0}, \mathbb{Q})$$

where $p_2^* \colon H^2(X'_{\mathsf{K}_0}, \mathbb{Q}) \to H^2(T', \mathbb{Q})$ with $T' = T_{RbR} \cap (X'_{\mathsf{K}_0} \times X'_{\mathsf{K}_0})$ is the usual pullback and p_{1*} is the trace $H^2(T',\mathbb{Q}) \to H^2(X',\mathbb{Q})$ obtained from $p_1^*: H_c^2(X'_{\mathsf{K}_0},\mathbb{Q}) \to H_c^2(T',\mathbb{Q})$ by Poincaré duality.

(To avoid problems with quotient singularities we could take a finite index normal subgroup K_1 of K_0 such that X'_{K_1} is smooth and consider $\mathsf{K}_0/\mathsf{K}_1$-invariant cohomology of X_{K_1}.)

One easily checks that $(T_{RbR})^*$ gives rise to an endomorphism of $\mathbb{H}^2(X_{\mathsf{K}_0},\mathbb{Q})$ which preserves the Hodge decomposition and the subspace W (see (1.2)). Similarly, we find an action of $(T_{RbR})_*$ on the dual of $\mathbb{H}^2(X_{\mathsf{K}_0},\mathbb{Q})$ (which is isomorphic to $\mathrm{Im}\{H_2(X'_{\mathsf{K}_0},\mathbb{Q}) \to H_2(Z_{\mathsf{K}_0},\mathbb{Q})\}$). One has

$$\int_{(T_{RbR})_*c} \omega = \int_c (T_{RbR})^*\omega \quad \text{for all } c \in H_2(X'_{\mathsf{K}_0},\mathbb{Q}) \qquad \omega \in H^2_{DR}(X'_{\mathsf{K}_0},\mathbb{Q})\,.$$

Moreover, one has

$$\langle \omega_1, (T_{RbR})^*\omega_1 \rangle = \langle (T_{RbR})^*\omega_1, \omega_2 \rangle \tag{2}$$

for all $\omega_1, \omega_2 \in H^2_{DR}(X'_{\mathsf{K}_0},\mathbb{Q})$. We thus find an action of the Hecke algebra A on the cohomology $\mathbb{H}^2(X_{\mathsf{K}_0},\mathbb{Q})$.

We can also define an action on modular forms. First we associate to a modular form $f = (f_1,\ldots,f_{h^+})$ on X_{K_0} of weight $\mathbf{k}$ a $\mathbb{C}$-valued function $\mathbf{f}$ on $\mathsf{G}(\mathbb{A}_\mathbb{Q})$ by

$$\mathbf{f}(\gamma g_j g) = (f_j|_k g_\infty)(\sqrt{-1},\ldots,\sqrt{-1})\,,$$

where $\gamma \in \mathsf{G}(\mathbb{Q})$, $g \in \mathsf{G}(\mathbb{R})^0 \times \mathsf{K}_0$, $g = (g_\infty g_f)$, cf. Ch.I (7.1). The function $\mathbf{f}$ determines the h^+-tuple $(f_1,\ldots,f_{h^+})$. Therefore, we sometimes identify $\mathbf{f}$ with the tuple $(f_1,\ldots, f_{h^+})$. We have $\mathbf{f}(sgr) = \mathbf{f}(g)$ for $r \in \mathsf{K}_0$, $s \in \mathsf{G}(\mathbb{Q})$. Let now $b \in B$ and write $RbR = \bigcup_l b_l R$ as a disjoint union of cosets. We assume $(b_l)_\infty = 1$. Then we define an action on modular forms by

$$(\mathbf{f}|RbR)(g) = \sum \mathbf{f}(gb_l) \qquad g \in \mathsf{G}(\mathbb{A}_\mathbb{Q})$$

The action of $S(\mathfrak{p})$ is given by $(\mathbf{f}|S(\mathfrak{p}))(g) = \mathbf{f}(g\pi_\mathfrak{p})$ with $\pi_\mathfrak{p}$ a prime element of $\mathfrak{o}_\mathfrak{p}$. So this action is trivial.

The action on the Fourier coefficients of $\mathbf{f}$ is as follows. Let $\mathbf{f} = (f_1,\ldots,f_{h^+})$ be of weight $\mathbf{k} = (k,\ldots,k)$. Each f_j has a Fourier expansion

$$f_j = \sum_{\substack{\nu=0 \\ \text{or } \nu \gg 0}} a_{\nu,j} e^{2\pi i \operatorname{Tr} \nu z}\,,$$

where $\nu \in (\mathfrak{a}_j^{-1})^\vee = \mathfrak{d}^{-1}\mathfrak{a}_j$. Note that $a_{\nu,j}\nu^{-k/2}$ depends only on the ideal $(\nu\mathfrak{d})$. Let $\mathfrak{m}$ be an integral ideal of $\mathfrak{o}$. Then $\mathfrak{m} = \nu\mathfrak{a}_j^{-1}\mathfrak{d}$ for some j, $\nu \in \mathfrak{d}^{-1}\mathfrak{a}_j$. We define

$$c(\mathfrak{m},\mathbf{f}) = \begin{cases} a_{\nu,j}\mathsf{N}(\mathfrak{a}_j\mathfrak{d}^{-1})^{-k/2} & \mathfrak{m} = \nu\mathfrak{a}_j^{-1}\mathfrak{d} \\ 0 & \text{otherwise}\,. \end{cases}$$

Then if we renormalize $T'(\mathfrak{m}) = \mathsf{N}(\mathfrak{m})^{(k-2)/2}T(\mathfrak{m})$ we have

$$c(\mathfrak{m},\mathbf{f}|T'(\mathfrak{m})) = \sum_{\mathfrak{m}+\mathfrak{n}\subset\mathfrak{a}} \mathsf{N}(\mathfrak{a})^{k-1} c(\mathfrak{m}\mathfrak{n}\mathfrak{a}^{-2},\mathbf{f})\,.$$

We consider the Dirichlet series

$$L(\mathbf{f}, s) = \sum_{\mathfrak{m}} c(\mathfrak{m}, \mathbf{f}) \mathsf{N}(\mathfrak{m})^{-s} \qquad \operatorname{Re}(s) > \frac{k}{2} + 1 \tag{3}$$

This converges for $\operatorname{Re}(s) > \frac{k}{2} + 1$ and can be continued meromorphically to the whole s-plane, cf. Ch. I.6. If $\mathbf{f}$ is a cusp form then $L(\mathbf{f}, s)$ is holomorphic for all s. If f is a common eigenform for all $T'(\mathfrak{p})$ then $L(\mathbf{f}, s)$ admits an Euler product using (1). We say that $\mathbf{f}$ is normalized if $c(\mathfrak{o}, \mathbf{f}) = 1$.

The action of the Hecke algebra on modular forms is compatible with the action in the cohomology.

(1.4) Definition. The curve T_N defines a class in $H_2(Z_{\mathsf{K}_0}, \mathbb{Q})$, hence by Poincaré duality and $j\colon X'_{\mathsf{K}_0} \hookrightarrow Z_{\mathsf{K}_0}$ a class t_N in $\mathbb{H}^2(X_{\mathsf{K}_0}, \mathbb{Q})$. The form $\frac{1}{4}(w_1 + w_2)$ on all components defines a class $t_0 \in \mathbb{H}^2(X_{\mathsf{K}_0}, \mathbb{Q})$. The classes in $\mathbb{H}^2(X_{\mathsf{K}_f}, \mathbb{Q})$ coming from components of skew-hermitian curves are called *Hirzebruch-Zagier classes.* The classes t_N generate a subspace $\mathbb{T}$ of $\mathbb{H}^2(X_{\mathsf{K}_0}, \mathbb{Q})$, stable under the Hecke algebra. The space of all Hirzebruch-Zagier classes is denoted $\mathbb{T}_{HZ}$.

We wish to know the action of the Hecke operators on $\mathbb{T}$.

(1.5) Proposition. *Let $\mathfrak{p}$ be a prime ideal of K lying over a prime p which does not divide the discriminant of K. Then if $p \nmid N$*

$$T(\mathfrak{p}) t_N = \begin{cases} t_{pN} & \text{if } p \text{ splits in } K \\ t_{p^2 N} + p t_N & \text{if } p \text{ is inert}\,. \end{cases} \tag{4}$$

Proof. Components of T_N are defined by homomorphisms $\varphi\colon \mathrm{GL}(2, \mathbb{Q}) \to \mathrm{GL}(2, K)$ or $Q^* \to \mathrm{GL}(2, K)$ where Q^* is the unit group of an indefinite quaternion algebra over $\mathbb{Q}$. Suppose now that p splits, $p = \mathfrak{p}\mathfrak{p}'$, and that T_N is not compact (i.e. passes through the cusps). Then $T(\mathfrak{p})$ has degree $p + 1$ and we can take for b_l $l = 1, \ldots, p + 1$ the matrices $\begin{pmatrix} \pi_\mathfrak{p} & 0 \\ 0 & 1 \end{pmatrix}, \begin{pmatrix} 1 & a \\ 0 & \pi_\mathfrak{p} \end{pmatrix}$ with $\pi_\mathfrak{p}$ a prime element and $a \in \mathfrak{o}_\mathfrak{p} \bmod \mathfrak{p}$. The homomorphism φ induces $\mathrm{GL}(2, \mathbb{A}_\mathbb{Q}) \to \mathsf{G}(\mathbb{A}_\mathbb{Q})$ which is the diagonal embedding followed by conjugation by an element g of B with $\mathsf{N}_{K/\mathbb{Q}}(\det g) = N$. Conjugation by b_l replaces g by an element g_1 with $\mathsf{N}(\det g_1) = Np$. The same holds for the case $Q^* \to \mathrm{GL}(2, K)$ (compact curves). If p is inert then $\deg T(\mathfrak{p}) = p^2 + 1$ and we can take as b_l: $\begin{pmatrix} p & 0 \\ 0 & 1 \end{pmatrix}, \begin{pmatrix} 1 & a \\ 0 & p \end{pmatrix}$ with $a \bmod p$. Again g is replaced by g_1 with $\mathsf{N}(\det g_1) = Np^2$ but if $a \in \mathbb{F}_p \subset \mathbb{F}_{p^2} = \mathfrak{o}_p/p$ then g_1 is divisible by p. The proposition now follows easily from this. □

[One can also prove an analogue of (1.5) for the classes $t_N^{(k)}$, $k > 2$ defined in Section 2. It says that under the assumption of (1.5)

$$T(\mathfrak{p}) t_N^{(k)} = \begin{cases} t_{pN}^{(k)} & \text{if } p = \mathfrak{p}\mathfrak{p}' \text{ splits} \\ t_{p^2 N}^{(k)} + p^{k-1} t_N^{(k)} & \text{if } p = \mathfrak{p} \text{ is inert}\,.] \end{cases} \tag{5}$$

Next, we study the action of the involution τ on the cohomology of X. Recall that τ is induced on a component $\mathrm{PGL}^+(\mathfrak{o} \oplus \mathfrak{a})\backslash\mathfrak{H}^2$ by $(z_1, z_2) \mapsto (-1/Az_2, -1/Az_1)$ with $A = \mathsf{N}(\mathfrak{a})$. The action on a modular form of weight k on $\mathrm{PGL}^+(\mathfrak{o} \oplus \mathfrak{a})$ is by

$$f(z_1, z_2) \mapsto \tau(f)(z_1, z_2) = (z_1 z_2)^{-k} f(-1/Az_2, -1/Az_1)$$

and on differential forms by

$$\tau^*\omega_f = -\omega_{\tau(f)}, \qquad \tau^*\eta_f = -\eta'_{\tau(f)}$$

$$\tau^* w_1 = w_2 .$$

We thus obtain an involution of $\mathbb{H}^2(X, \mathbb{Q})$ sending the cohomology $\mathbb{H}^2(X_\Gamma, \mathbb{Q})$ of a component to itself. One has

$$\mathbb{H}^2(X, \mathbb{Q})^\tau = \mathbb{H}^2(X/\tau, \mathbb{Q}) := j^* H^2(X/\tau, \mathbb{Q})$$

with $j\colon X \to X/\tau$ the natural map. We consider

$$\mathbb{H}^2_{\mathrm{sp}}(X, \mathbb{Q}) = \text{orthogonal complement of } W \text{ in } \mathbb{H}^2(X, \mathbb{Q}),$$

the *cuspidal* part of the cohomology. It carries a natural Hodge structure.

(1.6) Lemma. *Let $\mathbb{H}^2_{\mathrm{sym}}(X)$ be the $+1$-eigenspace of τ on $\mathbb{H}^2(X, \mathbb{C})$. Then $f \mapsto \eta_f + \tau^*\eta_f$ gives an isomorphism*

$$S_2 \xrightarrow{\sim} \mathbb{H}^2_{\mathrm{sym}}(X) \cap H^{1,1}_{\mathrm{sp}}(X, \mathbb{C}) .$$

(1.7) *Remark.* If one looks at a component X_Γ ($\Gamma = \mathrm{PGL}^+(\mathfrak{o} \oplus \mathfrak{a})$) instead of X then one must replace in (1.6) the space S_2 by the space $S_2^{+-}(\Gamma)$, i.e. the vector space of holomorphic functions $f\colon \mathfrak{H} \times \mathfrak{H}_- \to \mathbb{C}$ such that $f|\gamma(z) = (\det\gamma)^{-1}\mathsf{N}(cz+d)^2 f(\gamma z)$ for all $\gamma \in \Gamma$.

Now τ acts on S_2 and we have a commutative diagram

$$\begin{array}{ccl} S_2 & \to & V := \mathbb{H}^2_{\mathrm{sym}}(X) \cap H^{1,1}_{\mathrm{sp}}(X, \mathbb{C}) \\ \downarrow{\scriptstyle\tau} & & \downarrow{\scriptstyle\kappa} \\ S_2 & \to & V, \end{array}$$

where $\kappa = \varepsilon_1\varepsilon_2$ is the involution induced by complex conjugation $(z_1, z_2) \mapsto (\bar{z}_1, \bar{z}_2)$ ("Frobenius at the infinite places"). So we find an eigenspace decomposition

$$V \simeq V_+ \oplus V_- ,$$

where

$$V_+ \cap \mathbb{H}^2_{\mathrm{sp}}(X) \simeq S_2^{\tau=-1}$$

$$V_- \cap \mathbb{H}^2_{\mathrm{sp}}(X) \simeq S_2^{\tau=1} .$$

A formula for $\dim S_2$ was given in IV.2. The dimension of $S_2^{\tau=1}$ can be computed as follows. We have $\dim S_2^{\tau=1} = p_g(Y) - p_g(Y/\tau)$, hence it suffices to compute $p_g(Y/\tau)$. Now if S is a smooth surface with an involution τ we can blow up the *isolated* fixed points of τ to obtain a surface S' and an involution τ' (which lifts τ) such that τ' has no isolated fixed points. Then S'/τ' is smooth. One now has

$$\chi(S') = 2\chi(S'/\tau') + \tfrac{1}{4}\mathrm{K}_{S'} \cdot R \quad (\mathrm{K}_{S'}\text{: canonical divisor class})$$

with R the ramification divisor of $S' \to S'/\tau$, see [10]. In our situation going back to S we thus find

$$\chi(S) = 2\chi(S/\tau) + \frac{1}{4}\mathrm{K}_S \cdot F - \frac{m}{4}$$

with F the 1-dimensional fixed point set of τ and m the number of fixed points. E.g. if K has a prime discriminant p one has

$$\chi(Y) - 2\chi(Y/\tau) = \left[\frac{p-29}{24}\right]. \quad ([\ ] = \text{integer part})$$

For $\dim S_2^{\tau=1}$ see also VI.8.

2. The Dual of T_N

The arithmetic cycle T_N determines a homology class in $H_2(Z_\Gamma, \mathbb{Q})$, hence via Poincaré duality and $j\colon X_\Gamma^0 \hookrightarrow Z_\Gamma$ a cohomology class t_N in $j^*H^2(Z_\Gamma, \mathbb{Q}) = \mathbb{H}^2(X_\Gamma, \mathbb{Q})$. We define $t_0 \in \mathbb{H}^2(X_\Gamma, \mathbb{Q})$ as the cohomology class of $\frac{1}{4}(w_1 + w_2)$ and $t_N^0 \in \mathbb{H}^2_{\mathrm{sp}}(X_\Gamma, \mathbb{Q})$ as the unique $\mathbb{Q}$-linear combination $t_N + \alpha t_0$ which is orthogonal to W. (This exists since $\int_{T_N} w_1 = \int_{T_N} w_2$ and $t_0^2 \neq 0$.) One sees easily that $t_N^0 \in \mathbb{H}^2_{\mathrm{sym}}(X_\Gamma, \mathbb{Q}) \cap \mathbb{H}^{1,1}_{\mathrm{sp}}(X_\Gamma, \mathbb{Q})$. So by (1.6) of the preceding section t_N^0 corresponds to a Hilbert modular form $f \in S_2^{+-}(\Gamma)$ (i.e. t_N^0 is represented by $\eta_f + \bar{\eta}_f$). What is f?

Let F be a component of T_N given by $z_1 = Cz_2$. We assume for simplicity that F is non-singular and does not pass through singularities of $\Gamma\backslash\mathfrak{H}^2$. We have a diagram

$$\begin{array}{ccccc} z \in \mathfrak{H} & \to & G(C)\backslash\mathfrak{H} & & \\ \downarrow\ \ \downarrow & & \downarrow & & \\ (Cz, z) \in \mathfrak{H}^2 & \to & G(C)\backslash\mathfrak{H}^2 & \to & \Gamma\backslash\mathfrak{H}^2 \end{array}$$

with $G(C) = \{M \in \Gamma\colon M^*CM' = \pm C\}$, see V.1. Let L_F be the line bundle on $G(C)\backslash\mathfrak{H}^2$ corresponding to F. Let s be a section of L_F with divisor $G(C)\backslash\mathfrak{H} \subset G(C)\backslash\mathfrak{H}^2$. Define a hermitian metric on L_F by

$$\|s\|_{(z_1, z_2) \bmod G(C)} = \left|\frac{z_1 - Cz_2}{z_1 - C\bar{z}_2}\right|.$$

The Chern form is given by $\dfrac{1}{2\pi i}\partial\bar{\partial}\log\|s\|^2$ which is $\left(\text{if } C = \begin{pmatrix}\alpha & \beta \\ \gamma & \delta\end{pmatrix}\right)$ $\dfrac{1}{2\pi i}\{(\gamma z_1 \bar{z}_2 + \delta z_1 - \alpha\bar{z}_2 - \beta)^{-2} dz_1 \wedge d\bar{z}_2 + \text{complex conjugate}\}$.

This suggests the introduction for $N \in \mathbb{Z}_{>0}$, $k \in \mathbb{Z}_{>2}$, k even, of

$$\omega_N^{(k)}(z_1, z_2) = \sum \left(a\sqrt{D}z_1 z_2 - \lambda' z_1 + \lambda z_2 + \frac{b}{A}\sqrt{D}\right)^{-k}, \qquad (z_1, z_2) \in \mathfrak{H} \times \mathfrak{H}_-$$

where the summation is over all triples $(a, b, \lambda) \in \mathbb{Z}^2 \times \mathfrak{a}^{-1}$ satisfying $abD + \lambda\lambda' A = N$. If there are no such triples we put $\omega_N^{(k)} = 0$. One can check that $\omega_N^{(k)}$ is a Hilbert modular form of weight k in $S_k^{+-}(\Gamma)$.

For $k = 2$ one runs into convergence problems. So consider for $s \in \mathbb{C}$, $\mathrm{Re}(s) > 0$ the function $\omega_{N,s}$

$$\sum \left(a\sqrt{D}z_1 z_2 - \lambda' z_1 + \lambda z_2 + \frac{b}{A}\sqrt{D}\right)^{-2} \left|a\sqrt{D}z_1 z_2 - \lambda' z_1 + \lambda z_2 + \frac{b}{A}\sqrt{D}\right|^{-s}$$

and define $\omega_N = \omega_N^{(2)}$ as the limit for $s \to 0$. That this limit exists and is a form in $S_2^{+-}(\Gamma)$ can be shown as in [172], where D. Zagier introduced and studied a slightly different $\omega_N^{(k)}$.

(2.1) Theorem. *Under the isomorphism* $\mathbb{H}^2_{\mathrm{sym}}(X_\Gamma) \cap \mathbb{H}^{1,1}_{\mathrm{sp}}(X_\Gamma, \mathbb{C}) \simeq S_2^{+-}(\Gamma)$ *the class* t_N^0 *corresponds to* $\dfrac{N}{(2\pi i)^3}(\eta_f + \bar{\eta}_f)$ *with* $f = \omega_N \in S_2^{+-}(\Gamma)$.

Proof. We have to show

$$\int_{T_N} \eta_g = \frac{N}{(2\pi i)^3} \int_{\Gamma\backslash\mathfrak{H}^2} \eta_g \wedge \bar{\eta}_f \tag{1}$$

for all $g \in S_2^{+-}(\Gamma)$. For $C = \begin{pmatrix} \alpha & \beta \\ \gamma & \delta \end{pmatrix}$, $s \in \mathbb{C}$ we put

$$\phi_C(z_1, z_2) = (\gamma \bar{z}_2 + \delta)^{-2}(z_1 - C\bar{z}_2)^{-2} \quad \text{and} \quad \phi_{C,s} = \phi_C |\phi_C|^s .$$

We claim that for $\mathrm{Re}(s) > 0$

$$\int_{\Gamma\backslash\mathfrak{H}^2} \eta_g \wedge \bar{\phi}_{C,s}(z_1, z_2)(y_1 y_2)^s d\bar{z}_1 \wedge dz_2 = \frac{c_s}{N} \int_{G(C)\backslash\mathfrak{H}} |\gamma\bar{z} + \delta|^{-2s} \eta_g(Cz, z) , \tag{2}$$

where $c_s \to 2\pi i$ as $s \to 0$. Indeed, if $\mathscr{F}$ is a fundamental domain for the action of $G(C)$ on $\mathfrak{H}$ then $\mathfrak{H} \times \mathscr{F}$ is a fundamental domain for the action of $G(C)$ on $\mathfrak{H}^2$. Therefore we can split the left-hand side as

$$(2\pi i)^2 \int_{G(C)\backslash\mathfrak{H}} (\gamma z_2 + \delta)^{-2} \left(\int_{\mathfrak{H}} g(z_1, \bar{z}_2)(\bar{z}_1 - Cz_2)^{-2} |\bar{z}_1 - Cz_2|^{-2s} y_1^s (\mathrm{Im}\, Cz_2)^s dz_1 \wedge d\bar{z}_1 \right) dz_2 \wedge d\bar{z}_2 .$$

But the inner integral equals $c_s g(Cz_2, \bar{z}_2)$ with $c_s \to 2\pi i$ as $s \to 0$. This can be proved as Proposition (3.1) of Hejhal [68]. So if the matrices C_j define the different components of T_N we find

$$\begin{aligned}
&\int_{\Gamma\backslash\mathfrak{H}^2} \eta_g \wedge \omega_{N,s}(\bar z_1, z_2)(y_1 y_2)^2 \, d\bar z_1 \wedge dz_2 \\
&= \sum_j \int_{G(C_j)\backslash\mathfrak{H}} \eta_g \wedge \bar\phi_{C_{j,s}}(y_1 y_2)^s \, d\bar z_1 \wedge dz_2 \\
&= \frac{c_s}{N} \sum_j \int_{G(C_j)\backslash\mathfrak{H}} |\gamma_j \bar z + \delta_j|^{-2s} \eta_g(C_j z, z) \\
&= \frac{c_s}{N} \int_{T_N} |\gamma \bar z + \delta|^{-2s} \eta_g .
\end{aligned}$$

Taking the limit as $s \to 0$ we find (1). So $\eta_f + \bar\eta_f$ represents $\frac{(2\pi i)^3}{N}$ times the Poincaré dual of T_N on the orthogonal complement of $\mathbb{C}w_1 \oplus \mathbb{C}w_2$ in $\mathbb{H}^2(X_\Gamma, \mathbb{C})$. Since $\eta_f + \bar\eta_f$ is orthogonal to $E_\sigma^\vee$ for all σ this proves the theorem. □

The proof given here was inspired by an argument of Zagier given in [174], where he showed an analogue of (1) for forms of higher weight.

To treat the higher weight case we need some cohomology, cf. [165]. Let $r_{\mathbb{Q}}\colon \mathrm{GL}(2, \mathbb{Q}) \to V_{\mathbb{Q}}$ be the representation of $\mathrm{GL}(2, \mathbb{Q})$ on k-fold symmetric tensors on $\mathbb{Q}^2$. Via $\mathrm{GL}(2, K) \hookrightarrow \mathrm{GL}(2, \mathbb{R}) \times \mathrm{GL}(2, \mathbb{R})$ this defines a representation $\rho = \rho^{(k)}\colon \mathrm{GL}(2, K) \to V_{\mathbb{R}} \otimes V_{\mathbb{R}}$. This gives rise to a local system or vector bundle by $U = \Gamma\backslash(\mathfrak{H}^2 \times V_{\mathbb{R}} \otimes V_{\mathbb{R}})$. The complexification of U is again denoted by U. This U has a natural hermitian metric coming from a Γ-invariant symmetric bilinear form $P \otimes P$, where P is the unique symmetric bilinear form on $V_{\mathbb{Q}}$ with the property that

$$^t\left(\begin{pmatrix} a \\ b \end{pmatrix}^{\otimes k}\right) P \begin{pmatrix} c \\ d \end{pmatrix}^{\otimes k} = (ad - bc)^k \quad \text{for all } a, b, c, d \in \mathbb{Q}$$

cf. [150] p. 230. We can extend U over the cusp resolutions by noting for a cusp of type (M, V) that U is trivial over $\mathbb{C}^2/M$, hence can be extended trivially over Y^+ (see II.3 for notation) and descends to a line bundle on $Y(M, V) = Y^+/V$. We now define

$$\mathbb{H}^2(X_\Gamma, U) := j^* j_! H_c^2(\Gamma\backslash\mathfrak{H}^2, U)$$

with $j_!\colon H_c^2(\Gamma\backslash\mathfrak{H}^2, U) \to H^2(Y_\Gamma, U)$ and $j^*\colon H^2(Y_\Gamma, U) \to H^2(\Gamma\backslash\mathfrak{H}^2, U)$. This group can be interpreted as an intersection cohomology group, see [16]. Moreover, $\mathbb{H}^2(X_\Gamma, U)$ carries a homogeneous Hodge structure. A modular form $f \in S_{k+2}$ defines a class ω_f in $\mathbb{H}^{2,0}(X_\Gamma, U)$ by $(2\pi i)^2 f(z_1, z_2) \begin{pmatrix} z_1 \\ 1 \end{pmatrix}^{\otimes k} \otimes \begin{pmatrix} z_2 \\ 1 \end{pmatrix}^{\otimes k} dz_1 \wedge dz_2$ since

$$\begin{pmatrix} \gamma(z) \\ 1 \end{pmatrix}^{\otimes k} = (cz + d)^{-k} \rho^{(k)}(\gamma) \begin{pmatrix} z \\ 1 \end{pmatrix}^{\otimes k} \quad \text{for } z \in \mathfrak{H}, \quad \gamma \in \mathrm{GL}(2, \mathbb{R}).$$

One can show that

$$\begin{aligned}
\mathbb{H}^{2,0} &\cong S_{k+2}(\Gamma) \\
\mathbb{H}^{1,1} &\cong S_{k+2}^{+-}(\Gamma) + S_{k+2}^{-+}(\Gamma).
\end{aligned}$$

(We do not really need this. One could simply work in the subspace generated by cusp forms.) The hermitian metric on U together with the Hodge $*$-operator on Y_Γ gives rise to a $*$-operator from U-valued (p,q)-forms to U^*-valued $(2-p, 2-q)$-forms. A calculation now yields

$$\langle \eta_f, *\eta_g \rangle = 2^{2k}(-1)^k(2\pi i)^4 f(z_1, \bar{z}_2)\overline{g(z_1, \bar{z}_2)} y_1^k y_2^k \, dx_1 \, dy_1 \, dx_2 \, dy_2$$

for $f, g \in S_{k+2}^{+-}(\Gamma)$. (Here $\langle \ , \ \rangle$ denotes the trace $U \times U^* \to \mathcal{O}_{Y_\Gamma}$.)

We now associate to a modular curve F on X_Γ a section of $U|F$ following Tong [165]. Suppose that F is defined by $\{(z, Cz): z \in \mathfrak{H}\} \subset \mathfrak{H}^2$. We can view $P\rho(C)$ as a linear map $V \to V^*$, hence its inverse $(P\rho(C))^{-1}$ as an element of $\mathrm{Hom}(V^*, V) \cong V \otimes V$.

(2.2) Lemma. *The element $(P\rho(C))^{-1} \in V \otimes V$ is invariant under $G(C)$.*

Proof. We have

$$\begin{aligned} \rho(M)\rho(C^{-1})P^{-1}{}^t\rho(M') &= \rho(MC^{-1}M'^{-1})\rho(M')P^{-1}\rho({}^tM') \\ &= \rho(C^{-1})P^{-1} . \end{aligned}$$

for all $M \in G(C)$. □

This gives us the desired section s_F of U over F and a current $\hat{s}_F$ by

$$\hat{s}_F(\alpha) = \int_F \langle s_F, \alpha | F \rangle ,$$

where α is any compactly supported U^*-valued $(1,1)$-form. This again gives us a cohomology class $\mathbb{H}^{1,1}(X_\Gamma, U)$. The analogue of (2.1) is

(2.3) Proposition. *The cohomology class $t_N^{(k+2)} \in \mathbb{H}^{1,1}(X_\Gamma, U)$ defined by T_N corresponds to $c(\eta_f + \bar{\eta}_f)$ with $f = \omega_N^{(k+2)} \in S_{k+2}^{+-}(\Gamma)$ and with $c \in \mathbb{C}^*$ independent of N.*

The proof, which is similar to but simpler than that of (2.1), is left to the reader.

3. The Generating Series of the Modular Curves

In this section we study the generating series $\sum T_N q^N$ of the curves T_N. To get an idea what to expect we look at the "degenerate" case $X = X_\Gamma$ with $\Gamma = \mathrm{PSL}(2, \mathbb{Z}) \times \mathrm{PSL}(2, \mathbb{Z})$. Then $X = \mathbb{P}^1 \times \mathbb{P}^1$. The Hecke correspondence T_N (defined by equations $z_1 = (az_2 + b)/(cz_2 + d)$ with $ad - bc = N$) is symmetric and therefore its homology class is (v, v) in $H_2(X, \mathbb{Z}) \cong \mathbb{Z} \oplus \mathbb{Z}$ with v equal to the intersection number of T_N with $\mathbb{P}^1 \times \{\infty\}$. Note that T_N intersects $\mathbb{P}^1 \times \{\infty\}$ only at $\{\infty, \infty\}$. In local coordinates (ζ_1, ζ_2), $\zeta_j = \exp 2\pi i z_j$, T_N is given by equations $\zeta_1^a = \zeta_2^d$ with $ad = N$. Therefore $T_N \cdot \mathbb{P}^1 \times \{\infty\} = \sigma_1(N)$ (sum of divisors of N), and we get

$$T_N \sim \sigma_1(N)(P_1 + P_2) ,$$

where P_1 (resp. P_2) is a horizontal (resp. vertical fibre) of X. So the generating series

$$\sum [T_N] q^N \in R[[q]] \qquad q = \exp 2\pi i z$$

with $R = H_2(X, \mathbb{Z})$ is obviously linked to the Eisenstein series

$$\frac{1}{4}\zeta(-1) + \sum_{n \geqslant 1} \sigma_1(n) q^n$$

of weight 2 on $\mathrm{SL}(2, \mathbb{Z})$. (Unlike the Eisenstein series of higher weight this is not a modular form due to the conditional convergence of the usual sum defining the Eisenstein series, cf. I.6.)

By viewing a Hilbert modular surface as a "twisted" version of X and by Serre's remark relating the dimension of a part of the homology of Y_{Γ_K} to the dimension of a space of modular forms in one variable (see Introduction of this chapter) one is easily led to the idea that the generating series of the homology classes of the T_N on a Hilbert modular surface can be described in terms of modular forms of one variable. In particular the intersection numbers $T_N \cdot C$ (N varies and C is fixed) on Y_Γ should be the Fourier coefficients of a modular form. This was proved by Hirzebruch and Zagier using explicit formulas for the intersection numbers of T_N and T_M in [85].

Instead of studying $\sum T_N q^N$ one may as well look at the generating series of the ω_N. This was done by Zagier [172]. One of his results (adapted and generalized to our case) is the following.

For $k \in \mathbb{Z}_{\geqslant 2}$, k even we define

$$\Omega_\Gamma^{(k)}(z, \tau) = \sum_{N=1}^{\infty} N^{k-1} \omega_N^{(k)} e^{2\pi i N \tau} \qquad (z \in \mathfrak{H} \times \mathfrak{H}_-, \tau \in \mathfrak{H}).$$

(We shall often omit the Γ in $\Omega_\Gamma^{(k)}$.)

(3.1) Theorem. *For fixed $\tau \in \mathfrak{H}$ the function $\Omega^{(k)}(z, \tau)$ is a Hilbert modular cusp form of weight k in $S_k^{+-}(\Gamma)$. For fixed $z \in \mathfrak{H} \times \mathfrak{H}_-$ the function $\Omega^{(k)}(z, \tau)$ is a cusp form of weight k on $\Gamma_0(D)$ and character χ_D.*

The first assertion is clear. Zagier proved the second by expanding $\Omega^{(k)}$ as an infinite sum of Poincaré series. Another approach, suggested by Eichler and worked out by Vignéras, Oda, Kudla and others views $\Omega^{(k)}$ as a theta series of an indefinite quadratic form. E.g. one has:

(3.2) Theorem (Vignéras [168]). *Let q be a quadratic form of signature $(2, n-2)$ on $\mathbb{R}^n$ whose associated bilinear form is $\langle v, w \rangle = q(v+w) - q(v) - q(w)$. If $L \subset \mathbb{R}^n$ is a lattice on which q is integral and if $z \in \mathbb{C}^n$ with $q(z) = 0$, $\langle z, \bar{z} \rangle > 0$ then*

$$\sum_{v \in L} \frac{q(v)^{k-1}}{\langle v, z \rangle^k} e^{2\pi i q(v)}$$

is a modular form of weight k, whose level is that of q and whose character is χ_D with D the discriminant of q.

Theta series as in the theorem were introduced by Siegel. Their transformation behaviour can be proved by elementary means, but can be better understood using the Weil representation—a projective representation of $SL(2, \mathbb{R})$ on $L^2(\mathbb{R}^n)$, cf. [170]. An elementary proof is given in [168]; other references are [104, 109] and Oda's paper [122].

Theorem (3.1) follows from (3.2) by taking for L the lattice of skew-hermitian matrices which are integral w.r.t. $\mathfrak{a}$ with the quadratic form $A \cdot \det$ and $z = \begin{pmatrix} 1 & z_1 \\ z_2 & z_1 z_2 \end{pmatrix}$.

(3.3) Theorem. *For all $t \in \mathbb{H}^2(X_\Gamma, \mathbb{C})$ the generating series*

$$g_t = \langle t_0, t \rangle + \sum_{N=1}^{\infty} \langle t_N^0, t \rangle q^N ,$$

where $\langle , \rangle$ denotes the intersection pairing on $\mathbb{H}^2(X_\Gamma, \mathbb{C})$, is a modular form in M_2^γ with γ the genus of $\mathfrak{a}$.

Here M_2^γ is the subspace of $M_2(\Gamma_0(D), \chi_D)$ generated by modular forms whose Fourier coefficients are real and satisfy $a(N) = 0$ if there is a prime q with $q|D$ and $\chi_{D(q)}(N)(A, D)_q = -1$, see appendix to Ch. V.

Proof. Let t be class of type (1, 1) in $\mathbb{H}^2_{sp}(X_\Gamma, \mathbb{C})$ orthogonal to w_1 and w_2, say t is represented by $\eta_f + \eta'_g$, $f, g \in S_2^{+-}(\Gamma)$. Then

$$\langle t_N^0, t \rangle = * \int_{\Gamma \backslash \mathfrak{H}^2} (\eta_{\omega_N} + \eta'_{\omega_N}) \wedge (\eta_f + \eta'_g)$$

with $* \in \mathbb{C}^*$ independent of N and t, so $g_t = * \int (\eta_\Omega + \eta'_\Omega)(\eta_f + \eta'_g)$ is a modular form in $M_2(\Gamma_0(D), \chi_D)$. As we know from V.5

$$2\langle t_0, t_0 \rangle + 2 \sum_{N=1}^{\infty} \langle t_0, t_N \rangle q^N = \frac{1}{4} \operatorname{vol}(\Gamma \backslash \mathfrak{H}^2) + \sum_{N=1}^{\infty} \operatorname{vol}(T_N) q^N$$

belongs to $M_2(\Gamma_0(D), \chi_D)$ as well. It is not difficult to check that these forms lie in M_2^γ. □

Let V be a finite dimensional vector space and $V^\vee$ its dual. We call $\sum_{N=0} v_N q^N \in V[[q]]$ a *modular form with coefficients in V* if for every $\lambda \in V^\vee$ the Fourier series $\sum \lambda(v_N) q^N$ is a modular form (of the same level and character). In this sense $\sum t_N q^N$ is a modular form on $\Gamma_0(D)$ with character χ_D and with coefficients in $\mathbb{H}^2(X_\Gamma, \mathbb{Q})$. Likewise, $\Omega^{(k)}$ is a modular form with coefficients in $S_2^{+-}(\Gamma)$.

Let now $\mathbb{T}_\Gamma \subset \mathbb{H}^2(X_\Gamma, \mathbb{C})$ be the subspace generated by the classes t_N, $N \geqslant 0$. By (3.3) we get a linear map

$$g \colon \mathbb{T}_\Gamma \to M_2^\gamma \qquad t \mapsto g_t .$$

(3.4) Proposition. *g is injective.*

Proof. By the Hodge index theorem the intersection form on the Néron-Severi group has signature $(1+, m-)$. The plus sign is provided by t_0: $\langle t_0, t_0 \rangle > 0$. If $t \in \mathbb{T}_\Gamma$ and $\langle t, t_N \rangle = 0$ for all N then $\langle t, t \rangle < 0$ if $t \neq 0$, a contradiction. □

We shall see that g is an isomorphism in the next section.

(3.5). Several variations on this theme are possible. For example, suppose that $D = p$ is a prime discriminant and let $\Gamma = \mathrm{PSL}(2, \mathfrak{o})$. The curve F_{Np^2} has two components if it is non-empty, see Ch. V. If we write

$$a\sqrt{p}z_1z_2 - \lambda' z_1 + \lambda z_2 + b\sqrt{p} = 0\,,$$

then the sign of $\left(\frac{a}{p}\right) + \left(\frac{b}{p}\right)$ distinguishes the two components $F^{\pm}_{Np^2}$. We can consider (with $\omega_{N,\pm}$ the dual form of $F^{\pm}_{Np^2}$)

$$\Omega_0^{(k)}(z,\tau) = \sum_N N^{k-1}(\omega^{(k)}_{N,+} - \omega^{(k)}_{N,-})e^{2\pi i N\tau} \qquad (z \in \mathfrak{H}^2, \tau \in \mathfrak{H})\,.$$

Considered as a function of τ it is a cusp form of weight k on $\mathrm{SL}(2,\mathbb{Z})$ and as a function of z a Hilbert modular form of weight k on $\mathrm{SL}(2,\mathfrak{o})$, see [174, Thm. 7].

4. The Doi-Naganuma Lifting

In this section we use the description $X = \bigsqcup_{i=1}^{h} \mathrm{PGL}(\mathfrak{o} \oplus \mathfrak{a}_i)\backslash(\mathbb{C}\backslash\mathbb{R})^2$ rather than the description with h^+ components. We let $\mathbb{T}$ be the subspace of $\mathbb{H}^2(X, \mathbb{Q})$ generated by the t_N, $N \geqslant 0$. By (2.1) we know that t_N^0 (cuspidal part of t_N) corresponds to a form $\frac{N}{(2\pi i)^3}(\eta_f + \bar{\eta}_f)$ with $f = \omega_N \in S_2$. (Here we view Hilbert modular forms as h-tuples of functions on $(\mathbb{C}\backslash\mathbb{R})^2$ rather than as h^+-tuples of functions on $\mathfrak{H}^2$.)

The generating series $\Omega^{(k)} = \sum_{N=1}^{\infty} N^{k-1}\omega_N^{(k)}e^{2\pi i N\tau}$ is both a Hilbert modular form and a one variable modular form by (3.1). This can be used to establish a relationship between modular forms in one and two variables. The Petersson product $\langle\ , \Omega^{(k)}\rangle$ with $\Omega^{(k)}$ defines a linear map

$$\mathrm{DN}\colon S_k(\Gamma_0(D), \chi_D) \to S_k$$

which is called the *Doi-Naganuma lifting*. Using the Petersson product on S_k we have an adjoint linear map $\mathrm{DN}^*\colon S_k \to S_k(\Gamma_0(D), \chi_D)$ given by the same kernel function $\Omega^{(k)}$. Up to a constant this is the map g of (3.4).

Let

$$G_N^{(k)}(\tau) = \frac{(4\pi N)^{k-1}}{(k-2)!}\sum_\gamma \chi_D(a)(c\tau + d)^{-k}e^{2\pi i N\gamma(\tau)}\,, \qquad (k > 2)$$

where the sum is over all $\gamma = \begin{pmatrix} a & b \\ c & d \end{pmatrix}$ in $\Gamma_\infty\backslash\Gamma_0(D)$ with Γ_∞ the isotropy group in

$\Gamma_0(D)$ of the cusp at infinity. It is characterized by the property that for all $f \in S_k(\Gamma_0(D), \chi_D)$ we have $\langle f, G_N^{(k)} \rangle = N$-th Fourier coefficient of f. It follows that under DN the N-th Poincare-series $G_N^{(k)}$ is mapped to $N^{k-1}\omega_N^{(k)}$. For $k = 2$ there are again convergence problems. One takes as definition of $G_N^{(2)}$

$$G_N^{(2)} = \lim_{s \to 0} 4\pi N \sum_{\gamma \in \Gamma_\infty \backslash \Gamma_0(D)} \chi_D(a)(c\tau + d)^{-2}|c\tau + d|^{-2s} e^{2\pi i N \gamma(\tau)},$$

which turns out to be holomorphic.

Again $\langle\ , G_N^{(2)} \rangle$ gives the N-th Fourier coefficient. We shall write G_N for $G_N^{(2)}$.

(4.1) Lemma. DN *sends Hecke eigenforms to Hecke eigenforms.*

Proof. We prove this for $k = 2$ leaving $k > 2$ to the reader. We know $T(p)G_M = G_{Mp}$ for $(p, DM) = 1$. Moreover, by (1.5) we know the action of $T(\mathfrak{p})$ on the t_M^0, hence on the ω_M: $T(\mathfrak{p})\omega_M = p\omega_{Mp}$ if $p = \mathfrak{p}\mathfrak{p}'$ splits, and $p^2\omega_{Mp^2} + p\omega_M$ if $p = \mathfrak{p}$ is inert. If $f \in S_2(\Gamma_0(D), \chi_D)$ is a Hecke eigenform and $f = \sum \lambda_M G_M$ then, if p splits

$$\begin{aligned} T(\mathfrak{p})(\mathrm{DN}(f)) &= T(\mathfrak{p})(\textstyle\sum \lambda_M M \omega_M) \\ &= \textstyle\sum \lambda_M M p \omega_{Mp} \\ &= \mathrm{DN}(\textstyle\sum \lambda_M (T(p)G_M)) \\ &= \mathrm{DN}(T(p)f) = c_p(\mathrm{DN}(f)), \end{aligned}$$

while if p is inert

$$\begin{aligned} T(\mathfrak{p})(\mathrm{DN}(f)) &= T(\mathfrak{p})(\textstyle\sum \lambda_M M \omega_M) \\ &= \textstyle\sum \lambda_M (M p^2 \omega_{Mp^2} + p\omega_M) \\ &= \mathrm{DN}(\textstyle\sum \lambda_M (T(p^2)G_M + pG_M) \\ &= \mathrm{DN}((T(p^2) + p)f) = c_p' \mathrm{DN}(f). \end{aligned}$$

□

It follows from (4.1) that for each normalized Hecke eigenform $f \in S_2(\Gamma_0(D), \chi_D)$ the lift $\mathrm{DN}(f)$ is either zero or has the form $c_f \hat{f}$, with $\hat{f}$ a normalized Hilbert modular Hecke eigenform. By computing explicit expansions of $\Omega^{(k)}$, it was shown in [174] that the second alternative always occurs and that $c_f = \dfrac{(-1)^{k/2}\pi}{2^{k-3}(k-1)}$, independent of f. (This fact will not be proved here.)

To $\hat{f}$ we can associate the Dirichlet series

$$L(\hat{f}, s) = \sum_{\mathfrak{m}} c(\mathfrak{m}, \hat{f}) \mathrm{N}(\mathfrak{m})^{-s} \qquad \mathrm{Re}(s) > \frac{k}{2} + 1.$$

It has an analytic continuation to all of $\mathbb{C}$ and satisfies a functional equation: Put

$$\Lambda(\hat{f}, s) = D^s (2\pi)^{-s} \Gamma(s)^2 L(\hat{f}, s);$$

then $\Lambda(\hat{f}, s) = \Lambda(\hat{f}, k - s)$.

(4.2) Theorem. *If $f = \sum_{n=1}^{\infty} a(n)q^n \in S_k(\Gamma_0(D), \chi_D)$ is a normalized eigenform then*

$$L(\hat{f}, s) = L(f, s)L(f^\rho, s)$$

with $f^\rho = \sum \overline{a(n)} q^n$ and $L(f, s) = \sum a(n)n^{-s}$.

Proof. By the proof of (4.1) we know that for primes $p \nmid D$

$$a(\mathfrak{p}, \hat{f}) = \begin{cases} a(p) & \text{if } p = \mathfrak{p}\mathfrak{p}' \text{ splits} \\ a(p^2) + p^{k-1} & \text{if } p = \mathfrak{p} \text{ is inert} . \end{cases}$$

Using the Euler products

$$L(\hat{f}, s) = \prod_{\mathfrak{p}} [1 - a(\mathfrak{p}, \hat{f})N(\mathfrak{p})^{-s} + N(\mathfrak{p})^{k-1-2s}]^{-1}$$

$$L(f, s) = \prod_{p} [1 - a(p)p^{-s} + \chi_D(p)p^{k-1-2s}]^{-1}$$

and $\overline{a(p)} = \chi_D(p)a(p)$ we find that $L(\hat{f}, s)$ and $L(f, s) \cdot L(f^\rho, s)$ agree up to finitely many Euler factors. But they satisfy the same functional equation, hence they are equal. □

Let $M_2^+(\Gamma_0(D), \chi_D) = \{f = \sum a(n)q^n \in M_k(\Gamma_0(D), \chi_D): a(n) = 0$ for all n with $\chi_D(n) = -1\}$ and $S_k^+(\Gamma_0(D), \chi_D) = M_k^+(\Gamma_0(D), \chi_D) \cap S_k(\Gamma_0(D), \chi_D)$. We have a surjective map $S_k(\Gamma_0(D), \chi_D) \xrightarrow{\pi^+} S_k^+(\Gamma_0(D), \chi_D)$ given on Hecke eigenforms by $f = \sum a(n)q^n \to f^+ = \sum \mathrm{Re}(a(n))q^n = \frac{1}{2}(f + f^\rho)$. By (4.2), DN factors through π^+.

(4.3) Proposition. DN *restricted to $S_k^+(\Gamma_0(D), \chi_D)$ is injective.*

Proof. We have to prove that if f, g are normalized Hecke eigenforms in $S_k(\Gamma_0(D), \chi_D)$ and $\mathrm{DN}(f) = \mathrm{DN}(g)$, then $f = g$ or $f = g^\rho$. So let $f = \sum a(n)q^n$ and $g = \sum b(n)q^n$; then by the formulas

$$a(\mathfrak{p}, \hat{f}) = a(\mathfrak{p}, \hat{g}) = \begin{cases} a(p) & \text{if } p = \mathfrak{p}\mathfrak{p}' \text{ splits} \\ a(p)^2 + 2p^{k-1} & \text{if } p = \mathfrak{p} \text{ is inert} . \\ a(p) + \overline{a(p)} & \text{if } p = \mathfrak{p}^2 \text{ ramifies} \end{cases}$$

we see: $a(p) = b(p)$ for splitting p and $a(p)^2 = b(p)^2$ for inert p. Therefore, $a(p) = \pm b(p)$ for inert p. If we have always the same sign then $f = g$ or $f = g^\rho$. Suppose we do not have the same sign always. By replacing g by g^ρ if necessary, we may assume that for a set of primes with upper density $> 1/2$ (or even $\geqslant 3/4$) we have $a(p) = b(p)$. Associated to f and g we have ℓ-adic representations r_f, r_g: $\mathrm{Gal}(\bar{\mathbb{Q}}/\mathbb{Q}) \to \mathrm{GL}(2, \mathbb{Q}_\ell)$, see [136]. The traces of a Frobenius element at p agree if $a(p) = b(p)$, and if $r_f \not\cong r_g$ this can only happen for a set of primes of density $\leqslant 1/2$, by the Čebotarev density theorem (see [143]). □

(4.4) Theorem. *Let $\mathbb{T}_\mathbb{C} \subset \mathbb{H}^2(X, \mathbb{C})$ be the subspace generated by the classes t_N, $N \geqslant 0$. Then $t_N \mapsto N$-th Fourier coefficient defines an isomorphism*

$$g^\vee : \mathbb{T}_\mathbb{C} \to M^\vee \to (M_2^+(\Gamma_0(D), \chi_D))^\vee ,$$

where M is the linear subspace generated by $S_2^+(\Gamma_0(D),\chi_D)$ *and the Eisenstein series* $\sum_{D=D_1D_2} E_{D_1,D_2}$. *Moreover,* $g^\vee$ *maps* $\mathbb{T}_{\mathbb{C}} \subset \mathbb{H}^2_{sp}(X,\mathbb{C})$ *to* $(S_2^+(\Gamma_0(D),\chi_D))^\vee$.

Proof. It follows from Section 3 and V.5 that $g^\vee$ is injective. Since DN is injective on S_2^+ it follows that DN* is surjective onto S_2^+. Now DN* is induced by $t \to \sum_{N=1}^{\infty} \langle t, t_N^0\rangle q^N$ (up to a factor). By Poincaré duality this gives $\mathbb{T}_{\mathbb{C}} \cap \mathbb{H}^2_{sp} \to (S_2^+(\Gamma_0(D),\chi_D))^\vee$. The rest follows from V.5. □

(4.5) *Remark.* The codimension of $S_2^+(\Gamma_0(D),\chi_D)$ in $M_2^+(\Gamma_0(D),\chi_D)$ is 2^{t-1}. Define 2^{t-1} volume forms as follows. Fix a genus C and define $t_{0,C}$ to be $\frac{1}{4}(w_1 + w_2)$ on each component $\mathrm{PGL}^+(\mathfrak{o} \oplus \mathfrak{a})\backslash\mathfrak{H}^2$ with $\mathfrak{a} \in C$ and zero anywhere else. Let $\mathbb{T}'_{\mathbb{C}}$ be the subspace of $\mathbb{H}^2(X,\mathbb{C})$ generated by $\mathbb{T}$ and these $t_{0,C}$. Then $g^\vee$ extends to an isomorphism $\mathbb{T}'_{\mathbb{C}} \to (M_2^+(\Gamma_0(D),\chi_D))^\vee$.

Theorem (4.2) was proved by Zagier for D prime by expressing $\Omega_\Gamma^{(k)}$ as a sum of Poincaré series (in one variable). Theorem (4.4) was conjectured in [85] by Hirzebruch and Zagier and proved by Zagier in [174] and by Oda in [122]. That $g^\vee$ corresponds to DN* was conjectured in [85] and proved by Oda. Their proofs are different from those given here.

The idea of lifting modular forms goes back to Doi and Naganuma. In [27] they constructed a lifting from modular forms on $\mathrm{SL}(2,\mathbb{Z})$ to Hilbert modular forms on $\mathrm{SL}(2,\mathfrak{o}_K)$ by associating to $L(f,s)$ a Dirichlet series $L(f,s)L(f,s,\chi_D)$ and by observing a functional equation for this Dirichlet series. This implied that the form $\hat{f}$ defined as in (2) is a Hilbert modular form for $\mathrm{SL}(2,\mathfrak{o}_K)$ if $\mathfrak{o}_K$ is Euclidean. Their construction was used by Naganuma [119] to define a lifting from modular forms on $\Gamma_0(D)$ and "Nebentypus" to Hilbert modular forms. In [138] Saito computed the traces of the Hecke operators on spaces of "self-conjugate" Hilbert modular forms for cyclic field extensions $K/\mathbb{Q}$ of prime degree. By observing the similarity with the trace formulas for elliptic modular forms he showed that one can lift elliptic modular forms to Hilbert modular forms and that the image of this lifting is the space of "self-conjugate" Hilbert modular forms. Shintani then provided a representation theoretic interpretation of this lifting, cf. [155]. This was exploited by Langlands in a systematic way. For a description of the results we refer to [45].

The result of Hirzebruch and Zagier has inspired several people to look at geodesic cycles in quotients of other bounded symmetric domains. Kudla and Millson have set up a program to extend the result to e.g. $\mathrm{SO}(p,q)$, see [105]. Oda studied the case of $\mathrm{SO}(2,q)$, see [122].

(4.6) *Remark.* The function $\Omega_0^{(k)}$ given in (3.5) gives the lifting from $S_k(\mathrm{SL}(2,\mathbb{Z}))$ to Hilbert modular forms of weight k.

We add a lemma that is needed later on.

(4.7) Lemma. *Let* $\pi^+ : M_2(\Gamma_0(D),\chi_D) \to M_2^+(\Gamma_0(D),\chi_D)$ *be the projection given on eigenforms by* $f = \sum a(n)q^n \to f^+ = \sum \mathrm{Re}(a(n))q^n$. *Then*

$$\pi^+(T(M)g_{t_1}) = g_{t_M}.$$

Proof. In view of V (5.1) it suffices to prove this with t_1 and t_M replaced by t_1^0 and t_M^0. Since DN and DN* send eigenforms to eigenforms DN*DN commutes with the Hecke operators $T(M)$ and since DN*DN factors through π^+ it also commutes with $\pi^+(T(M))$. We find

$$\begin{aligned}\pi^+(T(M)g_{t_1^0}) &= \pi^+(T(M)(\mathrm{DN}^*\mathrm{DN}G_1)) \\ &= \mathrm{DN}^*\mathrm{DN}(\pi^+T(M)G_1) \\ &= \mathrm{DN}^*\mathrm{DN}(T(M)G_1)) \\ &= \mathrm{DN}^*\mathrm{DN}(G_M) \\ &= g_{t_M^0}\,.\end{aligned}$$

□

To a Hilbert modular cusp form f which is a normalized eigenform one can associate several L-series. The first one is

$$\text{i) } L(f,s) = \sum c(\mathfrak{a})\mathrm{N}(\mathfrak{a})^{-s}, \qquad \left(\mathrm{Re}(s) > \frac{k}{2} + 1\right)$$

where $\mathfrak{a}$ runs over the non-zero integral ideals and the $c(\mathfrak{a}) = c(\mathfrak{a}, f)$ are the Fourier coefficients of f, see VI.1 and I.6. A second one is the function

$$\text{ii) } G(f,s) = \zeta(2s-2k+2)\sum_{m=1}^{\infty} c((m))m^{-s} \qquad (\mathrm{Re}(s) > k+1)$$

studied by Asai [6].

Moreover, we can consider the convolution (cf. I.6)

$$\text{iii) } \zeta_K(2s-2k+2)L_{f,f^\rho}(s) = \zeta_K(2s-2k+2)\sum_{\mathfrak{a}} c(\mathfrak{a})c(\mathfrak{a}')\mathrm{N}(\mathfrak{a})^{-s}\,.$$

If we write

$$L(f,s) = \prod_{\mathfrak{p}} (1 - c(\mathfrak{p})\mathrm{N}(\mathfrak{p})^{-s} + \mathrm{N}(\mathfrak{p})^{k-1-2s})^{-1}$$

and we factor $X^2 - c(\mathfrak{p})X + \mathrm{N}(\mathfrak{p})^{k-1} = (X-\alpha_{\mathfrak{p}})(X-\beta_{\mathfrak{p}})$ the Euler factor at p of ii) is the inverse of

$$\begin{cases}(1-\alpha_{\mathfrak{p}}\alpha_{\mathfrak{p}'}p^{-s})(1-\alpha_{\mathfrak{p}}\beta_{\mathfrak{p}'}p^{-s})(1-\alpha_{\mathfrak{p}'}\beta_{\mathfrak{p}}p^{-s})(1-\beta_{\mathfrak{p}}\beta_{\mathfrak{p}'}p^{-s}) & (p) = \mathfrak{p}\mathfrak{p}' \\ (1-\alpha_{\mathfrak{p}}p^{-s})(1-\beta_{\mathfrak{p}}p^{-s})(1-p^{2k-2-2s}) & (p) = \mathfrak{p} \\ (1-\alpha_{\mathfrak{p}}^2p^{-s})(1-\beta_{\mathfrak{p}}^2p^{-s})(1-p^{k-1-2s}) & (p) = \mathfrak{p}^2\,.\end{cases}$$

We define for $\mathrm{Re}(s) > k+1$

$$G_\chi(f,s) = \zeta(2s-2k+2)\prod_{p|D}(1+p^{-s+k-1})\sum \chi_D(m)C((m))m^{-s}$$

with $\chi = \chi_D$ the quadratic character of $K/\mathbb{Q}$. By comparing Euler products, we find

$$G(f,s)G_\chi(f,s) = \zeta_K(2s-2k+2)L_{f,f^\rho}(s) \tag{1}$$

In the rest of this section we shall assume for simplicity that $h^+ = 1$. (We refer to Chapter XI for generalizations.)

(4.8) Theorem (Asai [6]). *Define* $G^*(f,s) = D^{s/2}(2\pi)^{-2s}\Gamma(s)\Gamma(s-k+1)G(f,s)$ *and* $G_\chi^*(f,s) = D^{3s/2}\Gamma(2\pi)^{-2s}\Gamma(s)\Gamma(s-k+1)G_\chi(f,s)$. *Then both* $G^*(f,s)$ *and* $G_\chi^*(f,s)$ *have an analytic continuation to the whole s-plane except* $s=k$, $s=k-1$ *where they can have simple poles. They satisfy the functional equations* $G^*(f,s) = G^*(2k-1-s)$ *and* $G_\chi^*(f,s) = G_\chi^*(2k-1-s)$.

Proof. For G^* one considers $g(z) = f(z,\bar{z})$. This is a real analytic function on $\mathfrak{H}$ satisfying $g(\gamma z) = |cz+d|^{2k}g(z)$ for all $\gamma = \begin{pmatrix} * & * \\ c & d \end{pmatrix} \in \mathrm{SL}(2,\mathbb{Z})$. We use Rankin's convolution: consider

$$J(s_1) = \int_{\mathrm{SL}(2,\mathbb{Z})\backslash\mathfrak{H}} y^{k-2}g(z)E^*(z,s_1)\,dx\,dy \quad \text{with } s_1 = s-k+1\,.$$

Here $E^*(z,s)$ is the non-holomorphic Eisenstein series on $\mathrm{SL}(2,\mathbb{Z})$ introduced in Chapter I.6. By Proposition (6.8) of Chapter I we get the analytic continuation and the functional equation of $J(s_1)$: $J(s_1) = J(1-s_1)$. Note that $J(s_1)$ has a pole at $s=0, 1$ if and only if

$$\int_{\mathrm{SL}(2,\mathbb{Z})\backslash\mathfrak{H}} g(z)y^{k-2}\,dx\,dy \neq 0 \tag{2}$$

If Γ_∞ is the isotropy group of ∞ in $\mathrm{SL}(2,\mathbb{Z})$ we find from the definition of $E(z,s)$:

$$\begin{aligned} J(s_1) &= \pi^{-s_1}\Gamma(s_1)\zeta(2s_1)\int_{\mathrm{SL}(2,\mathbb{Z})\backslash\mathfrak{H}} y^{k-2}\Bigl(\sum_{\gamma\in\Gamma_\infty\backslash\mathrm{SL}(2,\mathbb{Z})} \mathrm{Im}(\gamma z)^{s_1}\Bigr)g(z)\,dx\,dy \\ &= \pi^{-s_1}\Gamma(s_1)\zeta(2s_1)\int_{\Gamma_\infty\backslash\mathfrak{H}} y^{s_1+k-2}g(z)\,dx\,dy\,. \end{aligned}$$

Taking as a fundamental domain for Γ_∞ the set $\{z = x+iy \in \mathfrak{H} : 0 \leqslant x < 1\}$, we find

$$\begin{aligned} \int_{\Gamma_\infty\backslash\mathfrak{H}} y^{s_1+k-2}g(z)\,dx\,dy &= \int_0^\infty \Bigl(\int_0^1 g(z)\,dx\Bigr) y^{s_1+k-2}\,dy \\ &= \int_0^\infty \Bigl(\sum_{m=1}^\infty c(m)e^{4\pi my/\sqrt{D}}\Bigr)y^{s_1+k-2}\,dy\,, \end{aligned}$$

and so by working out the integral we get

$$J(s_1) = \pi^{k-1}G^*(f,s)\,.$$

The case $G_\chi^*(f,s)$ is similar: one takes $g(z) = f_\chi(z,\bar{z})$ with f_χ the twist of f by the quadratic character χ. (Since $\mathfrak{o}/\mathfrak{d} \cong \mathbb{Z}/D\mathbb{Z}$ we can view χ as a Dirichlet character on $\mathfrak{o}$.) One takes a Rankin type convolution for $\Gamma_0(D)$ instead of $\mathrm{SL}(2,\mathbb{Z})$. □

(4.9). Now take a Hecke eigenform f. Then $G(f,s)G_\chi(f,s)$ has a (simple) pole at $s=k$ if and only if $f = f^\rho$, by (1) and I. (6.9). Therefore we conclude: if f is a lift (either from $S_k(\mathrm{SL}(2,\mathbb{Z}))$ or from $S_k(\Gamma_0(D),\chi_D)$ then precisely one of $G(f,s)$ and $G_\chi(f,s)$ has a simple pole at $s=k$; the other one does not vanish at $s=k$.

(4.10) Corollary. *Let f be a Hecke eigenform of weight k. Then*
i) *$G(f,s)$ has a (simple) pole at $s=k$ if and only if $\int_{T_1}\eta_f^{(k)}\neq 0$,*
ii) *$G_\chi(f,s)$ has a (simple) pole at $s=k$ if and only if $\int_{F-F'}\eta_f^{(k)}\neq 0$ with F, F' the two components of F_{p^2}.*

Proof. i) $\int_{T_1}\eta_f^{(k)}$ equals the expression in (2) up to a non-zero multiplicative constant.
(ii) Use a similar but more complicated argument. (Or use the remark that in view of the kernel function for the lift from $\mathrm{SL}(2,\mathbb{Z})$ to $\mathrm{SL}(2,\mathfrak{o})$ we have: f is a lift from $S_k(\mathrm{SL}(2,\mathbb{Z}))$ if and only if $\int_{F-F'}\eta_f^{(k)}\neq 0$. Then we use (4.9) and i) to see that f is a lift from $\mathrm{SL}(2,\mathbb{Z})$ if and only if $G_\chi(f,s)$ has a pole at $s=k$. For weight 2 there is nothing to prove since there are no cusp forms of weight 2 on $\mathrm{SL}(2,\mathbb{Z})$.) □

(4.11) Corollary. *Let f be a Hecke eigenform of weight k. Then the following are equivalent:*
i) *either $G(f,s)$ or $G_\chi(f,s)$ has a simple pole at $s=k$,*
ii) *$L_{f,f^\rho}(s)$ has a simple pole at $s=k$,*
iii) *$f=f^\rho$.*

5. The Intersection Number of T_M and T_N

Let $X_m(D)=\bigcup_{\mathfrak{a}}X_{\Gamma_m(\mathfrak{o}\oplus\mathfrak{a})}$ be the disjoint union of 2^{t-1} surfaces, one for each genus of K. The curve T_N (=union of T_N on all components of $X_m(D)$) defines a class $t_N\in\mathbb{H}^2(X_m(D))=\bigoplus\mathbb{H}^2(X_{\Gamma_m(\mathfrak{o}\oplus\mathfrak{a})})$. We know that

$$g_{t_M}=\frac{1}{2}\mathrm{vol}(T_M)+\sum_{N=1}^{\infty}(t_M\cdot t_N)q^N \tag{1}$$

belongs to $M_2^+=M_2^+(\Gamma_0(D),\chi_D)$. We deduce an explicit formula for $t_M\cdot t_N$ from the identity (see (4.7))

$$g_{t_M}=\pi^+(T(M)g_{t_1}) \tag{2}$$

and the formula for g_{t_1} given in V.8. Here $T(M)$ is the M-th Hecke operator and $\pi^+\colon M_2\to M_2^+$ is the projection which is given *on eigenforms* by $\sum a(n)q^n\mapsto\sum_{n=0}^{\infty}\mathrm{Re}(a(n))q^n$. (Cf. appendix to Chapter V.)

We choose an integer $P=P(M,N)$ which is a multiple of (M,N) such that for every prime p dividing D

$$P/(M,N)\equiv\begin{cases}M/(M,N)\bmod D(p) & \text{if } \mathrm{ord}_p(M)\leqslant\mathrm{ord}_p(N)\\ N/(M,N)\bmod D(p) & \text{otherwise}.\end{cases} \tag{3}$$

(5.1) Theorem. $g_{t_M}=\sum_{N=0}^{\infty}\sum_{n|(M,N)}\frac{1}{2}(\chi_D(n)+\chi_D(P/n))(H_D(MN/n^2)+I_D(MN/n^2))q^N$.

Proof. In view of (2) this follows from V. Thm. (8.6) and the following lemma.

(5.2) Lemma. Let $f = \sum a(n)q^n \in M_2^\varepsilon(\Gamma_0(D), \chi_D)$ *with* ε *the principal genus. Then* $\pi^+(T(m)f) = \sum b(m,n)q^n$ *with*

$$b(m,n) = \frac{1}{2} \sum_{d|(m,n)} (\chi_D(d) + \chi_D(P/d))da(mn/d^2).$$

Proof. Let $\pi^\varepsilon\colon M_2 \to M_2^\varepsilon$ be the projection. We may assume that $f = \pi^\varepsilon g$ with g an eigenform since these forms generate M_2. If $g = \sum c(n)q^n$ then

$$\pi^\varepsilon g = 2^{-t} \sum_n \left(\sum_{D_1} c(n, D_1) \right) q^n,$$

where the inner sum is over discriminants dividing D and the $c(n, D_1)$ are defined by (Cf. appendix to Chapter V.):

$$\begin{aligned} &c(n,1) = c(n) \\ &c(n,D_1)c(m,D_1) = c(nm,D_1) && \text{if } (n,m) = 1 \\ &c(n,D_1D(p)) = \chi_{D(p)}(n)c(n,D_1) && \text{if } p \nmid n,\ p \nmid D_1 \\ &c(n,D_1) = \overline{c(n,D_2)} && \text{if } D = D_1D_2\,. \end{aligned} \tag{4}$$

If we write $g^{(D_1)} = \sum c(n, D_1)q^n$ then $g^{(D_1)}$ is a Hecke eigenform and we have

$$\begin{aligned} \pi^+(T(m)f) &= \pi^+(T(m)\pi^\varepsilon g) \\ &= \pi^+(T(m)2^{-t} \sum_{D_1} g^{(D_1)}) \\ &= 2^{-t} \sum_{D_1} c(m, D_1)\pi^+(g^{(D_1)}) \\ &= 2^{-t-1} \sum_{D_1} c(m, D_1) \sum_n (c(n, D_1) + \overline{c(n, D_1)})q^n\,. \end{aligned}$$

So we have to show that

$$\sum_{D_1} c(m, D_1)(c(n, D_1) + \overline{c(n, D_1)}) = \sum_{d|(m,n)} (\chi_D(d) + \chi_D(P(m,n)/d))d \sum_{D_1} c(mn/d^2, D_1)\,.$$

Since

$$c(m, D_1)c(n, D_1) = \sum_{d|(m,n)} \chi_D(d)dc(mn/d^2, D_1)\,, \tag{5}$$

we have to prove

$$\sum_{D_1} c(m, D_1)\overline{c(n, D_1)} = \sum_{D_1} \sum_{d|(m,n)} \chi_D(P(m,n)/d)dc(mn/d^2, D_1)\,. \tag{6}$$

For a prime p dividing D we put $r_p = \min(\operatorname{ord}_p(m), \operatorname{ord}_p(n))$, and we let $m = m_0 \prod p^{r_p}$, $n = n_0 \prod p^{r_p}$, $P = P_0 \prod p^{r_p}$. Then each divisor d of (m,n) can be written as $d_0 \prod p^{r_p}$ with $d_0|(m_0, n_0)$, so we find for the right-hand side of (6):

$$\sum_{D_1} \sum_{d_0|(m_0,n_0)} \chi_D(P_0/d_0) d_0 \prod p^{r_p} c(m_0 n_0/d_0^2, D_1)$$
$$= \sum_{D_1} (\prod p^{r_p}) \chi_D(P_0) \sum_{d_0} \chi_D(d_0) d_0 c(m_0 n_0/d_0^2, D_1)$$
$$= (\prod p^{r_p}) \sum_{D_1} \chi_D(P_0) c(m_0, D_1) c(n_0, D_1) \quad \text{by (5)}.$$

The left-hand side of (6) equals

$$\prod p^{r_p} \sum_{D_1} c(m_0, D_1) \overline{c(n_0, D_1)}$$

since $c(p^{r_p}, D_1)\overline{c(p^{r_p}, D_1)} = ((c(p, D_1)\overline{c(p, D_1)})^{r_p} = p^{r_p}$. So (6) follows from

$$\sum_{D_1} c(m_0, D_1)\overline{c(n_0, D_1)} = \sum_{D_1} \chi_D(P_0) c(m_0, D_1) c(n_0, D_1)$$

and this again follows from the observation that a term in the right-hand side equals $c(m_0, D_3) c(n_0, D_3)$ with $D_3 = D_1 \prod D(p)^{\pm}$, where the product runs over those primes p for which $\mathrm{ord}_p(m) \leqslant \mathrm{ord}_p(n)$ and with $\pm = +1$ if $D(p)$ does not occur in D_1, $\pm = -1$ otherwise. This follows from (4) and the definition of P. □

The intersection number $t_M \cdot t_N$ is a sum of contributions $(t_M \cdot t_N)_{\mathfrak{a}}$ given by the following corollary, whose proof is left to the reader.

(5.3) Corollary. *The intersection number* $(t_M \cdot t_N)_{\mathfrak{a}}$ *on* $X_{\Gamma_m(\mathfrak{o} \oplus \mathfrak{a})}$ *is given by*

$$(t_M \cdot t_N)_{\mathfrak{a}} = \sum_{n|(M,N)} n(H_D(MN/n^2) + I_D(MN/n^2)) \prod_{p|D} \frac{\chi_{D(p)}(n) + \chi_{D(p)}(PA/n)}{2}.$$

Theorem (5.1) was proved in [85] for D prime and in [62] for general D by counting special points, see Ch. V, 6 and 8.

6. The Action of the Hecke Algebra on the Cohomology

Using the description of $\mathbb{H}^2_{\mathrm{sp}}(X, \mathbb{Q})$ in terms of modular forms one can decompose $\mathbb{H}^2_{\mathrm{sp}}(X, \mathbb{C})$ as a direct product of Hodge structures of weight 2 corresponding to the eigenforms. Recall that $\mathbb{H}^2_{\mathrm{sp}}(X, \mathbb{Q})$ is the orthogonal complement of W. We have (with $X = X_{K_0}$):

$$\mathbb{H}^2(X, \mathbb{Q}) = \mathbb{H}^2_{\mathrm{sp}}(X, \mathbb{Q}) \oplus (\mathbb{Q}(-1))^{h^+},$$

where $\mathbb{Q}(-1)$ is the rational Hodge structure of type (1, 1) of $(2\pi i)^{-1}\mathbb{Q} \subset \mathbb{C}$, see [22] for notations and definitions. An important remark at this point is that we have a bilinear map

$$\Psi\colon \mathbb{H}^2_{\mathrm{sp}}(X, \mathbb{Q}) \times \mathbb{H}^2_{\mathrm{sp}}(X, \mathbb{Q}) \to H^4(X, \mathbb{Q}) \cong (\mathbb{Q}(-2))^{h^+}$$

which comes from Poincaré duality. By the symmetry of the Hecke correspondences we have for such a correspondence T

$$\Psi(T^*\omega_1,\omega_2) = \Psi(\omega_1, T^*\omega_2)\,. \tag{1}$$

We denote by E the commutative subring of $\mathrm{End}(\mathbb{H}^2_{\mathrm{sp}}(X,\mathbb{Q}))$ generated over $\mathbb{Q}$ by the Hecke operators $T(\mathfrak{m})$. This is finite-dimensional $\mathbb{Q}$-algebra.

(6.1) Lemma. *E is a semi-simple $\mathbb{Q}$-algebra.*

Proof. We show that $E \otimes_{\mathbb{Q}} \mathbb{C}$ is semi-simple. Note that $\mathbb{H}^2_{\mathrm{sp}}(X,\mathbb{C})$ is a direct sum of four $E \otimes \mathbb{C}$-modules isotypic to the $H^{2,0}$-component (which we identify with S_2), each of which is preserved by E. It is well-known that S_2 has a basis of eigenforms for all Hecke operators and that if two eigenforms have the same eigenvalues for all Hecke operators then one is a multiple of the other (Multiplicity One Principle, cf. [114]). So S_2 is an $E \otimes \mathbb{C}$-module of rank 1 and $E \otimes \mathbb{C}$ is semi-simple. □

By the lemma we can decompose

$$E = \bigoplus \mathscr{k}_i$$

as a direct sum of number fields. If $[\mathscr{k}_i : \mathbb{Q}] = d_i$ then $\sum d_i = \dim S_2$. By choosing a set of primitive idempotents $\{e_1, \ldots, e_n\}$ in E such that $\mathscr{k}_i = e_iE = Ee_i = e_iEe_i$ $(i = 1, \ldots, n)$ we can split up $\mathbb{H}^2_{\mathrm{sp}}(X,\mathbb{Q})$. A normalized eigenform f in e_iS_2 defines an embedding $\sigma\colon \mathscr{k}_i \to \mathbb{C}$ by $t(f) = \sigma(t)f$. Since $\mathscr{k}_i \otimes_{\mathbb{Q}} \mathbb{C} = \bigoplus_\sigma \mathbb{C}$ with the summands corresponding to the embeddings $\sigma\colon \mathscr{k}_i \to \mathbb{C}$ we have

$$e_iS_2 = \bigoplus_\sigma \mathbb{C}f^\sigma\,,$$

where the f^σ are the normalized eigenforms with $t(f^\sigma) = \sigma(t)f^\sigma$. We call the f^σ the *companions* of f.

We thus obtain a sub-Hodge structure $e_i\mathbb{H}^2_{\mathrm{sp}}(X,\mathbb{Q})$ of $\mathbb{H}^2_{\mathrm{sp}}(X,\mathbb{Q})$ on which the field $\mathscr{k}_i$ acts and which is polarized by the restriction of Ψ as follows from (1). In fact, since $e_i\mathbb{H}^2_{\mathrm{sp}}(X,\mathbb{Q})$ is an irreducible E-module Schur's lemma implies that Ψ induces $\Psi_i\colon e_i\mathbb{H}^2_{\mathrm{sp}}(X,\mathbb{Q}) \times e_i\mathbb{H}^2_{\mathrm{sp}}(X,\mathbb{Q}) \to \mathbb{Q}(-2)$. The action of $\mathscr{k}_i$ gives rise to

$$\theta\colon \mathscr{k}_i \to \mathrm{End}(e_i\mathbb{H}^2_{\mathrm{sp}}(X,\mathbb{Q}))\,. \tag{2}$$

Now we choose a set $F = \{f_1, \ldots, f_n\}$ of normalized eigenforms, one for each field $\mathscr{k}_i$ (i.e. $f_i \in e_iS_2$). If f is a normalized eigenform with $f \in e_iS_2$ then we shall write $e_i = e_f$, $\mathscr{k}_i = \mathscr{k}_f$ etc. We define for a normalized eigenform f

$$H^2(M_f,\mathbb{Q}) := e_f\mathbb{H}^2_{\mathrm{sp}}(X,\mathbb{Q})\,.$$

The symbol M_f suggests the conjectural motive associated to f of which we study here only the Hodge realization; it will be used only in the combination $H^2(M_f,\mathbb{Q})$. (See [24] and Ch. XI.)

(6.2) Theorem. *Corresponding to $E = \bigoplus \mathscr{k}_f$ there is a direct sum decomposition*

$$\mathbb{H}^2_{\mathrm{sp}}(X,\mathbb{Q}) = \bigoplus_{f\in F} H^2(M_f,\mathbb{Q})$$

of polarized Hodge structures. Moreover, if $\mathscr{k}_f$ acts on $H^2(M_f,\mathbb{Q})$ via θ then $\Psi_f(\theta(t)x,y) = \Psi_f(x,\theta(t)y)$ with Ψ_f the polarization on $H^2(M_f,\mathbb{Q})$.

The involutions ε_1 and ε_2 defined in section 1 induce an action of $\mathbb{Z}/2\mathbb{Z} \times \mathbb{Z}/2\mathbb{Z}$ on X and on Y, hence on $\mathbb{H}^2_{\mathrm{sp}}(X, \mathbb{Q})$. This commutes with the action of the Hecke operators and so we split

$$H^2(M_f, \mathbb{Q}) = \bigoplus_{s, s' \in \{+,-\}} H^2(M_f, \mathbb{Q})_{ss'},$$

where ε_1^* (resp. ε_2^*) acts on $H^2(M_f, \mathbb{Q})$ as $s \cdot \mathrm{id}$ (resp. $s' \cdot \mathrm{id}$).

(6.3) Proposition. *For every normalized eigenform $f \in S_2$ we have $\operatorname{rank}_{k_f} H^2(M_f, \mathbb{Q})_{ss'} = 1$. Furthermore, $\Psi_f = \operatorname{Tr}_{k_f/\mathbb{Q}} \psi_f$ with $\psi_f \colon H^2(M_f, \mathbb{Q}) \times H^2(M_f, \mathbb{Q}) \to k_f$ which with respect to a suitable basis $e_{ss'} \in H^2(M_f, \mathbb{Q})_{ss'}$ is of the form*

$$\begin{pmatrix} & & & 1 \\ & & 1 & \\ & 1 & & \\ 1 & & & \end{pmatrix}.$$

Proof. Since ε_1^* and ε_2^* commute with the Hecke operators each $H^2(M_f, \mathbb{Q})_{ss'}$ is a k_f-module. Now $H^2(M_f, \mathbb{Q})_{ss'} \otimes_{\mathbb{Q}} \mathbb{C}$ is generated by $\eta_g + s\varepsilon_1^*\eta_g + s'\varepsilon_2^*\eta_g + ss'\varepsilon_1^*\varepsilon_2^*\eta_g$ with g running through the companions of f, hence $\operatorname{rk}_{k_f} H^2(M_f, \mathbb{Q})_{ss'} = 1$. By $\Psi_f(\varepsilon_1^* x, \varepsilon_1^* y) = \Psi_f(\varepsilon_2^* x, \varepsilon_2^* y) = -\Psi_f(x, y)$ we have $\Psi_f(x, y) = 0$ if $x \in H^2(M_f, \mathbb{Q})_{ss'}$, $y \in H^2(M_f, \mathbb{Q})_{tt'}$ with $(st, s't') \neq (+,+)$. Choose a generator $e_{ss'} \in H^2(M_f, \mathbb{Q})_{ss'}$. The map $t \to \Psi_f(\theta(t)e_{++}, e_{--})$ from k_f to $\mathbb{Q}$ is $\mathbb{Q}$-linear, hence of the form $\operatorname{Tr}_{k_f/\mathbb{Q}}(\alpha t)$ for some $\alpha \in k_f$. Similarly, $\Psi_f(\theta(t)e_{+-}, e_{-+}) = \operatorname{Tr}_{k_f/\mathbb{Q}}(\beta t)$ for some β and all $t \in k_f$. □

Define

$$\mathbb{H}_2^{\mathrm{sp}}(X, \mathbb{Q}) = \left\{ \gamma \in \operatorname{Image}(H_2(X', \mathbb{Q}) \to H_2(Z, \mathbb{Q})) \colon \int_\gamma w = 0 \text{ all } w \in W \right\}.$$

It splits as a direct sum $\bigoplus_f H_2(M_f, \mathbb{Q})$ and each $H^2(M_f, \mathbb{Q})$ splits into four eigenspaces $H^2(M_f, \mathbb{Q})_{ss'}$. Choose $\gamma_{ss'} \in H^2(M_f, \mathbb{Q})_{ss'}$ dual to $e_{ss'}$ with $e_{ss'}$ as in (6.3).

(6.4) Definition. Let g be a companion of f. The periods of g are defined as

$$\omega_{ss'}(g) = \int_{\gamma_{ss'}} \omega_g .$$

The following period relation was observed by Oda:

(6.5) Theorem (Hodge-Riemann bilinear relation):

$$\omega_{++}(g)\omega_{--}(g) = -\omega_{+-}(g)\omega_{-+}(g) = \frac{1}{4}\int_{\Gamma\backslash\mathfrak{H}^2} \omega_g \wedge \bar{\omega}_g .$$

Proof. Let g correspond to $k_f \xrightarrow{\sigma} \mathbb{C}$. Denote the image of $e_{ss'}$ in $H^2(M_f, \mathbb{Q})_{ss'} \otimes_\sigma \mathbb{C}$ by $e^\sigma_{ss'}$. Then $\omega_g = \sum \omega_{-s-s'}(g) e^\sigma_{ss'}$, so

$$0 = \int \omega_g \wedge \omega_g = 2\omega_{++}(g)\omega_{--}(g) + 2\omega_{+-}(g)\omega_{-+}(g)$$

$$0 < \int \omega_g \wedge \bar{\omega}_g = 2\omega_{++}(g)\omega_{--}(g) - 2\omega_{+-}(g)\omega_{-+}(g)\,. \qquad \square$$

(6.6). For a Hecke eigenform f we have

$$\omega_f = \sum_{s,s'} \omega_{-s-s'} e_{ss'}\,.$$

For later use we give a change of basis:

$$\begin{pmatrix} \omega_f \\ \varepsilon_1^* \omega_f \\ \varepsilon_2^* \omega_f \\ \bar{\omega}_f \end{pmatrix} = \begin{pmatrix} \omega_{--} & \omega_{-+} & \omega_{+-} & \omega_{++} \\ -\omega_{--} & -\omega_{-+} & \omega_{+-} & \omega_{++} \\ -\omega_{--} & \omega_{-+} & -\omega_{+-} & \omega_{++} \\ \omega_{--} & -\omega_{-+} & -\omega_{+-} & \omega_{++} \end{pmatrix} \begin{pmatrix} e_{++} \\ e_{+-} \\ e_{-+} \\ e_{--} \end{pmatrix}.$$

We shall sometimes use (abuse) the notation $\omega_{ss'}^{\sigma}$ for the periods of the companion of f corresponding to $\sigma: k_f \hookrightarrow \mathbb{C}$.

7. The Periods of Eigenforms

For a product $X = X_1 \times X_2$ of two modular curves the cohomology group $H^2(X, \mathbb{C})$ can be written as a direct sum of terms $H^{p,q}(X_1) \otimes H^{r,s}(X_2)$ with $p + q + r + s = 2$. Since Hilbert modular surfaces are regular ($b_1 = 0$ see IV.6) there is no obvious way of using H^1 to study H^2. Nevertheless, if the conjecture mentioned in the introduction to this chapter is true then $H^2(M_f, \mathbb{Q})$ ought to be a tensor product of k_f-Hodge structures of weight 1. Oda showed in [124] that this is indeed the case.

If F is a totally real field of degree d over $\mathbb{Q}$ then an element $w \in \mathrm{GL}(2, F)\backslash(\mathbb{C} - \mathbb{R})^d$ defines a rational F-Hodge structure of weight 1 as follows. Let $V_{\mathbb{Q}} = F \oplus F$ (as F-module). A complex structure on $V_{\mathbb{R}}$ is defined by the embedding

$$F \oplus F \hookrightarrow \mathbb{C}^d \qquad (\alpha, \beta) \mapsto \Sigma(\alpha) + \Sigma(\beta) w\,,$$

where $\Sigma: F \to \mathbb{R}^d$ is defined using the d embedding of F into $\mathbb{R}$. A complex structure on $V_{\mathbb{R}}$ yields a direct sum decomposition $V_{\mathbb{C}} = V_{0,-1} \oplus V_{-1,0}$ with $V_{-1,0} = V_{\mathbb{R}} \subset V_{\mathbb{C}}$ and $\bar{V}_{-1,0} = V_{0,-1}$. The dual vector space carries a F-Hodge structure of weight $+1$.

(7.1) Theorem (Oda). *The rational k_f-Hodge structure $H^2(M_f, \mathbb{Q})$ is the tensor product of two rational k_f-Hodge structures of weight* 1:

$$H^2(M_f, \mathbb{Q}) \cong W^{(1)} \otimes_{k_f} W^{(2)}\,.$$

Proof. Let g_σ be a companion of f corresponding to $\sigma: k_f \hookrightarrow \mathbb{C}$. Consider the $\mathrm{GL}(2, k_f)$-orbits of

$$\omega_1 = (\omega_{+-}(g_\sigma)/\omega_{++}(g_\sigma))_{\sigma\in\Sigma} \qquad \Sigma = \{\sigma\colon k_f \hookrightarrow \mathbb{C}\}$$

$$\omega_2 = (\omega_{-+}(g_\sigma)/\omega_{++}(g_\sigma))_{\sigma\in\Sigma}\,.$$

We claim $\omega_1, \omega_2 \in \sqrt{-1}\mathbb{R}$. To see this, note that all Fourier coefficients of f are real (by $\langle T(\mathfrak{m})f, f\rangle = \langle f, T(\mathfrak{m})f\rangle$). So if $F_\infty = \varepsilon_1\varepsilon_2$ is the involution given by complex conjugation $(z_1, z_2)\mapsto(\bar{z}_1, \bar{z}_2)$ we find $F_\infty^*\eta_{g_\sigma} = \bar{\eta}_{g_\sigma}$ and $F_\infty(\gamma_{ss'}) = ss'\gamma_{ss'}$, hence

$$\int_{\gamma_{ss'}} F_\infty^*\eta_{g_\sigma} = ss'\int_{\gamma_{ss'}} \eta_{g_\sigma}\,,$$

implying that $\omega_{ss'}$ is real or pure imaginary depending on the sign of ss'. We deduce that ω_1, ω_2 define k_f-Hodge structures $W^{(1)}$, $W^{(2)}$ as above. Choose a k_f-bilinear form ψ_i on $W^{(i)}$ and a k_f-basis $e_+^{(i)}$, $e_-^{(i)}$ of $W_\mathbb{Q}^{(i)}$ such that $\psi_i(e_+^{(i)}, e_-^{(i)}) = +1$ and $\psi_i(e_s^{(i)}, e_s^{(i)}) = 0$. We define a k_f-module isomorphism

$$H^2(M_f, \mathbb{Q}) \xrightarrow{\sim} W_\mathbb{Q}^{(1)} \otimes_{k_f} W_\mathbb{Q}^{(2)}$$

by sending $e_{ss'}$ to $\pm e_s^{(1)}\otimes e_{s'}^{(2)}$ with the $(+)$-sign only if $s = s' = +$. We have to check that the Hodge decompositions of both sides match. By extension of scalars we have

$$H^2(M_f, \mathbb{Q})\otimes_\sigma \mathbb{C} \xrightarrow{\sim} (W_\mathbb{Q}^{(1)}\otimes_\sigma \mathbb{C})\otimes_\mathbb{C}(W_\mathbb{Q}^{(2)}\otimes_\sigma\mathbb{C})\,.$$

We check the $(2,0)$-component using (6.5). On the left-hand side this is

$$\mathbb{C}(\omega_{--}(g_\sigma)e_{++} + \omega_{-+}(g_\sigma)e_{+-} + \omega_{+-}(g_\sigma)e_{-+} + \omega_{++}(g_\sigma)e_{--})\,, \tag{1}$$

where we write e_{ss} instead of $e_{ss'}\otimes 1$. For the right-hand side note that the $(1,0)$-component of $W_\mathbb{Q}^{(i)}\otimes_\sigma\mathbb{C}$ equals (with $+\,-$ for $i=1$, and $-\,+$ for $i=2$)

$$\mathbb{C}(e_+^{(i)} + (\omega_{++}(g_\sigma)/\omega_{\pm\mp}(g_\sigma))e_-^{(i)})$$

so that the $(2,0)$-component is indeed (1). Now that the $(2,0)$-components match, the $(0,2)$-components do so too and consequently also the $(1,1)$-components must coincide. □

The involution τ of X_Γ induced on a component $\mathrm{PGL}^+(\mathfrak{o}\oplus\mathfrak{a})\backslash\mathfrak{H}^2$ by $(z_1, z_2)\mapsto(-1/Az_1, -1/Az_2)$ acts on modular forms and cohomology. It permutes the eigenforms and induces isomorphisms

$$H^2(M_f, \mathbb{Q}) \xrightarrow{\sim} H^2(M_{\tau(f)}, \mathbb{Q}) \qquad k_f \xrightarrow{\sim} k_{\tau(f)}$$

compatible with θ_f and $\theta_{\tau(f)}$. Since

$$\tau T(\mathfrak{m}) = T(\mathfrak{m}')\tau$$

as operators on modular forms, we have $k_f = k_{\tau(f)}$ and τ induces an involution $\rho = \rho_f$ of k_f by sending the Fourier coefficient $c(\mathfrak{m}, f)$ to $c(\mathfrak{m}', f)$.

We have two possibilities:

i) $\tau(f) = f$ (f is called *self-conjugate* in this case).

ii) $\tau(f) = f^\rho$, the companion of f associated to $k_f \xrightarrow{\rho} k_f \hookrightarrow \mathbb{C}$.

Indeed, if $\tau(f) \neq f$ then $\tau(f)$ and f^ρ have the same eigenvalues, hence by the

Multiplicity One Theorem $\tau(f) = f^\rho$. In the first case $\rho = \mathrm{id}_{k_f}$ while in the second case ρ is non-trivial. In any case τ induces

$$H^2(M_f, \mathbb{Q}) \xrightarrow{\sim} H^2(M_f, \mathbb{Q})^\rho ,$$

where the second k_f-Hodge structure is obtained from $H^2(M_f, \mathbb{Q})$ by composing θ with ρ: $\theta^\rho = \theta_f \rho$. Now τ interchanges $H^2(M_f, \mathbb{Q})_{+-}$ and $H^2(M_f, \mathbb{Q})_{-+}$, hence $\tau(\gamma_{+-}) = \theta(\alpha)\gamma_{-+}$ for some non-zero α. Moreover $\tau_*(\gamma_{++}) = \pm\gamma_{++}, \tau_*(\gamma_{--}) = \pm\gamma_{--}$, hence

$$\tau^*(\omega_1(g_\sigma)) = \pm\theta(\alpha)\omega_2(g_{\sigma\rho}) \tag{2}$$

(7.2) Proposition. *The involution τ induces an isomorphism of rational k_f-Hodge structures $W^{(1)} \xrightarrow{\sim} (W^{(2)})^\rho$.*

The k_f-Hodge structures just constructed are the rational Hodge structures of an isogeny class of abelian varieties over $\mathbb{C}$: Let $A^{(i)} = \mathbb{C}^d/\mathfrak{o}_{k_f} + \mathfrak{o}_{k_f}\omega_i$. This is a complex torus with real multiplication by k_f, hence an abelian variety (cf. IX.1). Indeed, the standard alternating form on $k_f \oplus k_f$ gives after multiplication by a suitable integer an integral valued form on the lattice which is the restriction of the real part of a positive definite hermitian form. By Lefschetz's criterium this shows that $A^{(i)}$ is a polarized abelian variety over $\mathbb{C}$ admitting an embedding $F \hookrightarrow \mathrm{End}(A^{(i)} \otimes \mathbb{Q})$. Unfortunately, we do not have information on the field of definition of these abelian varieties. Moreover, the choice of the lattice is arbitrary. But at least (7.2) agrees with the expectation that $A^{(1)}$ should be defined over K and that $A^{(2)}$ should be the conjugate (twisted with χ_D) form of $A^{(1)}$ w.r.t. $K/\mathbb{Q}$. It seems to be difficult to prove the existence of abelian varieties $A^{(i)}$ defined over K with the required properties.

For a Doi-Naganuma lift $\hat{g}$ of $g \in S_2(\Gamma_0(D), \chi_D)$ one expects a relation between the periods of $\hat{g}$ and g. Such a relation was proved by Oda in [124]. The proof is based on the relation $L(\hat{g}, s) = L(g, s)L(g^\rho, s)$ of (4.2) and the observation that the "critical values" $L(f, 1)$ are expressed by periods, cf. Deligne's philosophy in [24]. The basic example which expresses the value of an L-series as a period is the Mellin transform of a cusp form $f = \sum a(n)q^n$ on $\mathrm{SL}(2, \mathbb{Z})$:

$$\int_0^\infty f(iy)y^{s-1}\,dy = (2\pi)^{-s}\Gamma(s)\sum a(n)n^{-s} \qquad \mathrm{Re}(s) > \frac{k}{2} + 1 .$$

For simplicity we assume that $h^+ = 1$ in the rest of this section. (This assumption can be removed). If $f \in S_2$ with

$$f = \sum_{\substack{\nu \in \mathfrak{d}^{-1} \\ \nu \gg 0}} c((\nu)\mathfrak{d})e^{2\pi i \operatorname{Tr} \nu z}$$

$$L(f, s) = \sum c(\mathfrak{m})N(\mathfrak{m})^{-s} , \qquad \Lambda(f, s) = D^s(2\pi)^{-2s}\Gamma(s)^2 L(f, s)$$

we have similarly

$$\Lambda(f,s) = \int f(iy_1, iy_2)(y_1y_2)^{s-1}\,dy_1\,dy_2\,, \tag{3}$$

$$\begin{aligned}\Lambda(f,1) &= i^{-2}\int f(z_1,z_2)\,dz_1\wedge dz_2\\ &= (2\pi)^{-2}\int \omega_f\,,\end{aligned} \tag{4}$$

where the integrals are over $U^+\backslash\mathbb{R}^2_+$ and $\mathrm{Re}(s) > \frac{k}{2} + 1$. (Using (3) one obtains the analytic continuation and the functional equation $\Lambda(f,s) = \Lambda(f,k-s)$.)

The image of $U^+\backslash\mathbb{R}^2_+ \to X$, defined by $y \mapsto iy$ can be compactified by adding the cusp at ∞ to it. However, the image is a cycle that does not belong to the image of $H_2(X',\mathbb{Q}) \to H_2(X,\mathbb{Q})$. Indeed, by using the homotopy equivalence $X' \sim S_r^\circ$ of I.4 (S_r is the "part of X with distance $\geqslant r$ to the cusps") and the exact sequence

$$\to H_2(S_r^\circ,\mathbb{Q}) \to H_2(S_r,\partial S_r,\mathbb{Q}) \xrightarrow{\partial} H_1(\partial S_r,\mathbb{Q}) \to \tag{5}$$

it suffices to compute ∂ of it. But the image of $U^+\backslash\{y\in\mathbb{R}^2_+\colon \mathrm{N}(y) = r\}$ is a circle whose class is non-zero in $H_1(S_r,\mathbb{Q})$. To remedy this we look at sums of similar cycles weighted by a character.

Let $\mathfrak{n} = (v)$ be an integral ideal and let $V \subset U^+$ be a subgroup of finite index such that $V \subset 1 + \mathfrak{n}$. We define for $\alpha \in \mathfrak{o}$

$$D_{\alpha/v} = \{z \in \mathfrak{H}^2\colon \mathrm{Re}(z) = \alpha/v\}\,.$$

We have a map $i\colon V\backslash D_{\alpha/v} \to \Gamma\backslash\mathfrak{H}^2$, where V acts on $D_{\alpha/v}$ by $z \mapsto \varepsilon z + (1-\varepsilon)\alpha/v$. The quotient $V\backslash D_{\alpha/v}$ is topologically a cylinder $S^1 \times \mathbb{R}$ and its image in X_Γ can be compactified by adding the orbits of α/v and $\infty \in \mathbb{P}^1(\mathbb{Q})$. This gives a 2-cycle $\gamma_{\alpha/v,V}$ in X_Γ. (Replacing V by a subgroup V' multiplies this cycle by $[V:V']$.)

(7.3) Lemma. *If* $\chi\colon (\mathfrak{o}/\mathfrak{n})^* \to \mathbb{C}^*$ *is a non-trivial character then*

$$\gamma_{\mathfrak{n},V} = \sum_{\alpha \bmod \mathfrak{n}} \chi(\alpha)\gamma_{\alpha/v,V} \in \mathrm{Im}(H_2(X',\mathbb{Q}) \to H^2(X,\mathbb{Q}))\,.$$

Proof. By (5) it suffices to check $\partial\gamma_{\mathfrak{n},V} = 0$. Now $\partial(V\backslash D_{\alpha/v}) = x \in H_1(\partial S_r,\mathbb{Q})$ is independent of α. So $\partial\gamma = \sum_\alpha \chi(\alpha)x = 0$ if χ is non-trivial. □

Let χ be a Dirichlet character $(\bmod\,\mathfrak{n})$ and let $V \subset U^+$ of finite index with $V \subset 1 + \mathfrak{n}$. Then $\sum_{\alpha \bmod \mathfrak{n}} \chi(\alpha)\gamma_{\alpha/v,V}$ defines a class $\gamma_{\chi,V}$ in $H_2^{\mathrm{sp}}(X,\mathbb{Q})$ by projection.

(7.4) Lemma. *We have*

$$\gamma_{\chi,V} \in \begin{cases} H_2^{\mathrm{sp}}(X,\mathbb{Q})_{++} \oplus H_2^{\mathrm{sp}}(X,\mathbb{Q})_{--} & \text{if } \chi(-1) = +1\\ H_2^{\mathrm{sp}}(X,\mathbb{Q})_{+-} \oplus H_2^{\mathrm{sp}}(X,\mathbb{Q})_{-+} & \text{if } \chi(-1) = -1\,. \end{cases}$$

Moreover, $\gamma_{\chi,V} \in H_2(X,\mathbb{Q})_{ss'}$ if $\chi(\varepsilon_0) = s$, $\chi(\varepsilon_0') = s'$ *for a fundamental unit* ε_0 *with* $\varepsilon_0 > 0$, $\varepsilon_0' < 0$.

Let $G(\chi)$ be the Gauss sum

$$G(\chi) = \sum_{\alpha \bmod n} \chi(\alpha\sqrt{D})e^{2\pi i \operatorname{Tr}(\alpha/\nu)}$$

and define a twist of $L(f, s)$ by

$$L(f, s, \chi) = \sum_{\substack{\nu \in \mathfrak{d}^{-1} \\ \nu \gg 0}} \chi(\nu\sqrt{D})c(\nu\mathfrak{d})\mathrm{N}(\nu\mathfrak{d})^{-s} .$$

(7.5) Theorem (Oda). *Let $f \in S_2$ be a normalized eigenform. Then*

$$\Lambda(f, 1, \chi) = \frac{1}{[U^+ : V]} \frac{1}{(2\pi)^2} \frac{1}{G(\overline{\chi})} \int_{\gamma_{\overline{\chi},V}} \omega_f ,$$

where $V \subset U^+$ is of sufficiently large index and the orientation of $\gamma_{\overline{\chi},V}$ is chosen appropriately.

Proof. We have

$$\overline{\chi(\alpha)}G(\chi) = \sum_{\mu \bmod (\nu)} \chi(\mu\sqrt{D})e^{2\pi i \operatorname{Tr} \alpha\mu/\nu}$$

hence if we set

$$f_\chi = \sum_{\mu} \chi(\mu\sqrt{D})c(\mu\mathfrak{d})e^{2\pi i \operatorname{Tr} \mu z}$$

then

$$f_\chi(z) = G(\overline{\chi})^{-1} \sum_{\alpha \bmod (\nu)} \overline{\chi(\alpha)} f(z + \alpha/\nu) .$$

So by (3), (4)

$$\begin{aligned} \Lambda(f, 1, \chi) &= \frac{1}{G(\overline{\chi})} \sum_{\alpha \bmod n} \overline{\chi(\alpha)} \int_{U^+\backslash \mathbb{R}_+^2} f(\alpha/\nu + iy)\mathrm{N}(y)^{s-1}\, dy \\ &= \frac{1}{[U^+ : V]} \frac{1}{G(\overline{\chi})} \sum_{\alpha} \overline{\chi(\alpha)} \int_{V\backslash \mathbb{R}_+^2} f(\alpha/\nu + iy)\, dy \\ &= \frac{1}{[U^+ : V]} \frac{1}{G(\overline{\chi})} \sum_{\alpha} \overline{\chi(\alpha)} \int_{\gamma_{\alpha/\nu}} \frac{1}{(2\pi)^2} \omega_f . \end{aligned}$$

□

Next we define the periods of modular forms in $S_2(\Gamma_0(D), \chi_D)$. The kernel $\Gamma_0^\chi(D)$ of $\Gamma_0(D) \to \{\pm 1\}$, $\begin{pmatrix} a & b \\ c & d \end{pmatrix} \mapsto \chi_D(a)$ defines a modular curve $X_0^\chi(D)$ defined over $\mathbb{Q}$ with a natural morphism $X_0^\chi(D) \to X_0(D)$, also defined over $\mathbb{Q}$. The norm map of function fields $\mathbb{Q}(X_0^\chi(D)) \to \mathbb{Q}(X_0(D))$ induces a map Nm: $J_0^\chi(D) \to J_0(D)$ of the Jacobians. We denote the connected component of the kernel of Nm by P. It is an abelian subvariety of $J_0^\chi(D)$, called the *Prym variety* of $X_0^\chi(D) \to X_0(D)$. We have a Hodge-decomposition

$$H^1(P,\mathbb{Q})\otimes\mathbb{C}=\{\omega_g\colon g\in S_2(\Gamma_0(D),\chi_D)\}\oplus\{\bar\omega_g\colon g\in S_2(\Gamma_0(D),\chi_D)\},$$

where $\omega_g = 2\pi i g\, dz$.

The Hecke algebra acts on the whole situation and the subalgebra A of $\operatorname{End} H^1(P,\mathbb{Q})$ generated by the Hecke operators can be decomposed as a product of fields. For every normalized eigenform g there is a unique primitive idempotent e_g with $e_g A e_g = e_g A = A e_g = \mathscr{K}_g$, the field of Fourier coefficients of g. This gives us an abelian subvariety P_g defined by $P_g = e_g P$ (cf. Shimura [148]) and an associated rational $\mathscr{K}_g$-Hodge structure: $H^1(P_g,\mathbb{Q})$. The abelian variety P_g is of dimension $[\mathscr{K}_g:\mathbb{Q}]$ and is defined over $\mathbb{Q}$.

The action of K_g defines $\theta_g\colon \mathscr{K}_g \hookrightarrow \operatorname{End}(H^1(P_g,\mathbb{Q}))$. From $\langle T(n)g,h\rangle = \chi_D(n)\langle g, T(n)h\rangle$ it follows that $\mathscr{K}_g$ is a CM-field: a totally imaginary quadratic extension of a totally real field k_g. The intersection form on $H^1(X_0(D),\mathbb{Q})$ gives rise to a non-degenerate pairing

$$\Phi_g\colon H^1(P_g,\mathbb{Q})\times H^1(P_g,\mathbb{Q})\to\mathbb{Q}.$$

Since $\Phi_g(\theta_g(\alpha)x,y)=\Phi_g(x,\theta_g(\rho(\alpha))y)$, with ρ the (unique) complex conjugation of $\mathscr{K}_g$ (for all $\alpha\in A$, $x, y\in H^1(P_g,\mathbb{Q})$), we have $\Phi_g = \operatorname{Tr}_{\mathscr{K}_g/\mathbb{Q}}\phi_g$ for some ρ-skew symmetric form ϕ_g on $H^1(P_g,\mathbb{Q})$.

The fact that the Doi-Naganuma goes from S_2^+ to $S_2(\Gamma)$ suggests to look at the abelian variety

$$B_g=(1+W_D)P_g$$

with W_D the involution of $J_0(D)$ induced by $z\mapsto -1/Dz$. Shimura (see [150]) proved that B_g is defined over our quadratic field K. We have

$$W_D^* T(n) W_D^* = \chi_D(n) T(n),$$

so conjugation by W_D^* induces an involution of A which is complex conjugation ρ on $\mathscr{K}_g$. The form Φ_g restricts to a skew-symmetric bilinear form $\Phi_g^+ = \operatorname{Tr}_{k_g/\mathbb{Q}}\phi_g^+$ with ϕ_g^+ a k_g-valued k_g-bilinear alternating form. The Hodge decomposition for an embedding $\sigma\colon k_g\hookrightarrow\mathbb{C}$ is (with $h=g^\sigma$)

$$H^1(B_g,\mathbb{Q})\otimes_{k_g,\sigma}\mathbb{C}=\mathbb{C}(\omega_h+W_D^*(\omega_h))\oplus\mathbb{C}(\bar\omega_h+W_D^*(\bar\omega_h))$$

We know: $W_D^*\omega_g=\lambda_g\omega_{g^\rho}$ with $|\lambda_g|=1$, $\lambda_{g^\rho}=\bar\lambda_g$, cf. [150].

Complex conjugation $(z\mapsto -\bar z)$ induces an involution of $H^1(B_g,\mathbb{Q})$ whose $\pm$ eigenspaces $H^1(B_g,\mathbb{Q})_\pm$ are 1-dimensional k_g-modules. It also defines $\pm$ eigenspaces $H_1(B_g,\mathbb{Q})_\pm$ *of* $H_1(B_g,\mathbb{Q})$ and we choose $e_\pm\in H_1(B_g,\mathbb{Q})_\pm$ such that $\psi_g^+(e_+,e_-)=1$, where ψ_g^+ is a skew-symmetric k_g-bilinear form such that $\operatorname{Tr}_{k_g/\mathbb{Q}}\psi_g^+$ corresponds to Φ_g^+ under Poincaré duality. We define the *periods of a companion h of g* by

$$\omega_\pm(h)=\int_{e_\pm}\omega_h=\frac{1}{2}\int_{e_\pm}(\omega_h+W_D^*(\omega_h)) \tag{5}$$

(7.6) Proposition. i) $\frac{1}{2}\int_{X_0^\chi(D)}\omega_h\wedge\bar\omega_h=\overline{\omega_+(h)}\omega_-(h)-\omega_+(h)\overline{\omega_-(h)}$

ii) $\overline{\omega_\pm(h)}=\pm\bar\lambda_h\omega_\pm(h)$.

Proof. i) is trivial and ii) goes as follows

$$\begin{aligned}\overline{\omega_{\pm}(h)} &= \overline{\int_{e_{\pm}} \omega_h} \\ &= \pm \int_{e_{\pm}} \omega_{h^\rho} \quad \text{(since compl. conjugation sends } e_{\pm} \textit{ to } \pm e_{\pm}) \\ &= \pm \int_{e_{\pm}} \bar{\lambda}_h \omega_h \quad (\text{since } W_D^* e_{\pm} = e_{\pm} \textit{ and } W_D^*(\omega_h) = \lambda_h \omega_{h^\rho}) . \qquad \square\end{aligned}$$

Now we come to the relation between the periods of an eigenform $f \in S_2(\Gamma_0(D), \chi_D)$ and its Doi-Naganuma lift $\hat{f}$. Note that for any companion g of f we have $k_{\hat{g}} = k_g$ and let ρ_g be complex conjugation on $\mathcal{K}_g$.

(7.7) Theorem (Oda's period relation for the Doi-Naganuma lifting). *Let $f \in S_2(\Gamma_0(D), \chi_D)$ be a normalized eigenform and $\hat{f}$ its Doi-Naganuma lift. There exist elements $a_{ss'} \in k_{\hat{g}}$ such that for any companion $\hat{g}$ of $\hat{f}$ associated to σ: $k_{\hat{g}} \hookrightarrow \mathbb{C}$*

$$\omega_{ss'}(\hat{g}) = \lambda_g \sigma(a_{ss'}) \omega_s(g) \omega_{s'}(g) . \qquad s, s' \in \{+, -\}$$

(7.8) *Remark.* The fact that g and g^ρ have the same lift $\hat{g}$ is compatible with this relation since $\omega_{\pm}(h^\rho) = \bar{\lambda}_h \omega_{\pm}(h)$ and $\bar{\lambda}_h = \lambda_{h^\rho} = \lambda_h^{-1}$.

Proof. i) We first treat the case $s = s'$. Let $\mathfrak{n}$ be an integral ideal of K of norm n and let χ_0 be a Dirichlet character (mod n). Then the composition $\chi = \chi_0 N_{K/\mathbb{Q}}$ is a Dirichlet character (mod $\mathfrak{n}$). Moreover, $\chi(\varepsilon_0) = \chi_0(-1) = \pm 1$, so for suitable $V \subset U^+$ by (7.4) $\gamma_{\chi, V} \in H_2^{\mathrm{sp}}(M_{\hat{f}}, \mathbb{Q})_{\pm\pm}$. By (7.5)

$$\Lambda(\hat{g}, 1, \chi) = [U^+ : V]^{-1} (2\pi)^{-2} G(\bar{\chi})^{-1} \int_{\gamma_{\bar{\chi}, V}} \omega_{\hat{g}} . \tag{6}$$

Since $H_2(M_{\hat{f}}, \mathbb{Q})_{\pm\pm}$ has dimension 1 over $k_{\hat{f}}$ we have

$$\pm \int_{\gamma_{\bar{\chi}, V}} \omega_{\hat{g}} = \begin{cases} \sigma(\alpha) \omega_{++}(\hat{g}) & \text{if } \chi_0(-1) = +1 \\ \sigma(\alpha) \omega_{--}(\hat{g}) & \text{otherwise} \end{cases} \tag{7}$$

for some $\alpha \in k_{\hat{g}}$. For g we obtain similarly

$$\Lambda(g, 1, \chi) = (2\pi)^{-1} G(\bar{\chi}_0)^{-1} \cdot \begin{cases} \sigma(\beta) \omega_+(g) & \text{if } \chi_0(-1) = +1 \\ \sigma(\beta) \omega_-(g) & \text{otherwise} \end{cases} \tag{8}$$

for some $\beta \in \mathcal{K}_g$. Now we use (cf. (4.2))

$$L(\hat{g}, s, \chi) = L(g, s, \chi_0) L(g^\rho, s, \chi_0) .$$

Using (6), (7) and (8) this relation gives for $s = \pm$

$$\sigma(\alpha) \omega_{ss}(\hat{g}) = [U^+ : V] \frac{G(\bar{\chi})}{G(\bar{\chi}_0)^2} \sigma(\beta) \sigma(\rho_g(\beta)) \lambda_g \omega_s(g)^2 . \tag{9}$$

But $G(\chi) = G(\chi_0)^2$, so the case $s = s'$ is done if $\beta \neq 0$. According to a theorem of Shimura ([151], Thm. 2) we can choose χ_0 such that $L(1, g, \chi_0) \neq 0$ even if we require $\chi_0(-1) = +1$ or $\chi_0(-1) = -1$. By (8) this implies $\beta \neq 0$.

ii) Next we do the case $s \neq s'$. Since $\hat{f}$ is self-conjugate we find by applying τ

$$\omega_{+-}(\hat{g}) = \int_{e_{+-}} \omega_{\hat{g}} = \int_{\tau(e_{+-})} \tau^*(\omega_{\hat{g}}) = -\int_{\sigma(a)e_{+-}} \omega_{\hat{g}} = \sigma(-a)\omega_{-+}(\hat{g}) \tag{10}$$

for some $a \in k_{\hat{g}}$. We now claim:

$$\int_{\Gamma\backslash\mathfrak{H}^2} \omega_{\hat{g}} \wedge \bar{\omega}_{\hat{g}} = r\left(\int_{T_1} \eta_{\hat{g}}\right) \cdot \int_{\Gamma_0(D)\backslash\mathfrak{H}} \omega_g \wedge \bar{\omega}_g\,, \tag{11}$$

where r is a rational number. This will be proved later (7.9). Applying the period relations (6.5) and (7.6) to (11) we get

$$\omega_{++}(\hat{g})\omega_{--}(\hat{g}) = -\omega_{+-}(\hat{g})\omega_{-+}(\hat{g}) = r\left(\int_{T_1} \eta_{\hat{g}}\right)\omega_+(g)\omega_-(g) \tag{12}$$

In view of (10) and (12), it suffices to prove

$$\int_{T_1} \omega_{\hat{g}} = \sigma(b)\omega_{+-}(\hat{g})$$

for some $b \in k_{\hat{g}}$. But this follows from the fact that complex conjugation F_∞ changes the orientation of T_1, hence $t_1^0 \in H_2^{\text{sp}}(X, \mathbb{Q})_{+-} \oplus H_2^{\text{sp}}(X, \mathbb{Q})_{-+}$. □

(7.9) Proposition. *Let $g \in S_2(\Gamma_0(D), \chi_D)$ be a normalized Hecke eigenform and let $\hat{g}$ be its Doi-Naganuma lift. Then*

$$\int_{\Gamma\backslash\mathfrak{H}^2} \omega_{\hat{g}} \wedge \bar{\omega}_{\hat{g}} = r\left(\int_{T_1} \eta_{\hat{g}}\right) \int_{\Gamma_0(D)\backslash\mathfrak{H}} \omega_g \wedge \bar{\omega}_g$$

with $r \in \mathbb{Q}^*$.

Proof. We have $\mathrm{DN}^*\mathrm{DN}(g) = \mathrm{DN}^*(\hat{g}) = c_g^{-1}\langle \Omega, \hat{g}\rangle$. But $\langle \Omega, \hat{g}\rangle = \mu_g(g + g^\rho)$ with $\mu_g = \frac{1}{2}(2\pi i)^{-1} \int_{T_1} \eta_{\hat{g}}$ by (2.1). Now we have

$$\begin{aligned}
\int \omega_{\hat{g}} \wedge \bar{\omega}_{\hat{g}} &= -2^2(2\pi i)^4 \langle \hat{g}, \hat{g}\rangle \\
&= -2^2(2\pi i)^4 \langle \hat{g}, \mathrm{DN} g\rangle \\
&= -2^2(2\pi i)^4 \langle \mathrm{DN}^*\hat{g}, g\rangle \\
&= -2(2\pi i)^3 c_g^{-1}\left(\int_{T_1} \eta_{\hat{g}}\right)\langle g, g\rangle \\
&= i(2\pi i)c_g^{-1}\left(\int_{T_1} \eta_{\hat{g}}\right)\int \omega_g \wedge \bar{\omega}_g
\end{aligned}$$

and $c_g \in \pi^{-1} \cdot \mathbb{Q}^*$, see VI.4. □

8. The Contribution of an Eigenform to the Picard Number

If S is a smooth projective algebraic surface over $\mathbb{C}$ any compact algebraic curve on S defines a class in $H_2(S,\mathbb{Q})$ and by Poincaré duality a class in $H^2(S,\mathbb{Q})$. The classes in $H^2(S,\mathbb{Q})$ obtained in this way from the algebraic curves on S generate a $\mathbb{Q}$-linear subspace of $H^2(S,\mathbb{Q})$, the space of algebraic cycles. Its dimension is called the *Picard number* of S and denoted by $\rho = \rho(S)$. The difference $b_2(S) - \rho(S)$ is a birational invariant of S, called the Lefschetz number.

It is well-known that the space of algebraic cycles is the intersection of $H^{1,1}(S,\mathbb{C}) \cap H^2(S,\mathbb{Q})$ inside $H^2(S,\mathbb{C})$. This is an easy consequence of the standard exact sequence

$$\to H^1(S,\mathcal{O}_S^*) \to H^2(S,\mathbb{Z}) \to H^2(S,\mathcal{O}_S) \to$$

and the Hodge decomposition of $H^2(S,\mathbb{C})$.

In order to compute $\rho(Y)$ for a Hilbert modular surface $Y = Y(D)$ it suffices to compute the dimension of the $\mathbb{Q}$-linear subspace of $\mathbb{H}^2_{\mathrm{sp}}(X,\mathbb{Q})$ generated by the images of the algebraic cycles, since the rest of the cohomology is known. We call this subspace $\mathbb{H}^2_{\mathrm{sp}}(X,\mathbb{Q})_{\mathrm{alg}}$. We define

$$H^2(M_f,\mathbb{Q})_{\mathrm{alg}} = \mathbb{H}^2_{\mathrm{sp}}(X,\mathbb{Q})_{\mathrm{alg}} \cap H^2(M_f,\mathbb{Q})$$

and call its dimension the *Picard number* ρ_f of $H^2(M_f,\mathbb{Q})$. We have

$$\rho_f = \dim_{\mathbb{Q}}(H^2(M_f,\mathbb{Q}) \cap H^{1,1}).$$

(8.1) Proposition. $H^2(M_f,\mathbb{Q})_{\mathrm{alg}} \cong \mathrm{Hom}_{k_f}(W_f^{(1)}, W_f^{(2)})$ *(in the category of rational k_f-Hodge structures).*

Proof. We know

$$H^2(M_f,\mathbb{Q}) \simeq W_f^{(1)} \otimes W_f^{(2)}.$$

Now $(W_f^{(i)})^\vee = W_f^{(i)}(1)$ (shift by 1), hence

$$H^2(M_f,\mathbb{Q}) \simeq \mathit{Hom}_{k_f}((W_f^{(1)})^\vee, W_f^{(2)}) = \mathit{Hom}_{k_f}(W_f^{(1)}, W_f^{(2)})(-1).$$

Therefore,

$$\begin{aligned} H^2(M_f,\mathbb{Q})_{\mathrm{alg}} &= (0,0) - \text{part of } \mathit{Hom}_{k_f}(W_f^{(1)}, W_f^{(2)}) \quad (\text{in } -- \otimes \mathbb{C}) \\ &= \mathrm{Hom}_{k_f}(W_f^{(1)}, W_f^{(2)}). \end{aligned}$$ □

Note that (8.1) implies that ρ_f is a multiple of $d = [k_f : \mathbb{Q}]$.

Recall that a *CM*-field is a finite field extension L of $\mathbb{Q}$ such that for every embedding $\sigma\colon L \hookrightarrow \mathbb{C}$ complex conjugation of $\mathbb{C}$ defines a non-trivial automorphism of L which does not depend on σ. The fixed field of complex conjugation is a totally real subfield of which L is a totally imaginary quadratic extension.

(8.2) Corollary. i) $\rho_f = 0 \Leftrightarrow W_f^{(1)}$ *and* $W_f^{(2)}$ *are not isomorphic.*

ii) $\rho_f = d \Leftrightarrow W_f^{(1)} \simeq W_f^{(2)}$ *and* $\mathrm{End}_{k_f} W_f^{(i)} = k_f$.

iii) $\rho_f = 2d \Leftrightarrow W_f^{(1)} \simeq W_f^{(2)}$ *and* $\mathrm{End}_{k_f} W_f^{(i)} = L$, *where L is a CM-field whose totally real subfield is k_f.*

Proof. We have $\mathrm{Hom}_{\mathscr{k}_f}(W_f^{(1)}, W_f^{(2)}) \neq (0) \Leftrightarrow$ there exists $\gamma \in \mathrm{SL}(2, \mathscr{k}_f)$ with $\gamma\omega_1 = \omega_2 \Leftrightarrow W_f^{(1)}$ and $W_f^{(2)}$ are isomorphic (as rational $\mathscr{k}_f$-Hodge structures). Since the Hodge structures $W_f^{(i)}$ come from abelian varieties $A_f^{(i)}$ we have $\mathrm{End}_{\mathscr{k}_f} W_f^{(i)} = (\mathrm{End}_{\mathfrak{o}_{\mathscr{k}_f}} A_f^{(i)}) \otimes \mathbb{Q}$ and the result follows immediately from the possibilities for $\mathrm{End}_{\mathfrak{o}_{\mathscr{k}_f}}(A_f^{(i)})$, see e.g. [115]. □

This enables us to find part of the algebraic cycles.

(8.3) Theorem. *Let f be a self-conjugate eigenform. Then $H^2(M_f)^{\mathrm{sym}}$ is of rank* 1 *over $\mathscr{k}_f$ and is generated by algebraic cycles. (Here* sym *means τ acts as* 1.)

Proof. We know that τ interchanges $H_2(M_f)_{+-}$ and $H_2(M_f)_{-+}$ and acts as -1 on $H_2(M_f)_{++}$ and $H_2(M_f)_{--}$. Hence $\mathrm{rk}_{\mathscr{k}_f} H^2(M_f)^{\mathrm{sym}} = 1$. If g is a companion of f and if $\gamma \in H^2(M_f)^{\mathrm{sym}}$ then

$$\int_\gamma \omega_g = \int_{\tau(\gamma)} \omega_g = \int_\gamma \tau^*\omega_g = -\int_\gamma \omega_g = 0\,,$$

hence our space consists of type (1, 1)-classes. □

On the other hand we know that $\mathbb{T} \cap H^2(M_f) \subset H^2(M_f)^{\mathrm{sym}}$.

(8.4) Theorem. *If D is a product of primes $\equiv 1 \pmod 4$ then $\mathbb{T} \cap \mathbb{H}^2_{sp}(X) = \bigoplus H^2(M_f)^{\mathrm{sym}}$, where the sum is over a maximal set of self-conjugate normalized eigen forms which are not* $\mathrm{Gal}(\bar{\mathbb{Q}}/\mathbb{Q})$*-conjugate (i.e. not companions).*

Proof. Under the given condition $\mathrm{PGL}^+ = \mathrm{PSL}$ and the fixed point locus of τ is $F_D \cup F_1$, see (9.1). A computation as at the end of VI.1 tells us that $p_g(X) - 2p_g(X/\tau) = \dim S_2^+(\Gamma_0(D), \chi_D)$ is the dimension of the space generated by the self-conjugate forms. □

If D does not satisfy the conditions of (8.4) the situation is more difficult, cf. e.g. (9.1). We refrain from a discussion here since the subject is taken up again in Chapter XI in a more general—and less specific—setting.

In some cases one can exclude extra cycles in $H^2(M_f)$—i.e. those not coming from $\mathbb{T}$.

Let f be a self-conjugate form and assume as above in (8.4) that f is a lift $\hat{g}$ of g. Then the rational Hodge structure $W_f^{(i)}$ is that of a factor B_g of a Jacobian of a modular curve. Then if $\mathrm{End}_{\mathscr{k}_f} B_g$ has rank 2 then according to Shimura B_g is isogenous to a product of elliptic curves of CM-type, see [149, 147]. It then also follows that g is associated to a primitive grössencharacter ψ of an imaginary quadratic field. If the level of g is prime this cannot occur, see [147]. We find:

(8.5) Corollary (Oda). *If D is prime then $\rho_f = d = [\mathscr{k}_f : \mathbb{Q}]$ for all self-conjugate forms f.*

As mentioned before, one uses here only the Hodge realization of the "motive" M_f. More results can be obtained if one uses not only the infinite places but also the finite places. This will be done in Chapter XI.

(8.6) Corollary. *If D is a prime and all forms in S_2 are self-conjugate then the Lefschetz number $b_2 - \rho$ of $Y(D)$ equals $3p_g(Y(D))$.*

Proof. This follows immediately from (8.5). □

Examples. All cases where D is prime and where $p_g(Y_{\Gamma(\mathfrak{o}\oplus\mathfrak{a})}/\tau) = 0$ for all $\mathfrak{a}$: i.e. $29 \leqslant D \leqslant 181$, $D = 197, 229, 269, 293, 317$.

VII. The Classification of Hilbert Modular Surfaces

In the next two chapters the emphasis is on the geometry of Hilbert modular surfaces. Algebraic surfaces are classified by the growth of their plurigenera, which roughly speaking is a way of measuring the ampleness of the canonical divisor class. On Hilbert modular surfaces we have in some sense two "canonical divisor classes", the usual canonical class $c_1(\Omega^2)$ and the class $c_1(\Omega^2(\log D))$ with D the divisor resolving the cusps. The second one is more natural for Hilbert modular surfaces, but by using the first one the standard theory of algebraic surfaces becomes available. Since the difference D of $c_1(\Omega^2(\log D))$ and $c_1(\Omega^2)$ is relatively small if the volume of $\Gamma\backslash\mathfrak{H}^2$ is big one expects in general that the two classes behave similarly.

Algebraic surfaces with vanishing irregularity ($q = 0$) are divided into four classes according to the growth of their pluri-genera. These classes are: the rational surfaces, the blown-up $K3$ and Enriques surfaces, the blown up honestly elliptic surfaces and the surfaces of general type.

It was shown by Hirzebruch, Van de Ven and Zagier that for almost all Hilbert modular surfaces Y_Γ with $\Gamma = \mathrm{PSL}(\mathfrak{o} \oplus \mathfrak{a})$ and $\mathfrak{a}$ in the principal genus or in the genus of $(\sqrt{D})$ one has $\chi > 1$ and $c_1^2 > 0$ and this implies that these surfaces are of general type; they also showed how to classify the remaining surfaces. For this they used Castelnuovo's classical rationality criterion ($q = 0, P_2 = 0$) and the fact that certain configurations of non-singular rational curves with self-intersection number -2 are not admitted for surfaces in one of the classes.

We give the classification of the Hilbert modular surfaces Y_Γ with $\Gamma = \mathrm{PSL}(\mathfrak{o} \oplus \mathfrak{a})$ or a principal congruence subgroup of $\mathrm{PSL}(\mathfrak{o} \oplus \mathfrak{a})$.

If one deals with an algebraic surface it is important to know whether the surface is minimal or how a minimal model can be obtained. It turns out that the components of some of the modular curves F_N are exceptional curves. Blowing down these curves and then a few curves from the resolutions of the singularities that become exceptional yields (if Y_Γ is not rational) a non-singular surface Y_Γ°. It was conjectured in [82] (and [86]) that Y_Γ° is a minimal model of Y_Γ. This conjecture can be verified in many cases, e.g. for all surfaces Y_Γ for which Y_Γ is not of general type (by just showing that $c_1^2 = 0$) and for all surfaces Y_Γ associated to a sufficiently small principal congruence subgroup. One of the methods for showing that Y_Γ° is minimal consists in exhibiting an effective canonical divisor on Y_Γ. Such a canonical divisor contains all exceptional curves. We present two ways of constructing such canonical divisors, a geometric one which works for surfaces with $c_1^2 > 0$ and an arithmetic one which uses the isomorphism between the space $\mathbb{T}$ generated by the algebraic cycles t_N and $M_2^+(\Gamma_0(D), \chi_D)$. The canonical divisors that we thus find will be an important tool for the finer classification in the next chapter.

1. The Rough Classification of Algebraic Surfaces

Let S be a non-singular complex algebraic surface. The canonical line bundle of S will be denoted by K. Sometimes we also denote the canonical (divisor) class of S by K. Its homology class in $H^2(S, \mathbb{Z})$ equals $-c_1(S)$. As usual, $P_m = P_m(S) = \dim_{\mathbb{C}} \Gamma(S, \mathcal{O}(\mathrm{K}^{\otimes m}))$, $m \geqslant 1$ is the *m-th pluri-genus* of S. If $P_m \neq 0$, then let Φ_m be the rational map

$$S \to \mathbb{P}^{P_m - 1}$$

defined (up to projective equivalence) by choosing a basis $\varphi_1, \ldots, \varphi_{P_m}$ of $\Gamma(S, \mathcal{O}(\mathrm{K}^{\otimes m}))$ and setting

$$\Phi_m(x) = (\varphi_1(x) : \cdots : \varphi_{P_m}(x)),$$

whenever this is defined, i.e. x is not a base point of $\Gamma(S, \mathcal{O}(\mathrm{K}^{\otimes m}))$.

The *Kodaira dimension* $\kappa(S)$ of S is defined as follows.

$$\kappa(S) = -1 \text{ if } P_m = 0 \; \forall m \geqslant 1$$

$$\kappa(S) = \max_{m: P_m \neq 0} \dim \Phi_m(S) \quad \text{otherwise}.$$

Obviously, $-1 \leqslant \kappa(S) \leqslant 2$. Furthermore, we have

$$P_m(S) = O(m^{\kappa(S)}).$$

By an *exceptional curve* (of the first kind) on a surface S we mean a non-singular rational curve with self-intersection number -1. An irreducible curve E is an exceptional curve if and only if $E \cdot E = \mathrm{K} \cdot E = -1$. By blowing up a point of a non-singular surface S we obtain an exceptional curve. Exceptional curves play an important role in the theory of algebraic surfaces. It is well-known that an exceptional curve E on an algebraic surface S can be contracted (or blown down), i.e. there exists a non-singular algebraic surface S' and a birational morphism $\pi: S \to S'$ such that $\pi(E)$ is a point of S' and $\pi | S - E \to S' - \pi(E)$ is an isomorphism. In this case the canonical class K_S of S is related to that of S' by

$$\mathrm{K}_S = \pi^* \mathrm{K}_{S'} + E. \tag{1}$$

Consequently, we have

$$\mathrm{K}_S^2 = \mathrm{K}_{S'}^2 - 1. \tag{2}$$

It is a basic result in the theory of algebraic surfaces that any birational map $g: S \to S'$ of two (non-singular) algebraic surfaces can be factored into a finite number of blowing ups (of points) and blowing downs (of exceptional curves). In particular, if g is a morphism g can be represented as the composition of finitely many contractions of exceptional curves.

Let S and S' be (non-singular) algebraic surfaces. We say that S' is *minimal* (resp. a *minimal model* of S) if S' does not contain exceptional curves (resp. if S' is minimal and birationally equivalent to S). By blowing down an exceptional curve the rank of $H^2(-,\mathbb{Q})$ diminishes by 1. Hence a minimal model is obtained after finitely many steps. In general such a minimal model is not unique.

The classification of algebraic surfaces was shaped by Italian geometers around the turn of the century, esp. by Enriques. It was revised by Kodaira in the sixties who gave modern proofs and extended it to the non-algebraic case. It is by no means a classification up to isomorphism but rather lists properties of surfaces whose Kodaira dimension assumes a given value, and is usually called the rough or *Enriques-Kodaira* classification. We limit ourselves to algebraic surfaces with $q = 0$ since all Hilbert modular surfaces have this property. The classification of the *minimal* non-singular (compact complex) algebraic surfaces with vanishing irregularity ($q = 0$) can be summarized by the following diagram.

Kodaira dimension	type	χ	K^2	
$\kappa = -1$	rational	1	8	$S \cong \Sigma_n,\ n \neq 1$
			9	$S \cong \mathbb{P}^2$
$\kappa = 0$	$K3$	2	0	
	Enriques surface	2	0	
$\kappa = 1$	honestly elliptic	$\geqslant 1$	0	
$\kappa = 2$	general type	$\geqslant 1$	$\geqslant 1$	

We explain the terminology used in the table, but we must refer to [10] for an extensive explanation and proofs. A surface is called *rational* if it is birationally equivalent to $\mathbb{P}^2$.

The minimal rational surfaces are $\mathbb{P}^2$ and the surfaces Σ_n $n \in \mathbb{Z}_{\geqslant 0}$, $n \neq 1$. The surface Σ_n ($n \geqslant 0$) can be given by the equation

$$x_1 y_1^n - x_2 y_2^n = 0$$

in $\mathbb{P}^2 \times \mathbb{P}^1$ where $(y_1 : y_2)$ and $(x_0 : x_1 : x_2)$ denote the coordinates in $\mathbb{P}^1$ and $\mathbb{P}^2$. Equivalently, it can be described as the bundle space of the $\mathbb{P}^1$-bundle over $\mathbb{P}^1$ associated to $H \oplus H^{n+1}$ with H the bundle of degree 1 on $\mathbb{P}^1$. The surface Σ_n has a section whose divisor has self-intersection $-n$ (given by $x_1 = x_2 = 0$). Blowing down this section in Σ_1 gives $\mathbb{P}^2$.

A surface is called a *K*3 *surface* if $q = 0$ and if $K = 0$. A non-singular quartic surface in $\mathbb{P}^3$ is an example of a $K3$ surface. A $K3$ surface is simply-connected and all $K3$ surfaces are diffeomorphic.

An *Enriques surface* is a surface with $q = 0$, $2K = 0$ but $K \neq 0$. An Enriques surface admits as a double cover a $K3$ surface. It has a cyclic group of order 2 as its fundamental group.

A surface S is called *elliptic* if S admits an elliptic fibration, i.e. a morphism $p: S \to C$ onto a non-singular curve C such that all but finitely many fibres are non-singular elliptic curves. In this case the irregularity q of S equals the geometric genus of C if S is not isomorphic to $C \times E$ with E an elliptic curve. Kodaira has given a list of all possibilities for the fibres which are not elliptic curves. An elliptic surface is called *honestly elliptic* if $\kappa \geqslant 1$.

A surface is called of *general type* if Φ_m defines a birational map of S onto its image for large m. (This is easily seen to be equivalent to $\kappa = 2$.) For example, a non-singular surface of degree $\geqslant 5$ in $\mathbb{P}^3$ is of general type. If S is minimal, then for m large Φ_m is a birational morphism which contracts (trees of) non-singular rational curves with self-intersection -2 to singular points and is an isomorphism elsewhere. Bombieri proved that $m \geqslant 5$ suffices.

A surface S with $\kappa(S) \geqslant 0$ has a unique minimal model, whereas a surface with $q = 0$ and $\kappa(S) = -1$ has infinitely many minimal models: $\mathbb{P}^2$ and the surfaces Σ_n with $n \geqslant 0, n \neq 1$.

2. Configurations of Curves on Surfaces

The purpose of this section is to show how one can get information about the place of a surface in the Enriques-Kodaira classification using configurations of curves on it.

(2.1) Proposition. *Let S be a non-singular algebraic surface. If S contains two exceptional curves C_1, C_2 with $C_1 \cdot C_2 > 0$ or if S contains an irreducible curve with* $C \cdot C \geqslant 0$ and $\mathrm{K} \cdot C < 0$ *then all plurigenera of S vanish.*

Proof. Suppose we have a non-trivial holomorphic section of $\mathrm{K}^{\otimes m}$ with divisor Φ $(m \geqslant 1)$. Write

$$\Phi = aC_1 + bC_2 + R$$

with $a, b \geqslant 0$ and $R \cdot C_i \geqslant 0$. Then

$$-2m = \Phi \cdot (C_1 + C_2) \geqslant -a - b + (a + b) C_1 \cdot C_2 \geqslant 0\,,$$

a contradiction. Similarly, if $\Phi = aC + R$ with $a \geqslant 0$, $R \cdot C \geqslant 0$ we get

$$0 > \Phi \cdot C = aC^2 + R \cdot C \geqslant 0$$

again a contradiction. □

This proposition combined with the Enriques-Kodaira classification yields the following useful criterion.

(2.2) Rationality Criterion. *If S is a non-singular algebraic surface with $q = 0$ and if S contains either two intersecting exceptional curves or an irreducible curve C with $C^2 \geqslant 0$ and $\mathrm{K} \cdot C < 0$ then S is rational.*

In fact, one does not need the full strength of the Enriques-Kodaira classification here but *Castelnuovo's criterion*—a non-singular algebraic surface is rational if and only if $q(S) = P_2(S) = 0$—suffices.

(2.3) *Remark.* The factorization of birational maps in monoidal transformations and their inverses and the non-occurrence of intersecting exceptional curves on non-rational surfaces with $q = 0$ imply that a non-rational surface with $q = 0$ has a unique minimal model.

For a divisor D on an algebraic surface $p_a(D)$ is defined in such a way that the adjunction formula holds:

$$\mathrm{K}\cdot D + D\cdot D = 2p_a(D) - 2\,.$$

If D is effective then $p_a(D)$ coincides with the usual arithmetic genus $p_a = 1 - \chi(\mathcal{O}_D)$. So if D is an irreducible curve then $p_a(D) \geqslant g(D)$, the ordinary genus of D, equality holding only if D is non-singular.

Moreover, if D is irreducible, then $p_a = 0$ if and only $D \cong \mathbb{P}^1$ and $p_a = 1$ if and only if D is either a non-singular elliptic curve, a rational curve with one ordinary double point or a rational curve with one ordinary cusp.

Often one can derive information on the type of a surface by looking at (-2)-configurations, defined as follows.

(2.4) Definition. By a (-2)-curve on a surface S we mean a non-singular rational curve with self-intersection number -2. By a (-2)-configuration on S we mean a configuration formed by a finite number of (-2)-curves the union of which is connected.

(2.5) Lemma. *Let $\mathscr{C}$ be a (-2)-configuration on an algebraic surface S with $\kappa \geqslant 0$. Suppose that E is an exceptional curve on S which intersects $\mathscr{C}$. Then the configuration $\mathscr{C} \cup E$ looks like*

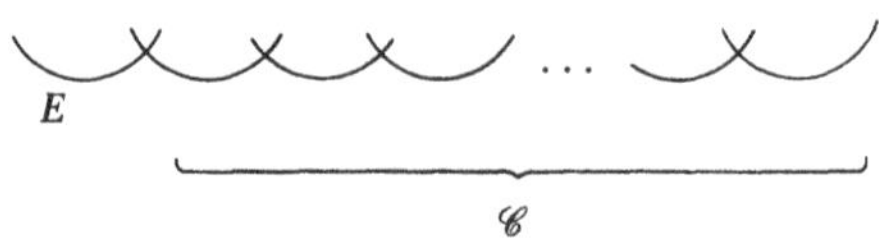

in particular, $\mathscr{C}$ is linear.

Proof. Suppose that $E\cdot C \geqslant 2$ for a (-2)-curve $C \in \mathscr{C}$. By blowing down E we obtain on the new surface a curve C' with $C'\cdot C' \geqslant 0$, $\mathrm{K}C' < 0$. If E intersects two (-2)-curves of $\mathscr{C}$ transversally then by blowing down E we get two intersecting exceptional curves. In both cases all pluri-genera have to vanish by (2.1) and this contradicts $\kappa \geqslant 0$. Therefore E intersects only one (-2)-curve of $\mathscr{C}$ transversally, say C_1. By blowing down E the curve C_1 becomes an exceptional curve which by the same argument can intersect only one (-2)-curve of $\mathscr{C}$, etc. □

(2.6) Proposition. *Let S be a minimal surface of general type. For any irreducible curve C on S one has $\mathrm{K}\cdot C \geqslant 0$ and $\mathrm{K}\cdot C = 0$ only if C is a (-2)-curve. Moreover,* *the number of* (-2)*-curves on* S *is finite and all of them are numerically independent on* S.

The proposition (for a proof see [10]) implies that on a surface of general type, not necessarily minimal, one always has $\mathrm{K}\cdot C \geqslant -1$ for each irreducible curve C on S. In fact, $\mathrm{K}\cdot C = -1$ implies that C is exceptional and $\mathrm{K}\cdot C = 0$ implies that C is a (-2)-curve.

Let S be a minimal surface of general type. If Φ_m is a birational morphism then its exceptional fibres are exactly the (-2)-configurations. Hence the (-2)-configurations on S are the resolutions of the singular points of $\Phi_m(S)$ and so by a theorem of Mumford they must have a negative definite intersection matrix.

(2.7) Proposition. *The only* (-2)*-configurations that can occur on an algebraic surface of general type are those whose dual graph is one of the following Dynkin diagrams.*

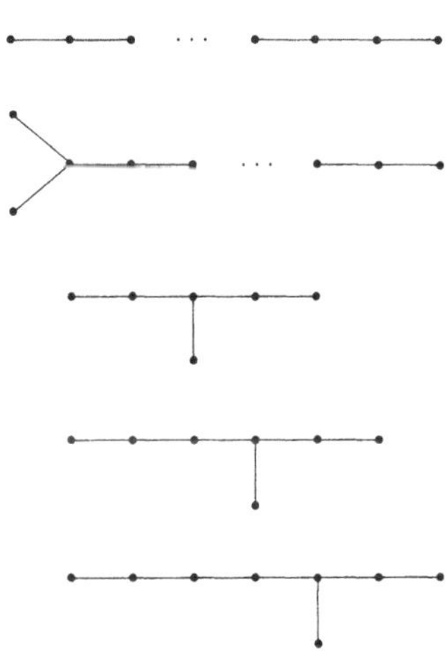

A_k

D_k

E_6

E_7

E_8

Proof. If S is minimal then the proposition is clear since these are the only configurations with negative definite intersection matrix. If we blow up a point of a curve C on a surface S then $\mathrm{K}\cdot C$ increases. Hence, if a new (-2)-curve arises by blowing up a point then this curve must have been an exceptional curve (for $\mathrm{K}\cdot C$ was -1). By Lemma (2.5) such an exceptional curve can only intersect an A_k-configuration and intersects no other exceptional curve. Blowing up a point of this exceptional curve leads only to configurations of type A_k. □

In his study of elliptic surfaces Kodaira made a list of the possible fibres of an elliptic surface. Such a fibre is a divisor D satisfying $D\cdot D = \mathrm{K}\cdot D = 0$.

Kodaira's list of elliptic fibres

Type

1. non-singular elliptic curve C with $C^2 = 0$

2. rational curve C with one ordinary cusp and $C^2 = 0$

3. two (-2)-curves intersecting with multiplicity 2 in one point.

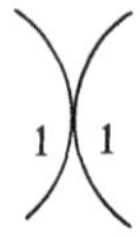

4. three (-2)-curves intersecting transversally in one point

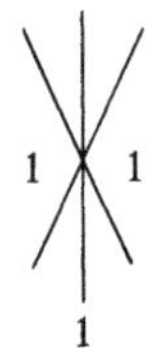

5. a cycle of $l \geqslant 2$ (-2)-curves or one rational curve C with one ordinary double point and $C^2 = 0$

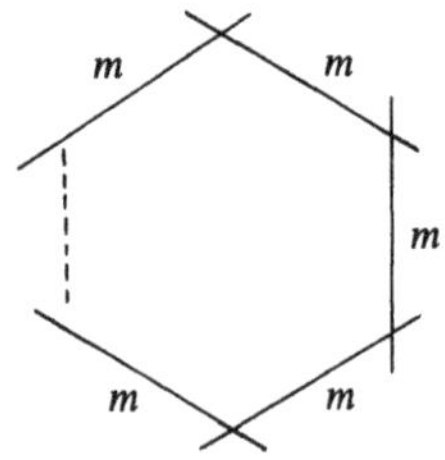

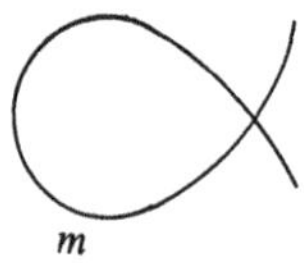

6. the (-2)-configuration E_6'

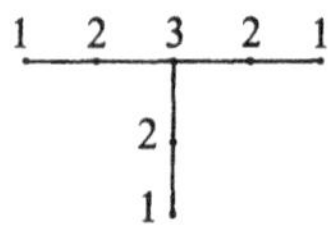

7. the (-2)-configuration E_7'

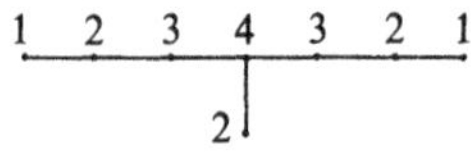

8. the (-2)-configuration E_8'

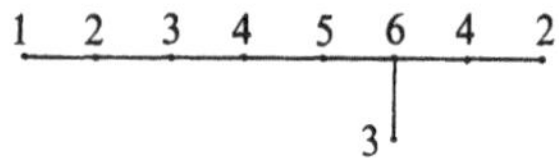

9. the (-2)-configuration D_{k-1}' ($k \geqslant 5$ curves)

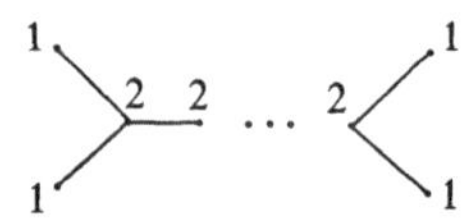

The multiplicities of the components of the fibres are given in the diagrams

(2.8) Definition. By an *elliptic configuration* on a surface S we mean a finite set of curves on S, the union of which is set-theoretically an elliptic fibre.

Let $\varphi: S \to B$ be an elliptic fibring of the algebraic surface S. Suppose that S is minimal. Then according to Kodaira the canonical class of S is given by

$$\mathrm{K}_S = \varphi^*(\mathbf{k}) + \sum (m_F - 1)F,$$

where the sum is over all fibres F with multiplicity (= g.c.d. of multiplicity of the components) $m_F \geqslant 2$ and where $\mathbf{k}$ is a divisor class on B of degree $p_g(S) + q(S) - 1$.

(2.9) Proposition. *Let S be a simply-connected non-rational algebraic surface. If S contains an elliptic configuration $\mathscr{C}$ then S is a blown up $K3$-surface or a blown up honestly elliptic surface. Moreover, if S also contains a (-2)-curve $D \notin \mathscr{C}$ such that $\mathscr{C} \cup D$ is connected, then S is a blown up (elliptic) $K3$-surface.*

Proof. If $\mathscr{C} = \{C_1, \ldots, C_m\}$ then $\mathrm{K} \cdot C_i = 0$ by the adjunction formula and Kodaira's list. If S is of general type this implies that $\mathscr{C}$ is a (-2)-configuration of the type mentioned in (2.7). However, none of these is an elliptic configuration. So S is either a blown up $K3$-surface or a blown up honestly elliptic surface. (It is not an Enriques surface since S is simply-connected.) Now assume that S also contains a (-2)-curve D which meets $\mathscr{C}$. From Kodaira's list it follows that $\mathscr{C} \cup D$ is not an elliptic configuration and is not of type A_k, so by Lemma (2.5) no exceptional curve intersects it. We thus may assume that S is minimal. Suppose that S admits an elliptic fibration $\phi: S \to \mathbb{P}^1$. At least one of the irreducible components of $\mathscr{C} \cup D$, say G, is not contained in a fibre of ϕ. The formula for the canonical class (K is a sum of fibres with non-negative coefficients) gives

$$0 = \mathrm{K} \cdot G = \sum a_i F_i \cdot G ,$$

where the F_i are fibres of ϕ. Since $F_i G > 0$ this implies $a_i = 0$ i.e. $\mathrm{K} = 0$ and S is a $K3$ surface. □

We did not prove that S is an *elliptic* $K3$-surface. Let us sketch how to do this. Assign to the curves of $\mathscr{C}$ their multiplicities such that $\mathscr{C}$ is a elliptic fibre from Kodaira's list. By the Riemann-Roch theorem one finds $h^0(S, \mathcal{O}(\mathscr{C})) \geqslant 2$. The linear system $|\mathscr{C}|$ does not have fixed components, for if we write $|\mathscr{C}| = F + |D|$ with F the fixed component, and $F \neq \varnothing$ than $D^2 < 0$ as follows from the shape of the intersection matrices of the elliptic fibres. Since $\mathscr{C}^2 = 0$ it also does not have isolated base points. Hence $|\mathscr{C}|$ contains a 1-dimensional linear system containing $\mathscr{C}$ giving a holomorphic map $S \to \mathbb{P}^1$ whose fibres are connected by Stein factorization and the fact that $\mathscr{C}$ is reduced and connected. By Bertini's theorem all but a finite number of these fibres are irreducible and non-singular, and by the adjunction formula their genus equals 1.

3. Classification Theorems

We give the classification of Hilbert modular surfaces Y_Γ with $\Gamma = \mathrm{PSL}(\mathfrak{o} \oplus \mathfrak{a})$ or a principal congruence subgroup of it.

The isomorphism class of $Y_{\Gamma(\mathfrak{o}\oplus\mathfrak{a})}$ only depends on the genus of $\mathfrak{a}$. If D is not a prime then there are at least two genera. We recall the notation for these genera:

(3.1) Notation. *We write $q_1, \ldots, q_t$ for the primes dividing D in increasing order (so $D = q_1^n q_2 \ldots q_t$ with $n = 1$ if $q_1 \neq 2$, $n = 2$ or 3 if $q_1 = 2$). We denote a genus γ by the t-tuple $(\pm, \ldots, \pm)$, where the i-th sign is that of* $(\mathrm{N}(\mathfrak{a}), D)_{q_i}$.

So the principal genus corresponds to $(+, \ldots, +)$. If $\mathfrak{a}$ belongs to the genus γ in the real quadratic field of discriminant D then the (isomorphism class of) $Y_{\mathrm{PSL}(\mathfrak{o}\oplus\mathfrak{a})}$ is denoted by $Y_\gamma(D)$.

In [82] the classification was given for D a prime discriminant. This was extended to arbitrary discriminants and $\gamma =$ the principal genus in [86]. There also the classification of the surfaces $\mathrm{PSL}(2, \mathfrak{o})\backslash\mathfrak{H} \times \mathfrak{H}_-$ with $\mathfrak{H}_- = \{z \in \mathbb{C}\colon \mathrm{Im}\, z < 0\}$

was given. By the map $\mathfrak{H} \times \mathfrak{H} \to \mathfrak{H} \times \mathfrak{H}_-$ given by $(z_1, z_2) \to (z_1/\sqrt{D}, -z_2/\sqrt{D})$ one finds an isomorphism $\mathrm{PSL}(2, \mathfrak{o})\backslash \mathfrak{H} \times \mathfrak{H}_- \to \Gamma(\mathfrak{o} \oplus (\sqrt{D}))\backslash \mathfrak{H}^2$. The genus of $(\sqrt{D})$ is given by

$$\gamma = (\ldots, \mathrm{sign}(-1, D)_{q_i}, \ldots).$$

In particular one has:

(3.2) Lemma. *The ideal $(\sqrt{D})$ lies in the principal genus if and only if K contains an element of norm -1.*

The condition in the lemma is equivalent with: the discriminant D is a sum of two squares (in $\mathbb{Z}$) or with: D has no prime factor $\equiv 3 \pmod 4$, cf. [15].

The classification is as follows:

(3.3) Theorem. *The Hilbert modular surface $Y_\gamma(D)$ is of general type except in the following cases:*

It is rational for

$D = 5, 8, 12, 13, 17, 21, 24, 28, 33, 60$	$\gamma =$ *principal genus*
$D = 12$	$\gamma = (--)$.

It is a blown up K3 surface for

$D = 29, 37, 40, 41, 44, 56, 57, 69, 105$	$\gamma =$ *principal genus*
$D = 21, 24, 28, 33, 40$	$\gamma = (--)$.

It is a blown up honestly elliptic surface for

$D = 53, 61, 65, 73, 76, 77, 85, 88, 92, 93, 120, 140, 165$	$\gamma =$ *principal genus*
$D = 44, 57, 65$	$\gamma = (--)$
$D = 60$	$\gamma = (+--)$ *or* $(-+-)$
$D = 105$	$\gamma = (--+)$.

Besides the groups $\Gamma(\mathfrak{o} \oplus \mathfrak{a})$ we consider the principal congruence subgroups. Let $\Gamma[\mathfrak{c}]$ be the image in $\Gamma(\mathfrak{o} \oplus \mathfrak{a})$ of the group

$$\left\{\begin{pmatrix} \alpha & \beta \\ \gamma & \delta \end{pmatrix} \in \mathrm{SL}(\mathfrak{o} \oplus \mathfrak{a}) : \alpha \equiv \delta \equiv 1 \pmod{\mathfrak{c}}, \gamma \in \mathfrak{c}\mathfrak{a}, \beta \in \mathfrak{c}\mathfrak{a}^{-1}\right\},$$

where $\mathfrak{c}$ is an ideal of $\mathfrak{o}$. It is called the principal congruence subgroup of $\Gamma(\mathfrak{o} \oplus \mathfrak{a})$ of level $\mathfrak{c}$, see IV.5. We denote the isomorphism class of the surface $Y_{\Gamma[\mathfrak{c}]}$ by $Y_\gamma(D, \mathfrak{c})$ with γ the genus of $\mathfrak{a}$.

(3.4) Theorem. *The Hilbert modular surfaces $Y_\gamma(D, \mathfrak{c})$ with $\mathfrak{c} \neq \mathfrak{o}$ are of general type except in the following cases:*

D	$\mathrm{N}(\mathfrak{c})$	γ	type of $Y_\gamma(D, \mathfrak{c})$
5	4	$+$	rational
	5	$+$	rational
8	2	$+$	rational
	4	$+$	blown up $K3$
12	2	$++$	blown up $K3$
	2	$--$	rational
	3	$++$	elliptic
	3	$--$	rational
	4	$++$	blown up $K3$
	6	$++$	elliptic
13	3	$+$	rational
17	2	$+$	rational
21	3	$--$	blown up $K3$
24	2	$--$	blown up $K3$
	3	$++$	elliptic
28	2	$++$	blown up $K3$
	3	$--$	blown up $K3$
33	2	$--$	blown up $K3$

These two theorems illustrate clearly the fact that if the volume of $\Gamma\backslash\mathfrak{H}^2$ becomes big the two classes $c_1(\Omega^2_{Y_\Gamma})$ and $c_1(\Omega^2_{Y_\Gamma}(\log D))$ differ little and the surfaces are of general type with respect to both classes.

We shall prove (3.3) in Section 5 and 6. Theorem (3.4) can be proved similarly. With the same methods one can classify other Hilbert modular surfaces, cf. [11].

4. Exceptional Curves on Hilbert Modular Surfaces

The results of the preceding sections show that it is important to know the exceptional curves on a Hilbert modular surface—or even better—a way of determining a minimal model. None of the curves in the resolutions of the singularities is exceptional, so the first candidates are the components of the F_N. In order to check them we need to know their genus and their intersection number with the canonical class.

The first Chern class of a Hilbert modular surface was given in Proposition (2.8) of Chapter IV. Let F_B be a component of F_N. It is the image under a birational map $\Gamma_B\backslash\mathfrak{H} \to Y$ induced by a volume preserving map $\mathfrak{H} \to \mathfrak{H}^2$. Here Γ_B is a discrete subgroup of $\mathrm{PSL}(2, \mathbb{R})$. As a corollary we have:

(4.1) Corollary. *The intersection number of F_B with the cohomology class c_1 $(= -\mathrm{K}_{Y_\Gamma})$ satisfies*

$$c_1 \cdot F_B = 2 \int \omega + \sum_x Z_x \cdot F_B ,$$

where $\omega = -\dfrac{1}{2\pi}\dfrac{dx \wedge dy}{y^2}$ *is integrated over* $\Gamma_B \backslash \mathfrak{H}$ *and where the sum is over the singularities of* X_Γ *and* Z_x *is the local Chern cycle of* x *(see* IV.2).

Let Δ be a discrete subgroup of $\mathrm{PSL}(2, \mathbb{R})$ such that $\Delta \backslash \mathfrak{H}$ has finite volume. Define an integer $c_1(\Delta)$ by

$$c_1(\Delta) = 2e(\overline{\Delta \backslash \mathfrak{H}}) - \sum a_r(\Delta) - \sigma(\Delta) . \tag{1}$$

Here $a_r(\Delta)$ is the number of Δ-inequivalent elliptic fixed points of order r, $\sigma(\Delta)$ is the number of cusps and $e(\overline{\Delta \backslash \mathfrak{H}})$ is the Euler number of $\overline{\Delta \backslash \mathfrak{H}}$.

(4.2) Lemma. *Let* F_B *be a component of a curve* F_N *on* Y_Γ. *If the quotient singularities on* F_B *are all of type* $(r, 1, 1)$, $r \geqslant 2$ *then*

$$c_1[F_B] \geqslant c_1(\Gamma_B) .$$

Proof. Every cusp of $\Gamma_B \backslash \mathfrak{H}$ maps to a point on the cusp resolutions and thus contributes at least 1 to $c_1[F_B]$. If γ is an element of order r in Δ generating the isotropy group of $z \in \mathfrak{H}$ in Δ then z maps to the unique fixed-point of $\gamma (\in \Gamma)$ in $\mathfrak{H}^2$, which is a quotient singularity of type $(r, 1, 1)$. The Chern cycle equals $\dfrac{r-2}{r}$ times the only resolution curve and contributes at least $\dfrac{r-2}{r}$ to $c_1[F_B]$. Since all contributions $Z_x \cdot F_B$ in (4.1) are non-negative we get

$$c_1[F_B] \geqslant 2 \int_{\Gamma_B \backslash \mathfrak{H}} \omega + \sum_r a_r(\Gamma_B) \frac{r-2}{r} + \sigma(\Gamma_B) .$$

On the other hand,

$$e(\overline{\Gamma_B \backslash \mathfrak{H}}) = \int_{\Gamma_B \backslash \mathfrak{H}} \omega + \sum_r a_r(\Gamma_B) \frac{r-1}{r} + \sigma(\Gamma_B) ,$$

which together with the inequality above gives the required result. □

We now look at a number of curves F_N on $Y_\gamma(D)$. Let $t = \#\{p : p | D, p \text{ prime}\}$. We shall assume that $(\mathrm{N}(\mathfrak{a}), D) = 1$ which we can achieve by altering $\mathfrak{a}$ inside its genus. As before we set $A = \mathrm{N}(\mathfrak{a})$.

N = 1. F_1 occurs only if γ is the principal genus and in this case it has 2^{t-1} components which are permuted transitively by the Hurwitz-Maass group. Each component is isomorphic to $\overline{\mathrm{PSL}(2, \mathbb{Z}) \backslash \mathfrak{H}}$ and intersects the cusp resolutions transversally in a simple point of the cycles. Since $a_2 = a_3 = 1$ we get 2^{t-1} configurations of $\mathbb{P}^1$'s

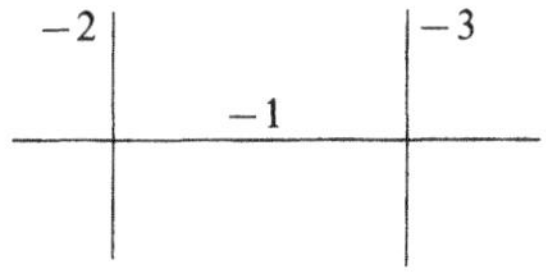

(2)

which are disjoint by the results of Chapter V. (We find $c_1[F_A] = 1$ for each component, hence every component is exceptional.)

N = 2. F_2 occurs only if 2 is the norm of an ideal from the genus γ. Then if $D \neq 8$ each of the

$$\frac{1}{2} \prod_{\substack{q|D \\ q \text{ prime}}} (1 + \chi_{D(q)}(2A))$$

components of F_2 is isomorphic to $\overline{\Gamma_0(2)\backslash\mathfrak{H}}$ and intersects the resolution cycles of the cusps in two points (which are not double points of the cycles) and passes through an order 2 quotient singularity. Each component is exceptional and gives rise to a configuration

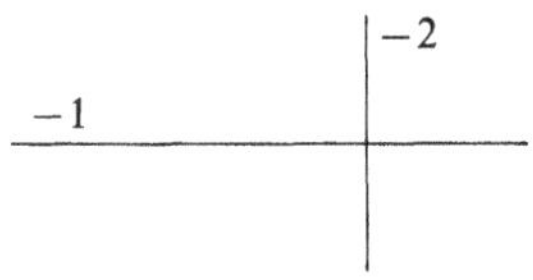

(3)

These configurations are disjoint and permuted transitively by the Hurwitz-Maass group.

N = 3. F_3 occurs only if 3 is the norm of an ideal in the genus γ. If $D \neq 12$ all of the

$$\frac{1}{2} \prod_{\substack{q|D \\ q \text{ prime}}} (1 + \chi_{D(q)}(3A))$$

components are isomorphic to $\overline{\Gamma_0(3)\backslash\mathfrak{H}}$ and intersects the resolution cycles transversally in two simple points. Since $a_3 = 1$ we find $c_1[F] = 1$ for every component of F_3.

N = 4. F_4 occurs if γ is the principal genus. If D is odd then F_4 has 2^{t-1} components, while if D is even, $D \neq 8$, F_4 has 2^{t-1} components. In both cases the Hurwitz-Maass group acts transitively on the set of components. If $\left(\frac{D}{2}\right) = +1$ then each component is isomorphic to $\overline{\Gamma_0(4)\backslash\mathfrak{H}}$, whereas if $\left(\frac{D}{2}\right) = -1$ we must replace $\Gamma_0(4)$ by a normal extension of $\Gamma_0(4)$ of index 3 generated by $\begin{pmatrix} 1 & -\frac{1}{2} \\ 2 & 0 \end{pmatrix}$. We then have $a_3 = 2$, $a_r = 0$

if $r \neq 3$ and $\sigma = 1$. F_4 always intersects the cusp resolutions transversally in simple points. If $2|D$, $D \neq 8$ then all components are isomorphic to $\overline{\Gamma_0(2)\backslash\mathfrak{H}}$. In this case we get 2^{t-2} configurations (3).

N = 9. F_9 occurs only if γ is the principal genus. If $3|D$, $D \neq 12$ then F_9 has 2^{t-1} components each isomorphic to $\overline{\Gamma_0(3)\backslash\mathfrak{H}}$. Since $a_3 \neq 1$ we have $c_1[F] \geqslant 1$ for every component F of F_9 in this case. We thus find 2^{t-1} times a configuration

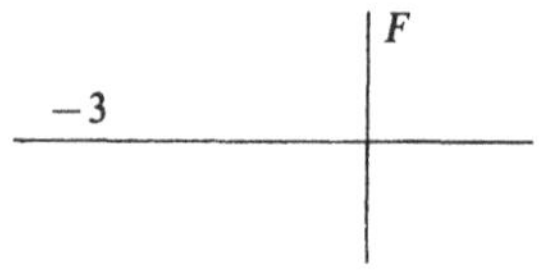

which are disjoint except for $D = 105$. There we get twice

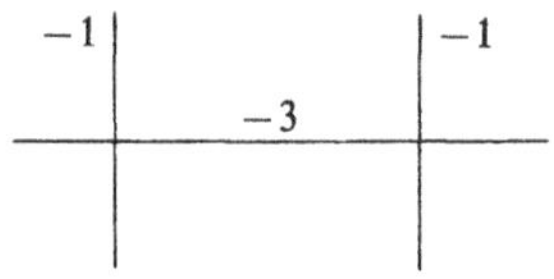

(since $4 \cdot 9 \cdot 9 - 3^2 = 3 \cdot 105$, cf. V.6.)

Suppose that Y is a regular surface which is *not* rational. Then if $\mathscr{C}$ is an irreducible curve on Y the inequality

$$c_1[\mathscr{C}] \geqslant 1$$

implies $c_1[\mathscr{C}] = 1$ and $p_a(\mathscr{C}) = 1$, i.e. $\mathscr{C}$ is non-singular and rational and hence an exceptional curve. Indeed, $c_1[\mathscr{C}] \geqslant 2$ would imply by the adjunction formula

$$\mathscr{C}^2 \geqslant c_1[\mathscr{C}] + 2p_a(\mathscr{C}) - 2 \geqslant 0\,,$$

hence by the rationality criterion (2.2) that Y is rational.

If Y_Γ is not rational (as follows from (3.3), this implies $D \neq 8, 12$) then intersecting exceptional curves do not occur on Y_Γ or on a surface obtained from Y_Γ by blowing down exceptional curves. Hence F_1, if present, gives rise to 3.2^{t-1} contractions of exceptional curves (cf. (2)) and F_4 gives rise to 2^{t-1} contractions, together 2^{t+1} contractions. Similarly, all components of F_2 give rise to 2 contractions, and every component of F_3 gives one contraction. Finally, F_9 if $3|D$, $D \neq 12$ gives rise to 2^{t-1} contractions if $\mathfrak{a}$ belongs to the principal genus. Moreover, if $D = 105$ after blowing down F_9 we can also blow down the images of the two (-3)-curves intersected twice by F_9. We thus obtain.

(4.3) Theorem. *Let $Y = Y_\Gamma$ with $\Gamma = \mathrm{PSL}(\mathfrak{o} \oplus \mathfrak{a})$ and $(A, D) = 1$. Suppose that Y is not a rational surface. Then we can blow down successively*

$$s = 2\prod_q (1 + \chi_{D(q)}(A)) + \prod_q (1 + \chi_{D(q)}(2A)) + \frac{1}{2}\prod_q (1 + \chi_{D(q)}(3A))$$

$$+ \left(1 - \left(\frac{D}{3}\right)^2\right)\prod_q (1 + \chi_{D(q)}(9A)) + \begin{cases} 2 & \text{if } D = 105, \mathfrak{a} \text{ in the principal genus} \\ 0 & \text{otherwise} \end{cases}$$

exceptional curves. (*The product is taken over all primes dividing D.*)

If Y_Γ is not rational we let Y_Γ° (or $Y_\gamma^\circ(D)$) be the surface obtained from Y_Γ by contracting these s exceptional curves.

(4.4) Conjecture. *Y_Γ° is the minimal model of Y_Γ.*

This conjecture—originally made by Hirzebruch and Van de Ven [82], see also [86]—can be checked in many cases, in particular, in all cases where Y_Γ is not of general type. Section 7 is devoted to this conjecture.

Suppose that $Y_\gamma(D)$ is not rational. The involution τ induces an involution τ° on the surface $Y_\gamma^\circ(D)$. We claim:

(4.5) Proposition. *The involution τ° acts without isolated fixed points.*

Proof. The isolated fixed points of τ on $Y_\gamma(D)$ occur in the resolutions of isolated fixed points of τ on $\Gamma\backslash\mathfrak{H}^2$, in the resolutions of elliptic fixed points on F_1 and F_3 and in the resolutions of the cusps. Isolated fixed points of τ on $\Gamma\backslash\mathfrak{H}^2$ can be read off from the tables [63]. They are the order 2 elliptic fixed points on F_2. They disappear in the transition from $Y_\gamma(D)$ to $Y_\gamma^\circ(D)$ since the image of F_2 on $Y_\gamma^\circ(D)$ intersects the 1-dimensional fixed points locus of τ° (the image of F_D or of $F_D \cup F_{D/4}$). Fixed points lying on the resolutions of elliptic fixed points on F_1 and F_3 also map to the 1-dimensional fixed point set of τ°. Isolated fixed points lying in the cusp resolutions lie on F_4 (cf. V.9). So also they do not lead to isolated fixed points on $Y_\gamma^\circ(D)$. Finally, the images of blown down curves all map to the 1-dimensional fixed point set of τ°. □

5. Estimates for the Numerical Invariants

If S is an algebraic surface with $q = 0$, $\chi > 1$ and $c_1^2 > 0$ then S is a surface of general type. The classification of Hilbert modular surfaces will be carried out by showing that for almost all Hilbert modular surfaces $Y_\gamma(D)$ we have $\chi > 1$ and $c_1^2 > 0$ and by treating the remaining cases separately.

(5.1) Proposition. *For $D > 500$ and all γ one has $c_1^2(Y_\gamma(D)) > 0$, $\chi(Y_\gamma(D)) > 1$.*

Proof. By (3.3) and (3.4) of Chapter IV it suffices to prove this for the principal genus. We have (cf. IV (2.5))

$$c_1^2(Y_\gamma(D)) = 4\zeta_K(-1) - \tfrac{1}{3}a_3^+ - \ell^- ,$$
$$c_2(Y_\gamma(D)) = 2\zeta_K(-1) + \tfrac{3}{2}a_2 + \tfrac{5}{3}a_3^+ + \tfrac{8}{3}a_3^- + \ell ,$$
$$12\chi(Y_\gamma(D)) = c_1^2(Y_\gamma(D)) + c_2(Y_\gamma(D)) ,$$

where $\ell = \sum \ell(M_\sigma)$, $\ell^- = \sum \ell(M_\sigma^\vee)$ and a_2, a_3^+, a_3^- are the numbers of quotient singularities of type (2; 1, 1), (3; 1, 1) and (3; 1, −1). (Note that these are the only ones that occur since $D > 500$.)

The dominant term in the formulas is $\zeta_K(-1)$ for which we have the estimate

$$\zeta_K(-1) > \frac{D^{3/2}}{360} \tag{1}$$

since $\zeta_K(-1) = \dfrac{D^{3/2}}{4\pi^4}\zeta_K(2)$ by the functional equation and the estimate

$$\zeta_K(2) = \sum_{\mathfrak{a}} \mathsf{N}(\mathfrak{a})^{-2} \geqslant \sum_{n=1}^{\infty} n^{-4} = \zeta(4) = \frac{\pi^4}{90} ,$$

where $\mathfrak{a}$ runs over the non-zero integral ideals. Hence we get

$$c_2(Y_\gamma(D)) > D^{3/2}/180 > 12 ,$$

so it suffices to show that $c_1^2 > 0$. For this we must combine (1) with estimates for a_3^+ and ℓ^- from above. It will be convenient to separate the cases D odd and D even and to notice that in the latter case (1) can be strengthened to

$$\zeta_K(-1) > \frac{D^{3/2}}{288} \qquad (D \equiv 0 \ (\mathrm{mod}\ 4)) \tag{1'}$$

This can be proved using the Euler product

$$\zeta_K(2) = \zeta(2) \prod_p \left(1 - \left(\frac{D}{p}\right)p^{-2}\right)^{-1} > \zeta(2) \prod_{p\neq 2} (1 + p^{-2})^{-1} = \frac{5}{4}\zeta(4) .$$

The number a_3^+ is given in terms of the class number of $\mathbb{Q}(\sqrt{-3D})$ (see I.5), so we first need an estimate for the class numbers of imaginary quadratic number fields.

(5.2) Lemma. *If $-\Delta < -4$ is a fundamental discriminant then* $h(-\Delta) \leqslant \dfrac{\sqrt{\Delta}}{\pi}\log \Delta$.

Remark. A slightly stronger statement than this was quoted incorrectly from the literature in [86], where also various other aspects of the estimates of c_1^2 were done in an unnecessarily complicated way; in penance for this, D. Zagier has written the present section.

Proof. Let $\chi(n) = \left(\dfrac{-\Delta}{n}\right)$ be the Dirichlet character associated to $-\Delta$. Then $h(-\Delta) = \dfrac{\sqrt{\Delta}}{\pi} L(1,\chi)$ and we must estimate $L(1,\chi)$. By partial summation

$$L(1,\chi)=\sum_{n=1}^{\infty}\chi(n)/n=\sum_{n=1}^{\infty}(s(n)-s(n-1))/n=\sum_{n=1}^{\infty}s(n)/n(n+1)\,,$$

where $s(n)=\chi(1)+\cdots+\chi(n)$. Since $s(n)-s(n-1)\leqslant 1$ for all n and $s(n)=0$ for $n\equiv 0 \pmod{\Delta}$, we have $s(n)\leqslant s^*(n):=\min(n,\delta)$ with $\delta=[\Delta/2]$. So using partial summation again we find

$$L(1,\chi)\leqslant\sum_{n=1}^{\infty}\frac{s^*(n)}{n(n+1)}=\sum_{n=1}^{\infty}(s^*(n)-s^*(n-1))/n=\sum_{n=1}^{\delta}\frac{1}{n}<\log\Delta \qquad (\Delta\geqslant 5)\,. \quad \square$$

Using this we find the estimate

$$\frac{1}{3}a_3^+\leqslant\frac{4}{3\pi}\sqrt{D/3}\log D/3<0.2451\sqrt{D}\log D\,. \tag{2}$$

Finally we must estimate ℓ^-. Since the kernel of $\mathrm{Sq}^+\colon \mathscr{C}\ell^+\to\mathscr{C}\ell^+$ has order 2^{t-1}, one has

$$\ell^-=\sum_{A\in\mathscr{C}\ell^+}\ell((\mathrm{Sq}^+(A)^\vee)=2^{t-1}\sum_{B\in\operatorname{Im}\mathrm{Sq}^+}\ell(B^\vee)\leqslant 2^{t-1}\ell(D)\,,$$

where

$$\ell(D)=\sum_{A\in\mathscr{C}\ell^+}\ell(A)=\frac{1}{2}\sum_{|x|<\sqrt{D}}\sigma_0\left(\frac{D-x^2}{4}\right)$$

(cf. IV.3(3)), with the usual convention that $\sigma_0(n)=0$ if n is not an integer.

(5.3) Lemma. *Let D be the discriminant of a real quadratic field. Then*

$$\sum_x\sigma_0\left(\frac{D-x^2}{4}\right)\leqslant\begin{cases}D^{1/2}\left(\dfrac{3}{2\pi^2}\log^2 D+1.05\log D\right) & \text{all } D\\[2ex] D^{1/2}\left(\dfrac{3}{4\pi^2}\log^2 D+0.8\log D\right) & D\equiv 0 \pmod 4\,.\end{cases}$$

Proof. We have

$$\sum_x\sigma_0\left(\frac{D-x^2}{4}\right)\leqslant 2\sum_{\substack{D=x^2+4mn\\ 1\leqslant n\leqslant m}}1\leqslant 2\sum_{1\leqslant n<\sqrt{D/4}}\ \sum_{\substack{|x|<\sqrt{D-4n^2}\\ x^2\equiv D \pmod{4n}}}1\,.$$

For each $n\geqslant 1$ we let

$$b(n)=\#\{x \pmod{2n}\colon x^2\equiv D \pmod{4n}\}\,.$$

One easily checks that $b(n)$ is a multiplicative function with $b(p^a)\leqslant 2$ for all prime powers (it is the coefficient of n^{-s} in $\zeta_K(s)/\zeta(2s)$), so

$$b(n)\leqslant\sigma_0^*(n):=\#\left\{d\colon d|n,\left(d,\frac{d}{n}\right)=1\right\}\,.$$

Hence

$$\sum_x\sigma_0\left(\frac{D-x^2}{4}\right)\leqslant 2\sum_{1\leqslant n<\sqrt{D/4}}b(n)\left[\frac{\sqrt{D-4n^2}}{n}+1\right]\leqslant 2F(\sqrt{D/4})$$

with

$$F(x) = \sum_{1 \leqslant n \leqslant x} \sigma_0^*(n) \left(\frac{2\sqrt{x^2 - n^2}}{n} + 1 \right) \qquad (x > 0) .$$

Standard methods of analytic number theory give

$$F(x) = \frac{6}{\pi^2} x \log x + O(x \log x) ,$$

and some work with a table computer (facilitated by the fact that $F(x)$ is monotone increasing) gives the explicit bound

$$F(x) \leqslant \frac{6}{\pi^2} x \log^2(2x) + 2.1 x \log(2x) \qquad (x > 125) .$$

This proves the first bound. The second goes similarly if we note that $b(2n) \leqslant \sigma_0^*(n)$, $b(4n) = 0$ when $D \equiv 0 \pmod 4$. □

Combining (5.3) with (2) and (1) or (1′), we see that c_1^2 is certainly positive if

$$\begin{aligned} \frac{D}{72} &\geqslant 2^{t-2}\left(\frac{3}{4\pi^2}\log^2 D + \log D\right) + 0.2451 \log D \qquad (D \equiv 0 \pmod 4)) \\ \frac{D}{90} &\geqslant 2^{t-2}\left(\frac{3}{2\pi^2}\log^2 D + \log D\right) + 0.2451 \log D \qquad (D \equiv 1 \pmod 4)) \end{aligned} \qquad (3)$$

These equalities are satisfied for $D \geqslant D_0(t)$ and $D \geqslant D_1(t)$, resp., where the first few values of $D_0(t)$ and $D_1(t)$ are given by

t	1	2	3	4	5
D_0	<500		895	1875	3992
D_1			1555	3268	6971

But the smallest discriminant which is even and has 5 prime factors is $4 \cdot 3 \cdot 5 \cdot 7 \cdot 11 \geqslant D_0(5)$ and the smallest odd one with 5 prime factors is $3 \cdot 5 \cdot 7 \cdot 11 \cdot 19 \geqslant D_1(5)$, so our inequality holds whenever $t = 5$. If $t \geqslant 6$, then it holds even more since increasing $t\ (\geqslant 5)$ by 1 multiplies the minimal discriminant by at least 13 and barely more than doubles the right hand side of (3). Hence it suffices to check (3) for the discriminants D with $t \leqslant 4$ and $D \leqslant D_0(t)$ (D even) or $D \leqslant D_1(t)$ (D odd). This is quickly done on a table computer (the only cases >1875 which must be checked are $3 \cdot 5 \cdot 11 \cdot 13$, $3 \cdot 5 \cdot 11 \cdot 17$ and $3 \cdot 5 \cdot 7 \cdot 29$).

6. Proof of the Classification

We now come to the proofs of Theorem (3.3). To begin with, since $c_1^2 > 0$, $\chi > 1$ by (5.1) for all $Y_\gamma(D)$ with $D > 500$ these surfaces are of general type. For $D < 500$ the values of c_1^2 and χ and if $\chi \neq 1$ for $c_1^2(Y_\gamma^\circ(D))$ are listed in table on p. 269. There are 11 cases where $\chi = 1$.

(6.1) Proposition. *If $\chi = 1$, i.e. if $D = 5, 8, 12, 13, 17, 21, 24, 28, 33, 60$ and γ the principal genus or if $D = 12, \gamma = (--)$ then $Y_\gamma(D)$ is rational.*

We use Criterion (2.2). In eight cases we produce intersecting exceptional curves. For these surfaces we indicate an intersection diagram of non-singular rational curves with their self-intersection numbers. Blowing down exceptional curves in these diagrams leads to intersecting exceptional curves which guarantees rationality.

$D = 5$

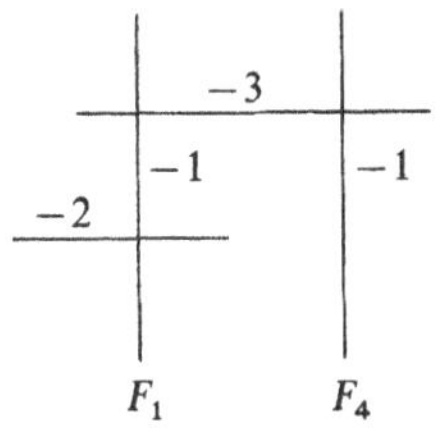

$D = 8$

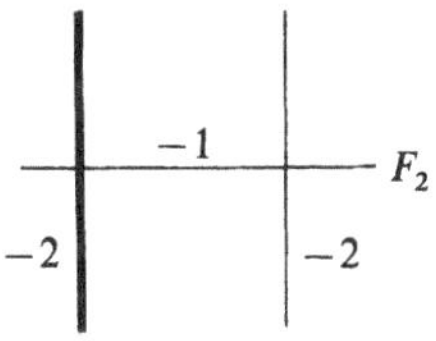

$D = 13$

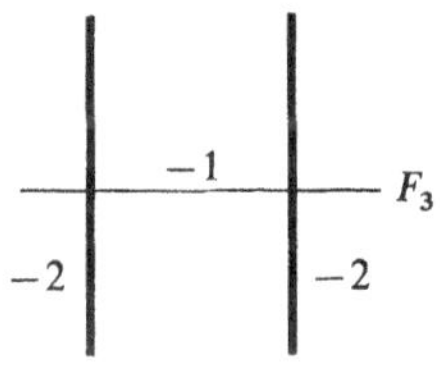

$D = 12, \gamma = (--)$

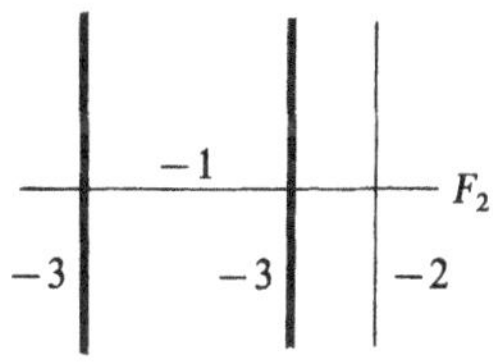

$D = 17$

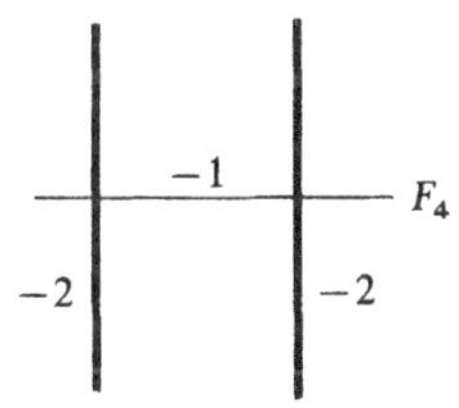

$D = 24, \gamma = (++)$

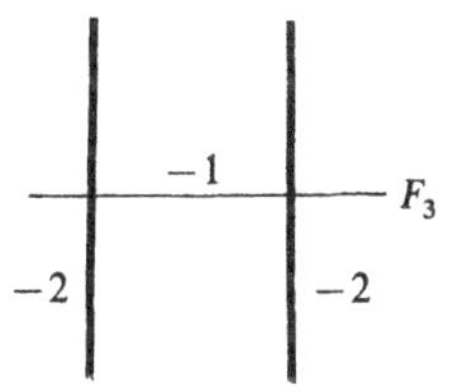

$D = 28, \gamma = (++)$

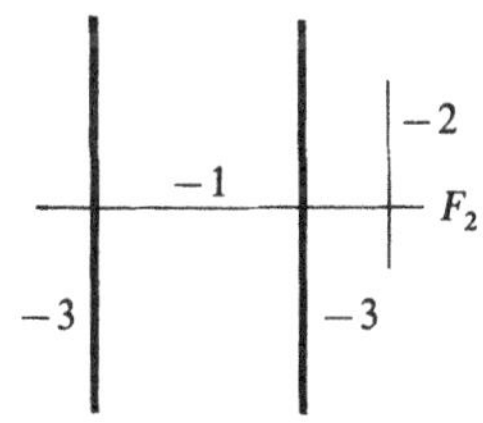

$D = 33, \gamma = (++)$

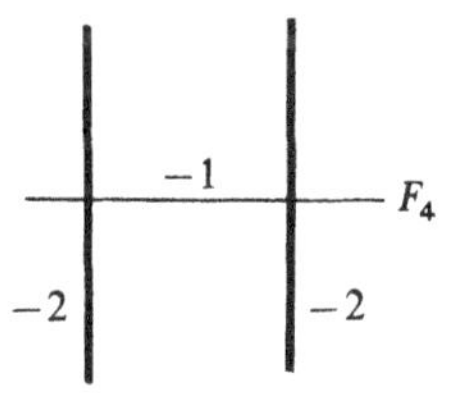

In these figures heavily drawn curves belong to the cusp resolutions

For a change we produce in the remaining two cases an irreducible curve C with $C^2 \geqslant 0$, $KC < 0$, which by (2.2) also suffices. For $D = 12$, $\gamma = (++)$ take a component C of F_{13}; using (4.1) one finds $KC = -2$. For $D = 60$, $\gamma = (++)$ take F_{10} (one component). It also satisfies $K \cdot F_{10} = -2$. By the adjunction formula $C^2 \geqslant 0$ follows. For $D = 21$ one takes F_7. This proves the proposition.

(6.2) Proposition. *If* $\chi = 2$, *i.e.* $D = 29, 37, 40, 41, 44, 56, 57, 69, 105$ *and* $\gamma =$ *principal genus or if* $D = 21, 24, 28, 33, 40$, $\gamma = (--)$, *then* $Y_\gamma^\circ(D)$ *is a* $K3$ *surface.*

Look at the principal cusp (i.e. the one corresponding to $\mathfrak{o}$) for γ equal to the principal genus. If $\mathfrak{o} = \mathbb{Z}1 + \mathbb{Z}w_0$ with w_0 reduced and if $w_0 = b_0 - \cfrac{1}{b_1 - \cdots}$ is the continued fraction then F_1 intersects a curve S_0 in the resolution with self-intersection number $-b_0$.

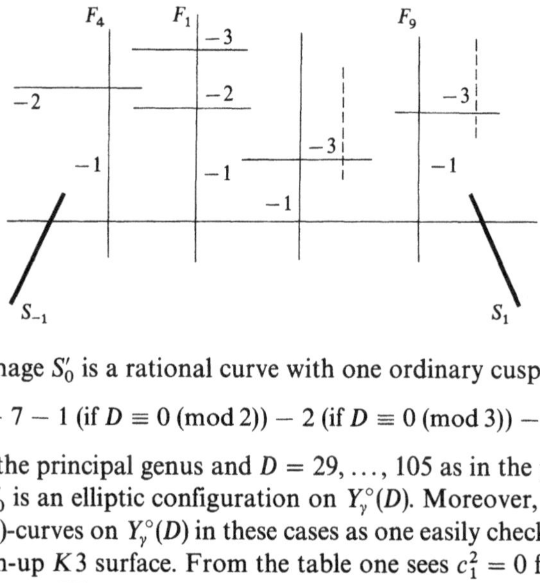

On $Y_\gamma^\circ(D)$ the image S_0' is a rational curve with one ordinary cusp and it satisfies

$$K_{Y^\circ} \cdot S_0' = b_0 - 7 - 1 \text{ (if } D \equiv 0 \pmod 2) - 2 \text{ (if } D \equiv 0 \pmod 3) - 2 \text{ (if } D = 105).$$

(cf. (4.3)). If γ is the principal genus and $D = 29, \ldots, 105$ as in the proposition then $K_{Y^\circ} S_0' = 0$, so S_0' is an elliptic configuration on $Y_\gamma^\circ(D)$. Moreover, the images of S_1 and S_{-1} are (-2)-curves on $Y_\gamma^\circ(D)$ in these cases as one easily checks. By (2.9) $Y_\gamma^\circ(D)$ must be a blown-up $K3$ surface. From the table one sees $c_1^2 = 0$ for $Y_\gamma^\circ(D)$ for our γ and D, so these are $K3$.

If $D = 21, 24, 40$, $\gamma = (--)$ we have the following configuration

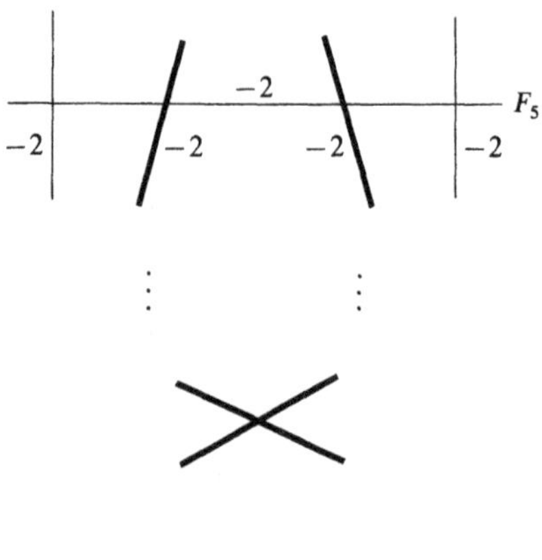

which proves that $Y_\gamma(D)$ must be blown-up $K3$. Checking $c_1^2 = 0$ for $Y_\gamma^\circ(D)$ then suffices.

Finally, for the remaining cases we have the configurations

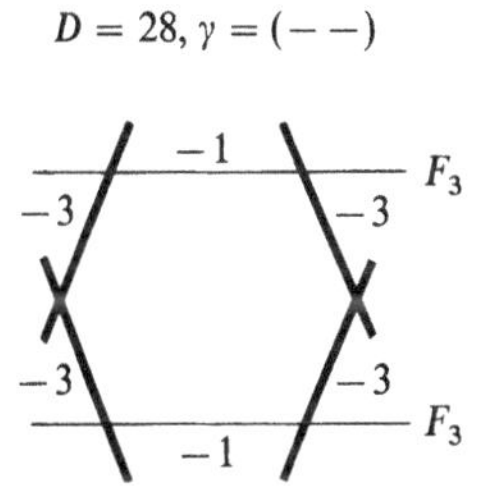

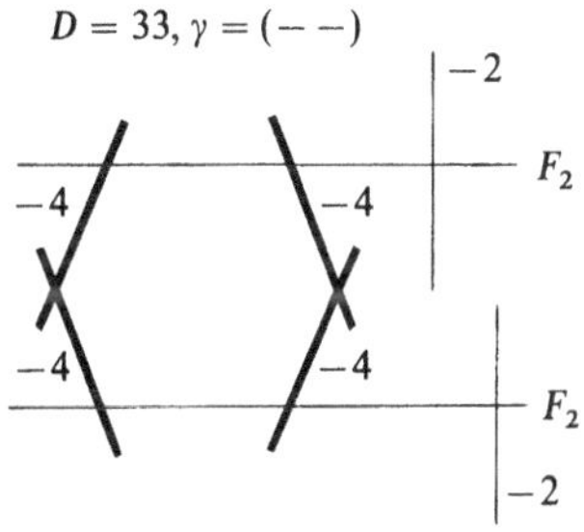

Blowing down leads to (-2)-configurations which imply (together with $c_1^2 = 0$) that $Y_\gamma^\circ(D)$ is $K3$ in these cases.

(6.3) Proposition. *The surface $Y_\gamma^\circ(D)$ is a honestly elliptic surface for $D = 53$, 61, 65, 73, 76, 77, 85, 88, 92, 93, 120, 140, 165, $\gamma = (+\cdots+)$, for $D = 44, 57, \gamma = (--)$, for $D = 60$ $\gamma = (-+-)$, $\gamma = (+--)$, for $D = 65$, $\gamma = (--)$ and $D = 105$, $\gamma = (--+)$.*

We use again (2.9). In many cases one curve F_N and a number of (-2)-curves in the cusp resolutions give an elliptic (-2)-configuration. For example we have for $N = 5$, $N = 6$ (if F_6 is not compact)

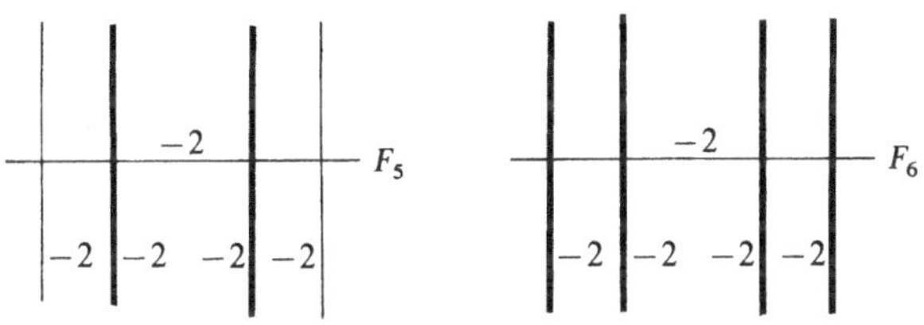

This works for $\gamma =$ principal genus and $D = 53$, 93 (with F_7), $D = 61$, 85, 88 (with F_9) or $D = 73$ (with F_6 after blowing down F_4), and also for $D = 44$, $\gamma = (--)$ and $D = 60, \gamma = (-+-)$ (with F_7), $D = 57, \gamma = (--)$, $D = 105, \gamma = (--+)$ (with F_8), $D = 65$, $\gamma = (--)$ (with F_7 and F_8). After blowing down F_4 on $Y_{++}(65)$ a number of curves coming from the cusp resolution form an elliptic configuration. On $Y_{++}(92)$ we have a configuration

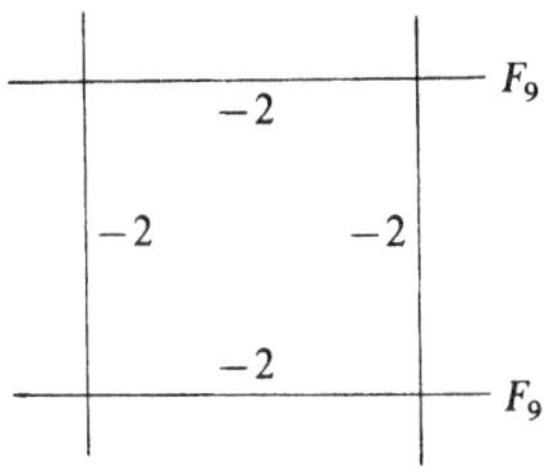

Finally, for $D = 77, D = 140, D = 165, \gamma = (+ \cdots +)$ we find an elliptic curve C with $C^2 = 0$, namely F_{11}, F_{14} and F_{15} resp. This proves (6.3) and hence (3.3), since the estimate of (5.1) and the table on p. 269 show that $c_1^2 > 0, \chi > 1$ in the remaining cases.

7. Canonical Divisors

If Γ is a discrete subgroup of $\mathrm{PSL}(2, \mathbb{R})^2$ such that $\Gamma\backslash\mathfrak{H}^2$ is *compact* and non-singular, then the canonical class $-c_1$ of $\Gamma\backslash\mathfrak{H}^2$ is ample and this prevents $\Gamma\backslash\mathfrak{H}^2$ from having non-singular rational or elliptic curves. This fact can be proved in the same way as the following analogue for Hilbert modular surfaces.

(7.1) Proposition. *Let C be an irreducible curve of arithmetic genus* 0 (*resp.* 1) *on a Hilbert modular surface Y. Then C has at least* 3 (*resp.* 1) *points in common with the resolution of the cusp and quotient singularities.*

Proof. Suppose the genus is zero. Then C is a non-singular rational curve. If C has less than 3 points in common with the resolutions then the universal cover of $C - C \cap (\text{resolutions})$ is $\mathbb{P}^1$, $\mathbb{C}$ or $\mathbb{C}^*$. We thus find a non-constant holomorphic map of the universal cover to $\mathfrak{H}^2$. By composing with the projections $\mathfrak{H}^2 \to \mathfrak{H}$ one gets holomorphic maps from $\mathbb{P}^1$, $\mathbb{C}$ or $\mathbb{C}^*$ to $\mathfrak{H}$, which by Picard's little theorem have to be constant, a contradiction. The genus 1 case is similar. □

If the volume of $\Gamma\backslash\mathfrak{H}^2$ is big then $\Gamma\backslash\mathfrak{H}^2$ is very similar to a compact quotient. Therefore one expects that a Hilbert modular surface in general does not have too many non-singular rational or elliptic curves apart from the rational curves in the resolution of singularities.

Let us start with the rational curves. If S is a regular non-rational algebraic surface and C is a non-singular rational curve on S then $\mathrm{K} \cdot C \geqslant -1$ by (2.2) and the adjunction formula. Moreover, $\mathrm{K} \cdot C = -1$ if and only if C is an exceptional curve.

A conjecture on the number of exceptional curves on $Y_\gamma(D)$ was given in (4.4). A regular surface which is neither rational nor of general type is minimal if and only if $c_1^2(S) = 0$. Using the tables at p. 269 one can easily check:

(7.2) Proposition. *If $Y_\gamma(D)$ is neither rational nor of general type then $Y_\gamma^\circ(D)$ is the minimal model of $Y_\gamma(D)$.*

If $Y_\gamma(D)$ is of general type we do not have such an easy way to check the truth of (4.4). One can construct a minimal model of a surface of general type if one knows an effective canonical divisor on it. An effective canonical divisor D on a surface of general type contains all exceptional curves since $\mathrm{K} \cdot E = D \cdot E = -1 < 0$ for all exceptional curves E. If E is an exceptional curve on S we blow it down to obtain a new surface S' and a new effective canonical divisor D' and S'. By repeating this process we obtain in finitely many steps a minimal model. The following proposition, which emerged in [41, 36, 83], gives a way of constructing such effective canonical divisors.

(7.3) Proposition. *Let S be a compact algebraic surface and L a divisor class on S with* $L \cdot L > 0$. *Suppose that D is a divisor on S with* $D \cdot D = D \cdot L = L \cdot L$ *then* $L - D$ *is zero in the rational homology of S.*

Proof. We have $(D - L) \cdot L = 0$, hence by the algebraic index theorem either $D - L$ is zero (in rational homology) or $(D - L)^2 < 0$. Since we know $(D - L)^2 = 0$, D is homologous to L. □

(7.4) *Remark.* If $H_1(S, \mathbb{Z}) = 0$ then $H_2(S, \mathbb{Z})$ is without torsion, so L is homologous to D. Since $q = 0$ we have the exact sequence

$$0 = H^1(S, \mathcal{O}_S) \to H^1(S, \mathcal{O}_S^*) \to H^2(S, \mathbb{Z}),$$

that is, homology implies linear equivalence. So D is also linearly equivalent to L.

(7.5) *Example.* Let $\Gamma = \Gamma[3]$ be the principal congruence subgroup of Γ_K of level (3) with $K = \mathbb{Q}(\sqrt{5})$. It acts freely on $\mathfrak{H}^2$ and it is of index $360 = \mathrm{PSL}(2, \mathbb{F}_9)$ in Γ_K, so $\mathrm{vol}(\Gamma \backslash \mathfrak{H}^2) = \frac{1}{15} 360 = 24$. The ten cusps are each resolved by a cycle $(-3, -3, -3, -3)$:

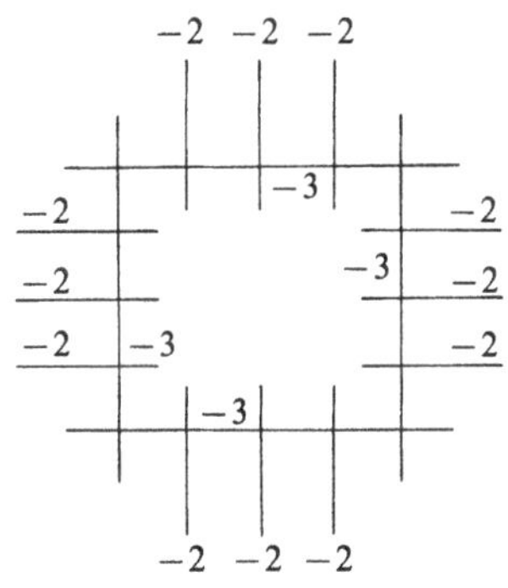

We find $\chi(Y_\Gamma) = 6$, $e(Y_\Gamma) = 24 + 10 \cdot 4 = 64$, hence $p_g = 5$, $c_1^2 = 8$. The curve F_1 has 30 disjoint components ((-2)-curves, each isomorphic to $\overline{\Gamma(3) \backslash \mathfrak{H}}$) and it intersects each resolution curve S_i in three points. Take the divisor with rational coefficients

$$D' = \frac{1}{5}\left(\sum S_j + 2F_1\right).$$

We have $D' \cdot D' = \mathrm{K} \cdot D' = 8$, i.e. $D' = \mathrm{K}$ in $H_2(Y_\Gamma, \mathbb{Q})$. Since $q = 0$ this implies that $D = 2F_1 + \sum S_j$ is linearly equivalent to 5K. The effective 5-canonical divisor D contains no exceptional curves; therefore Y_Γ is a *minimal* surface of general type.

(7.6) *Example.* $Y_{--+}(60)$ is a surface of general type with invariants $\chi = 5$, $c_1^2 = 4$. The divisor $2F_6$ + (cusp resolutions) is a canonical divisor on $Y_{--+}(60)$ containing no exceptional curves. So $Y^{\circ}_{--+}(60) = Y_{--+}(60)$ is a minimal surface.

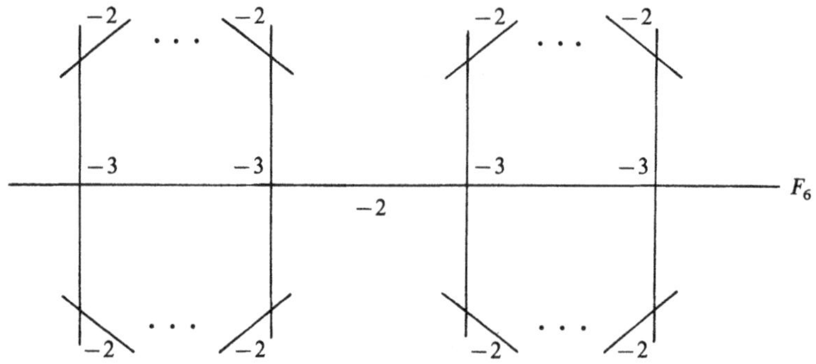

This method can be used to establish the minimality of $Y_\gamma^\circ(D)$ in many cases, see e.g. [81] or the examples in the next chapter. In [41] the minimality of $Y_\gamma^\circ(D)$ was established for γ = principal genus and D a prime discriminant with $89 \leqslant D \leqslant 317$ using slightly different methods.

If one knows an effective canonical divisor C on $Y_\gamma^\circ(D)/\tau^\circ$ then $C + F_D$ ($D \equiv 1 \pmod 4$) or else $C + F_D + F_{D/4}$ is an effective canonical divisor on $Y_\gamma^\circ(D)$.

(7.7) *Example.* $Y_+^\circ(193)/\tau^\circ$ is a blown-up $K3$ surface. A minimal model can be obtained by blowing down successively (the images of) components of the curves T_N and resolution curves. This gives a canonical divisor on $Y_+^\circ(193)$ containing no exceptional curves. Hence $Y_+^\circ(193)$ is minimal.

Before we proceed we give a lemma.

(7.8) Lemma ([74]). *Let τ be an involution on a compact complex surface S. Let C be a non-singular rational curve on S with $\tau(C) = C$ and which is not pointwise fixed by τ. Then τ has two fixed points on C. If C^2 is even then either both of them or none of them are isolated fixed points of τ on S. If C^2 is odd then exactly one of them is an isolated fixed point of τ on S.*

Proof. Choose a meromorphic section s of the normal bundle of C with $\tau^*s = \pm s$ and s not identically zero. If $\tau^*s = +s$ (resp. $-s$) then it follows from local considerations that the order of s in an isolated fixed point of τ is odd (resp. even) and that it is even (resp. odd) in an intersection point of C with 1-dimensional fixed point set of τ. Hence the parity of $\deg s = C^2$ equals that of the number of isolated fixed points of τ on C. □

(7.9) Lemma. *If E is an exceptional curve not pointwise fixed on $Y_\gamma^\circ(D)$ by τ° then $\tau^\circ(E) \cap E = \varnothing$. Moreover for each exceptional curve on $Y_\gamma(D)$ or $Y_\gamma^\circ(D)$ we have $\alpha(E) \cap E = \varnothing$ for all $\alpha \in \Gamma_m/\Gamma$, $\alpha \neq \mathrm{id}$.*

Proof. By (7.8) and the absence of isolated fixed points of τ° one has $\tau^\circ(E) \neq E$. Since $Y^\circ(D)$ is not rational E and $\tau^\circ(E)$ cannot intersect, see (2.2). Since α has only isolated fixed points we cannot have $\alpha(E) = E$ by (7.8). Again by (2.2) $\alpha(E)$ and E cannot intersect. □

(7.10) Proposition. *Let E be a non-singular rational curve on $Y_\gamma(D)$ which is not contained in the exceptional divisor of π: $Y_\gamma(D) \to Y_\gamma^\circ(D)$. Then E is exceptional if and only if $\pi(E)$ is exceptional.*

Proof. If E is exceptional it cannot intersect any other exceptional curve, hence $\pi(E)$ is also exceptional. If E is not exceptional but $\pi(E)$ is exceptional then E intersects the exceptional divisor of π and hence $\pi(E)$ intersects the 1-dimensional fixed point locus of τ°, see (4.5). By (7.9) $\pi(E)$ must be pointwise fixed, i.e. $\pi(E)$ belongs to F_D or $F_{D/4}$. But one can easily check that these curves do not contain exceptional components. □

(7.11) Theorem. *Let $N \neq 1, 2, 3, 4$ or 9 (if $3|D$) and suppose that $Y_\gamma(D)$ is not rational. Then F_N does not contain exceptional curves.*

Proof. If D is a prime discriminant the argument is very short and we treat this case first. If D is prime then F_N has 1 or 2 components, and each component of F_N is fixed by τ. An exceptional component gives rise to an exceptional curve on $Y_\gamma^\circ(D)$. By (7.9) an exceptional component of F_N transformed into itself by τ must be pointwise fixed. But it is easy to see that the fixed point set of τ° does not contain exceptional curves.

In the general case the argument is similar but more complicated. We claim

(i) If F_N contains an exceptional curve then F_N has 2^t components which are permuted transitively by $\Gamma_m \cup \Gamma_m\tau/\Gamma$.

Proof of (i). By (7.9) an exceptional component F of F_N gives rise to 2^{t-1} components under the action of Γ_m/Γ. Next, if $\alpha_w\tau(F) = F$ then by (7.8) $\alpha_w\tau$ has at least 2^{t-1} isolated fixed points. The isolated fixed points of $\alpha_w\tau$ are known. They lie on the intersection on F_{4w} with the cusps or (if $D \equiv 1 \pmod 4$) on the resolutions of quotient singularities corresponding to quadratic forms $[2w, 2w, (D/w + w)/2]$:

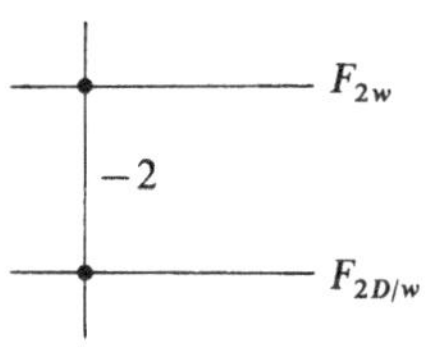

and lie on no other F_M. If $w \neq 1, D$ then F_{2w} and F_{4w} do not contain exceptional curves.

Next we show:

(ii) There exists a square-free divisor w of D such that $\alpha_w\tau$ acts trivially on the set of components of F_N if F_N has 2^t components. Proof of (ii): By V (3.3) one deduces that $R_1 = \emptyset$, $D^2|N$ and $(-1, D)_p = +1$ for all primes $p|D$. Then there is a ramified ideal with norm w in the genus of γ. By V.9 this implies that $\alpha_w \tau$ acts trivially.

Together (i) and (ii) give a contradiction if F_N contains an exceptional component.

□

(7.12) Theorem. *Suppose we can find on $Y_\gamma(D)$ an effective pluri-canonical divisor consisting only of components of the modular curves and the resolution curves. Then $Y_\gamma^\circ(D)$ is a minimal model of $Y_\gamma(D)$.*

Proof. Any effective pluri-canonical divisor contains all exceptional curves. Using (7.10) and (7.11) we see that the given pluri-canonical divisor on $Y_\gamma(D)$ gives rise to a pluri-canonical divisor on $Y_\gamma^\circ(D)$ containing no exceptional curves. □

(7.13) Conjecture. *Suppose that $Y_\gamma(D)$ is not rational. Then there exists an effective pluri-canonical divisor on $Y_\gamma(D)$ consisting only of resolution curves and of components of the modular curves.*

Remark. By (7.12) this conjecture implies the minimality of $Y_\gamma^\circ(D)$.

This conjecture can be reformulated as follows. By IV (2.8) the canonical class of $Y_\gamma(D)$ equals

$$-\frac{1}{4}t_0 - \sum_k s_k - \sum_1 \frac{1}{3} g_1,$$

where the s_k are the Poincare-duals of the homology classes of the resolution curves S_k of the cusps and the g_1 are the Poincare-duals of the homology classes of the resolution curves G_1 of the a_3^+ quotient singularities of type (3; 1, 1). It follows that the canonical class in $H^2(Y_\gamma(D), \mathbb{Q})$ lies in the $\mathbb{Q}$-linear subspace V of $H^2(Y_\gamma(D), \mathbb{Q})$ generated by the classes of the irreducible components of the curves T_N and the resolution curves. So (7.13) is equivalent to the conjecture that the canonical class of $Y_\gamma(D)$ lies in the positive cone V^+ inside V spanned by the classes of the components of the T_N and the resolution curves.

We now show that $-t_0$ belongs to V^+.

(7.14) Lemma. *A linear relation $\sum \lambda_{n_i} t_{n_i} = 0$ in V is equivalent to a linear relation $\sum \lambda_{n_i} a(n_i) = 0$ for all modular forms $f = \sum a(n) q^n$ in $M_2^\gamma(\Gamma_0(D), \chi_D)$.*

Proof. Evident from VI (4.4).

(7.15) Proposition. $-t_0 \in V^+$.

Proof. Note that a linear relation $\sum \lambda_{n_i} a(n_i) = 0$ for all $f \in S_2^\gamma$ with all $\lambda_{n_i} > 0$ implies a relation $\lambda_0 a(0) + \sum \lambda_{n_i} a(n_i) = 0$ with $\lambda_0 > 0$ since the Eisenstein series in M_2^γ has $a(0) < 0$ and all other coefficients non-negative. As D. Zagier pointed out to me a cusp form f cannot have only non-negative Fourier coefficients because then the L-series $\sum a(n) n^{-s}$ of f would have a pole (for $s = k$), contradicting the well-known holomorphy of the L-series. □

(7.16) Corollary. *There exists a modular form of even weight on Γ whose divisor in X_Γ consists only of curves T_N.*

Proof. Let L be the line bundle whose sections are the modular forms of weight 2. Its Chern class is represented by $-\frac{1}{4}t_0$, see IV (2.8). Hence (7.15) means that there exists an effective divisor consisting of components of the F_N and of the resolution curves only which belongs to the linear system defined by a power of L. This is a pull-back from X_Γ. □

We gave evidence for (7.13) in [81]. Indeed, we gave examples of Hilbert modular surfaces with lots of canonical divisors of the required form.

Hermann proved Conjecture (7.13) for an infinite set of $Y_\gamma(D)$ by constructing cusp forms of weight 2 with the help of theta series. In his case γ is the genus of $(\sqrt{D})$, i.e. $Y_\gamma(D)$ = resolution of the Baily-Borel compactification of $\mathrm{SL}(2,\mathfrak{o})\backslash\mathfrak{H}\times\mathfrak{H}_-$. Then we can interpret $Y_\gamma(D)$ as a moduli space of principally polarized abelian surfaces and we have thus a map $\mathrm{SL}(2,\mathfrak{o})\backslash\mathfrak{H}\times\mathfrak{H}_-\to\mathrm{Sp}(4,\mathbb{Z})\backslash\mathfrak{H}_2$ with $\mathfrak{H}_2$ the Siegel upper half space of degree 2, see Ch. IX. On $\mathrm{Sp}(4,\mathbb{Z})\backslash\mathfrak{H}_2$ we have the usual theta functions with half-integer characteristics. Using their restrictions to $\mathfrak{H}\times\mathfrak{H}_-$ one can construct modular forms of weight 1/2 with a character on $\mathrm{SL}(2,\mathfrak{o})$. By taking suitable products of four such theta functions we get cusp forms of weight 2. Under certain conditions the character is trivial. The zero sets of the theta functions with half integer characteristics in $\mathrm{Sp}(4,\mathbb{Z})\backslash\mathfrak{H}_2$ are known. Along these lines one finds:

(7.17) Theorem (Hermann [70]). *Let γ be the genus of $(\sqrt{D})$. Then there exists on $Y_\gamma(D)$ a canonical divisor consisting only of components of the curves F_N and resolution curves if D satisfies one of the following two conditions:*

i) *D has a divisor a with $a\not\equiv 1\ (\mathrm{mod}\ 8)$*

ii) *there are two integers $m, n>0$ with $m\equiv 7\ (\mathrm{mod}\ 8)$ and $D=(m^2-8)/n^2$.*

In particular, $Y_\gamma^\circ(D)$ is minimal in these cases.

Next we show that $Y_\gamma(D,\mathfrak{c})$ is minimal if $\mathrm{N}(\mathfrak{c})$ is big.

(7.18) Proposition. *Suppose that $Y_\gamma(D,\mathfrak{c})$ is of general type and $Y_\gamma(D)$ is not rational. If E is an exceptional curve on $Y_\gamma(D,\mathfrak{c})$ and E' is the irreducible curve on $Y_\gamma(D)$ which is the image under the natural rational map $Y_\gamma(D,\mathfrak{c})\to Y_\gamma(D)$, then E' is an exceptional curve.*

Proof. We show that $\mathrm{K}_{Y_\gamma(D)}E'<0$. By (2.2) this implies that $E'^2<0$, hence that E' is exceptional. Let $E_1=E,\ldots,E_r$ be the transforms of E under the action of $\Gamma/\Gamma[\mathfrak{c}]$. These are disjoint exceptional curves by (2.2) again. We have

$$\begin{aligned}\mathrm{K}_{Y_\gamma(D,\mathfrak{c})}(E_1+\cdots+E_r)&=-r\\&=-2\int_{\sum E_i}\omega-\sum E_i\Delta\,,\end{aligned}$$

with Δ the divisor resolving the cusps on $Y_\gamma(D,\mathfrak{c})$. Let $d=[\Gamma:\Gamma[\mathfrak{c}]]$. Then

$$\int_{\sum E_i} \omega = d \int_{E'} \omega \tag{1}$$

Moreover,

$$\Delta \sum E_i \leqslant d\Delta' E' , \tag{2}$$

with Δ' the divisor resolving the cusps of $Y_\gamma(D)$. This can be seen as follows. Let $G = \Gamma/\Gamma[\mathfrak{c}]$ operate on $Y_\gamma(D, \mathfrak{c})$. The quotient has only finitely many singularities. Using intersection theory on $Y_\gamma(D, \mathfrak{c})/G$ we have

$$\Delta \cdot \sum E_i = (\# G)\Delta'' \cdot E'' \quad \text{with}$$

Δ'', E'' the images of Δ, E on $Y_\gamma(D, \mathfrak{c})/G$. If we resolve the quotient singularities (which gives a surface Y_1) we get a surface which is a blow up of $Y_\gamma(D)$ in the cusps. We then have (writing g for $Y_1 \to Y_\gamma(D)$)

$$(g^*\Delta')(g^*E') = (\Delta'' + H_1)(E'' + H_2) ,$$

where H_1, H_2 are divisors (with non-negative rational coefficients) with support in the cusps. Moreover, $H_2(\Delta'' + H_1) = 0$, hence

$$\Delta'E' \geqslant (\Delta'' + H_1)E'' \geqslant \Delta''E'' .$$

Together (1) and (2) give

$$\mathrm{K}_{Y_\gamma(D)}E' = -2 \int_{E'} \omega - E'\Delta' - E'A ,$$

with $-A$ the Chern cycle of the quotient singularities. So

$$\mathrm{K}_{Y_\gamma(D)}E' \leqslant -\frac{r}{d}$$

and this is negative. □

(7.19) Theorem ([37]). *There exists an effectively computable constant $C = C(D)$ such that if $\mathrm{N}(\mathfrak{c}) \geqslant C$ then $Y_\gamma(D, \mathfrak{c})$ is minimal.*

Proof. Let E be an exceptional curve on $Y(D, \mathfrak{c})$. Keeping the notation of the proof of (7.18) we find for non-rational $Y_\gamma(D)$

$$\begin{aligned} -2 \int_{E'} \omega - E'\Delta' - E'A &= -1 \\ -2 \int_{E_i} \omega - E_i\Delta &= -1 . \end{aligned} \tag{3}$$

So

$$\begin{aligned} \sum E_i\Delta &= r - 2 \int_{\sum E_i} \omega \\ &= r - 2d \int_{E'} \omega \\ &= r - d + d(E'\Delta' + E'A') . \end{aligned}$$

By (7.1) we have $E'\Delta' + E'A' \geqslant 4/3$ (note that by (7.1) the E_i pass through the cusps, hence also E' does), so

$$\sum E_i \Delta \geqslant r + d/3 . \tag{4}$$

We claim: there exists a constant C' such that

$$\mathrm{K}_{Y_\gamma(D,\mathfrak{c})}\Delta \leqslant \frac{d}{\mathsf{N}(\mathfrak{c})} C' \tag{5}$$

Assuming this we see that by blowing down all E_i we get on the resulting surface Y'

$$\mathrm{K}_{Y'}\Delta' < 0 \tag{6}$$

if $r + d/3 > dC/\mathsf{N}(\mathfrak{c})$, i.e. if $\mathsf{N}(\mathfrak{c}) \geqslant 3C' =: C$. However, this implies that on a minimal model Y'' of Y' the image Δ_1 of Δ' satisfies $\mathrm{K}_{Y''}\Delta_1 < 0$. (Since Δ consists of cycles it can not be blown down completely.) This is impossible since on a minimal surface of general type $\mathrm{K}F \geqslant 0$ for every curve F.

To prove (5) one observes that the cusp resolutions on $Y_\gamma(D,\mathfrak{c})$ consist of $\dfrac{d}{\mathsf{N}(\mathfrak{c})}$ or $\dfrac{2d}{\mathsf{N}(\mathfrak{c})}$ times the primitive cycles associated to the modules $\mathfrak{a}^{-2}\mathfrak{c}$, where $\mathfrak{a}$ runs through $\mathscr{Cl}(K)$. (Cf. IV, Lemma (5.2)) So

$$\mathrm{K}_{Y_\gamma(D,\mathfrak{c})}\Delta \leqslant \frac{d}{\mathsf{N}(\mathfrak{c})} C'$$

with $C' = \sum (b_i - 2)$ (sum over all cycles associated to $\mathfrak{a}^{-2}\mathfrak{c}$ with $\mathfrak{a}$ running through $\mathscr{Cl}(K)$), see Ch. II.

If $Y_\gamma(D)$ is rational then we have

$$-2\int_{E'} \omega - E'\Delta' - E'A = -m \quad \text{with } m > 0$$

and so

$$\sum E_i \Delta = r - dm + d(E'\Delta' + E'A) ,$$

but by (7.1) $\sum E_i \Delta > 3r$, hence $(-m + E''\Delta' + E'A) > 0$ and we can replace (4) by $\sum E_i \Delta \geqslant de$ with e a positive rational number. The proof then goes the same way. □

If we just want to restrict to the case $\mathfrak{c} = (n)$ with $n \in \mathbb{Z}_{>1}$, then everything is much simpler.

(7.20) Lemma. *Let $\Gamma = \Gamma(\mathfrak{o} \oplus \mathfrak{a})$. If $f \in S_{2k}(\Gamma)$ then $\omega = f(dz_1 \wedge dz_2)^k$ defines a holomorphic section of $\mathrm{K}^{\otimes k}_{Y_{\Gamma[(n)]}}$ for $n \geqslant \max(k, 3)$.*

Proof. Since ω is Γ-invariant and since $\Gamma[(n)]$ acts freely for $n \geqslant 3$ we have to check only that ω extends over the cusps. Write

$$f = a_0 + \sum_{\nu \in M_+^\vee} a_\nu e^{2\pi i \operatorname{Tr} \nu z} \quad \text{(near a cusp)} .$$

Then ω extends over this cusp if

$$a_\nu = 0 \quad \text{for all } \nu \in M_+^\vee \cup \{0\} \text{ with } \min_{y \in M_+} \operatorname{Tr} \nu y < k\,,$$

see III.3. By replacing M by nM this condition is satisfied. □

(7.21) Corollary (Freitag [34]). *If n is big than $Y_\gamma(D,(n))$ does not contain non-singular elliptic nor non-singular rational curves other than those in the cusp resolutions.*

Proof. Suppose the cusp forms define an embedding of $\Gamma\backslash\mathfrak{H}^2$ into projective space. (They do for k big.) All elements of $S_{2k}(\Gamma)$ define pluri-canonical forms $\omega = f(dz_1 \wedge dz_2)^k$ on $Y_{\Gamma[(n)]}$ if $n \geqslant k$. Suppose that C is a non-singular rational or elliptic curve on $Y_{\Gamma[(n)]}$ not contained in the cusp resolutions. Restricting ω to C we find a section of $\mathrm{k}_C^{\otimes k}$ with k_C the canonical bundle of C (Use the adjunction formula $\mathrm{k}_C = \mathcal{O}(\mathrm{K} + C)|_C$.) This section is not constant if we choose f suitably. (The f's separate points.) This contradicts the assumption on the genus of C. □

(7.22) Corollary (Freitag [34]). *For n big we have*

$$\operatorname{Aut}(Y_\gamma(D,(n))) = \textit{Normalizer of } \Gamma/\Gamma[(n)] \textit{ in } \operatorname{Aut}(\mathfrak{H}^2)\,.$$

In particular, all automorphisms of $Y_\gamma(D,(n))$ are modular, they come from $\mathfrak{H}^2$.

Proof. By (7.21) every automorphism must preserve the cusp resolutions, i.e. induces an automorphism of $\Gamma\backslash\mathfrak{H}^2$. □

VIII. Examples of Hilbert Modular Surfaces

As an illustration of the previous chapters we now give examples. Reading this chapter will resemble a visit to a zoological garden: each of the surfaces presented here is a world in itself that can be explored as far as one wishes, either for its own peculiarities or to test general principles.

Some of the examples that we give can be identified with classical surfaces. As a consequence of this the properties of these classical surfaces acquire a new meaning by the modular interpretation of Hilbert modular surfaces. Other examples might serve to illustrate general facts of the theory of algebraic surfaces or to construct new examples.

The most natural approach to the geometry of examples of Hilbert modular surfaces is studying the maps defined by Hilbert modular forms. However, even for small discriminants the graduated ring of modular forms for a Hilbert modular group is already quite complicated and we can use this method only in very few cases. In most cases we use the (pluri-)-canonical maps to study Hilbert modular surfaces of general type. In general not much is known about the structure of the canonical map, but in a few cases where the numerical invariants are small something is known. Moreover, often we are able to construct effective canonical divisors on Hilbert modular surfaces which gives a lot of information that usually is not available.

In Section 1 we give some auxiliary remarks on rings of Hilbert modular forms and on linear systems on surfaces. The examples are given in Section 2 ordered by the discriminant of the quadratic number field to which they belong.

1. Preliminaries

To a Hilbert modular group Γ acting on $\mathfrak{H}^n$ one can associate the ring of modular forms

$$M(\Gamma) = \bigoplus_k M_k(\Gamma)$$

which is graduated by the weight k. This ring is finitely generated and integrally closed in its quotient field. Let $M_{\mathrm{ev}} = \bigoplus_k M_{2k}(\Gamma)$. A formula for the dimension of $\dim M_{2k}$ was given in Chapter IV. For $n = 2$, $D \neq 5, 12$ and $k > 0$ one has

$$\dim S_{2k}(\Gamma) = k(k-1)\operatorname{vol}(\Gamma\backslash\mathfrak{H}^2) + \chi(Y_\Gamma) - \begin{cases} \frac{1}{3}a_3^+ & k \equiv 2 \pmod 3 \\ 0 & \text{otherwise} \end{cases}$$

and $\dim M_{2k}(\Gamma) = \dim S_{2k}(\Gamma) + h$ with h = number of cusps. This implies

(1.1) Proposition. *Let $\Gamma = \Gamma(\mathfrak{o} \oplus \mathfrak{a})$ and let $m_\Gamma = \sum \dim M_{2k}(\Gamma) t^k$ be the Hilbert series of $M_{ev}(\Gamma)$. If $D \neq 5$, 12 we have*

$$m_\Gamma = \left(\sum_{i=0}^{6} b_i t^i\right)(1-t)^{-2}(1-t^3)^{-1},$$

where $b_0 = b_6 = 1$, $b_1 = b_5 = \chi(Y_\Gamma) + h - 3$, $b_2 = b_4 = 4\zeta_K(-1) - \chi - \frac{1}{3}a_3^+ - h + 3$ and $b_3 = 4\zeta_K(-1) + \frac{2}{3}a_3^+ - 2$. Furthermore

$$\begin{aligned} m_\Gamma = {} & (1+t^{10})(1-t)^{-1}(1-t^3)^{-1}(1-t^5)^{-1} && D = 5 \\ & (1-t^6)(1-t^8)(1-t)^{-1}(1-t^2)^{-1}(1-t^3)^{-2}(1-t^4)^{-1} && D = 12, \gamma = ++ \\ & (1-t^4)(1-t^{10})(1-t)^{-1}(1-t^2)^{-2}(1-t^3)^{-1}(1-t^5)^{-1} && D = 12, \gamma = -- . \end{aligned}$$

We shall study the examples by using linear systems. Therefore we now recall some notions related to linear systems on surfaces. Let D be an effective divisor on the compact algebraic surface S. Then D defines a rational map

$$\Phi_D: S \to \mathbb{P}^n \quad \text{with } n+1 = \dim H^0(S, \mathcal{O}_S(D)).$$

We can write $|D| = F + |M|$, where F is the fixed part of $|D|$ and M is the moving part. Then $\Phi_M = \Phi_D$. In general $|M|$ has base points and in such points Φ_M is not defined. Suppose $n \geqslant 2$. Then the image of S under Φ_D is either a curve or a surface. In the first case we say that $|D|$ is composed with a pencil. In the second case Φ_D is generically finite and

$$D^2 \geqslant M^2 \geqslant (\deg \Phi_D)(\deg \Sigma) \tag{1}$$

with Σ the image surface.

We are interested in the case where $D = \mathrm{K}_S$, the canonical divisor class of S. Moreover, we shall assume that S is a surface of general type with $p_g \geqslant 3$ for which the image is a surface. We shall write Φ_m for $\Phi_{m\mathrm{K}_S}$. All exceptional curves (of the first, i.e. (-1)-curves) belong to the fixed part of $|\mathrm{K}|$. Therefore, it is often useful to assume that S is a minimal model. All (-2)-curves are contracted to points under Φ_m.

Not much is known for the canonical map Φ_1, but one has the following result due to Beauville [12].

(1.2) Proposition. *Let S be a surface of general type such that the image of Φ_{K_S} is a surface Σ in $\mathbb{P}^{p_g-1}$. Then either $p_g(\Sigma) = 0$ or Σ is a surface of general type embedded in $\mathbb{P}^{p_g-1}$ by its canonical system.*

If the image surface of S under Φ_1 is a surface Σ one has very little information on $\deg \Phi_1$. However, one has the following result also due to Beauville l.c.:

(1.3) Proposition. *If S is a minimal surface of general type with* $K_S^2 \leqslant 3p_g(S) - 7$ *then* Φ_1 *is of degree* 2 *onto a ruled surface.*

As to the invariants for a surface of general type one has the estimates

$$K^2 \leqslant 9\chi \quad (\text{or equivalently } c_1^2 \leqslant 3c_2) \quad (\text{Miyaoka—Bogomolov})$$

$$K^2 \geqslant 2p_g - 4 \quad \text{if } S \text{ is minimal} \quad (\text{Noether}).$$

But for a Hilbert modular surface of general type one has

$$K^2 \leqslant 8\chi \quad (\text{equivalently, } c_1^2 \leqslant 2c_2),$$

see Ch. IV.

Minimal surfaces of general type with K^2 near the Noether bound of $2p_g - 4$ were studied by Horikawa. To state some of his results, recall that by Σ_d we denote a $\mathbb{P}^1$-bundle over $\mathbb{P}^1$ with a section Δ with $\Delta^2 = -d$ (see VII.1). We denote by F a fibre of this $\mathbb{P}^1$-bundle.

(1.4) Proposition ([87]). *Let* $n \in \mathbb{Z}_{\geqslant 2}$, *and let S be a minimal algebraic surface with* $p_g = n + 1$ *and* $c_1^2 = 2n - 2$. *Then S is the minimal resolution of singularities of one of the following normal surfaces*:

i) *double coverings of* $\mathbb{P}^2$ *with branch locus of degree* 8 ($n = 2$)

ii) *double coverings of* $\mathbb{P}^2$ *with branch locus of degree* 10 ($n = 5$)

iii) *double coverings of* Σ_d *whose branch locus is linearly equivalent to* $6\Delta + (n + 3 + 3d)F$ *with* $n \geqslant \max(d + 3, 2d - 3)$ *and* $n - d$ *odd*

iv) *double coverings of* Σ_d *whose branch locus is linearly equivalent to* $6\Delta + 4nF$ *with* $n = 3, 4$ *or* 5.

2. The Examples

D = 5. We shall treat three surfaces associated to $\mathbb{Q}(\sqrt{5})$ in detail.

Let $\Gamma = \Gamma[2]$ be the principal congruence subgroup of level (2) in Γ_K acting on $\mathfrak{H}^2$. The corresponding Hilbert modular surface Y_Γ has numerical invariants $c_1^2 = -7$, $c_2 = 19$. The group $\Gamma_K/\Gamma[2] \cong \mathrm{PSL}(2, \mathbb{F}_4) \cong \mathfrak{A}_5$ acts on it. There are five cusps, which we number 1, ..., 5. The action of $\Gamma_K/\Gamma[2]$ permutes the cusps and this establishes the isomorphism with $\mathfrak{A}_5$. The resolution cycle of a cusp is (3, 3, 3).

The curve F_1 consists of 10 components, each of which is isomorphic to $\overline{\Gamma(3)\backslash\mathfrak{H}}$ with $\Gamma(3) = \ker\{\mathrm{PSL}(2, \mathbb{Z}) \to \mathrm{PSL}(2, \mathbb{Z}/3\mathbb{Z})\}$. Each component intersects three cusp resolutions transversally in a simple point. We number the components of F_1 accordingly: $F_1^{ij} = F_1^{ji}$ $(1 \leqslant i < j \leqslant 5)$ does not intersect cusps i and j. The configuration formed by F_1 and the resolution cycles is indicated in the following diagram

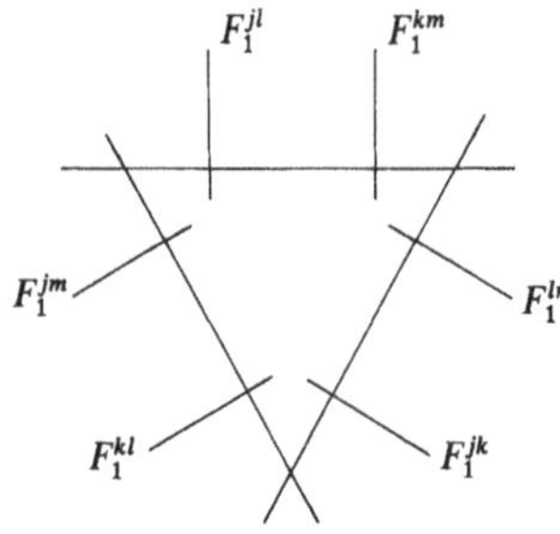

cusp i $\{1,\ldots,5\} = \{i,j,k,l,m\}$

These facts are easy to check.

Denote by Y_Γ° the surface obtained by blowing down the ten components of F_1 on Y_Γ. These are exceptional curves.

(2.1) Theorem. *Let $D = 5$ and let $\Gamma = \Gamma[2] = \ker\{\mathrm{PSL}(2,\mathfrak{o}) \to \mathrm{PSL}(2,\mathfrak{o}/2\mathfrak{o})\}$. Then*

i) *X_Γ is isomorphic to the surface in $\mathbb{P}^4$ given by $\sigma_2 = \sigma_4 = 0$ with σ_i the i-th elementary symmetric function in the coordinates of $\mathbb{P}^4$.*

ii) *Y_Γ° is isomorphic to the surface in $\mathbb{P}^4$ given by $\sigma_1 = \sigma_3 = 0$.*

Proof. Let L be the line bundle on Y_Γ whose sections are the modular forms of weight 2. Using VII (7.3) one checks that the following divisor is the divisor of a section of L:

$$G_i = \sum_{\substack{j=1 \\ j\neq i}}^{5} F_1^{ij} + H_i$$

with H_i the sum of the curves occuring in the cusp resolutions different from cusp i. We thus find five sections E_i. Since E_i vanishes in four cusps and has the value 1 in the remaining cusp (after normalization) and since there are no cusp forms of weight 2, E_i is an Eisenstein series:

$$E_0 = \sum_{\substack{(\alpha,\beta)=\mathfrak{o} \\ \alpha\equiv 0\,(2) \\ \beta\equiv 1\,(2)}} \mathrm{N}(\alpha z + \beta)^{-2}$$

Obviously, the E_i have no common zeroes and thus define a morphism $Y_\Gamma \to \mathbb{P}^4$ which factors through $X_\Gamma \to \mathbb{P}^4$. The involution $(z_1, z_2) \to (z_2, z_1)$ of $\mathfrak{H}^2$ induces an involution of Y_Γ and together with $\mathfrak{A}_5$ this generates the group $\mathfrak{S}_5$. Our morphism is $\mathfrak{S}_5$-equivariant. One easily checks that the image of our surface is again a surface. The E_i do not satisfy non-trivial linear relations nor non-trivial quadratic relations different from the relation $\sigma_2(E_1,\ldots,E_5) = 0$. This relation follows from the observation that $\sigma_2(E_1,\ldots,E_5)$ is a cusp form of weight 4 and by the fact that $\dim S_4(\Gamma) = 0$, cf. IV. As to cubic relations, since $(1:0:\cdots:0)$ lies on our surface a relation $a\sigma_1^3 + b\sigma_3 = 0$ implies $a = 0$. Since $\sigma_3(E_1,\ldots,E_5)|F_1^{ij} = E_k E_l E_m | F_1^{ij}$, which does not vanish identically, there is no relation $\sigma_3 = 0$. The dimension formula of IV.4 now implies a quartic relation $a\sigma_1^4 + b\sigma_1\sigma_3 + c\sigma_4 = 0 \pmod{\sigma_2}$. Again $a = 0$ as above. Restriction to F_1^{ij} gives $b = 0$. Therefore the image surface is a surface of degree 8. Since $L^2 = 8$ it follows that the morphism is of degree 1. Since L is ample by Baily-Borel our morphism is an isomorphism of X_Γ to the image surface. This proves *i*).

ii) On the surface Y_Γ° the image of the resolution cycle of one cusp is an anti-canonical divisor. (Use VII (7.3) again.) There are five such divisors. The divisor of E_k^{-1} is $-\sum_{k \neq j} F_1^{kj} - H_k$ on Y_Γ; hence E_k^{-1} defines a holomorphic section of $L^{-1} \otimes \mathcal{O}(F_1 + D)$ with D the divisor consisting of all cusp resolutions. Now $L = \mathrm{K}_{Y_\Gamma} \otimes \mathcal{O}(D) = \pi^* \mathrm{K}_{Y_\Gamma^\circ} \otimes \mathcal{O}(F_1 + D)$ with $\pi\colon Y_\Gamma \to Y_\Gamma^\circ$ the natural map, so $y_k = E_k^{-1}$ defines a section of $\mathrm{K}_{Y_\Gamma^\circ}^{-1}$ whose divisor is one of the divisors obtained above. These five divisors have no point in common, so $y_1, \ldots, y_5$ define a morphism $Y_\Gamma^\circ \to \mathbb{P}^4$. Using i) one finds the relations $\sigma_1 = \sigma_3 = 0$ and it is clear that the morphism is birational. Comparing Euler numbers one can easily check that it is an isomorphism onto its image.

(2.2) Corollary. i) $M_{\mathrm{ev}}(\Gamma[2]) = \mathbb{C}[x_1, \ldots, x_5]/(\sigma_2, \sigma_4)$, *where under the isomorphism* x_i *corresponds to the Eisenstein series* E_i.

ii) $M_{\mathrm{ev}}(\Gamma_K) = \mathbb{C}[\sigma_1, \sigma_3, \sigma_5, \Delta]/(\Delta^2 - F)$, *where*

$$F = \sigma_5(3125\sigma_5^3 + 2000\sigma_5^2\sigma_3\sigma_1^2 + 256\sigma_5^2\sigma_1^5 - 900\sigma_5\sigma_3^3\sigma_1 - 128\sigma_5\sigma_3^2\sigma_1^4 + 16\sigma_3^4\sigma_1^3 + 108\sigma_3^5)\,.$$

Proof. We have already proved the first assertion. The second follows because $M_{\mathrm{ev}}(\Gamma_K) = M_{\mathrm{ev}}(\Gamma[2])^{\mathfrak{A}_5}$; this is a quadratic extension of $M_{\mathrm{ev}}(\Gamma[2])^{\mathfrak{S}_5} = \mathbb{C}[\sigma_1, \sigma_3, \sigma_5]$ given by $\Delta = \prod_{1 \leqslant i < j \leqslant 5} (E_i - E_j)$ of weight 20 and Δ^2 as given. □

(2.3) *Remarks.* 1. The surface given by $\sigma_1 = \sigma_3 = 0$ is a classical one. It was studied by Clebsch and Klein. Its equation can be written as

$$\sum_{i=0}^{4} x_i = 0 \qquad \sum_{i=0}^{4} x_i^3 = 0$$

which explains the name *diagonal surface*. Hirzebruch discovered that it was a model for Y_Γ, see [76].

2. The threefold

$$\sum_{i=0}^{5} x_i = 0 \qquad \sum_{i=0}^{5} x_i^3 = 0$$

in $\mathbb{P}^5$ is a model for a Siegel modular variety. The six coordinate hyperplanes cut out on this threefold six surfaces, each of which is a "Humbert surface" isomorphic to Y_Γ°. This will be explained in Chapter IX.

3. The result

$$M_{\mathrm{ev}}^{\mathrm{sym}}(\Gamma_K) = M_{\mathrm{ev}}(\Gamma[2])^{\mathfrak{S}_5} = \mathbb{C}[\sigma_1, \sigma_3, \sigma_5]\,,$$

is due to Gundlach [52, 53]. The divisor of σ_5 in $\mathfrak{H}^2$ is divisible by 2, so there is a modular form Θ of weight 5 (with character) whose square is a multiple of σ_5. One can show that $\Theta = \prod_{1 \leqslant i < j \leqslant 5} \theta_{ij}$ with θ_{ij} a theta function whose fourth power equals $E_i + E_j$ on $\mathfrak{H}^2$. Using the dimension formula one thus finds

$$M(\Gamma_K, \chi) = \mathbb{C}[\sigma_1, \sigma_3, \Theta, \Delta/\Theta],$$

a result due to Gundlach and Resnikoff ([135]).

4. It was proved by Nagaoka that the graduated ring of symmetric even weight Hilbert modular forms with rational integral Fourier coefficients is generated by four elements, of weights 2, 6, 10, 12, see [120].

In our next example we take $\Gamma = \Gamma[(\sqrt{5})]$. In this case we also have an action of the icosahedral group $\mathfrak{A}_5 = \mathrm{PSL}(2, \mathbb{F}_5)$ on X_Γ and Y_Γ. The surface X has six cusps (numbered 1 to 6) each resolved by a cycle (3, 3). The curve F_1 has one component isomorphic to $\overline{\Gamma(5)\backslash\mathfrak{H}}$. It passes transversally through the double points of the six resolution cycles. The curve F_5 has 15 components F_5^{ij}, $1 \leqslant j < j \leqslant 6$, each of which intersects two cusps. The configuration on Y_Γ is as follows.

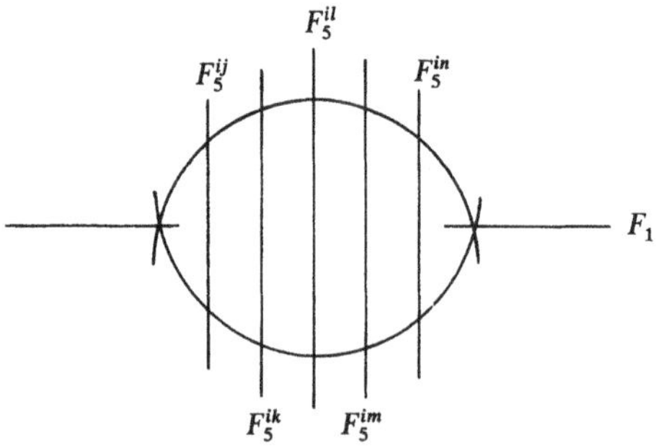

The action of $\mathfrak{A}_5$ on the set of the cusps is the same as that of $\mathfrak{A}_5$ on the six axes of an icosahedron.

The involution τ on Y_Γ (or X_Γ) induced by $(z_1, z_2) \to (z_2, z_1)$ on $\mathfrak{H}^2$ permutes the two resolution curves of each cycle. The fixed point set of τ on Y_Γ is F_1 (use V.9). Each of the six resolution cycles and F_1 gives rise to a configuration

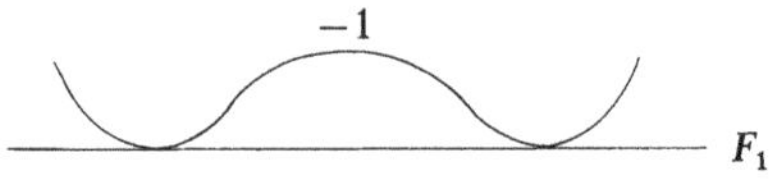

(1)

on the non-singular surface Y_Γ/τ. Blowing down the resulting six exceptional curves yields a surface isomorphic to X_Γ/τ. Since now $c_1^2 = 9$, $c_2 = 3$ and Y_Γ/τ is rational we have $X_\Gamma/\tau = \mathbb{P}^2$. The group $\mathfrak{A}_5$ acts on X_Γ/τ and the map $X_\Gamma \to X_\Gamma/\tau$ is $\mathfrak{A}_5$-equivariant.

The ring of invariants $\mathbb{C}[x_0, x_1, x_2]^{\mathfrak{A}_5}$ for the action of $\mathfrak{A}_5$ on $\mathbb{P}^2$ is generated by polynomials A, B, C and D of degree 2, 6, 10 and 15. We have $A = x_0^2 + x_1^2 + x_2^2$. There is one orbit of six points. One normalizes B by requiring that $B = 0$ defines the unique invariant curve of order 6 passing through these six points. Similarly, there is a unique invariant curve $C = 0$ of degree 10 passing through these six points with multiplicity >2. It has double cusps in these points. There is an invariant curve of degree 15 namely the union of the 15 lines connecting these six points. There is one relation R of degree 30, see Klein's Werke II, p. 354 for the formulas. The following theorem is due to Hirzebruch ([77]):

(2.4) Theorem. *Let $D = 5$ and Γ the principal congruence subgroup of level $(\sqrt{5})$ of Γ_K. The double cover π: $X_\Gamma \to X_\Gamma/\tau$ is $\mathfrak{A}_5$-equivariantly isomorphic to the double cover of $\mathbb{P}^2$ branched along $C = 0$. The six cusps of X_Γ correspond to the six singularities of $C = 0$.*

(2.5) Corollary. $M(\Gamma) = \mathbb{C}[x_0, x_1, x_2, c]/(c^2 = C)$
$M^{\text{sym}}(\Gamma) = \mathbb{C}[x_0, x_1, x_2]$
$M(\Gamma_K) = \mathbb{C}[A, B, c, D]/(R(A, B, c^2, D))$.

Proof. A computation of the invariants of F_1 shows that F_1 maps to an invariant curve of degree 10 in $\mathbb{P}^2$. This curve has double cusps in the 6-point orbit as (1) shows, hence F_1 maps to $C = 0$. The corollary follows from $L = (\pi')^*\mathcal{O}_{\mathbb{P}^2}(2)$ with π': $Y_\Gamma \to X_\Gamma \to X_\Gamma/\tau = \mathbb{P}^2$, a consequence of $L = \pi^*(\mathrm{K}_{Y_\Gamma} + \sum S_k)$ and $\mathrm{K}_{Y_\Gamma} = \pi^*(\mathrm{K}_{\mathbb{P}^2}) + F_1$. □

In the last example for this field we let $\Gamma = \Gamma[(3)]$, a case studied by Hirzebruch and the author. Here the group $\mathrm{PSL}(2, \mathbb{F}_9) = \mathfrak{A}_6$ acts. The surface Y_Γ has 10 cusp resolutions of type $(3, 3, 3, 3)$, numbered by the partitions $(ijk)(lmn)$ of the set $\{1, 2, \ldots, 6\}$. F_1 has 30 components $F_1^{ij,kl,mn}$, each a (-2)-curve. The configuration was indicated in VII (7.5).

We have $p_g = 5$. By considerations as in IV.5 the representation of $\mathfrak{A}_6$ on $S_2(\Gamma)$ must be the one given by the permutation representation on $\{(x_0, \ldots, x_5) \in \mathbb{C}^6: \sum x_i = 0\}$. Choose a basis s_i of $S_2(\Gamma)$ corresponding to the x_i. Then $\sigma_1(s_0, \ldots, s_5)$, $\sigma_2(s_0, \ldots, s_5)$ and $\sigma_4(s_0, \ldots, s_5)$ are cusp forms of weight 2, 4, and 8 on Γ_K, hence zero.

(2.6) Theorem. *Let $D = 5$ and $\Gamma = \Gamma[(3)]$. The cusp forms s_i $i = 0, \ldots, 5$ of weight 2 define an $\mathfrak{A}_6$-equivariant morphism of Y_Γ onto the surface given by $\sigma_1 = \sigma_2 = \sigma_4 = 0$ in $\mathbb{P}^5$. (Here $\mathfrak{A}_6$ acts on the coordinates by permuting them and σ_i is the i-th elementary symmetric function in the s_i.) The morphism is of degree 1 and it contracts the 30 components of F_1 to the 30 double points that make up the $\mathfrak{A}_6$-orbit of $(1 : 1 : \rho : \rho : \rho^2 : \rho^2)$ with $\rho = \exp(2\pi i/3)$.*

Proof. Using VII (7.3) one can show that twice a cusp resolution plus the components of F_1 it intersects is an effective canonical divisor. We get in this way 10 divisors $(s_i + s_j + s_k) = (s_l + s_m + s_n)$. This shows that the canonical map $Y_\Gamma \to \mathbb{P}^5$ is well defined. The image is a surface of degree 8. Since $c_1^2 = 8$ it follows that the morphism is of degree 1. The (-2)-curves map to the singularities of $\sigma_1 = \sigma_2 = \sigma_4 = 0$. The rest easily follows.

(2.7) *Remark.* The variety $\sigma_1 = \sigma_4 = 0$ in $\mathbb{P}^5$ is a model of a Siegel modular variety and $\sigma_2 = 0$ defines a "Humbert surface" in it, see Ch. IX and [39].

D = 8. Our first example is related to $\Gamma = \Gamma[2]$. We set $\hat{\Gamma} = \Gamma[2] \cup \Gamma[2]\alpha$ with α defined by $\begin{pmatrix} 1 + \sqrt{2} & 0 \\ 0 & -1 + \sqrt{2} \end{pmatrix}$. We denote by τ the involution on X_Γ (or $X_{\hat{\Gamma}}$, Y_Γ, $Y_{\hat{\Gamma}}$) induced by $(z_1, z_2) \mapsto (z_2, z_1)$. We have the following diagram of surfaces

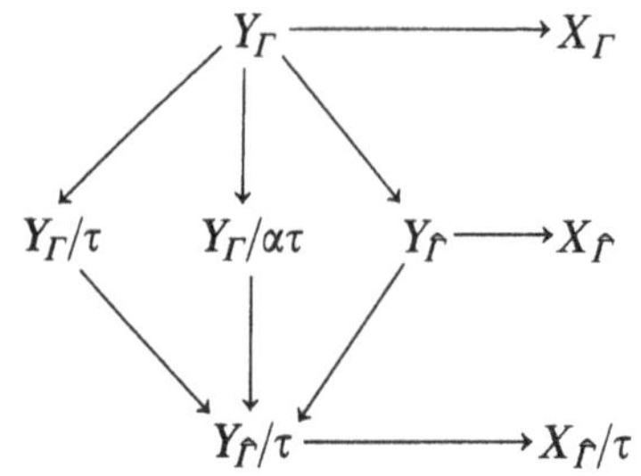

The group $\mathrm{PSL}(2, \mathbb{Z}/2\mathbb{Z})$, which is a central extension of $\mathfrak{S}_4$ by $\mathbb{Z}/2\mathbb{Z}$ (α defines the non-zero element of $\mathbb{Z}/2\mathbb{Z}$), acts on X_Γ, Y_Γ and Y_Γ/τ. It induces an action of $\mathfrak{S}_4$ on $Y_{\hat\Gamma}$, $X_{\hat\Gamma}$, $Y_{\hat\Gamma}/\tau$ and on $X_{\hat\Gamma}/\tau$.

Define a representation of $\mathfrak{S}_4$ on $V = \{(x_1, \ldots, x_4) \in \mathbb{C}^4 : \sum x_i = 0\}$ by $\sigma(x_i) = \operatorname{sgn} \sigma x_{\sigma(i)}$. Let $\mathbb{P}^2 = \mathbb{P}(V)$ and put

$$Q_1 = x_1x_2 + x_3x_4\,, \qquad Q_2 = x_1x_3 + x_2x_4\,, \qquad Q_3 = x_1x_4 + x_2x_3\,.$$

(2.8) Theorem. i) *$X_{\hat\Gamma}/\tau$ is isomorphic to $\mathbb{P}^2$ (with $\mathfrak{S}_4$-action). The cusps correspond to the orbit of* $(0:0:1:-1)$.

ii) *$Y_{\hat\Gamma}/\tau$ is isomorphic to the surface obtained by blowing up $\mathbb{P}^2$ twice in the orbit of* $(0:0:1:-1)$ *such that the strict transforms of $Q_1 = 0$, $Q_2 = 0$ and $Q_3 = 0$ intersect transversally.*

iii) *$X_{\hat\Gamma}$ is isomorphic to the double covering of $\mathbb{P}^2$ branched along the curves $\sigma_4 = 0$ and $Q_1Q_2Q_3 = 0$. It has six singular points.*

iv) *$Y_{\hat\Gamma}$ is the minimal resolution of singularities of $X_{\hat\Gamma}$. It is an Enriques surface blown up in four points.*

v) *Y_Γ is an unramified double cover of $Y_{\hat\Gamma}$. It is a $K3$ surface blown up in eight points.*

vi) *Y_Γ/τ is a $K3$ surface and is the minimal resolution of singularities of the double cover of $\mathbb{P}^2$ branched along $Q_1Q_2Q_3 = 0$.*

vii) *$Y_\Gamma/\alpha\tau$ is the minimal resolution of singularities of the double cover of $\mathbb{P}^2$ branched along $\sigma_4 = 0$.*

viii) *X_{Γ_K}/τ is isomorphic to $\mathbb{P}^2/\mathfrak{S}_4$.*

Proof. To each cusp (γ, δ) for Γ one can associate an Eisenstein series $E_{\gamma,\delta} = \lim_{s\to 0} \sum_{(\alpha,\beta)\equiv(\gamma,\delta) \bmod 2} \mathrm{N}(\alpha z + \beta)^{-1} |\mathrm{N}(\alpha z + \beta)|^{-s}$. These yield modular forms of weight 1 and character χ, where χ fits in the exact sequence $1 \to \Gamma \to \hat\Gamma \xrightarrow{\chi} \{+1, -1\} \to 1$. They are invariant under τ and span a 3-dimensional space E. The representation of $\mathfrak{S}_4$ on E is just the one defined on V.

The surface $X_{\hat\Gamma}/\tau$ is a non-singular regular surface with $c_1^2 = 9$, $c_2 = 3$, i.e. $X_{\hat\Gamma}/\tau$ is a complex projective plane. One can check that L_{X_Γ} is the pullback of $\mathcal{O}_{\mathbb{P}^2}(2)$. Altogether this proves $X_{\hat\Gamma}/\tau \cong_{\mathfrak{S}_4} \mathbb{P}(E)$.

The curve F_1 gives rise to 4 lines in $\mathbb{P}^2$. They are permuted by $\mathfrak{S}_4$. Choose coordinates such that F_1 is given by $\sigma_4 = 0$. Then i), ii) and viii) follow. Since the fixed point set of τ on $X_{\hat\Gamma}$ is $F_1 \cup F_2$ and F_2 corresponds to $Q_1Q_2Q_3 = 0$ iii) follows. By blowing down F_1 on Y_Γ one gets a $K3$ surface on which α acts freely. The rest is an easy consequence. □

(2.9) Theorem. i) $M(\Gamma) \cong \mathbb{C}[x_1, x_2, x_3, x_4, f, g]/(\sigma_1, f^2 - \sigma_4, g^2 - \sigma_3^2 + 4\sigma_2\sigma_4)$ ii) *If χ is the character of $\hat{\Gamma}$ defined by the exact sequence* $1 \to \Gamma \to \hat{\Gamma} \to \{1, -1\} \to 1$ *then* $M(\hat{\Gamma}, \chi) \cong \mathbb{C}[x_1, \ldots, x_4, h]/(h^2 - \sigma_4(\sigma_3^2 - 4\sigma_2\sigma_4))$. *Here x_i, f, g, $h = fg$ are modular forms of weights* 1, 2, 3, 5.

Proof. Since Y_Γ is a blown-up $K3$ surface with canonical divisor F_1 there exists a cusp form $f \neq 0$ of weight 2 on Γ vanishing on F_1. This implies f^2 is a multiple of σ_4. The dimension formula $\dim S_k(\Gamma) = 2k(k-2) + 2$ $(k \geqslant 3)$ predicts a cusp form g of weight 3 not divisible by a modular form of weight 1 or 2. This implies $\tau^* g = g$ (otherwise $f | g$) and $\dim S_3(\Gamma, \chi) = 4$ implies $\alpha^* g = -g$. Hence $(\alpha\tau)^* g = -g$, so g vanishes on the fixed point set F_2 of $\alpha\tau$. We see that $\alpha^*(fg) = fg$ since $\alpha^* f = -f$ ($Y_{\hat{\Gamma}}$ is an Enriques surface), so $h = fg$ lives on $\hat{\Gamma}$ and h^2 vanishes on $F_1 \cup F_2$. The dimension formulas now tell us that the elements thus found generate the whole rings. □

(2.10) Corollary. i) $M(\Gamma_K) \simeq \mathbb{C}[\sigma_2, \sigma_4, \sigma_3^2, \sigma_3\Delta, k]/(R, k^2 - \sigma_4\sigma_3^2)$, *where* $\Delta = \prod(x_i - x_j)$ *and* $R = 27(\sigma_3\Delta)^2 + [-4(\sigma_2^2 + 12\sigma_4)^3 + (27\sigma_3^2 + 2\sigma_2^3 - 72\sigma_2\sigma_4)^2]\sigma_3^2$.
ii) $M^{\mathrm{sym}}(\Gamma_K) \simeq \mathbb{C}[\sigma_2, \sigma_4, \sigma_3^2, \sigma_3\Delta]/(R)$.

Proof. One checks that τ acts on σ_3 and Δ by a change of sign. Thus ii) follows from i). By the dimension formula one knows that there is a skew-symmetric modular form of weight 5 on Γ_K vanishing on F_1 $(\sigma_4 = 0)$ and F_4 $(\sigma_3^2 = 0)$. □

A large part of this case is due to Hirzebruch; some results on the rings of modular forms were obtained by Gundlach [53]. In [120] Nagaoka proved that over $\mathbb{Z}$: $M_{\mathrm{ev}}^{\mathrm{sym}}(\Gamma_K)/\mathbb{Z} \cong \mathbb{Z}[x_2, x_4, x_6]$ (subscript = weight).

The second example for $D = 8$ is $\Gamma = \Gamma[\mathfrak{p}_7]$ with $\mathfrak{p}_7$ a prime ideal of norm 7. The eight cusps of X_Γ are resolved by cycles $(4, 2, 4, 2, 4, 2)$ giving rise to configurations

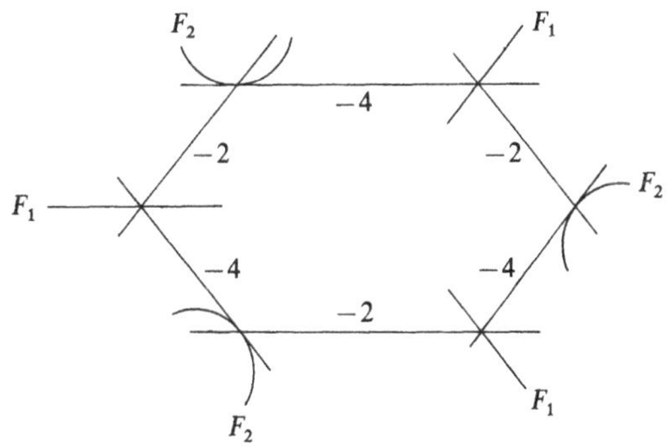

The curve F_1 is irreducible $(\simeq \overline{\Gamma(7)\backslash\mathfrak{H}})$ and has 24 cusps. The invariants are $q = 0$, $p_g = 6$, $c_1^2 = 8$. The surface admits an action by $G_{168} = \mathrm{PSL}(2, \mathbb{F}_7)$. The divisor $G = 2F_1 +$ all (-2)-curves of the cusps is a 2-canonical divisor.

(2.11) Theorem ([77]). *Y_Γ is the desingularization of the double cover of $\mathbb{P}^2$ ramified along the two unique curves A and B of degree 4 and 6 in $\mathbb{P}^2$ which are invariant under the non-trivial action of G_{168} on $\mathbb{P}^2$. The canonical map is the composition of $Y_\Gamma \to \mathbb{P}^2$ with the Veronese map $\mathbb{P}^2 \to \mathbb{P}^5$ given by conics.*

Proof. Y_Γ is minimal (use the divisor above). By Horikawa's results the canonical map ϕ_K maps Y_Γ with degree 2 onto a surface of degree 4 in $\mathbb{P}^5$, i.e. either the Veronese surface or a ruled surface, see [87]. But ruled surfaces do not admit a G_{168}-action. The curves F_1 and F_2 map to invariant curves of degree 4 and 6 since $KF_1 = 8$, $KF_4 = 12$. Since G contains F_1 twice F_1 is contained in the ramification locus, hence the branch locus is reducible. The divisor $2F_2$ + all (-2)-curves in the cusps is 3-canonical, hence also F_2 is contained in the ramification locus. □

This example is remarkable since Y_Γ carries an involution which is not modular, i.e. does not come from $\mathfrak{H}^2$. Indeed, the involution corresponding to Φ_K does not map the cusp resolutions to cusp resolutions. This phenomenon also occurs for modular curves, cf. Ogg [126].

D = 12. Let $\Gamma = \Gamma[\mathfrak{p}_2] \subset \mathrm{PSL}(2, \mathfrak{o})$ with $\mathfrak{p}_2$ the prime ideal of norm 2. The surface Y_Γ is a blown-up $K3$ surface with $c_1^2 = -2$. The two components of F_1 are exceptional curves. The maps $\mathfrak{H}^2 \to \mathfrak{H}^2$ given by $(z_1, z_2) \mapsto ((2+\sqrt{3})z_1, (2-\sqrt{3})z_2)$ and $(z_1, z_2) \mapsto (z_2, z_1)$ define two involutions ε and τ on Y_Γ which commute with the action of $\mathfrak{S}_3 \cong \mathrm{PSL}(2, \mathfrak{o}/\mathfrak{p}_2)$.

The divisor F_1 + (all cusp resolutions) is the divisor of a cusp form y of weight 2, cf. Ch. VII. There are three Eisenstein series x_1, x_2, x_3 of weight 2 and they are permuted by $\mathfrak{S}_3$. Moreover, $\sigma \in \mathfrak{S}_3$ acts on y by $y \mapsto \mathrm{sgn}(\sigma) y$. One sees immediately the $\tau^* x_i = x_i$. Since $Y_\Gamma/\varepsilon\tau$ is rational (the cusp resolutions and F_1 yield intersecting exceptional curves) the differential form $y\, dz_1 \wedge dz_2$ is not invariant, hence $(\varepsilon\tau)^* y = y$.

Consider now the rational map ϕ: $Y_\Gamma \to \mathbb{P}^3$ defined by x_1, x_2, x_3, y. It factors through $\varepsilon\tau$. One easily checks that the image is a surface, and since $c_1(L)^2 = 4$ it must be a quadric, invariant under $\mathfrak{S}_3$ and containing the $\mathfrak{S}_3$-orbit of $(0:0:1:0)$, i.e. given by $y^2 = \sigma_2$ (σ_i: i-th symmetric function of x_i). This quadric is a double cover of $\mathbb{P}^2$ ramified along $\sigma_2 = 0$, the image of F_1.

The forms x_1, x_2, x_3 and y generate a subring of $M_{\mathrm{ev}}(\Gamma)$ whose graded part of weight $2k$ has dimension $(k+1)^2$. Since we know that $\dim M_{2k}(\Gamma) = (k+1)^2 + (k-2)^2$ a new cusp form appears in $M_6(\Gamma)$, call it g. The forms x_1, x_2, x_3, y and g generate a subring whose weight $2k$-part has dimension $\geqslant (k+1)^2 + (k-2)^2$; hence they generate $M_{\mathrm{ev}}(\Gamma)$.

(2.12) Theorem. *The surface Y_Γ is the minimal resolution of singularities of the double cover of $\mathbb{P}^1 \times \mathbb{P}^1$ ramified along a curve of type (6,6), the image of F_6. The ring $M_{\mathrm{ev}}(\Gamma)$ equals $\mathbb{C}[x_1, x_2, x_3, y, g]/(y^2 - \sigma_2, g^2 - G)$, where G is a symmetric polynomial of degree 6 in x_1, x_2, x_3.*

(2.13) Corollary. *$M_{ev}(\Gamma_K) \cong \mathbb{C}[\sigma_1, \sigma_2, \sigma_3, g, \Delta y]/(g^2 - G, \Delta^2 y^2 - \sigma_2 H)$, where $\Delta = \prod_{i<j}(x_i - x_j)$ and $\Delta^2 = H$ is the usual discriminant relation.*

Note that this checks with Proposition (1.1). Note also that by VII (7.3), $F_6 +$ 3((-2)-curves in cusps) + 2((-3)-curves in cusps) + ((-2)-curves from quotient singularities on F_6) belongs to $|3L|$.

Next, consider the action of $\Gamma = \Gamma[2]$ on $\mathfrak{H} \times \mathfrak{H}_-$. Now Y_Γ is a minimal surface of general type with $c_1^2 = 8$, $p_g = 5$. The matrix $\begin{pmatrix} 2+\sqrt{3} & 0 \\ 0 & 2-\sqrt{3} \end{pmatrix}$ defines an extension $\hat{\Gamma}$ of Γ and an involution α on Y_Γ with 8 isolated fixed points. The resolved quotient $Y_{\hat{\Gamma}}$ is a surface with $c_1^2 = 4$, $p_g = 3$.

Consider in $\mathbb{P}^2 = \{x \in \mathbb{P}^3 : \sum x_i = 0\}$ with $\mathfrak{S}_4$-action by permutation of coordinates the four conics:

$$Q_i\colon x_i^2 + x_j x_k + x_j x_l + x_k x_l = 0 \qquad i = 1, \ldots, 4,$$
$$\{i, j, k, l\} = \{1, 2, 3, 4\}$$

In the double cover $\mathbb{P}^1 \times \mathbb{P}^1$ of $\mathbb{P}^2$ ramified along σ_2 there lie 8 components over the Q_i. The next theorem is due to Hirzebruch [80].

(2.14) Theorem. *The canonical system on $Y_{\hat{\Gamma}}$ defines a $\mathfrak{S}_4$-equivariant 4 : 1-morphism of $Y_{\hat{\Gamma}}$ to $X_{\hat{\Gamma}_\tau}/\beta = \mathbb{P}^2$ which factors through $X_{\hat{\Gamma}_\tau} \cong \mathbb{P}^1 \times \mathbb{P}^1$. The surface $Y_{\hat{\Gamma}}$ is the minimal resolution of the double cover of $X_{\hat{\Gamma}_\tau}$ ramified along F_3 (= inverse image of* the Q_i). Here β is induced by: $z \to ((1 + \sqrt{3})z + 2)/(-2z + 1 - \sqrt{3})$.

D = 13. We consider Y_Γ with $\Gamma = \ker[\mathrm{PSL}(2, \mathfrak{o}) \to \mathrm{PSL}(2, \mathfrak{o}/2\mathfrak{o})]$. This surface was studied in detail in [42], so we omit detailed proofs. It is a surface of general type with invariants $c_1^2 = -5$, $\chi = 5$, $q = 0$ with an action of the alternation group $\mathfrak{A}_5$ which permutes the 5 cusps. Each cusp yields a configuration

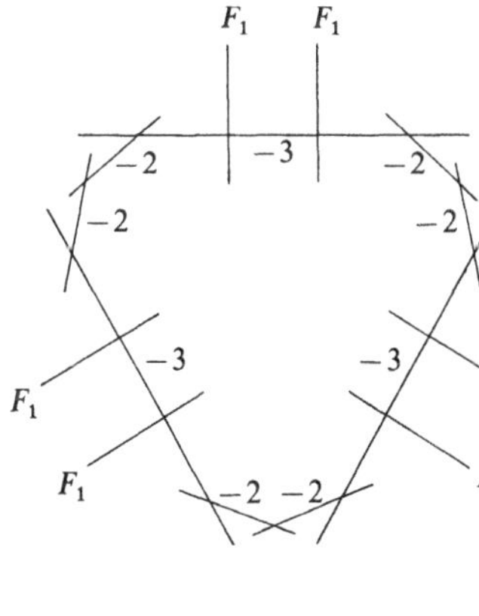

By blowing down F_1 (10 components) one gets a surface Y_Γ^0 with $c_1^2 = 5$. The curve F_4 has 5 components, each of which intersects only one cusp. A component of F_4 and all curves of the cusp resolution it intersects form a canonical divisor. We thus

find five canonical divisors, hence five cusp forms s_i $i = 1, \ldots, 5$ of weight 2 which are permuted by $\mathfrak{A}_5$.

(2.15) Theorem. *The cusp forms* s_i $i = 1, \ldots, 5$ *of weight* 2 *define a morphism of degree* 1 *of* Y_Γ^0 *onto a quintic surface in* $\mathbb{P}^4$ *given by* $\sum_{i=1}^5 x_i = 0$, $12\sum_{i=1}^5 x_i^5 = 5\sum_{i=0}^5 x_i^2 \sum_{j=0}^5 x_j^3$.

Using this model the ring of modular forms for Γ and for Γ_K was determined completely, e.g.

$$M_{ev}(\mathrm{SL}(2,\mathfrak{o})) \cong \mathbb{C}[X_2, X_4, X_6, Y_6, X_8]/(X_4^3 - X_6 Y_6, X_8^2 - P),$$

where subscripts denote the weight and P is a polynomial (given explicitly) in X_2, X_4, X_6, Y_6.

D = 17. The surface $Y_{\Gamma[(2)]}$ is a surface of general type with $c_1^2 = 3$, $\chi = 6$, admitting an action of $G = \mathfrak{S}_3 \times \mathfrak{S}_3$. We have a diagram

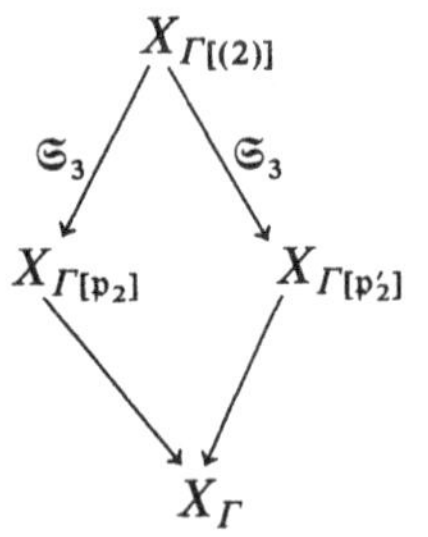

where $(2) = \mathfrak{p}_2\mathfrak{p}_2'$ and $\mathfrak{S}_3$ denotes a Galois group. The involution τ acts on $X_{\Gamma[(2)]}$ and on this diagram by interchanging $X_{\Gamma[\mathfrak{p}_2]}$ and $X_{\Gamma[\mathfrak{p}_2']}$. $X_{\Gamma[\mathfrak{p}_2]}$ (resp. $X'_{\Gamma[\mathfrak{p}_2']}$) has 3 cusps indexed by $1 \leqslant i \leqslant 3$ (resp. by $1 \leqslant j \leqslant 3$). The cusps of $X_{\Gamma[2]}$ are then indexed by (ij) with G acting in the obvious way.

On $Y_{\Gamma[(2)]}$ the six components of F_1 are exceptional curves and one obtains a minimal surface Y° with $c_1^2 = 9$ by blowing them down. There exists a canonical divisor on Y° consisting of (the image under $Y_{\Gamma[(2)]} \to Y^\circ$ of) all curves in cusp resolutions (11), (22) and (33) and the 6 components of F_2 intersected by them.

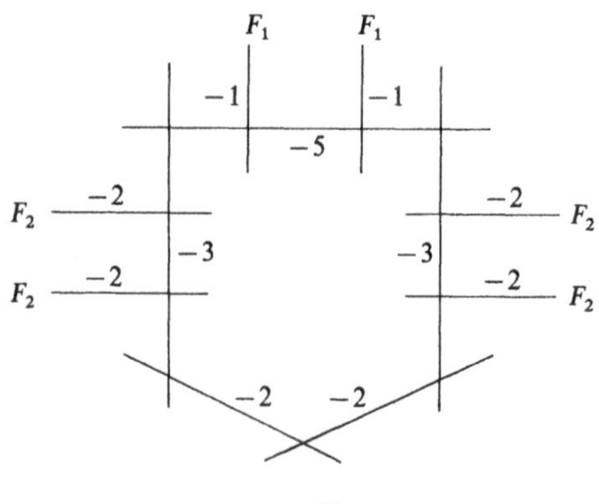

cusp 11

Let s be a corresponding section of $H^\circ(Y^\circ, K_{Y^\circ})$. Under the action of $\mathfrak{S}_3 \times (e) \subset G$ one obtains 6 sections s_σ with $\sum s_\sigma = 0$ since $p_g(Y_{\Gamma_K}) = 0$. The space $H^\circ(Y^\circ, K_{Y^\circ})$ decomposes as a sum $V_1 \oplus V_2 \oplus W$ of irreducible invariant subspaces under $\mathfrak{S}_3$ with $\dim V_i = 2$, $\dim W = 1$. Moreover, $\tau^* V_1 = V_2$. We choose $x_1, x_2, x_3 \in V_1 - \{0\}$ with $\sum x_i = 0$ which are permuted by $\mathfrak{S}_3$. Let σ_j be the j-th elementary symmetric function in the x_i and the $\sigma'_j = \tau^*\sigma_j$. Choose also $t \in W - \{0\}$. Then σ_2, σ'_2, σ_3, σ'_3 are G-invariants, whereas t and $\Delta = \prod_{i<j}(x_i - x_j)$, $\Delta' = \tau^*\Delta$ are semi-invariants.

(2.16) Theorem. *The canonical map $\phi\colon Y^\circ \to \mathbb{P}^4$ is a morphism of Y° onto the intersection of two cubic threefolds in $\mathbb{P}^4$. The image surface has nine singularities of type A_2 and nine of type A_1. It can be defined by equations*

$$\lambda t^3 + \mu t(\sigma_2 + \sigma'_2) + \Delta + \Delta' = 0$$

$$\sigma_3 + \sigma'_3 = 0$$

for some $\lambda, \mu \in \mathbb{C}$.

D = 21. Let $\Gamma = \Gamma[(2)] \cup \Gamma[(2)]\varepsilon$, where $\varepsilon = 55 + 12\sqrt{21}$ acts by $(z_1, z_2) \mapsto (\varepsilon z_1, \varepsilon' z_2)$. The surface Y_Γ is not minimal, by blowing down F_1 one obtains a minimal surface Y° of general type with $c_1^2 = 5$, $\chi = 5$, $q = 0$ with an action of $\mathfrak{S}_5$ (generated by $\hat{\Gamma}_K/\Gamma$ and τ). We proved in [42]:

(2.17) Theorem. *The canonical map $Y_\Gamma^\circ \to \mathbb{P}^3$ is a $\mathfrak{S}_5$-equivariant morphism of degree 1 onto the quintic surface*

$$\{(x_1 : \cdots : x_5) \in \mathbb{P}^4 \colon \sigma_1 = 0, 2\sigma_5 + \sigma_2\sigma_3 = 0\}$$

(σ_i: elem. symm. function in the x_i; $\mathfrak{S}_5$ acts by permuting the coordinates). This quintic has 20 ordinary double points corresponding to the 20 isolated fixed points of ε on $Y_{\Gamma[(2)]}$.

The canonical map of $Y_{\Gamma[(2)]}$ factors through that of Y_Γ. (Note that $p_g(Y_{\Gamma[(2)]}) = p_g(Y_\Gamma) = 4$.)

D = 24. We choose for $\Gamma \subset \mathrm{PSL}(2, \mathfrak{o})$ the principal congruence subgroup associated to the ideal of norm 2. Let Y_Γ° be the surface obtained from Y_Γ by blowing down F_1. We proved in [36]:

(2.18) Theorem. *Y_Γ° is a minimal surface of general type with $c_1^2 = 2$, $p_g = 3$ and $q = 0$. It is isomorphic to the minimal resolution of singularities of the double cover of $\mathbb{P}^2 = \{(x_0 : x_1 : x_2 : y) \in \mathbb{P}^3 \colon \sum x_i = 0\}$ branched along the curve $(\sigma_2 + y^2)(\sigma_3^2 + \sigma_2 y^4 + y^6) = 0$, the image of F_3 and F_{12}. (σ_i: elem. symm. function in x_i.)*

In [36] we deduced from this result the structure of $M(\mathrm{SL}(2, \mathfrak{o}))$. This ring is already quite complicated.

Next, we let $\Gamma = \Gamma[\mathfrak{p}_3]$ with $\mathfrak{p}_3 = (3 + \sqrt{6})$.

(2.19) Theorem. *The Hilbert modular surface Y_Γ associated to the action of Γ on $\mathfrak{H} \times \mathfrak{H}_-$ is a minimal surface of general type with $c_1^2 = 8$, $\chi = 7$, $q = 0$. Its canonical map is a morphism of degree 2 onto the Veronese surface $\mathbb{P}^2 \subset \mathbb{P}^5$. Y_Γ is the minimal resolution of singularities of the double cover of $\mathbb{P}^2 = \{(x_1, \ldots, x_4) \in \mathbb{P}^3 : \sum x_i = 0\}$ branched along the curve $\sigma_2(9\sigma_3^2\sigma_2 - \sigma_4\sigma_2^2 + 12\sigma_4^2) = 0$, where σ_i is the i-th elem. symmetric function in $x_1, \ldots, x_4$.*

Y_Γ has four cusps, which we number 1, 2, 3, 4. The curve F_6 has 6 components F_6^{ij} $1 \leqslant i < j \leqslant 4$ intersecting cusp i as follows:

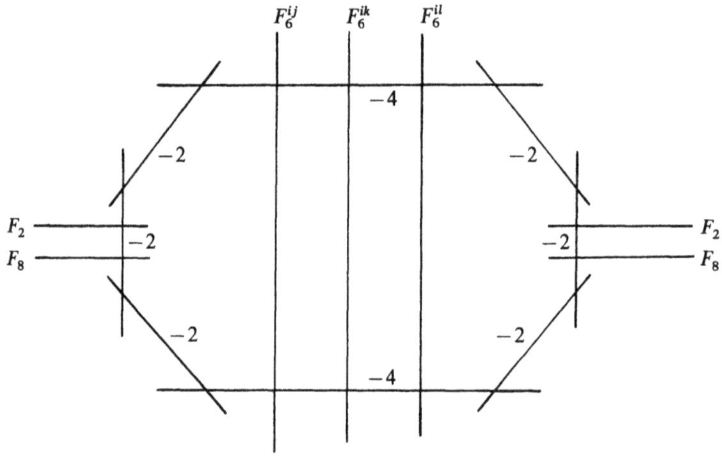

cusp i $\{i, j, k, l\} = \{1, 2, 3, 4\}$

One checks that

$$2 \sum_{i,j \neq 3,4} F_6^{ij} + (\text{curves of cusp resolution 1 and 2 with multiplicity 1})$$

is a canonical divisor, say the zero divisor of $s_{12} \in \Gamma(Y_\Gamma, K_{Y_\Gamma})$. Using the action of $\mathfrak{A}_4$ one obtains six sections s_{ij} of $K_{Y_\Gamma} = \phi^*\mathcal{O}(2)$, with $\phi\colon Y_\Gamma \to \mathbb{P}^2$ the canonical map. Let $\mathcal{L} = \phi^*\mathcal{O}(1)$. We can choose $s_i \in \Gamma(Y_\Gamma, \mathcal{L})$ such that $s_i s_j = s_{ij}$. Let σ_i be the i-th elementary symmetric function in the s_i. Then one can prove (cf. [36])

(2.20) Theorem. *The graded ring of Hilbert modular forms for the action of Γ on $\mathfrak{H} \times \mathfrak{H}_-$ is generated by the forms $\sigma_3/\sqrt{\sigma_4}$, σ_2, $\delta/\sqrt{\sigma_4}$, σ_4 and Δ of weights 1, 2, 3, 4 and 6, respectively, satisfying the relations implied by*

$$\Delta^2 = 256\sigma_4^3 - 128\sigma_4^2\sigma_2^2 + 16\sigma_4\sigma_2^4 + 144\sigma_4\sigma_3^2\sigma_2 - 27\sigma_3^4 - 4\sigma_3^2\sigma_2^3$$

$$\delta^2 = \sigma_2(9\sigma_3^2\sigma_2 - \sigma_4\sigma_2^2 + 12\sigma_4^2)$$

and by the notation.

D = 28. Let $\Gamma = \ker\{\mathrm{PGL}^+(2, \mathfrak{o}) \to \mathrm{PGL}(\mathfrak{o}/\mathfrak{p}_2)\}$ with $\mathfrak{p}_2$ the prime ideal of norm 2 be the extended congruence subgroup of level $\mathfrak{p}_2$.

(2.21) Theorem. *The Hilbert modular surface Y_Γ associated to the action of Γ on $\mathfrak{H} \times \mathfrak{H}_-$ is the minimal desingularization of the double cover of $\mathbb{P}^2$ branched along four conics given by*

$$(\sigma_1^2 - 4\sigma_2)(4\sigma_2^2\sigma_1^2 + 2\sigma_1^2\sigma_3 + \sigma_3^2) = 0 .$$

Y_Γ carries three cusp resolutions which look like

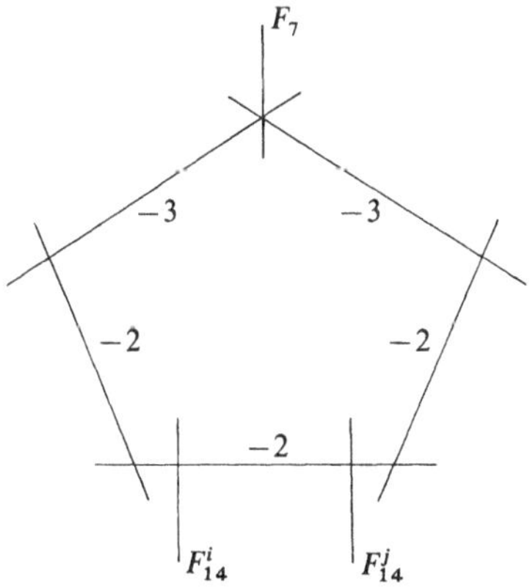

cusp k $\{i, j, k\} = \{0, 1, 2\}$

Y_Γ is minimal and has $c_1^2 = 2$, $p_g = 3$. The canonical map is holomorphic of degree 2 onto $\mathbb{P}^2$ ramified along F_7 and F_{14}. F_{14} consists of 3 components F_{14}^i with $KF_{14}^i = 2$. Also $KF_7 = 2$. The configuration now suffices to determine the equation.

D = 40. Let Γ be the principal congruence subgroup of $SL(2, \mathfrak{o})$ of level $\mathfrak{p}$, the prime ideal of norm 2. Then Y_Γ is a surface of general type with $\chi = 5$, $c_2 = 56$, $c_1^2 = 4$. There are six cusps, each resolved by a cycle

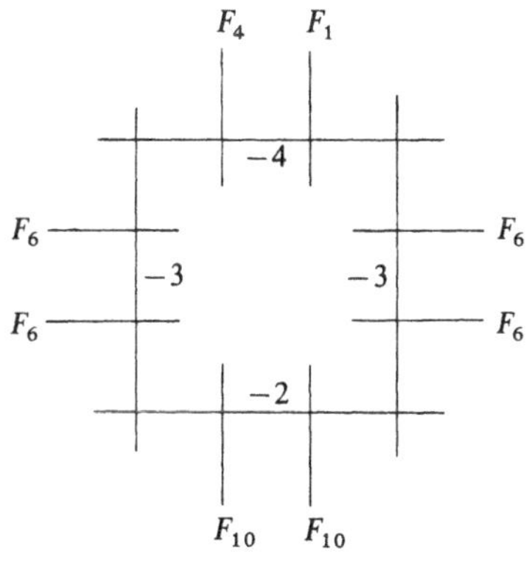

F_1 has two components and these are exceptional curves. Blowing them down yields a minimal surface Y° with $c_1^2 = 6$. There exists a canonical divisor

$$2F_4^{(1)} + \sum_{i=1}^{4} F_6^{(i)} + R \qquad (*)$$

with $F_4^{(1)}$ one of the three components of F_4, $F_6^{(1)}, \ldots, F_6^{(4)}$ the four components of F_6 intersecting the same two cusps as $F_4^{(1)}$ does and R the sum of the resolution curves of the two cusps intersected by $F_4^{(1)}$ and the two (-2)-curves resolving the quotient singularities on $F_4^{(1)}$. The group $\mathrm{SL}(2, \mathfrak{o}/\mathfrak{p}) = \mathfrak{S}_3$ acts on Y°. Let y_1 be a section of the canonical bundle of Y° having $(*)$ as divisor. The action of the alternating group $\mathfrak{A}_3$ yields three sections y_1, y_2, y_3 that are linearly independent. There is a further section of K_{Y° spanning a $\mathfrak{S}_3$-invariant subspace and linearly independent of y_1, y_2, y_3. Call it t. Let α be a generator of the Hurwitz-Maass extension of Γ. Then α is an involution with twelve isolated fixed points on Y°. The quotient Y°/α has as minimal resolution of singularities a surface with $p_g = c_1^2 = 3$. Using the trace formula one finds $y_i^\alpha = y_i$, $t^\alpha = -t$. One checks $y_i^\tau = -y_i$, $t^\tau = t$. Hence $-\alpha\tau$ is the identity on $H^\circ(Y^\circ, \mathrm{K}_{Y^\circ})$.

(2.22) Theorem. *The canonical map $Y^\circ \to \mathbb{P}^3$ given by the cusp forms of weight 2: $x \mapsto (y_1(x) : y_2(x) : y_3(x) : t(x))$ is a holomorphic map of degree 2 onto the cubic surface given by*

$$\sigma_1^3 - 4\sigma_1\sigma_2 + 8\sigma_3 - 4\sigma_1 t^2 = 0$$

with σ_i the i-th elementary symmetric function in the y_i. This cubic has three ordinary double points (the orbit of $(0:1:1:0)$) and contains twelve lines, the image of F_{10} ($t = 0$, three lines) and the image of the cusps ($\sigma_3 = 0$, nine lines).

D = 44. Let $\Gamma = \ker\{\mathrm{PGL}^+(2, \mathfrak{o}) \to \mathrm{PGL}(2, \mathfrak{o}/\mathfrak{p}_2)\}$ with $\mathfrak{p}_2$ the prime ideal of norm 2, act on $\mathfrak{H} \times \mathfrak{H}_-$. The minimal resolution of singularities Y of the quotient contains three exceptional curves; the components of F_2. After blowing them down we have the following configuration coming from the cusp resolutions on the resulting surface Y°

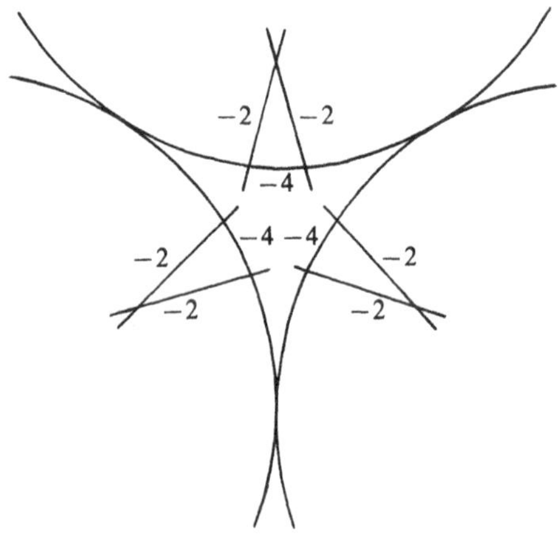

(2.23) Theorem. *Y° is a minimal surface of general type with $c_1^2 = 2$, $p_g = 3$. It is the minimal resolution of singularities of the double cover of $\mathbb{P}^2$ branched along the curve given by*

$$\sigma_2(\sigma_1^6 - 4\sigma_1^4\sigma_2 + 8\sigma_1^3\sigma_3 + 16\sigma_3^2) = 0\,,$$

where σ_i is the i-th elementary symmetric function in the coordinates x_0, x_1, x_2 of $\mathbb{P}^2$.

In all cases where the surface is a minimal surface of general type with $c_1^2 = 2p_g - 4$ we can apply Horikawa's results. If $p_g = 3$ this implies that the canonical map is a holomorphic map of degree 2 onto $\mathbb{P}^2$ which contracts all trees of (-2)-curves. The branch locus is of degree 8, in our case it is the image of F_4 and F_{11}.

There are many more cases with $c_1^2 = 2$, $p_g = 3$, where one can apply these methods.

D = 113. $Y^\circ(113)$ is a surface of general type with $p_g = 4$, $c_1^2 = 6$. It is also minimal. According to Horikawa the canonical map is a morphism of degree 2 onto a cubic surface in $\mathbb{P}^3$; it factors through τ. We have a configuration

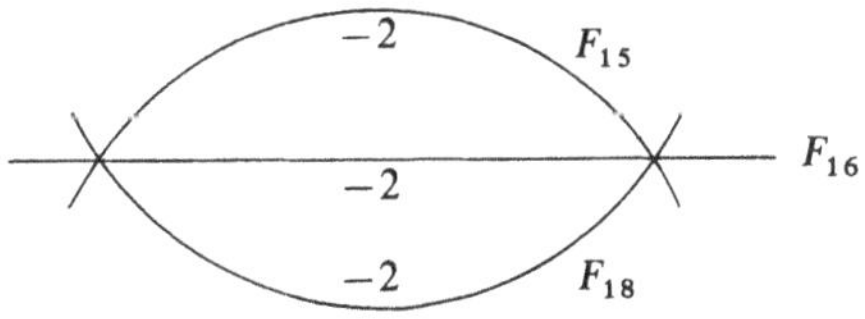

of three elliptic curves. The intersection points have quadratic form $15x^2 + 13xy + 16y^2$. This is a configuration D with $D^2 = \mathrm{K}D = \mathrm{K}^2 = 6$ (showing that $Y^\circ(113)$ is minimal). The following theorem is due to Hirzebruch:

(2.24) Theorem ([79]). *The canonical map of $Y^\circ(113)$ to $\mathbb{P}^3$ is a morphism of degree 2 onto the cubic surface*

$$xyw + x^3 - xz^2 - yz^2 + y^3 = 0\,.$$

Since the canonical map factors through τ and contracts all (-2)-curves it is easy to see that this cubic has a singularity of type A_3 (whose counter image is a set of two strings of 3 (-2)-curves in the cusp). Moreover, this cubic also has an Eckardt point (three lines passing through a point). This determines the cubic, see [79].

In this case one knows on $Y^\circ(113)$ all (-2)-curves, and all (-3)-curves (because they map to lines). They are all "known" curves, i.e. coming from F_N or the resolutions of singularities.

To conclude this section we give some examples—due to Bassendowski—for rather big discriminants.

(2.25) Theorem. *Let $D = 401$ or 541 and let Y_τ° be the minimal model of $Y(D)/\tau$. The canonical map is a morphism of degree 1 onto a sextic surface*

$$(x_1x_2x_3)^2 + Ax_0x_1x_2x_3 + Bx_0^2 = 0$$

with $\deg A = 2$, $\deg B = 4$ *in* x_0, x_1, x_2, x_3. *This sextic has a double curve consisting of three lines. All other singularities are rational double points.*

The theorem is proved using Horikawa's results and explicit canonical divisors. E.g. for $D = 401$ the counter image of the three lines consists of $F_{57} + F_{46} + F_{41} + F_{38}$ + resolution curves on Y_τ°, see [11].

IX. Humbert Surfaces

This chapter is devoted to complex abelian surfaces whose endomorphism ring contains an order from a real quadratic field. The moduli spaces of such abelian surfaces are Hilbert modular surfaces. Since the moduli spaces of polarized complex abelian varieties are Siegel modular varieties we find natural maps of Hilbert modular surfaces to Siegel modular threefolds. The images are called Humbert surfaces, after G. Humbert, who studied abelian surfaces with real multiplication, i.e. with an endomorphism ring that contains an order from a real quadratic field.

Section 1 describes the moduli of complex abelian varieties with real multiplication and the maps to the Siegel modular varieties. In Section 2 we consider the Humbert surfaces in more detail. In Section 3 we give examples of Siegel modular varieties and the Humbert surfaces in them. In the last section we make some remarks about curves of genus 2 whose Jacobian has real multiplication. The moduli of abelian schemes with real multiplication will be considered in detail in the next chapter.

1. Modular Embeddings

Take an abelian variety A over $\mathbb{C}$. The underlying complex manifold of A is a complex torus $T = V/\Lambda$, with V an n-dimensional complex vector space and $\Lambda \subset V$ a lattice. Since A is algebraic there exists an ample line bundle L on A whose first Chern class (in $H^2(A, \mathbb{Z}) = \bigwedge^2 \operatorname{Hom}(\Lambda, \mathbb{Z})$) gives a non-degenerate alternating bilinear form $E: \Lambda \times \Lambda \to \mathbb{Z}$. The $\mathbb{R}$-linear extension of E represents the image of E under $H^2(A, \mathbb{Z}) \hookrightarrow H^2(A, \mathbb{C})$, where the latter group has been identified with $\bigwedge^2 \operatorname{Hom}_{\mathbb{R}}(V, \mathbb{C}) = \bigwedge^2(W \oplus \overline{W})$ with $W = \operatorname{Hom}_{\mathbb{C}}(V, \mathbb{C})$. Since E lies in the (1,1)-part of H^2, i.e. in $W \otimes \overline{W}$ it follows that $E(ix, iy) = E(x, y)$ for all $x, y \in V$. It is well-known that E is ample if and only if $E(ix, x) > 0$ for all non-zero $x \in V$.

(1.1) Definition. Let $T = V/\Lambda$ be a complex torus. A Riemann form on T is a non-degenerate alternating bilinear form $\Lambda \times \Lambda \to \mathbb{Z}$ whose $\mathbb{R}$-linear extension to $V \times V$ satisfies

i) $E(ix, iy) = E(x, y)$ for all $x, y \in V$

ii) $E(ix, x) > 0$ for all $x \neq 0$, $x \in V$.

Thus the first Chern class of an ample line bundle on T defines a Riemann form. Conversely, every Riemann form occurs as the first Chern class of an ample line bundle on T. The existence of a Riemann form is a necessary and sufficient condition for T to be algebraic.

By a polarized abelian variety we shall mean in this chapter a complex torus provided with a Riemann form.

We recall the standard description of the moduli of n-dimensional polarized abelian varieties.

Take a vector space $W = \mathbb{Q}^{2n}$ of dimension $2n$ over $\mathbb{Q}$ and equip it with an alternating form

$$J = \begin{pmatrix} 0 & 1_n \\ -1_n & 0 \end{pmatrix}.$$

Let

$$\mathrm{Sp}(2n, \mathbb{Q}) = \{g \in \mathrm{GL}(2n, \mathbb{Q}) \colon {}^t g J g = J\}$$

be the symplectic group fixing this form. If we write $\alpha \in \mathrm{GL}(2n, \mathbb{Q})$ as 2×2 matrix of $n \times n$ blocks

$$\alpha = \begin{pmatrix} A & B \\ C & D \end{pmatrix}$$

then the condition says ${}^t AC = {}^t CA$, ${}^t AD - {}^t CB = 1$, ${}^t BD = {}^t DB$. Choose positive integers $d_1, \ldots, d_n$ with $d_i | d_{i+1}$ and set

$$\begin{aligned} \mathfrak{L} &= \mathbb{Z}^n \times \prod_{i=1}^{n} \mathbb{Z} d_i \subset W \\ \mathrm{Sp}(\mathfrak{L}) &= \{g \in \mathrm{Sp}(2n, \mathbb{Q}) \colon g\mathfrak{L} = \mathfrak{L}\}\,. \end{aligned} \tag{1}$$

Furthermore, define

$$\mathfrak{H}_n = \{\tau \in M_n(\mathbb{C}) \colon {}^t\tau = \tau, \operatorname{Im}\tau \text{ positive definite}\}\,,$$

the *Siegel upper half plane* of degree n. The group $\mathrm{Sp}(2n, \mathbb{R})$ acts on $\mathfrak{H}_n$ by

$$\tau' \mapsto \alpha\tau = (A\tau + B)(C\tau + D)^{-1} \qquad \alpha = \begin{pmatrix} A & B \\ C & D \end{pmatrix}.$$

This action is not effective, but $\mathrm{PSp}(2n, \mathbb{R}) = \mathrm{Sp}(2n, \mathbb{R})/\{\pm 1\}$ acts effectively and it is the full group of analytic automorphisms of the complex domain $\mathfrak{H}_n$. The groups $\mathrm{PSp}(\mathfrak{L})$ = image of $\mathrm{Sp}(\mathfrak{L})$ in $\mathrm{PSp}(2n, \mathbb{R})$ act properly discontinuously on $\mathfrak{H}_n$, but in general not freely.

The meaning of the quotient space is:

$$\mathrm{PSp}(\mathfrak{L})\backslash\mathfrak{H}_n = \begin{cases} \text{moduli space of pairs } (A, E) \text{ of complex} \\ n\text{-dimensional abelian varieties } A \text{ with Riemann} \\ \text{form } E \text{ such that } (H_1(A, \mathbb{Z}), E) \simeq (\mathfrak{L}, J|\mathfrak{L})\,. \end{cases}$$

For an abelian variety we let $\mathrm{End}(A)$ be the endomorphism ring of A and we put $\mathrm{End}^\circ(A) = \mathrm{End}(A) \otimes_{\mathbb{Z}} \mathbb{Q}$. If A is simple then $\mathrm{End}^\circ(A)$ is a division algebra over

$\mathbb{Q}$. A polarization (i.e. Riemann form) on A defines an involution on $\mathrm{End}^\circ(A)$, $a \mapsto a^\circ$, called the *Rosati involution*. It is determined by $E(ax, y) = E(x, a^\circ y)$. The important fact is that it is positive: $\mathrm{Tr}(aa^\circ) > 0$ (if $a \neq 0$). Division algebras with positive involution have been classified. In the surface case ($n = 2$) this yields:

(1.2) Proposition. *If A is a simple abelian surface then* $\mathrm{End}(A)$ *is one of the following*: i) $\mathbb{Z}$, ii) *an order in a real quadratic field*, iii) *an order in an indefinite quaternion (division) algebra over* $\mathbb{Q}$, iv) *an order in a quartic CM-field* (*i.e. a totally imaginary extension of a real quadratic field*). *If A is isogenous to a product of elliptic curves $E_1 \times E_2$ then* $\mathrm{End}(A) = \mathbb{Z}$ *if E_1 and E_2 are not isogenous. If they are isogenous, then* $\mathrm{End}(A)$ *is an order in $M_2(F)$ with $F =$* $\mathrm{End}^\circ(E_i)$. ($F = \mathbb{Q}$ *or an imaginary quadratic field.*)

For an abelian variety A we write

$$\mathrm{NS}(A) = \mathrm{Im}\{H^1(A, \mathcal{O}_A^*) \to H^2(A, \mathbb{Z})\}$$

for the Néron-Severi group of A. Its rank ρ is called the Picard number of A. We have $1 \leqq \rho \leqq h^{1,1}(A) = n^2$ ($n = \dim(A)$). Put $\mathrm{NS}^\circ(A) = \mathrm{NS}(A) \otimes \mathbb{Q}$. For a polarized abelian variety we have an identification

$$\{a \in \mathrm{End}^\circ(A)\colon a = a^\circ\} \xrightarrow{\sim} \mathrm{NS}^\circ(A) \subset H^2(A, \mathbb{Q})$$

$$a \mapsto E(a-, -) \in \mathrm{Hom}(\textstyle\bigwedge^2 \Lambda, \mathbb{Q}) .$$

(1.3) Definition. Let A be an n-dimensional complex torus. We say that A has real multiplication if there exists a totally real number field of degree n over $\mathbb{Q}$ and a $\mathbb{Q}$-algebra homomorphism $K \to \mathrm{End}^\circ(A)$. Then $\mathfrak{o} = K \cap \mathrm{End}(A)$ is an order in K and we say that A has multiplication by $\mathfrak{o}$.

(1.4) Lemma. *A complex torus with real multiplication is an abelian variety.*

Proof. Suppose $A = V/\Lambda$ has multiplication by $\mathfrak{o} \subset K$. Then $\Lambda \otimes \mathbb{Q}$ is a projective K-module of rank 2. Choose an isomorphism $\varphi\colon \Lambda \otimes \mathbb{Q} \xrightarrow{\sim} K \oplus K$ of K-modules. Then φ determines a $\mathbb{C}$-vector space isomorphism $V \xrightarrow{\sim} \mathbb{C}^n$ such that $\varphi^{-1}(0, \beta) \in \Lambda \otimes \mathbb{Q} \subset V$ maps to $(\beta^{(1)}, \ldots, \beta^{(n)}) \in \mathbb{R}^n$ with $x \to x^{(i)}$ the i-th embedding of K into $\mathbb{R}$. Then $\varphi^{-1}(\alpha, 0)$ maps to $(\alpha^{(1)} z_1, \ldots, \alpha^{(n)} z_n)$ for some $z = (z_1, \ldots, z_n) \in \mathbb{C}^n$, $z_i \notin \mathbb{R}$. By changing φ if necessary we may assume $z_i \in \mathfrak{H}$. Consider an alternating form on $K \oplus K$ and take its trace to $\mathbb{Q}$. After multiplication with a suitable integer (>0) this form is integral on Λ. If we take e.g. the standard alternating form on $K \oplus K$ we find (up to a multiple)

$$E(\zeta, \eta) = \sum_{i=1}^{n} (\mathrm{Im}\, z_i)^{-1} \,\mathrm{Im}(\zeta_i \bar{\eta}_i) . \tag{2}$$

One checks immediately that this gives a Riemann form. □

We now assume that $\mathfrak{o}$ is the ring of integers in a totally real field K of degree n and consider n-dimensional abelian varieties with multiplication by $\mathfrak{o}$. In order to get the right moduli space we do not fix *one* polarization but a class of polarizations:

(1.5) Definition. The *polarization module* of an abelian variety with multiplication by $\mathfrak{o}$ is

$$P = \bigwedge\nolimits_{\mathfrak{o}}^{2} H_1(A,\mathbb{Z})^{\vee} ,$$

where $^{\vee}$ means the $\mathbb{Z}$-dual of an $\mathfrak{o}$-module (e.g. $\mathfrak{o}^{\vee}$ is the inverse different). The ordered polarization module is the pair (P, P_+) with

$$P_+ = \{\sigma \in P\colon \sigma(Ix \wedge x) > 0 \text{ for all } x \in H_1(A,\mathbb{R}), x \neq 0\}$$

with I the endomorphism defining the complex structure on $H_1(A,\mathbb{R})$. The elements of P_+ are called the totally positive elements.

Since $\mathfrak{o}$ is a Dedekind domain any projective $\mathfrak{o}$-module Λ of rank 2 is isomorphic to $\mathfrak{o} \oplus \mathfrak{a}$ for some fractional ideal $\mathfrak{a}$ and then $\bigwedge^2 \Lambda \simeq \mathfrak{a}$. Therefore, the isomorphism class of the rank one $\mathfrak{o}$-module P determines that of the rank 2 module Λ. We can associate to a pair (P, P_+) an element of $\mathscr{C}\ell^+(K)$.

Now let a projective $\mathfrak{o}$-module Λ be given. A complex structure on $\Lambda \otimes \mathbb{R}$ (which always exists) uniquely determines an abelian variety A with multiplication by $\mathfrak{o}$ and $H_1(A,\mathbb{Z}) = \Lambda$. Hence the isomorphism classes of complex abelian varieties with multiplication by $\mathfrak{o}$ and with $H_1(A,\mathbb{Z})$ given correspond to the complex structures on $\Lambda \otimes \mathbb{R}$. These are parametrized by

$$\mathrm{GL}_{\mathfrak{o}}(\Lambda)\backslash \mathrm{GL}_{\mathfrak{o}\otimes\mathbb{R}}(\Lambda \otimes \mathbb{R})/\mathbf{k} ,$$

where $\mathbf{k}$ is the centralizer of one complex structure. We now bring the positivity P_+ on P into play. We look at complex abelian varieties with multiplication by $\mathfrak{o}$ with $H_1(A,\mathbb{Z}) = \Lambda$ and with $P_+ \subset P$ given, i.e. for each embedding $K \to \mathbb{R}$ an ordering of $(\bigwedge^2{}_{\mathfrak{o}}\Lambda^{\vee}) \otimes \mathbb{R}$ is given. It then makes sense to speak of the totally positive elements $(\bigwedge^2{}_{\mathfrak{o}}\Lambda^{\vee})_+$ of $\bigwedge^2{}_{\mathfrak{o}}\Lambda^{\vee}$. The following is then clear:

(1.6) Proposition. *The isomorphism classes of complex abelian varieties A with multiplication by $\mathfrak{o}$ with $H_1(A,\mathbb{Z}) = \Lambda$ and with $P_+ = (\bigwedge^2{}_{\mathfrak{o}}\Lambda^{\vee})_+$ are parametrized by*

$$\mathrm{GL}_{\mathfrak{o}}^{+}(\Lambda)\backslash \mathrm{GL}_{\mathfrak{o}\otimes\mathbb{R}}^{+}(\Lambda \otimes \mathbb{R})/\mathbf{k} , \tag{3}$$

where $\mathrm{GL}_{\mathfrak{o}}^{+}(\Lambda)$ is the group of $\mathfrak{o}$-automorphisms of Λ preserving the set of totally positive elements of $\bigwedge^2 \Lambda^{\vee}$ and where $\mathbf{k}$ is the centralizer of one complex structure.

Assuming $\Lambda = \mathfrak{o} \oplus \mathfrak{a}$ for some ideal $\mathfrak{a}$ with its standard order $(\bigwedge^2 \Lambda^{\vee})_+ \subset \bigwedge^2 \Lambda^{\vee}$ we can identify (3) with $\mathrm{PGL}^+(\mathfrak{o} \oplus \mathfrak{a})\backslash \mathfrak{H}^n$. Explicitly, to $z \in \mathfrak{H}^n$ we associate the complex torus $T_z = \mathbb{C}^n/(\mathfrak{o}z + \mathfrak{a})$ with $\mathfrak{o}z + \mathfrak{a}$ embedded in $\mathbb{C}^n$ using $K \to \mathbb{R}^n$.

If one fixes a polarization instead of (P, P_+) then the situation is different. In fact, given an abelian variety with $j\colon \mathfrak{o} \to \mathrm{End}(A)$ and a polarization there exists an isomorphism $\varphi\colon \Lambda \xrightarrow{\sim} \mathfrak{o} \oplus \mathfrak{a}$ for some $\mathfrak{a}$ such that the polarization induces the standard alternating form on $\mathfrak{o} \oplus \mathfrak{a}$ (i.e. $\mathrm{Tr}_{K/\mathbb{Q}}$ of $\begin{pmatrix} 0 & 1 \\ -1 & 0 \end{pmatrix}$). The various φ now differ by elements of $\mathrm{PSL}(\mathfrak{o} \oplus \mathfrak{a})$. So $\mathrm{PSL}(\mathfrak{o} \oplus \mathfrak{a})\backslash \mathfrak{H}^n$ is the moduli space of triples (A, j, r), A a polarized n-dimensional complex abelian variety, $j\colon \mathfrak{o} \to \mathrm{End}(A)$ and r

an $\mathfrak{o}$-module isomorphism $H_1(A, \mathbb{Z}) \xrightarrow{\sim} \mathfrak{o} \oplus \mathfrak{a}$ carrying the Riemann form to the standard form. The map $\mathfrak{H}^n \to \mathfrak{H}^n$, $z \to \lambda z$ for totally positive λ induces an isomorphism $\mathrm{PSL}(\mathfrak{o} \oplus \mathfrak{a})\backslash\mathfrak{H}^n \xrightarrow{\sim} \mathrm{PSL}(\mathfrak{o} \oplus \lambda\mathfrak{a})\backslash\mathfrak{H}^n$ but such a λ changes the polarization! The polarization given by the standard form on $\mathfrak{o} \oplus \mathfrak{a}$ is a principal polarization if and only if $\mathfrak{a} = \mathfrak{o}^\vee$. In general, this Riemann form is equivalent to

$$\begin{bmatrix} & & d_1 & \\ 0 & & & \ddots \\ & & & & d_n \\ -d_1 & & & & \\ & \ddots & & 0 & \\ & & -d_n & & \end{bmatrix}, \tag{4}$$

where $d_1, \ldots, d_n$ with $d_i | d_{i+1}$ are the elementary divisors of the abelian group $\mathfrak{o}^\vee/\mathfrak{a}$ (note that $\mathfrak{a} \subset \mathfrak{o}^\vee$ since the form is integral.)

In order to compactify moduli spaces of abelian varieties one has to look at (mild) degenerations of abelian varieties: degenerations consisting of an extension of an abelian variety with a multiplicative group

$$1 \to (\mathbb{C}^*)^s \to A \to B \to 0 .$$

Multiplication by $\mathfrak{o}$ makes it necessary to require such sequences to be $\mathfrak{o}$-module sequences. But an abelian variety B with multiplication by $\mathfrak{o}$ must have dimension $\geqq [K : \mathbb{Q}]$, see [115]. So the degenerations in our case will be necessarily completely of multiplicative type. Such degenerations are up to isomorphism of the form: an algebraic torus with character group a projective $\mathfrak{o}$-module of rank 1. We shall treat degenerating families of abelian varieties with real multiplication in the next chapter in order to construct a compactification of the moduli spaces of abelian varieties with real multiplication.

Since we can interpret the varieties $\mathrm{PSL}(\mathfrak{o} \oplus \mathfrak{a})\backslash\mathfrak{H}^n$ as moduli spaces of polarized complex abelian varieties with some additional structure there must exist "forgetful" maps

$$\mathrm{PSL}(\mathfrak{o} \oplus \mathfrak{a})\backslash\mathfrak{H}^n \to \mathrm{PSp}(\mathfrak{L})\backslash\mathfrak{H}_n \tag{5}$$

(with $\mathfrak{L}$ determined by $\mathfrak{a}$) which are called *modular embeddings*. We now describe these explicitly. Such a map is induced by an identification of the $\mathfrak{o}$-module $\mathfrak{o} \oplus \mathfrak{a}$ with alternating form with the symplectic lattice $\mathfrak{L}$. The standard alternating form $\mathrm{Tr}_{K/\mathbb{Q}}(ad - bc)$ on $\mathfrak{o} \oplus \mathfrak{a}$ is $\mathbb{Z}$-equivalent to (4)$\Leftrightarrow \mathfrak{a} \subset \mathfrak{o}^\vee$ and $\mathfrak{o}^\vee/\mathfrak{a}$ has elementary divisors d_i. Now (5) is induced by $\mathrm{SL}(\mathfrak{o} \oplus \mathfrak{a}) \to \mathrm{Sp}(\mathfrak{L})$ as follows. Choose $R \in \mathrm{GL}(n, \mathbb{R})$ such that

$$\mathfrak{L} = \begin{pmatrix} R & 0 \\ 0 & {}^tR^{-1} \end{pmatrix} \mathfrak{o} \oplus \mathfrak{a} ,$$

where we view $\mathfrak{o} \oplus \mathfrak{a}$ as embedded in $\mathbb{R}^{2n}$ using $K \to \mathbb{R}^n$. For $r \in \mathbb{R}^n$ we write $r^* = \mathrm{diag}(r_1, \ldots, r_n) \in M_n(\mathbb{R})$. The map

$$\begin{pmatrix} a & b \\ c & d \end{pmatrix} \mapsto \begin{pmatrix} R & 0 \\ 0 & {}^tR^{-1} \end{pmatrix} \begin{pmatrix} a^* & b^* \\ c^* & d^* \end{pmatrix} \begin{pmatrix} R^{-1} & 0 \\ 0 & {}^tR \end{pmatrix}$$

induces $\mathfrak{H}^n \to \mathfrak{H}_n$ given by $z \mapsto Rz^{*t}R$. Together these describe (5) explicitly.

2. Humbert Surfaces

Here we study Hilbert modular surfaces as subvarieties of Siegel modular varieties.

We start by giving an alternative description of the moduli of complex abelian surfaces. If A is a complex abelian surface then $H^2(A, \mathbb{C})$ carries a Hodge structure with Hodge numbers $h^{0,2} = h^{2,0} = 1$, $h^{1,1} = 4$. A polarization is given by $\xi \in H^2(A, \mathbb{Z}) \cap H^{1,1}$. Let U be the orthogonal complement of ξ in $H^2(A, \mathbb{R})$ with respect to the cup product. The restriction of the cup product to U is of signature (2, 3). The subspace $U \cap H^{1,1}$ of U is 3-dimensional and the cup product is negative definite on it. We thus find a 5-dimensional vector space V with a quadratic form Δ of signature (3, 2) (after a change of sign).

Let us compute Δ explicitly. Write $A = \mathbb{C}^2/L_\tau$ with L_τ the lattice spanned by the columns of $(T\sigma)$ with $\tau = {}^t\sigma T^{-1} \in \mathfrak{H}_2$, $T = \begin{pmatrix} d_1 & 0 \\ 0 & d_2 \end{pmatrix}$, $d_1 | d_2$, $d_i \in \mathbb{Z}_{>0}$. We have natural real coordinates $x_1, \ldots, x_4$ satisfying $\begin{pmatrix} z_1 \\ z_2 \end{pmatrix} = (T \ \sigma)\begin{pmatrix} x_1 \\ \vdots \\ x_4 \end{pmatrix}$. The Riemann form is supposed to be given by $\xi = dx_1 \wedge dx_3 + dx_2 \wedge dx_4$. With these notations we have:

(2.1) Lemma. *If $\omega = \sum a_{ij} dx_i \wedge dx_j$ with (a_{ij}) alternating then $\omega \in H^{1,1} \cap U$ if and only if*

$$a_{13} + a_{24} = 0 \qquad (1)$$

$$-a_{14}\tau_1 + (a_{13} - a_{24})\tau_2 + a_{23}\tau_3 + a_{12}(\tau_1\tau_3 - \tau_2^2) + a_{34} = 0\,, \qquad (2)$$

where $\tau = \begin{pmatrix} \tau_1 & \tau_2 \\ \tau_2 & \tau_3 \end{pmatrix}$. In particular, $U \cap H^{1,1}$ is in bijection with the set of vectors $(-a_{14}, a_{13} - a_{24}, a_{23}, a_{12}, a_{34})$ satisfying (1) *and* (2).

For an abelian surface $A = \mathbb{C}^2/L_\tau$ the equations (1) and (2) describe the Néron-Severi group of A. Indeed the map

$$(r_{ij}) \mapsto \sum_{i<j} r_{ij}\, dx_i \wedge dx_j$$

gives a bijection between the additive group

$$M_\tau = \{R = (r_{ij}) : {}^tR = -R, (\tau \quad 1_2)R^t(\tau \quad 1_2) = 0\,,$$
$$r_{12}, d_1 r_{13}, d_1 r_{23}, d_2 r_{14}, d_2 r_{24}, d_1 d_2 r_{34} \in \mathbb{Z}\}$$

and $H^{1,1}(A) \cap H^2(a, \mathbb{Z}) = \mathrm{NS}(A)$. In order to get $U \cap \mathrm{NS}(A)$ add the condition

$r_{13} + r_{24} = 0$. The quadratic form on this lattice is $2(b^2 - ac - de)$. The dual lattice carries the form $(b^2 - 4ac - 4de)$. We thus find: *Picard number of A equals rank* M_τ.

We see that we are led to the following data: a 5-dimensional vector space $V/\mathbb{Q}$ equipped with a quadratic form Δ of signature (3,2). We consider the (dual) form given by

$$(a,b,c,d,e) \mapsto b^2 - 4ac - 4de\,.$$

In order to simplify the exposition we shall assume $d_1 = 1$, and put $t = d_2/d_1$. (This is not really a restriction of generality.) Now we look at

$$N_t = \{(a,b,c,d,e) \in \mathbb{Z}^5 \colon c, d \equiv 0 \,(\mathrm{mod}\, t)\} \subset V.$$

To an element $x \in V$, $x \neq 0$ we associate a subset $\mathfrak{H}_x$ of $\mathfrak{H}_2$(if $x = (a, \ldots, e)$):

$$\{\tau \in \mathfrak{H}_2 \colon a\tau_1 + b\tau_2 + c\tau_3 + d(\tau_2^2 - \tau_1\tau_3) + e = 0\}\,.$$

One easily checks: $\mathfrak{H}_x \neq \emptyset \Leftrightarrow \Delta(x) > 0$. If $\mathfrak{H}_x$ is not empty then it is a copy of $\mathfrak{H}^2$ embedded in $\mathfrak{H}_2$.

Let $\mathrm{Gr}^+(3, V)$ be the Grassmann variety of 3-dimensional linear subspaces of V on which Δ is positive definite.

(2.2) Proposition. *There is a bijection* $\mathrm{Gr}^+(3, V) \leftrightarrow \mathfrak{H}_2$ *given by* $W \mapsto \tau_w = \bigcap_{w \in W} \mathfrak{H}_w$.

We leave the proof to the reader. The inverse is given by $\tau \mapsto W = \{(a,b,c,d,e)\colon a\tau_1 + b\tau_2 + c\tau_3 + d(\tau_2^2 - \tau_1\tau_3) + e = 0\}$. Using this proposition we obtain a new description of the moduli of complex abelian surfaces.

Fix now $t \in \mathbb{Z}_{>0}$ and $\mathfrak{L}_t = \mathbb{Z}^3 \oplus \mathbb{Z}t$. Let Δ be a positive integer. Define a surface in $\mathrm{Sp}(\mathfrak{L}_t)\backslash\mathfrak{H}_2$ by

$$H_\Delta := \text{image under } \mathfrak{H}_2 \to \mathrm{Sp}(\mathfrak{L}_t)\backslash\mathfrak{H}_2 \text{ of all } \mathfrak{H}_x \text{ with } x \in N_t$$
$$\text{and } \Delta(x) = \Delta \text{ and } x \text{ primitive}\,.$$

The primitivity means that $x \in N_t$ is not divisible by an integer >1. Since $b^2 - 4ac - 4de$ assumes all positive values $\Delta \equiv$ square (mod $4t$) we see: $H_\Delta \neq \emptyset \Leftrightarrow \Delta > 0$, $\Delta \equiv$ square (mod $4t$). If H_Δ is not empty it is a (possibly reducible) complex surface (in general singular). We call H_Δ the *Humbert surface* of invariant Δ in $\mathrm{Sp}(\mathfrak{L}_t)\backslash\mathfrak{H}_2$. It is the analogue of the curve F_N on a Hilbert modular surface. We also define the analogue of the T_N:

$$G_\Delta = \bigcup_{r^2 | \Delta} H_{\Delta/r^2}\,,$$

where r runs over the integers $\geqslant 1$. The equations defining Humbert surfaces occur in the papers of Humbert, and in Hecke's thesis [64], cf. also Franke [32].

(2.3) Proposition. *Let* $\mathfrak{o}_\Delta$ *be the order of discriminant* Δ *in the real quadratic field* $\mathbb{Q}(\sqrt{\Delta})$. *Let* $A = \mathbb{C}^2/L_\tau$ *be an abelian surface. Then*

$$\mathrm{End}(A) \textit{ contains } \mathfrak{o}_\Delta \Leftrightarrow \exists \omega \in H^2(A, \mathbb{Z}) \cap U \quad \textit{with } \omega \wedge \omega = -2\Delta$$
$$\Leftrightarrow \tau \in G_\Delta\,.$$

Proof. The equivalence of both right hand sides is easily seen by associating to an equation

$$a\tau_1 + b\tau_2 + c\tau_3 + d(\tau_2^2 - \tau_1\tau_3) + e = 0 \tag{3}$$

the form $\omega = -2d\,dx_1 \wedge dx_2 + b(dx_1 \wedge dx_3 - dx_2 \wedge dx_4) - 2a\,dx_1 \wedge dx_4 + 2c\,dx_2 \wedge dx_3 + 2e\,dx_3 \wedge dx_4$, see (2.1). Then $\omega \in H^{1,1} \cap U$ and $\omega \wedge \omega = -2\Delta$. To prove the equivalence $\mathfrak{o}_\Delta \subset \operatorname{End}(A) \Leftrightarrow \tau \in G_\Delta$, suppose $\alpha \in \operatorname{End}(A)$, given by ${}^t\sigma\alpha = A^t\sigma + BT$, $T\alpha = C^t\sigma + DT$. Elimination of α gives $\tau C\tau + \tau D - A\tau - B = 0$. Subtraction of this relation from its transpose yields the desired equation with $\Delta =$ discriminant of minimum polynomial of α over $\mathbb{Q}$, etc. □

(2.4) Theorem. *Every irreducible component of H_Δ in $\operatorname{Sp}(\mathfrak{L}_t)\backslash\mathfrak{H}_2$ can be represented in $\mathfrak{H}_2$ by an equation $\tau_1 + b\tau_2 + ct\tau_3 = 0$ with $b^2 - 4ct = \Delta, 0 \leqq b < 2t$. The number of irreducible components of H_Δ is $\#\{b \,(\mathrm{mod}\, 2t)\colon b^2 \equiv \Delta \,(\mathrm{mod}\, 4t)\}$.*

In case $t = 1$ the number of components was determined by Humbert [88], cf. also Franke. To prove (2.4) one uses either generators of $\operatorname{Sp}(\mathfrak{L}_t)$, (see Rosati [137]) to bring an Equation (3) in standard form or one uses (2.3) (e.g. if Δ is a fundamental discriminant each component of H_Δ corresponds to a strict ideal class $[\mathfrak{b}]$ of $\mathfrak{o}_\Delta$ containing an ideal $\mathfrak{b} \in \delta^{-1}$ with $\delta^{-1}/\mathfrak{b} \cong \mathbb{Z}/t\mathbb{Z}$ (as abelian groups), see Section 1.).

Now write $\Delta = Df^2$ with D a fundamental discriminant. Then

$$\#\{b \,(\mathrm{mod}\, 2t)\colon b^2 \equiv \Delta \,(\mathrm{mod}\, 4t)\} = r_D^*(f,t)\,,$$

with $r_D^*(f,t) =$ number of $\operatorname{SL}(2,\mathbb{Z})$-inequivalent primitive representations of t by all binary quadratic forms of discriminant Δ ([85]). The function $r_D^*(f,t)$ is simultaneously multiplicative in t and in f. We have for a prime q:

$$\sum_{m=0}^{\infty}\sum_{n=0}^{\infty} r_D^*(q^m,q^n)X^mY^n = \frac{(1+Y)(1-\chi_D(q)XY)}{(1-X)(1-\chi_D(q)Y)(1-qXY^2)}\,.$$

In particular, $r_D^*(q^m,q^n) = r_D^*(q^{m-1},q^{n-2})$, $r_D^*(q^m,1) = r_D^*(q^m,q) = 1$.

In nature of these irreducible components is as follows.

(2.5) Proposition. *If Δ is not a square then the component of H_Δ defined by $\tau_1 + b\tau_2 + ct\tau_3 = 0$ is the image of a degree 1 morphism $\operatorname{PSL}(\mathfrak{o} \oplus \mathfrak{a})\backslash\mathfrak{H}^2 \to \operatorname{PSp}(\mathfrak{L}_t)\backslash\mathfrak{H}_2$ if $b \not\equiv 0 \,(\mathrm{mod}\, t)$. If $b \equiv 0 \,(\mathrm{mod}\, t)$ then it is the image of a symmetric Hilbert modular surface $\Gamma \cup \Gamma\tau\backslash\mathfrak{H}^2$ with $\Gamma = \operatorname{PSL}(\mathfrak{o} \oplus \mathfrak{a})$ and τ the involution defined in* V.9.

Proof. Let $S = \begin{pmatrix} 1 & \dfrac{b+\sqrt{\Delta}}{2} \\ 1 & \dfrac{b-\sqrt{\Delta}}{2} \end{pmatrix}$. Then the image of $\mathfrak{H}^2 \hookrightarrow \mathfrak{H}_2$, $(z_1,z_2) \mapsto S^{-1}\begin{pmatrix} z_1 & 0 \\ 0 & z_2 \end{pmatrix}{}^tS^{-1}$ is the surface defined by $\tau_1 + b\tau_2 + ct\tau_3 = 0$. One then computes the stabilizer. □

If Δ is a square then one gets a similar result. One has to replace PSL(2, K) by PSL(2, $\mathbb{Q}$)^2.

The quotient $\mathrm{Sp}(\mathfrak{L}_t)\backslash\mathfrak{H}_2$ admits a natural compactification, the *Satake* compactification. To describe it we map $\mathfrak{H}_2$ onto a bounded symmetric domain. Let

$$D_n = \{V \in M_n(\mathbb{C}) \colon V\bar{V} < 1_n\} .$$

By the Cayley transformation $\tau \mapsto (1 + i\tau)(1 - i\tau)^{-1}$ we can identify $\mathfrak{H}_2$ with D_2. The action of $\mathrm{Sp}(\mathfrak{L}_t)$ on $\mathfrak{H}_2$ goes over into an action on D_2 which can be extended to $\bar{D}_2$. Define for $r \leqq n$ $D_r \hookrightarrow \bar{D}_n$ by $\tau_1 \mapsto \begin{pmatrix} \tau_1 & 0 \\ 0 & 1_{n-r} \end{pmatrix}$. Then

$$\bar{D}_n = \bigcup_{r \leqq n} \bigcup_{g \in \mathrm{Sp}(2n, \mathbb{R})} gD_r .$$

We let

$$D_n^* = \bigcup_{r \leqq n} \bigcup_{g \in \mathrm{Sp}(2n, \mathbb{Q})} gD_r .$$

The Satake compactification of $\mathrm{Sp}(\mathfrak{L}_t)\backslash D_2$ is (as a set) $\mathrm{Sp}(\mathfrak{L}_t)\backslash D_2^*$. One can define the structure of a normal analytic space on it. One defines an algebraic structure on it using modular forms. A Siegel modular form (of weight k on $\mathrm{Sp}(\mathfrak{L}_t)$) is a holomorphic function $f \colon \mathfrak{H}_2 \to \mathbb{C}$ satisfying $f(\gamma\tau) = \det(c\tau + d)^k f(\tau)$ for all $\gamma = \begin{pmatrix} a & b \\ c & d \end{pmatrix} \in \mathrm{Sp}(\mathfrak{L}_t)$. Let $M^*(\mathrm{Sp}(\mathfrak{L}_t)) = \bigoplus_k M_k(\mathrm{Sp}(\mathfrak{L}_t))$ be the graded ring of Siegel modular forms on $\mathrm{Sp}(\mathfrak{L}_t)$. One can show that the elements of $M_k(\mathrm{Sp}(\mathfrak{L}_t))$ define for k large an embedding of $\mathrm{Sp}(\mathfrak{L}_t)\backslash\mathfrak{H}_2 \cong \mathrm{Sp}(\mathfrak{L}_t)\backslash D_2$ into a projective space. The closure of the image can be identified with $\mathrm{Sp}(\mathfrak{L}_t)\backslash D_2^*$, i.e. the Satake compactification is the analytic space V_t of $\mathrm{Proj}\, M^*(\mathrm{Sp}(\mathfrak{L}_t))$.

The Humbert surfaces H_Δ, G_Δ admit a natural compactification in V_t. We are interested in the cohomology classes that they define.

Let $\mathbb{H}^i(V_t, \mathbb{Q})$ $i = 0, \ldots, 6$ be the intersection homology groups of V_t. (Note that V_t is singular.) It has been shown that the intersection cohomology $\mathbb{H}^*(V_t, \mathbb{Q})$ coincides with the L^2-cohomology, see [179]. The Humbert surfaces G_Δ define classes g_Δ in $\mathbb{H}^2(V_t, \mathbb{Q})$. They are similar to the classes t_N defined in Chapter VI. One defines the class g_0 by the $\mathrm{Sp}(4, \mathbb{R})$-invariant form

$$\mathrm{Tr}(Y^{-1}\, d\tau\, Y^{-1}\, d\tau) .$$

The analogue of VI (3.1) is

(2.6) Theorem. *The series $\sum_{\Delta \geqq 0} g_\Delta q^\Delta \in \mathbb{H}^2[[q]]$ is the Fourier series of a modular form of weight 5/2 and with Fourier coefficients in $\mathbb{H}^2(V_t, \mathbb{Q})$ on $\Gamma_0(4t)$.*

We refer to [101] for modular forms of half-integral weight on $\Gamma_0(4)$.

The invariant metric on $\mathfrak{H}_2$ given by

$$ds^2 = \mathrm{Tr}(y^{-1}\, d\tau y^{-1}\, d\tau) \qquad y = \mathrm{Im}(\tau)$$

defines Chern forms c_1, c_2, c_3.

Evaluating c_1^2 on G gives the volume of G_Δ. The volume of H_Δ can be computed. For $t = 1$ it equals, if $\Delta = Df^2$ with D a fundamental discriminant,

$$2f^3 \sum_{\substack{p|f \\ p\,\text{prime}}} \left(1 - \left(\frac{\Delta}{p}\right)p^{-2}\right)\zeta_K(-1)$$

with $K = \mathbb{Q}(\sqrt{D})$. So for $t = 1$ this implies that $\sum g_\Delta q^\Delta$ can be identified with the modular form $\sum_{N \equiv 0,1 \pmod 4} H(2, N)q^N$ studied by Cohen (cf. [18]):

$$H(2, Df^2) = L\left(-1, \left(\frac{D}{\cdot}\right)\right) \sum_{d|f} \mu(d)\left(\frac{D}{d}\right) d\sigma_3(f/d)\,.$$

Next we shall look at the intersections of Humbert surfaces. Let $U \subset V_{\mathbb{Q}}$ be a 2-dimensional subspace or a rank 2 sublattice. We define

$$\mathfrak{H}_U = \{\tau \in \mathfrak{H}_2 \colon a\tau_1 + \cdots + e = 0 \text{ for all } (a, \ldots, e) \in U\}\,.$$

We have

$$\mathfrak{H}_U \neq \varnothing \quad \text{if and only if } \Delta | U > 0$$

and if $\mathfrak{H}_U$ is not empty then it is a copy of $\mathfrak{H}$ embedded in $\mathfrak{H}_2$. We define C_U as the image of $\mathfrak{H}_U$ in $V_t = \mathrm{Sp}(\mathfrak{L}_t)\backslash\mathfrak{H}_2$. If $\Delta | U > 0$ then it is a positive definite quadratic form. If we take U to be a rank 2 sublattice of N_t then $\Delta | U$ is also integral and the action of $\mathrm{Sp}(\mathfrak{L}_t)$ fixed the $\mathrm{GL}(2, \mathbb{Z})$-equivalence class of $\Delta | U$. Note that in this case $\phi = \Delta | U$ has the additional properties:

$$\begin{aligned} &\text{i) } \phi \text{ represents only quadratic residues mod } 4t\,, \\ &\text{ii) the discriminant } d_\phi \text{ of } \phi \text{ is divisible by } 16t\,. \end{aligned} \tag{4}$$

If we identify for a polarized abelian variety A_τ of dimension 2 the orthogonal complement of the polarization class $\xi \in H^2(A_\tau, \mathbb{Q}) = \bigwedge^2 \mathfrak{L}_t$ with $V_{\mathbb{Q}}$ as above then the curve C_U in V_t has the following interpretation.

(2.7) Lemma. *Let A_τ be an abelian surface corresponding to $\tau \in \mathrm{Sp}(\mathfrak{L}_t)\backslash\mathfrak{H}_2$. Then $\tau \in C_U$ if and only if $\mathrm{NS}^\circ(A_\tau)$ contains the quadratic space (U, ϕ).*

As a corollary we find for generic $\tau \in C_U$: $\mathrm{End}^\circ(A_\tau) =$ the Clifford algebra of (U, ϕ).

Take a primitive vector $v \in N_t$. It defines a component S of $H_{\Delta(v)}$. Let W be the orthogonal complement of $v \in V$. We have a diagram (see (2.2) and V.4)

$$\begin{array}{ccccc} \mathrm{Gr}^+(2, W) & \hookrightarrow & \mathrm{Gr}^+(3, V) & & U \mapsto \langle U, v\rangle \subset V \\ \wr\wr & & \wr\wr & & \\ \mathfrak{H}_2 & \hookrightarrow & \mathfrak{H}_2 & & \end{array} \tag{5}$$

which induces the inclusion of this component in V_t.

Suppose that $\Delta(v) = D$ is a fundamental discriminant. We then know that S is the image of a Hilbert modular surface X (see (2.5)). We have a map

$$j^*: \mathbb{H}^4(V_t, \mathbb{Q}) \xrightarrow{\cap g_D} \mathbb{H}^2(X, \mathbb{Q}) .$$

(2.8) Proposition. *The natural map j^* maps G, the subspace generated by the g_Δ, to $\mathbb{T}$. More precisely,*

$$j^* g_N = \sum_{x \in \mathbb{Z}} t_{(DN-x^2)/4t} \, .$$

Proof. The embedding (4) $\mathfrak{H}^2 \to \mathfrak{H}_2$ is induced by an identification of symplectic lattices $\mathfrak{o} \oplus \mathfrak{a} \to \mathfrak{L}_t$ with $\mathfrak{o} \subset \mathbb{Q}(\sqrt{D})$ the ring of integers and $\mathfrak{a}$ some ideal. Then N_t corresponds to a $\mathbb{Z}$-module in $\bigwedge^2(\mathfrak{o} \oplus \mathfrak{a})$ and the orthogonal complement of $v \in N_t$ defines a lattice of $\mathbb{Z}$-rank 4 in it. This is a lattice of hermitian matrices (w.r.t. $\mathfrak{a}$) as studied in Chapter V. The elements $w \in N_t$ defining G_N correspond to the elements $w' \in v^\perp$ defining $T_{(DN-x^2)/4t}$. By the adjunction formula this is also true for $N = D$. □

As an application take $t = 1$. The volume of G_Δ is proportional to the volume of the pull-back to H_1 since $M^+_{5/2}$ is one-dimensional. The pull-back is $\sum t_{(n-x^2)/4}$. This gives:

(2.9) Theorem (Siegel [159], Ch. 6). *Suppose that Δ is not a square. Then*

$$H(2, \Delta) = -\frac{1}{5} \sum_{\substack{x \in \mathbb{Z} \\ x \equiv \Delta \,(\mathrm{mod}\ 2)}} \sigma_1((\Delta - x^2)/4) ,$$

where $H(2, \Delta)$ is defined above.

Note that if Δ is square-free this gives a formula for $-12\zeta_{\mathbb{Q}(\sqrt{\Delta})}(-1)$, see I (6.5).

3. Examples

In this section we give two examples of Siegel modular 3-folds with some of the Humbert surfaces in them.

Let $\Gamma[n] = \ker\{\mathrm{PSp}(2, \mathbb{Z}) \to \mathrm{PSp}(\mathbb{Z}/n\mathbb{Z})\}$ be the principal congruence subgroup of $\mathrm{PSp}(2, \mathbb{Z})$. The quotient $\Gamma[n]\backslash\mathfrak{H}_2$ can be compactified by adding the $\Gamma[n]$-orbits of $\bigcup_{g \in \mathrm{Sp}(2,\mathbb{Q})} gD_1$ and $\bigcup_{g \in \mathrm{Sp}(2,\mathbb{Q})} gD_0$, see Section 2. The action of $\Gamma[n]$ on these sets of boundary components is not transitive.

(3.1) Lemma. *Let $n = p$ be a prime. The 1-dimensional (resp. 0-dim.) boundary components of $\Gamma[p]\backslash D_2^*$ correspond bijectively to the totally isotropic 1-dimensional (resp. 2-dimensional) linear subspaces of a fixed 4-dimensional symplectic vector space over $\mathbb{F}_p$.*

We have an incidence relation on the set of boundary components: a 1-dimensional and a 0-dimensional boundary component are incident if the 0-dimensional boundary component occurs as a boundary component of the 1-dimensional one.

We shall treat two cases: $n = 2$, $n = 3$.

n = 2. The 15 1-dimensional and the 15 0-dimensional boundary components form a $(15_3, 15_3)$-configuration. There are six (maximal) sets of five disjoint 1-dimensional boundary components and the action of $\mathrm{Sp}(4, \mathbb{F}_2)$ on them defines an isomorphism $\mathrm{Sp}(4, \mathbb{F}_2) \cong \mathfrak{S}_6$, (symmetric group).

If we number these six sets we have a numbering (ij) $1 \leqslant i, j \leqslant 6$, $i \neq j$ for the 1-dimensional boundary components. Call these components λ_{ij}. The configuration is given by the following diagram:

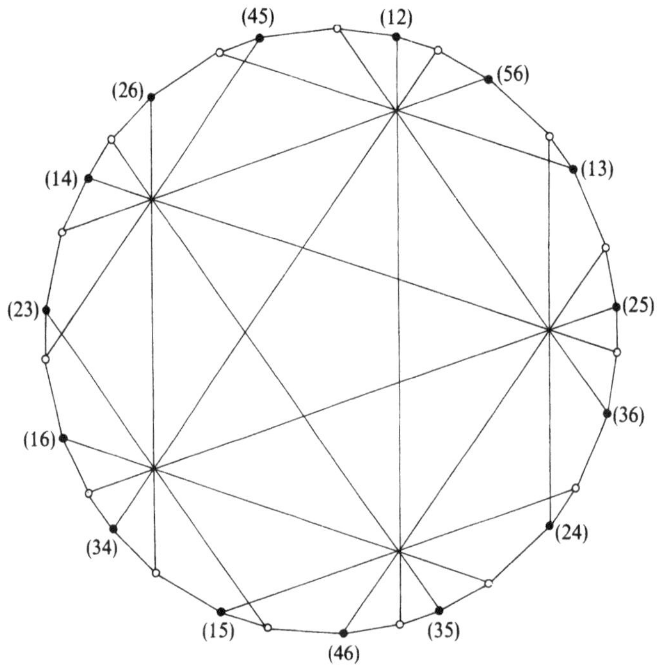

Consider now

$$\theta[m](\tau, \zeta) = \sum_{n \in \mathbb{Z}^2} \exp 2\pi i \left(\frac{1}{2}\left(n + \frac{m'}{2}\right)\tau\, {}^t\!\left(n + \frac{m'}{2}\right) + \left(n + \frac{m'}{2}\right)\left(\zeta + \frac{m''}{2}\right)\right)$$

where $m = \begin{pmatrix} m' \\ m'' \end{pmatrix}$ with m', $m'' \in \mathbb{Z}^2$. These are 16 functions on $\mathfrak{H}_2 \times \mathbb{C}^2$ ($\tau \in \mathfrak{H}_2$, $\zeta \in \mathbb{C}^2$). As function of ζ $\theta[m](\tau, \zeta)$ is either even or odd. The group $\mathrm{Sp}(4, \mathbb{F}_2)$ acts on these functions; it permutes the 6 odd ones giving thus an isomorphism $\mathrm{Sp}(4, \mathbb{F}_2) \cong \mathfrak{S}_6$ which differs from the one above by an outer automorphism.

Consider the 10 even $\theta[m] = \theta[m](\tau, 0)$ for $\zeta = 0$. These are modular forms of weight 1/2 on some subgroup of $\Gamma[2]$. Their fourth powers $\theta[m]^4$ are modular forms of weight 2 on $\Gamma[2]$.

It is well-known that $\prod_{\text{even}} \theta[m]$ does not vanish outside the Humbert surface H_1, cf. [38]. In fact, the zero set of a $\theta^4[m]$ is a component of H_1 in $X[2]$, the Satake compactification of $\Gamma[2]\backslash\mathfrak{H}_2$. Each component of H_1 is isomorphic to $(\Gamma(2)\backslash\mathfrak{H})^2$ and contains 6 λ_{ij}: we can set up a $1-1$ correspondence between the components of H_1 and the partitions of $\{1,\ldots,6\}$ in two sets of 3 elements. The component corresponding to (123)(456) contains the following configuration

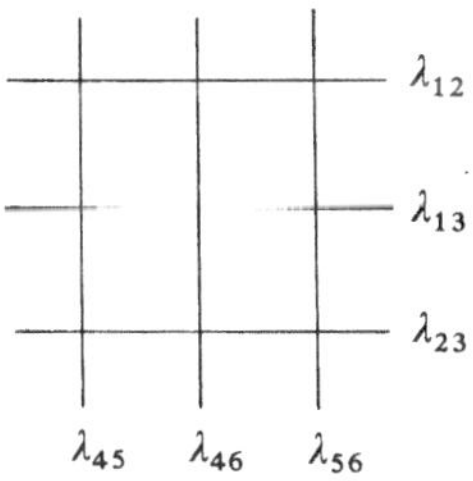

This gives the action of $\mathrm{Sp}(4,\mathbb{F}_2)$ on the $\theta[m]^4$ up to signs.

The $\theta[m]^4$ satisfy linear relations, e.g.

$$\theta^4\begin{bmatrix}0&0\\0&0\end{bmatrix}-\theta^4\begin{bmatrix}0&0\\0&1\end{bmatrix}-\theta^4\begin{bmatrix}0&1\\1&0\end{bmatrix}-\theta^4\begin{bmatrix}1&1\\0&0\end{bmatrix}=0\,.$$

All other linear relations can be obtained by letting $\mathrm{Sp}(4,\mathbb{F}_2)$ act.

(3.2) Theorem. *The* 10 *modular forms* $\theta^4[m]$ *span a 5-dimensional vector space and define an embedding*

$$X[2]\to\mathbb{P}^4\subset\mathbb{P}^9$$

and the image is the quartic 3*-fold defined by*

$$t_4^2-4t_8=0$$

where $t_4=\sum\theta^8[m]$, $t_8=\sum\theta^{16}[m]$.

If one changes the action of $\mathfrak{S}_6$ by an outer automorphism we find as equations for $X[2]\subset\mathbb{P}^5$

$$\begin{aligned} s_1&=0\\ s_2^2-4s_4&=0\,, \end{aligned}\tag{1}$$

where the s_i is the i-th power sum in the coordinates $x_j\, j=1,\ldots,6$ of $\mathbb{P}^5$ on which $\mathfrak{S}_6$ acts by the usual permutation. We refer to [38] for a more explicit description.

It is easy to see that the intersection homology group $\mathbb{H}^4(X[2],\mathbb{Q})$ is 1-dimensional.

(3.3) Proposition. *The Humbert surfaces* G_Δ *in* $X[2]$ *is the zero divisor of a modular form on* $\mathrm{Sp}(4,\mathbb{Z})$ *of weight* $-60H(2,\Delta)$.

Proof. Pullback to a non-singular model $\tilde{X}$ of $X[2]$ and use the exact sequence ($X[2]$ is simply-connected)

$$0 \to H^1(\tilde{X}, \mathcal{O}^*) \to H^2(\tilde{X}, \mathbb{Z}) \to$$

to see that homology implies linear equivalence. Since G_1 is the zero divisor of $\chi_{10} = \prod \theta^2[m]$ the proposition follows from (2.6). □

The ring of modular forms on $\mathrm{Sp}(4, \mathbb{Z})$ was determined by Igusa [90]. The subring of modular forms of even weight is a polynomial ring

$$\mathbb{C}[\psi_4, \psi_6, \chi_{10}, \chi_{12}],$$

where the subscript denotes the weight.

One can determine the equations for several G_Δ:

$$G_1\colon \chi_{10} = 0$$

$$G_4\colon \chi_{35}^2 = 0 \quad \text{(Igusa's notation [90])}$$

The components of H_4 are given by $x_i = x_j$ $i \neq j$ in terms of (1).

$$G_5\colon (\chi_{12} - 2^{-12}3^{-6}(\psi_6^2 + \psi_4^3))^2 - \psi_4(2 \cdot 3^{-1}\chi_{10} - 2^{11}3^{-6}\psi_6\psi_4)^2 = 0$$

cf. [38].

The Humbert surface G_5 has 6 components in $X[2]$. The i-th component $(1 \leqslant i \leqslant 6)$ is given by

$$2\rho_2 + \rho_1^2 = 0,$$

where ρ_k is the k-th elementary symmetric function in $x_1, \ldots, \hat{x}_i, \ldots, x_6$ (x_i deleted). If we write $u_j = x_j - \frac{1}{2}\rho_1$ $(j = 1, \ldots, 5)$ we find for the 6-th component the equations

$$\tau_2 = 0 \qquad \tau_4 = 0$$

in the coordinates $u_1, \ldots, u_5$ of $\mathbb{P}^4$. This is exactly the equation for the Hilbert modular surface of discriminant 5 and level 2 found in VIII.2. There we also found another equation. This can be obtained as follows:

Consider the projective dual variety Y of $X[2]$: it is given by:

$$\sum x_i = 0$$

$$\sum x_i^3 = 0$$

in $\mathbb{P}^5$ with coordinates $x_1, \ldots, x_6$. The hyperplanes $x_i = 0$ cut out on Y six surfaces, each of which is isomorphic to the cubic diagonal surface. These six surfaces correspond to the six components of G_5.

n = 3. Here the Satake compactification $X[3]$ has 40 boundary components of dimension 1 (resp. 0) forming a $(40_4, 40_4)$-configuration. One obtains a non-singular model by blowing up $X[3]$ along the ideal of cusp forms (see [89]) or by the theory of toroidal compactification. Under $Y[3] \to X[3]$ the fibre over a 1-dimensional boundary component is the universal elliptic curve of level 3 over $\Gamma(3)\backslash\mathfrak{H}$; the fibre over a 0-dimensional boundary component is a configuration of $\mathbb{P}^1$'s forming a tetrahedron:

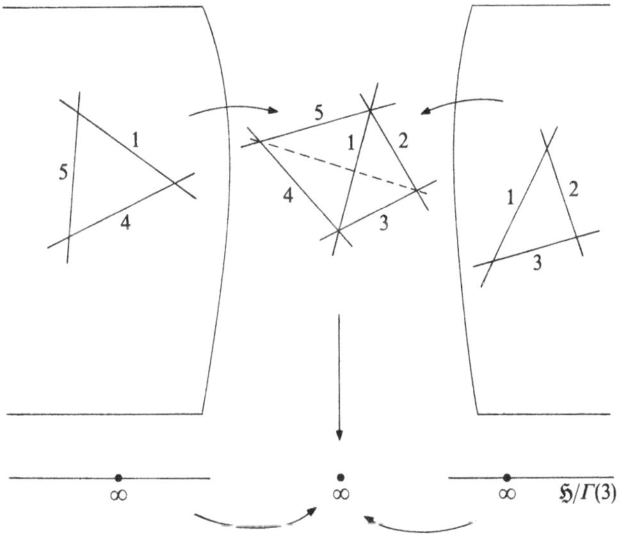

We have an action of $PSp(4, \mathbb{F}_3)$ on $X[3]$ and on $Y[3]$. This is a simple group of order 25,920.

Let L be the line-bundle on $X[3]$ (or $Y[3]$) defined by the factor of automorphy $\det(c\tau + d)$ and let D be the divisor on $Y[3]$ at infinity. One can show that the linear system M with $M = H_1 + D - 3L$ defines a morphism $Y[3] \xrightarrow{\phi} \mathbb{P}^4$ which is equivariant with respect to the group $PSp(4, \mathbb{F}_3)$. This group contains $\mathfrak{S}_6$ as a non-normal subgroup. Then we have:

(3.4) Theorem. *The morphism ϕ gives a birational morphism of the Siegel modular variety $Y[3]$ to its image in $\mathbb{P}^4 \subset \mathbb{P}^5$ which contracts the* 45 *components of H_1. The image of $Y[3]$ is given by*

$$\sigma_1 = 0\,, \qquad \sigma_4 = 0\,,$$

where σ_i is the i-th elementary symmetric function in x_j, $j = 1, \ldots, 6$. The 40 *universal elliptic curves in the boundary are mapped to the* 40 *elliptic pencils through* 9 *points in $\mathbb{P}^2$.*

The variety defined by these equations occurs already in a paper by Burkhardt (Math. Ann. 38). For the proof we refer to [39].

The Humbert surface G_5 has 36 components. One of these components is given by

$$\sigma_1 = 0\,, \qquad \sigma_2 = 0\,, \qquad \sigma_4 = 0\,.$$

This is the equation obtained for the Hilbert modular surface $\Gamma[3]\backslash\mathfrak{H}^2$ for discriminant 5. See VIII.1.

4. Jacobians with Real Multiplication

Let C be a complete non-singular algebraic curve of genus 2 with $J = \mathrm{Jac}(C)$ its Jacobian. Then J is a principally polarized abelian variety of dimension 2. When does J have real multiplication by (a given order in) a real quadratic field? We shall assume that C is defined over $\mathbb{C}$.

C is determined by the six branch points in $\mathbb{P}^1$ of the canonical morphism $C \to \mathbb{P}^1$. Let $p_i = (a_i : b_i)$ $i = 1, \ldots, 6$ be these six points and let (ij) be the determinant $a_i b_j - a_j b_i$. We define

$$y_1 = (12)(34)(56) + (13)(25)(46) + (24)(36)(15) + (16)(23)(45) + (14)(26)(35)\,.$$

By the action of the symmetric group $\mathfrak{S}_6$ we obtain six expressions y_i. They satisfy

$$\sum_{i=1}^{6} y_i = 0$$

$$\sum_{i=1}^{6} y_i^3 = 0\,.$$

These are the equations of the projective dual Y of the model for $X[2]$, the Siegel modular variety of level 2 and degree 2 obtained above. We have a birational morphism

$$Y \leftarrow \Gamma_2(2)\backslash\mathfrak{H}_2 \subset X[2] \subset \mathbb{P}^4 \tag{1}$$

(by associating to a point the tangent space at this point to $X[2]$) which contracts the ten components of the Humbert surface H_1. Using (1) and the equations given in Section 3 for some Humbert surfaces one gets quite explicit conditions on $p_1, \ldots, p_6$ such that J has real multiplication by a given order.

For example, J has multiplication by $\mathfrak{o}_{\mathbb{Q}(\sqrt{5})}$ if for some ordering of $p_1, \ldots, p_6$ we have $y_1 = 0$. An example of such a curve is provided by

$$y^2 = x \prod_{i=0}^{4} (x - \zeta^i)$$

with ζ a primitive 5-th root of 1. It corresponds to a point in the orbit under $\mathrm{Sp}(4, \mathbb{F}_2)$ of $(0 : 1 : \zeta : \zeta^2 : \zeta^3 : \zeta^4) \in Y$. The corresponding Jacobian has an automorphism group of order 10. The corresponding point on $\mathrm{SL}(2, \mathfrak{o}_{\mathbb{Q}(\sqrt{5})})\backslash\mathfrak{H}^2$ is a quotient singularity of order 5.

There is also a geometric characterization of principally polarized abelian surfaces which admit multiplication by $\mathfrak{o}_{\mathbb{Q}(\sqrt{5})}$. This characterization is due to Humbert [88].

We assume that $J = \mathrm{Jac}(C)$ is the Jacobian of a curve C. The bicanonical map $C \to \mathbb{P}^2$ sends C 2 : 1 to a conic Q. Let $p_1, \ldots, p_6$ be the branch points on Q of this map.

(4.1) Theorem (Humbert). Jac(C) *has real multiplication by* $\mathfrak{o}_{\mathbb{Q}(\sqrt{5})}$ *if and only if for a suitable labeling of the branch points there exists a conic D which is inscribed to the pentagon $p_1, \ldots, p_5$ and passes through p_6.*

Proof. To get the configuration for a Jacobian J with multiplication by $\mathfrak{o}_{\mathbb{Q}(\sqrt{5})}$ one takes an effective divisor H on J giving a principal polarization. Then if $\varepsilon = (1+\sqrt{5})/2$ we have for the intersection number

$$(2H, \varepsilon^* H) = 2\,\mathrm{Tr}_{K/\mathbb{Q}}(\varepsilon^2) = 6\,. \tag{2}$$

One uses $|2H|$ to map J to its Kummer surface $K(J)$, a quartic surface in $\mathbb{P}^3$ with sixteen singular points. Then by (2) there exists a curve E on $K(C)$ of degree 3 and E passes through six singular points of $K(C)$. Project now from such a singular point p onto a $\mathbb{P}^2$. The six standard conics through p (corresponding to singular points of the dual of $K(C)$, see [48]) map to lines $l_1, \ldots, l_6$ in $\mathbb{P}^2$. The closure of the image of E in $\mathbb{P}^2$ is a conic D passing through five intersection points of the l_i. This gives a configuration dual to the one described in the theorem.

Conversely, the configuration gives rise to a curve E of genus 2 on J with $(H, E) = 3$. Using the isomorphism $\{a \in \mathrm{End}^\circ(J)\colon a = a^\circ\} \simeq \mathrm{NS}^\circ(J)$ and that

$$\tfrac{1}{2}(\alpha H + \beta E, \alpha H + \beta E) = \alpha^2 + 3\alpha\beta + \beta^2$$

is a quadratic form of discriminant 5 one sees easily that J has real multiplication by $\mathfrak{o}_{\mathbb{Q}(\sqrt{5})}$, cf. Prop. (2.3). □

Humbert gives geometric characterizations for other discriminants too, see [88].

X. Moduli of Abelian Schemes with Real Multiplication

Hilbert modular varieties are the moduli spaces of abelian schemes with real multiplication. This interpretation makes it possible to construct models of Hilbert modular varieties over (schemes of) number rings. These models live in a somewhat larger category than that of schemes: they are algebraic stacks. They are however not proper over their base schemes. This is due to the fact that the abelian varieties which we consider can degenerate. Mumford has given a very explicit construction of degenerating abelian schemes, thus generalizing the idea of the Tate curve for elliptic curves. Using this one can construct smooth compactifications of moduli schemes of abelian schemes. The construction was carried out by Rapoport for Hilbert modular varieties over $\mathbb{Z}$. Faltings has extended the construction and applied it to the construction of a compactification of A_g, the moduli scheme of principally polarized abelian schemes over $\mathbb{Z}$. Faltings has also shown how to obtain a minimal (Baily-Borel) compactification over $\mathbb{Z}$.

The contents of this Chapter are the following. In Section 1 we define the notion of abelian schemes with real multiplication and study some of its properties. In Section 2 we define the objects that we want to classify; they are abelian schemes with real multiplication and a class of polarizations. By a theorem of M. Artin the corresponding functor is representable by an algebraic stack. We then describe how smooth compactifications of such a stack are constructed. We follow the method used by Faltings for A_g. Since the construction is rather involved our description is somewhat sketchy. If the reader wishes to see more details he should work them out himself or consult Faltings paper for an analogous situation. The construction is based on a choice of a cone decomposition for each cusp and therefore it is not unique. (For dimension two there is however a canonical choice.)

In Section 3 we give the applications to Hilbert modular forms. We state the result on the minimal compactification and derive some consequences. Since the minimal compactification is unique (though not smooth) one should try to use it instead of a smooth compactification whenever possible. The reader should notice how similar the final situation over $\mathbb{Z}$ is to that over $\mathbb{C}$.

The contents of the first three sections are not used in the rest of this book except for the fact that certain Hilbert modular varieties are defined over $\mathbb{Q}$.

Finally, in Section 4 we give the action of $\mathrm{Gal}(\bar{\mathbb{Q}}/\mathbb{Q})$ on the set of the components of certain Hilbert modular varieties.

1. Abelian Schemes with Real Multiplication

We discuss some basic properties of abelian schemes with real multiplication. A reference here is Rapoport's thesis [133], § 1, which we partly follow.

As before, K is a fixed totally real field of degree n over $\mathbb{Q}$ with ring of integers $\mathfrak{o} = \mathfrak{o}_K$. Let S be an arbitrary scheme. An abelian scheme A/S is a commutative group scheme such that $A \to S$ is separated, proper and smooth, has connected fibres and is of finite presentation. If $s: S \to A$ is the zero section then $\mathfrak{t}_A = \mathrm{Lie}(A)$ is the dual of $s^*\Omega^1_{A/S}$ (as locally free $\mathcal{O}_S$-module).

(1.1) Definition. We say that an abelian scheme A/S has real multiplication by $\mathfrak{o}$ if a homomorphism $m: \mathfrak{o} \to \mathrm{End}_S(A)$ with $m(1) = id$ is given such that $\mathfrak{t}_A$ is a locally (on S) free $\mathfrak{o} \otimes \mathcal{O}_S$-module of rank 1.

The definition implies that the relative dimension of A/S is n. Comparing with our (provisional) definition in Chapter IX we see that the condition on $\mathfrak{t}_A$ has been added. This rules out pathologies and will ensure that an abelian variety over a field of characteristic $p > 0$ with real multiplication can be lifted to characteristic 0. However, over $\mathbb{C}$ the condition is automatically fulfilled, as we see from:

(1.2) Lemma. *Let A be an abelian variety over a field k and let $m: \mathfrak{o} \to \mathrm{End}_k(A)$ be a homomorphism with $m(1) = id$. Then (A, m) has real multiplication by $\mathfrak{o}$ if and only if $\mathfrak{o} \otimes k$ acts faithfully on $\mathfrak{t}_A$.*

Proof. The ring $R = \mathfrak{o} \otimes k$ is a product of local Artin rings R_i with maximal ideals $\mathfrak{m}_i$. The localization of $\mathfrak{t}_A$ at $\mathfrak{m}_i$ is a direct sum of cyclic R_i-modules (of the form $R_i/\mathfrak{m}_i^{a_i}$ with $a_i \geqslant 0$). The action of R_j is faithful if and only if R_j occurs among these cyclic R_j-modules. So $\mathfrak{o} \otimes k$ acts faithfully if and only if R occurs in $\mathfrak{t}_A$ as a summand, but then for dimension reasons $R \cong \mathfrak{t}_A$. □

We refer to [133], p. 274 for an example of an abelian surface A over $k = \overline{\mathbb{F}}_p$ with $\mathfrak{o}_K \hookrightarrow \mathrm{End}(A)$ for a real quadratic field K for which $\mathfrak{t}_A$ is not free over $\mathfrak{o}_K \otimes k$.

Before we proceed we recall some facts about abelian varieties over a perfect field k of $\mathrm{char}(k) = p > 0$, needed in the proof of the next lemma.

a) Let $(\Omega^{\cdot}_{A/k}, d)$ be the complex of Kähler differentials and let $H^1_{DR}(A)$ be the n-th hypercohomology group of this complex (DR = De Rham). We have an exact sequence:

$$0 \to H^0(A, \Omega_{A/k}) \to H^1_{DR}(A) \to H^1(A, \mathcal{O}_A) \to 0 .$$

b) $H^1_{DR}(A)$ is the reduction mod p of the (contravariant) Dieudonné module of the p-divisible (Barsotti-Tate) group of A. The Dieudonné module M is a free module of rank equal to $2\dim(A)$ over the ring $W = W(k)$ of Witt vectors of k (= unique complete discrete valuation ring of characteristic 0 with residue field k and maximal ideal (p)). It is known (see [169]) for A with real multiplication by $\mathfrak{o}$ that $\mathfrak{o} \otimes \mathbb{Z}_p$ injects in $\mathrm{End}_{\mathbb{Z}_p}(M)$, hence M is torsion free over $\mathfrak{o} \otimes W(k)$.

(1.3) Lemma. *If A/S has real multiplication by $\mathfrak{o}$, then $H^1(A, \mathcal{O}_A)$ is an S-locally free $\mathfrak{o} \otimes \mathcal{O}_S$-module of rank* 1.

Proof. We have $\mathfrak{t}_{A^\vee} \cong H^1(A, \mathcal{O}_A)$. Note that if $\mathfrak{t}_A$ is free over $\mathfrak{o} \otimes k(x)$ for a closed point $x \in S$, then $\mathfrak{t}_A$ is locally free over $\mathfrak{o} \otimes \mathcal{O}_S$ in a neighbourhood of x. We thus may assume that $S = \mathrm{Spec}(k)$ with k a perfect field of characteristic $p > 0$.

M is torsion-free, hence projective over the Dedekind ring $\mathfrak{o} \otimes W(k)$. Consider $M \otimes \mathbb{Q}_p$ as $K \otimes W(k)$-module. Over an algebraic closure $\overline{W}$ of $Q(W(k))$ we see (if $K \otimes \overline{W} = \bigoplus_\sigma \overline{W}$)

$$M \otimes \mathbb{Q}_p \otimes \overline{W} \simeq \bigoplus_\sigma \overline{W}^{n_\sigma}, \qquad (n_\sigma \in \mathbb{Z}_{\geqslant 0})$$

where σ runs over the embeddings of K into $\overline{W}$. Obviously, this module $M \otimes \mathbb{Q}_p$ is free if and only if all n_σ are equal.

An endomorphism of A gives rise to an endomorphism of M with trace in $\mathbb{Z}$. But in order to get $\sum n_\sigma \sigma(\alpha) \in \mathbb{Q}$ for all $\alpha \in K$ all n_σ must be equal. So $M \otimes \mathbb{Q}_p$ is free over $K \otimes W(k)$. Since $\mathfrak{o} \otimes W(k)$ is a local ring this shows that M is free over $\mathfrak{o} \otimes W(k)$, hence $H^1_{DR}(A) = M/pM$ is free over $\mathfrak{o} \otimes k$.

By symmetry, $H^1_{DR}(A^\vee)$ is also free over $\mathfrak{o} \otimes k$. By the exact sequence

$$0 \to H^1(A^\vee, \Omega^1_{A^\vee/k}) \to H^1_{DR}(A^\vee) \to H^1(A^\vee, \mathcal{O}_{A^\vee}) \to 0$$
$$\wr\|$$
$$\mathfrak{t}_A$$

and the assumption on $\mathfrak{t}_A$ we see that $H^0(A^\vee, \Omega^1_{A^\vee/k})$ is free over $\mathfrak{o} \otimes k$. But this is $\mathfrak{t}^\vee_{A^\vee} = \mathrm{Hom}_{\mathfrak{o}\otimes k}(\mathfrak{t}_{A^\vee}, \mathfrak{o} \otimes k) \otimes \mathfrak{d}^{-1}$ with $\mathfrak{d}^{-1}$ the inverse different. So $\mathfrak{t}_{A^\vee}$ is free over $\mathfrak{o} \otimes k$. □

(1.4) Corollary. *The dual $A^\vee$ of an abelian scheme A with real multiplication m has real multiplication (by $m^\vee$).*

A polarization $\lambda\colon A \to A^\vee$ of an abelian scheme A/S with real multiplication m is called $\mathfrak{o}$-*linear* if $m^\vee \circ \lambda = \lambda \circ m$.

(1.5) Proposition. *An abelian variety A over a field k with real multiplication m admits an $\mathfrak{o}$-linear polarization.*

Proof. (We showed this already in Chapter IX for $\mathrm{char}(k) = 0$.) We may assume: A is isogenous to B^r with B a simple abelian variety. Put $E = \mathrm{End}^\circ(A)$. The Néron-Severi group tensored with the reals, $J = \mathrm{NS}^\circ(A) \otimes \mathbb{R}$, is a formally real Jordan algebra and the images of the ample line bundles are the totally positive elements of J, cf. [115].

Using the Rosati-involution $x \mapsto x'$ on E defined by an arbitrary polarization we can identify $\mathrm{NS}^\circ(A)$ with $\{x \in E\colon x' = x\}$. We have 3 cases: $\mathrm{End}^\circ(B)$ is

i) a totally real field F, or

ii) a quaternion algebra over a totally real field, or

iii) a quaternion algebra over a CM-field.

K sits in $\mathrm{End}^\circ(A) = M_r(\mathrm{End}^\circ(B))$ and we have to find a K-linear totally positive element of J. We treat case i), leaving ii), iii) to the reader. View $E \otimes \mathbb{R}$ as $\mathrm{End}_{F_\mathbb{R}}(V)$ for some $F_\mathbb{R}$-module V of finite type. The totally positive elements of J correspond to the $F_\mathbb{R}$-bilinear symmetric forms ϕ which are totally positive. We are looking for such a ϕ with $\phi(\lambda x, y) = \phi(x, \lambda y)$ for all $\lambda \in K$. But $(F \otimes K)_\mathbb{R} = \mathbb{R}^a$ and such a ϕ obviously exists. □

(1.6) Lemma. *Let S be a connected scheme of finite type over $\mathbb{Z}$. If $f: A \to B$ is a $\mathfrak{o}$-linear homomorphism of abelian schemes over S with real multiplication by $\mathfrak{o}$ then $f = 0$ or f is finite locally free* (i.e. *an isogeny*).

Proof. (Using Drinfeld's trick, cf. [94].) $\mathrm{Hom}_S(A, B)$ is represented by a separated scheme over S. Hence $f = 0$ defines a closed subscheme C of the base. Let $s \in C$ be a closed point such that $f = 0$ in s. If we show that $f = 0$ in $\hat{\mathcal{O}}_s$, then C is also open, so $C = S$. It suffices to show: $f = 0$ in $\hat{\mathcal{O}}_s/\mathfrak{m}_s^n$ for all $n \geqslant 1$. By our assumption on S (of finite type/$\mathbb{Z}$) $\hat{\mathcal{O}}_s/\mathfrak{m}_s^n$ is finite, hence a prime p is nilpotent. So we are in the situation: R a ring, $I \subset R$ an ideal such that (I, p) is nilpotent. By induction over $I^{(0)} = I, \ldots, I^{(n)} = ((I^{(n-1)})^2, pI^{(n-1)})$ we may assume $I^2 = pI = 0$. If $f = 0 \bmod I$ we now show $pf = 0$. For any R-algebra D the kernel of $B(D) \to B(D/I)$ lies in the formal group of B. Choose formal parameters $X_1, \ldots, X_g$ for this formal group and let $P \in B(D)$ with $f(P) \equiv 0 \pmod{I}$. Then $X_i(f(P)) \in ID$ and since $[p]^* X_i \equiv pX_i \pmod{I^2}$ we see $[p]f(P) = 0$. Since multiplication $[p]$ by p is surjective in the f.p.p.f. topology it follows that $f(P) = 0$. Since f is $\mathfrak{o}$-linear f is fibre by fibre finite if $f \not\equiv 0$; so f is then a proper map with finite fibres, hence finite. It is flat as a finite morphism of smooth schemes of the same dimension, cf. [1]. By assumption S is noetherian, so if $f \neq 0$, then f is locally free. □

Recall that we say that a projective $\mathfrak{o}$-module P of rank 1 is *ordered* if for all $\sigma: K \hookrightarrow \mathbb{R}$ an order on $P \otimes_\sigma \mathbb{R}$ is given.

Let A/S be an abelian scheme with real multiplication by $\mathfrak{o}$. A homomorphism $\lambda: A \to A^\vee$ is called *symmetric* if

$$\begin{array}{ccc} A & \xrightarrow{\ \lambda\ } & \\ \| & & A \\ A^{\vee\vee} & \xrightarrow{\ \hat{\lambda}\ } & \end{array} \qquad \text{commutes .}$$

We associate to A/S an $\mathfrak{o}$-module and a subset of this $\mathfrak{o}$-module as follows:

$$P(A/S) = \{\lambda \in \mathrm{Hom}_S(A, A^\vee) \colon \lambda \ \mathfrak{o}\text{-linear and symmetric}\}$$

$$P^+(A/S) = \{\lambda \in P(A/S) \colon \lambda \text{ a polarization}\} .$$

We call the pair (P, P^+) the *ordered polarization module.* By (1.6) P is a torsion free $\mathfrak{o}$-module for S connected and of finite type over $\mathbb{Z}$.

(1.7) Lemma. *If k is a perfect field then (P, P^+) (A/k) is an ordered $\mathfrak{o}$-module of rank* 1.

Proof. By (1.5) P is non-zero. We have an injection $P \to \bigwedge_{\mathfrak{o}}^2 H^1(A, \mathbb{Z}_\ell)$ for $\ell \neq p$, cf. [115], hence rk $P = 1$. We have to check that P^+ is the set of totally positive elements for some choice of orders on $P \otimes_\sigma \mathbb{R}$. We show that U^+ (the totally positive units of $\mathfrak{o}$) preserves P^+. If $\alpha \in \mathfrak{o}$ then $m(\alpha)^*L$ is ample for L ample; this corresponds to $\lambda \mapsto \alpha^2\lambda$, hence α^2 preserves P^+. But any totally positive element $\alpha \in \mathfrak{o}$ is of the form $\sum_{i=1}^r \beta_i^2$, $\beta_i \in K$, hence $n^2\alpha = \sum_{i=1}^r \gamma_i^2$ with $\gamma_i \in \mathfrak{o}$ for some n, see [106, p. 274]. Hence $n^2\alpha$ preserves P^+ and so does α. □

We show that elements of $P(P^+)$ can be lifted to formal deformations of A/k. The result is that the functor (P, P^+) defined on the category of abelian schemes with real multiplication is formally étale: if R is a ring and $I \subset R$ a nilpotent ideal, $R_0 = R/I$, then $P(A/R) = P(A/R_0)$.

Grothendieck proved a local-Torelli-type theorem for nilpotent fattenings, see [50]. Let R be a ring, I an ideal with $I^2 = 0$ and set $R_0 = R/I$, $S = \mathrm{Spec}(R)$, $S_0 = \mathrm{Spec}(R_0)$. According to Grothendieck one can associate to an abelian scheme A_0/R_0 an R-module M which is locally free of rank $2\dim A_0$ and such that (canonically) $M \otimes R_0 \xrightarrow{\sim} H^1_{DR}(A_0)$.

If R_0 is a perfect field of characteristic $p > 0$ then M is the contravariant Dieudonné module of A_0. Then $M \otimes R_0$ is provided with a filtration $F_0 \subset M \otimes R_0$ coming from $\mathrm{t}_{A_0}^\vee \subset H^1_{DR}(A_0)$. This gives rise to an equivalence of categories:

$$\begin{aligned}&\{\text{abelian schemes}/R \text{ with an isomorphism } A \otimes R_0 \xrightarrow{\sim} A_0\}\\ \simeq\ &\{\text{filtrations } F \subset M \text{ of } R\text{-modules such that } F \text{ is locally}\\ &\quad \text{a direct summand which extends } F_0\}\end{aligned}$$

If A_0 has real multiplication by $\mathfrak{o}$ then $F_0 \subset M \otimes R_0$ is a filtration of $\mathfrak{o} \otimes \mathcal{O}_S$-modules. Under the above equivalence abelian schemes/R with real multiplication by $\mathfrak{o}$ and with $A \otimes R_0 \xrightarrow{\sim} A_0$ correspond to those filtrations that are filtrations of $\mathfrak{o} \otimes \mathcal{O}_S$-modules.

A consequence is that we can lift abelian schemes over R_0 with real multiplication by $\mathfrak{o}$ to abelian schemes over R with real multiplication by $\mathfrak{o}$: since $\mathrm{t}_{A_0}^\vee$ is locally (on S_0) free of rank 1 we can choose locally a generator g_0. Choose an element $g \in M$ which reduces to g_0 on S_0. Then g generates (locally) a free $\mathfrak{o} \otimes \mathcal{O}_S$-module $F \subset M$ which reduces to F_0. This gives our lift A/R of A_0/R_0.

Let A/S be an abelian scheme with real multiplication by $\mathfrak{o}$. A polarization $\lambda\colon A_0 \to A_0^\vee$ induces an alternating bilinear form ϕ_λ on M for which F_0 is totally isotropic. Then λ extends to A/S if and only if F is isotropic for ϕ_λ. If moreover, λ is $\mathfrak{o}$-linear, then $\phi_\lambda = \mathrm{Tr}_{K/\mathbb{Q}}\,\Phi_\lambda$ where Φ_λ is a uniquely determined $\mathfrak{o}$-bilinear form: $M \times M \to \mathfrak{d}^{-1} \otimes \mathcal{O}_S$. Since F is an $\mathfrak{o}$-module of rank 1 λ must be zero on F. Hence any $\mathfrak{o}$-linear polarization λ of A_0/S_0 extends to an $\mathfrak{o}$-linear polarization.

(1.8) Corollary. *The functor* (P, P^+) *is formally étale, i.e. it does not see nilpotents*: $(P, P^+)(A/S) = (P, P^+)(A_0/S_0)$.

Applying Grothendieck's theory one sees that there are no obstructions to deforming an abelian variety with real multiplication by $\mathfrak{o}$ and moreover that the tangent space to the deformation space is $\mathrm{Hom}_{\mathfrak{o}\otimes\mathcal{O}_{S_0}}(F \to M/F)$, i.e.

$$\mathrm{Hom}_{\mathfrak{o}\otimes\mathcal{O}_{S_0}}(\mathfrak{t}_A^\vee, \mathfrak{t}_{A^\vee}) = \mathfrak{t}_{A^\vee} \otimes_{\mathfrak{o}\otimes\mathcal{O}_{S_0}} \mathfrak{t}_A \otimes \mathfrak{d}$$

with $\mathfrak{d}$ the different of K. Moreover, Schlessinger's theory [140] tells us that an abelian variety with real multiplication over an algebraically closed field k of characteristic $p > 0$ has a versal deformation over $W(k)[[t_1, \ldots, t_g]]$.

2. Modular Stacks

The best one can hope for in dealing with the moduli spaces for abelian schemes with real multiplication by a given order in a totally real field and a class of polarizations is to have a scheme $\mathcal{M}$ together with a (universal) abelian scheme $\mathcal{A}$ with real multiplication over $\mathcal{M}$ such that for every abelian scheme with real multiplication of the given type A/S there exists a unique morphism $S \xrightarrow{\phi} \mathcal{M}$ such that $A = \mathcal{A} \otimes_\phi S$; in other words, that $\mathcal{M}$ represents the functor which associates to a scheme S the isomorphism classes of such abelian schemes with real multiplication over S. However, we are asking to much. For one thing, abelian schemes with real multiplication can have automorphisms, so $S \to \mathcal{M}$ cannot be unique. One can try to repair this by imposing some extra structure on the abelian schemes we are dealing with to rule out non-trivial automorphisms. Instead of $\mathcal{M}$ we then expect a pair $\mathcal{S} = (S_0, S_1)$ of schemes with an abelian scheme with real multiplication over S_0 such that S_1 gives a kind of equivalence relation which tells us which identifications have to be made when we forget the extra structure. Such an object $\mathcal{S}$ is called an algebraic stack. And instead of morphisms $T \to \mathcal{M}$ we now consider descent data for $\mathrm{Hom}(T, \mathcal{S})$: an étale covering $\{T_i\}$ of T and morphisms $\phi_i\colon T_i \to S_0$ which are in some sense compatible with the equivalence relation S_1. When made precise the result is then almost as good as the one we hoped for.

We now start by explaining the notion of an algebraic stack, cf. [26].

Let A be the category in which the objects are the abelian schemes $(A/S, m)$ with real multiplication $m\colon \mathfrak{o} \hookrightarrow \mathrm{End}_S(A)$ by $\mathfrak{o}$ over variable base schemes S and in which the morphisms are Cartesian diagrams:

$$\begin{array}{ccc} A_1 & \longrightarrow & A_2 \\ \downarrow & & \downarrow \\ S_1 & \longrightarrow & S_2 \end{array}$$

whose upper horizontal arrow is $\mathfrak{o}$-linear. This category is fibered over the category Sch of schemes in groupoids, i.e. one has

i) for any scheme $T \to S$ over S and any object A/S in A there is a morphism in our category

$$\begin{array}{ccc} A' & \longrightarrow & A \\ \downarrow & & \downarrow \\ T & \longrightarrow & S \end{array}$$

ii) given a diagram

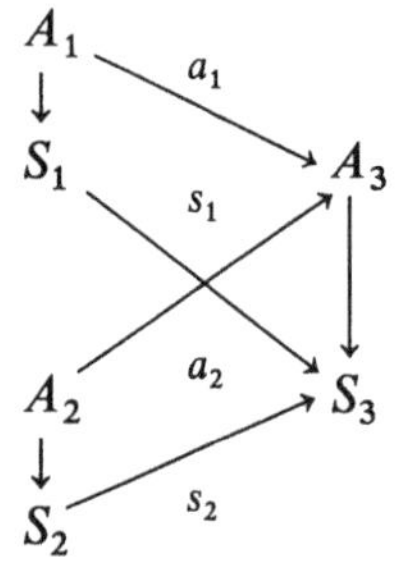

and a morphism $s_{12}\colon S_1 \to S_2$ that $s_1 = s_2 \cdot s_{12}$ there is a diagram

$$\begin{array}{ccc} A_1 & \xrightarrow{a_{12}} & A_2 \\ \downarrow & & \downarrow \\ S_1 & \xrightarrow[s_{12}]{} & S_2 \end{array}$$

such that $a_1 = a_2 a_{12}$ in A.

But our category satisfies more:

iii) Let S be a scheme. The functor which associates to an étale $T \xrightarrow{\phi} S$ the set $\mathrm{Hom}_T(\phi^* A_1, \phi^* A_2)$ is a sheaf in the étale topology of S.

iv) given an étale covering $T_i \to S$, objects A_i/T_i in A and descent data (i.e. isomorphisms $f_{ij}\colon A_i \xrightarrow{\sim} A_j$ over $T_i \times_S T_j$ satisfying a cocycle condition over $T_i \times T_j \times T_k$) there is an object A over S which gives by pullback A_i/T_i.

Conditions i), ... iv) say that our category A is a *stack* (for the étale topology over Sch; "champ" in french).

Instead of A we consider the substack $\mathsf{B} = \mathsf{B}_{C,C^+}$ whose objects are the triples $(A/S, m, \alpha)$ such that $(A/S, m)$ is in A and α is an isomorphism of sheaves of ordered $\mathfrak{o}$-modules (of rank 1): $(P_1 P^+) \to (C, C^+)$. Here (C, C^+) is the constant sheaf with value a fixed ordered projective $\mathfrak{o}$-module (on connected schemes); morphisms are cartesian diagrams compatible with the α's.

It would be desirable to have a final object in B: i.e. a ("universal") abelian scheme $\mathscr{A}$ over a base scheme $\mathscr{M}$ with multiplication by $\mathfrak{o}$ and with an α such that any triple $(A/S, m, \alpha)$ in B is obtained via pullback over a *unique* $S \to \mathscr{M}$. Since our objects can have non-trivial automorphisms this cannot exist. One could then hope to have a *coarse moduli scheme* M, i.e. if $F\colon \mathsf{Sch} \to \mathsf{Sets}$ is the functor which associates to S the set of isomorphism classes of tuples $(A/S, m, \alpha)$ in B one would like to have a scheme M and a morphism $F \to h_M := \mathrm{Hom}(-, M)$ such that

i) if k is an algebraically closed field $F(\mathrm{Spec}(k)) \to h_M(\mathrm{Spec}(k))$ is a bijection

ii) $F \to h_M$ is universal: for any scheme N and a morphism $F \to h_N$ there is a unique factorization via h_M ($\mathrm{Hom}(F, h_N) = \mathrm{Hom}(M, N)$).

By ii) M is then unique. An intuitive idea for the construction of M is: start with a triple $(A/S, m, \alpha)$ over $S = \mathrm{Spec}(k)$ with k an algebraically closed field and consider deformations of this triple. This should give an étale formal neighbourhood of a geometric point $\mathrm{Spec}(k) \to M$. Then one should glue together the algebraizations of these pieces to get a scheme.

However, what ensures that we get a scheme? What one can expect is an object of the following kind:

(2.1) Definition. An *algebraic stack* $\mathscr{S}$ is a diagram of schemes and morphisms

$$\begin{array}{ccc} S_1 \times_{S_0} S_1 & \xrightarrow{d_0} & S_1 \\ {\scriptstyle d_1}\downarrow & \square & \downarrow{\scriptstyle e_1} \\ S_1 & \xrightarrow[e_0]{} & S_0 \end{array} \qquad S_1 \times_{S_0} S_1 \xrightarrow{d_2} S_1$$

such that a) S_0 is quasi-compact for thc ćtalc topology

b) $e_i\colon S_1 \to S_0$ $i = 0, 1$ is étale

c) $(e_0, e_1)\colon S_1 \to S_0 \times S_0$ is finite

d) for every scheme T the following is a category all whose arrows are invertible: the objects are the elements of $\mathrm{Hom}(T, S_0)$, the arrows are the elements of $\mathrm{Hom}(T, S_1)$ and the composition of arrows is giving by d_2: if $f, g \in \mathrm{Hom}(T, S_1)$ with $e_0(f) = e_1(g)$ then there is a unique $h \in \mathrm{Hom}(T, S_1 \times_{S_0} S_1)$ with $d_1(h) = f$, $d_0(h) = g$. The composition is then $d_2(h)$.

One should view this object as a kind of quotient S_0/S_1 for the étale equivalence relation S_1. If $S_1 \to S_0 \times S_0$ is a closed immersion then $\mathscr{S}$ is called an *algebraic space*. An algebraic stack $\mathscr{S}$ as in (2.1) gives a stack in the sense above: the objects over a scheme T are the *descent data for* $\mathrm{Hom}(T, \mathscr{S})$: an étale covering $\{T_i\}$ of T with morphisms $T_i \xrightarrow{\phi_i} S_0$, $T_i \times T_j \xrightarrow{\phi_{ij}} S_1$ with $e_0(\phi_{ij}) = \phi_i$, $e_1(\phi_{ij}) = \phi_j$ such that the ϕ_{ij} satisfy a cocycle relation on $T_i \times T_j \times T_k$. One checks that the properties i), ii), iii), iv) above are satisfied. Algebraic spaces are studied in [3, 4].

We want to construct an algebraic stack $\mathscr{S}$ with

1) S_0 is of finite type and smooth $/\mathbb{Z}$

2) over S_0 there exists an abelian variety $\mathscr{A}$ with real multiplication m by $\mathfrak{o}$ and an isomorphism of sheaves for the étale topology

$$(P, P^+) \xrightarrow{\sim} (C, C^+)$$

such that $\mathscr{S}$ represents the functor $F = F_{C,C^+}\colon \mathsf{Sch} \to \mathsf{Sets}$ which associates to a scheme T the set of isomorphism classes of triples $(A/S, m, \alpha)$ (i.e. such that we have a functorial bijection $F(T) \leftrightarrow$ set of descent data of T for $\mathrm{Hom}(-, \mathscr{S})$).

Fix now C, C^+.

(2.2) Theorem (cf. [133]). *There exists a smooth algebraic stack $\mathscr{S}(C, C^+)$ of dimension g over $\mathbb{Z}$ which represents F_{C,C^+}.*

The proof of this theorem uses the criterion of M. Artin for the representability of functors by algebraic spaces or algebraic stacks [3, 4]. That the conditions given by Artin hold follows immediately from the literature and the properties of abelian schemes with real multiplication given in Section 1, cf. [133].

The fibre $\mathscr{S}(C, C^+) \otimes \mathbb{C}$ can be identified—in a non-canonical way—with a quotient $\mathrm{GL}_{\mathfrak{o}}^+(\Lambda) \backslash \mathrm{GL}_{\mathfrak{o}\otimes\mathbb{R}}^+(\Lambda \otimes \mathbb{R})/\mathbf{k}$ described in (1.6) of Chapter IX.

The Hilbert modular stacks found in (2.2) are not proper over $\mathbb{Z}$. To remedy this we must consider degenerating abelian varieties with real multiplication. Tate gave a description of a degenerating elliptic curve over a complete discrete valuation ring R with field of quotients $F = Q(R)$: the Tate curve $F^*/q^{\mathbb{Z}}$.

This construction can be generalized to degenerating abelian varieties by using the rigid analytic spaces of non-archimedean analysis, see [134]. Mumford gave a different construction of the Tate curve and generalized this to a construction of degenerating abelian varieties over complete local rings, [116]. Rapoport used this construction to obtain a compactification of Hilbert modular stacks in [133]. Faltings used an extension of Mumford's construction to compactify the modular stack A_g of principally polarized abelian varieties of dimension g. He also obtained a minimal ("Satake-Baily-Borel") compactification over $\mathbb{Z}$.

Here we give the relevant facts for the construction of a smooth compactification of Hilbert modular stacks over $\mathbb{Z}$ but refer to [30, 133] for details. We also indicate how to obtain a minimal compactification.

(2.3) Definition. A semi-abelian scheme G/S is a smooth algebraic group scheme of finite type over S with connected fibres which are extensions of abelian varieties by algebraic tori. If $m\colon \mathfrak{o} \hookrightarrow \mathrm{End}(G/S)$ with $m(1) = \mathrm{id}$ is given such that $\mathfrak{t}_G$ is a locally (on S) free $\mathcal{O}_S \otimes \mathfrak{o}$-module of rank 1, then we say that G/S has real multiplication by $\mathfrak{o}$.

Suppose that R is a normal complete local ring with maximal ideal $\mathfrak{m}$ and field of quotients $F = Q(R)$. Let G be a semi-abelian scheme over R with real multiplication by $\mathfrak{o}$ whose special fibre G_s is an extension of an abelian variety by a torus T. We may assume that $T \cong \mathbb{G}_m^r$ (otherwise extend R). The character group of T is a torsion-free, hence projective, $\mathfrak{o}$-module of rank 0 or 1. If G_s is not abelian this rank equals 1. If $\mathfrak{a}$ is a projective $\mathfrak{o}$-module of rank 1 we denote by

$$\mathfrak{a} \otimes \mathbb{G}_m$$

the algebraic torus with character group $\mathfrak{a}^\vee$, the dual of $\mathfrak{a}$ w.r.t. Tr.

Now assume that R is a discrete valuation ring and that G/S, $S = \mathrm{Spec}(R)$, is semi-abelian with as special fibre G_s a split torus. According to Raynaud [134] we can present G as a rigid analytic quotient of a split torus by a discrete subgroup. If G has real multiplication by $\mathfrak{o}$ then the rigid analytic space G^{an} of G is of the form

$$G^{\mathrm{an}} = (\mathfrak{a} \otimes \mathbb{G}_m)^{\mathrm{an}}/i(\mathfrak{b}),$$

where $\mathfrak{a}$, $\mathfrak{b}$ are projective $\mathfrak{o}$-modules of rank 1 and

$$i\colon \mathfrak{b} \to \mathfrak{a} \otimes \mathbb{G}_m(F) \tag{1}$$

is $\mathfrak{o}$-linear.

Symmetric $\mathfrak{o}$-linear homomorphisms $\lambda\colon G \to G^\vee$ correspond to $\mathfrak{o}$-linear maps $\mathfrak{b} \to \mathfrak{a}^\vee$. We thus can identify the polarization module P of G with $(\mathfrak{a} \otimes \mathfrak{b})^\vee$. Those λ which define a polarization on G/F correspond to the totally positive elements of $(\mathfrak{a} \otimes \mathfrak{b})^\vee$ for some order on $(\mathfrak{a} \otimes \mathfrak{b})^\vee$.

Let $B: \mathfrak{a}^\vee \times \mathfrak{b} \to F^*$ be the $\mathfrak{o}$-bilinear form which gives (1) over F. Then

i) B is "symmetric": for all $\lambda \in P$ viewed as $\lambda: \mathfrak{b} \to \mathfrak{a}^\vee$ we have $B(\lambda(\beta_1), \beta_2) = B(\lambda(\beta_2), \beta_1)$ for all $\beta_1, \beta_2 \in \mathfrak{b}$,

ii) for all $\lambda \in P^+$ and $\beta \in \mathfrak{b}$, $\beta \neq 0$ we have $B(\lambda(\beta), \beta) \in \mathfrak{m}$.

Now the idea is to construct conversely a semi-abelian scheme with real multiplication out of such data. The construction is due to Mumford and was applied to the case of semi-abelian schemes with real multiplication by Rapoport.

So, let R be an excellent normal ring, complete in the I-adic topology for some ideal I. Denote its quotient field $Q(R)$ by F and set $S = \mathrm{Spec}(R)$, $S_0 = \mathrm{Spec}(R/I)$. The adaptation of Mumford's theorem that we need is:

(2.4) Theorem. *Let $\mathfrak{a}$, $\mathfrak{b}$ be projective $\mathfrak{o}$-modules of rank 1 and let $P = (\mathfrak{a} \otimes \mathfrak{b})^\vee$ be ordered (by $P^+ \subset P$). Suppose we are given a $\mathfrak{o}$-bilinear form $B: \mathfrak{a}^\vee \times \mathfrak{b} \to F^*$ such that*

i) $B(\lambda(\beta_1), \beta_2) = B(\lambda(\beta_2), \beta_1)$ for all $\beta_1, \beta_2 \in \mathfrak{b}$, $\lambda \in \mathfrak{b} \to \mathfrak{a}^\vee$ $\lambda \in P$,

ii) for all $\lambda \in P^+$ and all $\beta \in \mathfrak{b}$, $\beta \neq 0$ we have $B(\lambda(\beta), \beta) \in I$. Then there exists a functorial construction which associates to these data a semi-abelian scheme G/S with real multiplication by $\mathfrak{o}$ with generic fibre an abelian variety over F and with special fibre $G \otimes S_0$ the algebraic torus $\mathfrak{a} \otimes \mathbb{G}_m$. If one interchanges $\mathfrak{a}$ and $\mathfrak{b}$ the construction yields a semi-abelian scheme G' which over F is the dual $\mathrm{Pic}^\circ(G)$ *of G. Moreover, the polarization module of G over F is (P, P^+): every $\lambda \in P$ defines a $\mathfrak{o}$-linear homomorphism $\mathfrak{a} \otimes \mathbb{G}_m \to \mathfrak{b}^\vee \otimes \mathbb{G}_m$ which gives rise to $G \to G'/F$. This is an $\mathfrak{o}$-linear polarization if and only if $\lambda \in P^+$.*

In the following we view $\mathscr{C}\ell^+(K)$ as the group of isomorphism classes of ordered projective $\mathfrak{o}$-modules of rank 1. Let $C \in \mathscr{C}\ell^+(K)$ be fixed.

(2.5) Definition. A cusp (for C) is an exact sequence

$$0 \to \mathfrak{a} \to \mathfrak{L} \to \mathfrak{b} \to 0$$

of projective $\mathfrak{o}$-modules of rank 1, 2, 1 respectively, plus an order on $(\bigwedge^2 \mathfrak{L})^\vee$ such that $(\bigwedge^2 \mathfrak{L})^\vee \in C$.

Two cusps are called equivalent if there is a commutative diagram

$$\begin{array}{ccccccccc} 0 & \to & \mathfrak{a}_1 & \to & \mathfrak{L}_1 & \to & \mathfrak{b}_1 & \to & 0 \\ & & \downarrow & & \downarrow & & \downarrow & & \\ 0 & \to & \mathfrak{a}_2 & \to & \mathfrak{L}_2 & \to & \mathfrak{b}_2 & \to & 0 \end{array}$$

such that the induced $(\bigwedge^2 \mathfrak{L}_1)^\vee \cong (\bigwedge^2 \mathfrak{L}_2)^\vee$ is an isomorphism of ordered $\mathfrak{o}$-modules.

(The link with the notion of cusp introduced in Chapter I is as follows.

Let $\mathfrak{c} \subset \mathfrak{o}$ be an ideal. Then $\mathfrak{c}^\vee \subset K$ is ordered in a natural way. To the cusp at ∞ for $\mathrm{GL}^+(\mathfrak{o} \oplus \mathfrak{c})$—in the old sense—we associate the sequence $0 \to \mathfrak{o} \to \mathfrak{o} \oplus \mathfrak{c} \to \mathfrak{c} \to 0$. Any old sense cusp is obtained by applying an element $A \in \mathrm{SL}(2, K)$. The linear map $K \oplus K \xrightarrow{A} K \oplus K$ transforms our sequence into a new one.)

As in Chapter II the compactification is based on rational polyhedral decompositions of $(M \otimes \mathbb{R})_+$ for certain ordered projective $\mathfrak{o}$-modules M of rank 1, one for each cusp.

Indeed, if $0 \to \mathfrak{a} \to \mathfrak{L} \to \mathfrak{b} \to 0$ is a cusp with $P = (\bigwedge^2 \mathfrak{L})^\vee$ ordered then M is the module of $\mathfrak{o}$-bilinear forms B on $\mathfrak{a}^\vee \times \mathfrak{b}$ which are symmetric, i.e. for every $\lambda \in P$ viewed as $\mathfrak{b} \xrightarrow{\lambda} \mathfrak{a}^\vee$ we have $B(\lambda(\beta_1), \beta_2) = B(\lambda(\beta_2), \beta_1)$ for all $\beta_1, \beta_2 \in \mathfrak{b}$. Since $M = (\bigwedge^2 \mathfrak{L})^\vee \cdot \mathfrak{b}^2$ it is ordered in a natural way. We denote by M_+ the set of totally positive elements of M and let V be (a subgroup of finite index of) the group of totally positive units of $\mathfrak{o}$.

We choose for each equivalence class of cusps a decomposition

$$(M \otimes \mathbb{R})_+ = \bigcup_{\sigma \in \Sigma} \sigma ,$$

where the σ are rational polyhedral cones (convex hulls of finitely many half-lines) such that

i) the decomposition is V-invariant and modulo V there are only finitely many cones,

ii) if $\sigma \in \Sigma$ then every face of σ is in Σ,

iii) if $\sigma, \tau \in \Sigma$ then $\sigma \cap \tau$ is a common face of σ and τ (if $\sigma \neq \tau$)

iv) every cone is spanned by a subset of a basis of M.

It is known that one can obtain such a cone decomposition by suitable subdivision, cf. Ch. II and [28, 8]. The finiteness in i) is a consequence of Dirichlet's theorem on units. Condition iv) is not indispensable but will assure that our compactification is smooth. In general there are many decompositions satisfying the requirements and no one seems canonical if $[K:\mathbb{Q}] \geqslant 3$. If $[K:\mathbb{Q}] = 2$ a canonical 'minimal' decomposition is specified by the boundary points of the convex hull of M_+ in $(M \otimes \mathbb{R})_+$, see Ch. II.

To the decomposition Σ of $(M \otimes \mathbb{R})_+$ we can associate a toroidal embedding as follows. Let S be the algebraic torus with character group $N = M^\vee$: $S = \operatorname{Spec}(\mathbb{Z}[X^n : n \in N]/(X^{n_1} X^{n_2} = X^{n_1+n_2}, X^0 = 1))$. Every $\sigma \in \Sigma$ defines a torus embedding $S_\sigma = \operatorname{Spec}(R_\sigma)$

$$R_\sigma = \mathbb{Z}[X^n : n \in \check{\sigma} \cap N] \quad \text{with } \check{\sigma} = \{l \in N_{\mathbb{R}} : l(m) \geqslant 0 \text{ for all } m \in \sigma\} .$$

S_σ is an affine scheme with an action of S which extends the natural action of S as a group scheme on itself by translation. Under this action there is only one closed orbit with stabilizer the subtorus belonging to the sublattice spanned by σ in M. If τ is a face of σ then S_τ is an open set in S_σ.

One can glue the S_σ's: glue S_{σ_1} with S_{σ_2} along $S_{\sigma_1 \cap \sigma_2}$ via $R_{\sigma_i} \subset R_{\sigma_1 \cap \sigma_2}$. One thus obtains a scheme S_Σ—a "toroidal embedding"—which is locally of finite type and contains S and on which S acts such that this extends the action of S on itself. The group V acts on S_Σ.

An elementary case of this construction was carried out in Chapter II, 2 over $\mathbb{C}$ instead of $\mathbb{Z}$. In that case every 2-dimensional σ_α determines $S_{\sigma_\alpha} = \operatorname{Spec}(\mathbb{Z}[u_\alpha, v_\alpha])$. If $g = \begin{pmatrix} a & b \\ c & d \end{pmatrix}$ transforms σ_α into σ_β then $\operatorname{Spec}(\mathbb{Z}[u_\alpha, v_\alpha])$ and $\operatorname{Spec}(\mathbb{Z}[u_\beta, v_\beta])$ are glued via $u_\beta = u_\alpha^a v_\alpha^b$, $v_\beta = u_\alpha^c v_\alpha^d$. This gives an infinite string of $\mathbb{P}^1$'s:

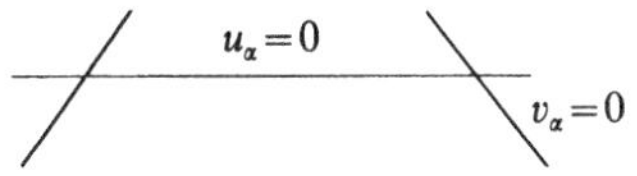

After taking the quotient by V we get a cycle of $\mathbb{P}^1$'s.

Given a cusp $0 \to \mathfrak{a} \to \mathfrak{L} \to \mathfrak{b} \to 0$ and a cone decomposition of $(M \otimes \mathbb{R})_+$ with M the ordered module of symmetric $\mathfrak{o}$-bilinear forms (in the sense defined above) on $\mathfrak{a}^\vee \times \mathfrak{b}$, we have a universal bilinear form

$$\phi\colon \mathfrak{a}^\vee \times \mathfrak{b} \to N \qquad (N = M^\vee)$$

given by $(\alpha, \beta) \mapsto (l \mapsto l(\alpha, \beta))$ for $l \in M$. This defines a bilinear form for every cone $\sigma \in \Sigma$

$$B_\sigma\colon \mathfrak{a}^\vee \times \mathfrak{b} \to F_\sigma^* \qquad (\alpha, \beta) \mapsto X^{\phi(\alpha,\beta)} \tag{2}$$

with $F_\sigma^* = Q(R_\sigma)$ the field of quotients of $R_\sigma = \mathbb{Z}[X^n\colon n \in \check{\sigma} \cap N]$.

Claim. i) $B_\sigma(\lambda(\beta), \beta) \in R_\sigma$ for all $\lambda \in P^+$, $\beta \in \mathfrak{b}$
ii) $B_\sigma(\lambda(\beta), \beta) = 0$ on the closed orbit of S_σ if $\beta \neq 0$.
The claim follows from $\phi(\alpha, \beta) \in \check{\sigma}$.

Now we indicate how the local pieces (in the étale topology) of our compactification are constructed.

Take a cone $\sigma \in \Sigma$. If $\sigma = (0)$ then we consider a closed point s of $\mathrm{Spec}(\mathbb{Z})$ and an abelian variety A with real multiplication by $\mathfrak{o}$ and polarization module (C, C^+) over $\bar{k}(s)$, the algebraic closure of the residue field of s. This abelian variety has versal deformation defined over $W(\bar{k}(s))[[t_1, \ldots, t_n]]$ by the deformation theory of abelian varieties, cf. Sec. 1. Let $R = W(\bar{k}(s))[[t_1, \ldots, t_n]]$ and $U = \mathrm{Spec}(R)$. This yields a piece of the algebraic stack $\mathcal{S}(C, C^+)$ defined in (2.3). If $\sigma \neq (0)$ we consider a closed point from the closed orbit of S_σ and let R be the strict henselization of the local ring of S_σ in s. Set $\mathfrak{m}$ for the maximal ideal of R and $\hat{R}$ for the $\mathfrak{m}$-adic completion of R. By (2) we have a bilinear form $B_\sigma\colon \mathfrak{a}^\vee \times \mathfrak{b} \to F_\sigma^*$. Mumford's construction (2.4) yields a semi-abelian scheme $\mathcal{G}$ with real multiplication by $\mathfrak{o}$ and polarization module (C, C^+) over $\hat{R}$ with special fibre $\mathfrak{a} \otimes \mathbb{G}_m$. By Artin's approximation theorem [3, 4] we see that $\mathcal{G}$ lives over R.

By the action of the torus $S = \mathrm{Spec}(\mathbb{Z}[X^n\colon n \in N])$ we get a stratification of $\mathrm{Spec}(R)$ indexed by the faces of σ. The construction is compatible with this stratification. R is an inductive limit of rings of finite type over $\mathbb{Z}$ and can be replaced by a suitable ring of finite type over $\mathbb{Z}$ over which $\mathcal{G}$ lives. We set $U = \mathrm{Spec}(R)$. Of course, U depends on σ, s and if $\sigma = (0)$ on A.

Let B be the modular stack of all semi-abelian schemes $G \to S$ with real multiplication by $\mathfrak{o}$ and with polarization module $P \simeq C$. There is an étale map: $U \to S_\sigma \times B$. The algebraic stack we want to construct will be a finite union of such U's. Indeed, for equivalence class of cusps and for each $\sigma \in \Sigma$ we take finitely many U's which cover (closed orbit of S_σ) $\times$ B. This is possible since B is quasi-compact.

(2.6) Theorem. *A choice of cone decompositions for the modules M of the cusps determines a smooth algebraic stack* $\tilde{\mathscr{S}} = (\tilde{S}_0, \tilde{S}_1)(C, C^+)$ *of finite type over* $\mathbb{Z}$ *together with a semi-abelian scheme* $\mathscr{G}$ *over* $\tilde{S}_0$ *with real multiplication which is abelian in the generic point and has polarization module* $(P, P^+) \cong (C, C^+)$ *as ordered* $\mathfrak{o}$*-module. It has the following properties:*

1) $\tilde{\mathscr{S}}$ *is proper over* $\mathbb{Z}$.

2) *Let* $\tilde{\mathscr{S}}^0$ *be the dense open subset where* $\mathscr{G}$ *is abelian. Then* $e_0^{-1}(\tilde{S}_0^0) = e_1^{-1}(\tilde{S}_0^0) \cong \mathrm{Isom}(e_0^*\mathscr{G}, e_1^*\mathscr{G})$ *and* $(\tilde{S}_0^0, \tilde{S}_1^0)$ *can be identified with the algebraic stack* (S_0, S_1) *of (2.4).*

3) $\tilde{\mathscr{S}} - \tilde{\mathscr{S}}^0$ *has a stratification* $\tilde{S}_0 - \tilde{S}_0^0 = \bigcup \tilde{S}_{0,\sigma}$, $\tilde{S}_1 - \tilde{S}_1^0 = \bigcup \tilde{S}_{1,\sigma}$, *where* σ *runs through the non-zero cones modulo totally positive units in the cone decompositions associated to the equivalence classes of cusps. The strict henselization of* $\tilde{\mathscr{S}}$ *in a closed point of* $\tilde{S}_{0,\sigma}$ *is isomorphic to the strict henselization of the torus embedding* S_σ *in a closed point of the closed orbit.*

Isom stands for the scheme of $\mathfrak{o}$-linear isomorphisms that preserve the ordered polarization modules. We refer to Rapoport's paper [133] and Faltings' paper [30] for more details and especially to the latter for a proof in an analogous situation.

(2.7) Corollary. *The geometric fibres of* $\mathscr{S}(C, C^+)$ *are irreducible.*

Proof. The complex fibre $\tilde{\mathscr{S}} \otimes \mathbb{C}$ is irreducible as we already know. By Zariski's main theorem all geometric fibres are connected. Since $\tilde{\mathscr{S}}$ is regular the geometric fibres of $\tilde{\mathscr{S}}$ and hence of $\mathscr{S}$ are irreducible. □

3. Hilbert Modular Forms

In this section we define the concept of a Hilbert modular form in the new setting of moduli stacks. As in the preceding section our treatment is sketchy. The theorems are stated for the sake of reference.

Let $\tilde{\mathscr{S}}$ be a stack as described in Theorem (2.6). A line bundle on such a stack is given by a line bundle L on $\tilde{S}_0$ with an isomorphism $e_0^*L \cong e_1^*L$ satisfying some natural compatibility conditions.

The determinant of the cotangent bundle $t^\vee_{\mathscr{G}/\tilde{\mathscr{S}}}$ gives rise to a line bundle on $\tilde{\mathscr{S}}$:

$$\omega = \mathrm{Nm}\, t^\vee_{\mathscr{G}/\tilde{\mathscr{S}}},$$

where Nm is the norm map from $\mathfrak{o} \otimes \mathcal{O}_{\tilde{\mathscr{S}}}$ to $\mathcal{O}_{\tilde{\mathscr{S}}}$.

(3.1) Definition. Let R be a commutative ring. An *R-valued Hilbert modular form* of weight k on $\tilde{\mathscr{S}}$ is a global section of $\omega^{\otimes k}$ on $\tilde{\mathscr{S}} \otimes_{\mathbb{Z}} R$.

If R is clear we simply speak of Hilbert modular forms.

Over $\mathscr{S}$—where $\mathscr{G}$ is abelian—a R-valued Hilbert modular form is given by associating to each R-scheme T and each abelian scheme A over T with real multiplication and polarization module $(P, P^+) \cong (C, C^+)$ a section over T of $\omega_{A/T}^{\otimes k}$ such that this is compatible with all morphisms

$$\begin{array}{ccc} A_1 & \longrightarrow & A_2 \\ \downarrow & & \downarrow \\ T_1 & \longrightarrow & T_2 \end{array}$$

in our stack over R.

Over $\mathbb{C}$ one has a canonical trivialization of ω over $\mathfrak{H}^n$ with the section $d\zeta_1 \wedge \cdots \wedge d\zeta_n$. Sections of $\omega^{\otimes k}$ over $\mathscr{S} \otimes \mathbb{C}$ correspond to forms $f(z)\, d\zeta_1 \wedge \cdots \wedge d\zeta_n$ with f satisfying $f(\gamma z) = \mathrm{N}(cz + d)^k f(z)$ for all $\gamma \in \mathrm{GL}^+(\mathfrak{o} \oplus \mathfrak{a})$ for some suitable $\mathfrak{a}$.

Modular forms over $\mathbb{C}$ have Fourier expansions at the cusps as we saw in Chapter I and II. These are expansions

$$f = \sum_{\substack{\nu \in N \\ \nu = 0 \text{ or} \\ \nu \gg 0}} a_\nu e^{2\pi i \operatorname{Tr} \nu z} \tag{1}$$

In the 2-dimensional case ($[K : \mathbb{Q}] = 2$) we have rewritten (1) in a local coordinate system (u_k, v_k), (cf. Ch. II)

$$f = a_0 + \sum_{\nu \gg 0} a_\nu u_k^{\operatorname{Tr} \nu A_{k-1}} v_k^{\operatorname{Tr} \nu A_k}.$$

The analogue of this is obtained as follows.

Given a cone decomposition Σ of the module M associated to a cusp we constructed a torus embedding S_Σ. Let $\sigma \in \Sigma$ be a maximal dimensional cone and S_σ the associated torus embedding. We denote by V the ring of the completion of S_σ in the closed orbit of S_σ: $V = \mathbb{Z}[[X^n : n \in N \cap \check{\sigma}]]$. E.g. for $n = 2$ $V = \mathbb{Z}[[u_k, v_k]]$.

Given a bilinear form B: $\mathfrak{a}^\vee \times \mathfrak{b} \to L^*$ with $L = Q(V)$ satisfying Condition i), ii) of (2.4) we find by (2.4) a degenerating family of abelian varieties G over V. From the way these families are constructed it follows that t_G is canonically trivialized. (In fact $\mathrm{t}_G \cong \mathrm{t}_T$ with $T = \mathfrak{a} \otimes \mathbb{G}_m$.) G is abelian over an open $\operatorname{Spec}(V[1/w])$ with w a monomial in V.

Suppose that f is a (R-valued) Hilbert modular form of weight k on $\mathscr{S}$. Since ω is free over V such an f defines an element of $V[1/w]$, i.e. we can write

$$f = \sum_{\nu \in N} a_\nu X^\nu \tag{2}$$

It can be extended to a section of $\omega^{\otimes k}$ over V if $a_\nu \neq 0$ implies $\nu \in \check{\sigma}$.

(3.2) Lemma (Koecher principle). *Suppose $n \geqslant 2$. Any section f of $\omega^{\otimes k}$ over $\mathscr{S}$ extends to a section of $\omega^{\otimes k}$ over $\tilde{\mathscr{S}}$.*

Proof. Since $f \in V[1/w]$ we have $w^l f \in V$ for some $l \geqslant 0$. Hence $a_\nu = 0$ if ν is not contained in some translate of $\check{\sigma}$. On the other hand, $a_{\varepsilon\nu} = a_\nu$ for all totally positive units ε. This easily implies: $a_\nu \neq 0 \Rightarrow \nu = 0$ or $\nu \gg 0$. (Consider $a_{\varepsilon^r \nu}$ for suitable ε and large r.)

(3.3) Theorem (q-expansion principle). *Let f be an R-valued Hilbert modular form on $\tilde{\mathscr{S}}$. Then*

1) *$f = 0$ if and only if $a_\nu = 0$ for all $\nu = 0$ or $\nu \gg 0$ in an expansion* (2).

2) *Let $R' \subset R$ be a subring. Then f is an R'-valued modular form on $\tilde{\mathscr{S}}$ if and only if $a_\nu \in R'$ for $\nu = 0$ or $\nu \gg 0$ in* (1).

Proof. Let A be an abelian group. By an A-valued modular form we mean a section of $\omega^k \otimes A$ on $\tilde{\mathscr{S}}$. One then has Fourier expansions in $A \otimes \mathbb{Z}[[X^n: n \in \check{\sigma} \cap N]]$. If A is an inductive limit $\varinjlim A_i$ then $\Gamma(\tilde{\mathscr{S}}, \omega^{\otimes k} \otimes A) = \varinjlim \Gamma(\mathscr{S}, \omega^{\otimes k} \otimes A_i)$, hence we can restrict ourselves to finitely generated rings.

i) If the Fourier expansion is zero for a σ then f vanishes in the completion of the local ring at a point. Since $A \otimes \omega^{\otimes k}$ is locally on $\tilde{\mathscr{S}}$ pullback from a module on $\mathbb{Z}$ it follows that f vanishes in all geometric fibres, since these geometric fibres are irreducible.

ii) Let $A = R/R'$ and apply i). □

Faltings has constructed a Baily-Borel compactification of $A_g/\mathbb{Z}$. One can adapt his methods to construct Baily-Borel compactifications of Hilbert modular varieties.

(3.4) Theorem. *The global sections of $\omega^{\otimes k}$ over $\tilde{\mathscr{S}} = \tilde{\mathscr{S}}(C, C^+)$ generate $\omega^{\otimes k}$ and define a morphism $\tilde{\mathscr{S}} \to \mathbb{P}^m_{\mathbb{Z}}$ which induces an embedding of $\mathscr{S}$. The image of $\tilde{\mathscr{S}} - \mathscr{S}$ consists of finitely many points ($/\mathbb{Z}$) and these are disjoint from the image of $\mathscr{S}$. The image of $\tilde{\mathscr{S}}$ is a normal projective $\mathbb{Z}$-scheme.*

For the proof in an analogous situation we refer to [31]. The idea is to construct sections of $\omega^{\otimes k}$ by means of theta functions (symmetric theta divisors). Alternatively, one could use embeddings of Hilbert modular stacks (with level structures) into A_g.

(3.5) Corollary. *The graded ring of Hilbert modular forms $\bigoplus_{k \geqslant 0} \Gamma(\mathscr{S}, \omega^{\otimes k})$ is a finitely generated $\mathbb{Z}$-algebra.*

Remark. The Hilbert modular forms in (3.5) are precisely the Hilbert modular forms with integral Fourier coefficients.

4. The Galois Action on the Set of Components

We go back to the adelic formulation introduced in I.7. This is the most appropriate language for arithmetic problems.

Let $\mathsf{G} = \mathsf{R}_{K/\mathbb{Q}}\mathrm{GL}(2)$, the Weil restriction of scalars, i.e. $\mathsf{G}(\mathbb{A}_\mathbb{Q}) = \mathrm{GL}(2, \mathbb{A}_K)$. We set $\mathsf{G}_f = \mathsf{G}(\mathbb{A}^f_\mathbb{Q})$. To any compact open subgroup K_f of G_f we associate the quotient

$$X'_{\mathsf{K}_f} = \mathsf{G}(\mathbb{Q})\backslash\mathsf{G}(\mathbb{A}_\mathbb{Q})/\mathsf{K}_f\mathsf{K}_\infty .$$

This is a (separated) scheme over $\mathbb{C}$. If $\mathsf{K}'_f \subseteq \mathsf{K}_f$ we have a natural morphism

$X'_{K'_f} \to X'_{K_f}$. If we let K_f vary we obtain a projective system of X_{K_f}'s and finite morphisms and we can form the projective limit

$$X' = \varprojlim_{K_f} X'_{K_f}.$$

This is separated quasi-compact scheme over $\mathbb{C}$. It is not of finite type.

For $K_f = K_0 = \prod_{v \text{ finite}} GL(2, \mathfrak{o}_v)$ we have given a modular interpretation of X'_{K_f}, see Ch. IX and above in this Chapter. One can give such an interpretation for all K_f. Let $K(n) = \{g \in G_f : g \equiv 1 \pmod{n}\}$. Suppose that $K_f \supseteq K(n)$. Then the points of X'_{K_f} are in 1-1-correspondence with the 4-tuples $(A, m, (P, P^+), k)$, where A is an abelian variety $/\mathbb{C}$ with real multiplication m, with ordered polarization module (P, P^+) and with a class $k \bmod K_f/K(n)$ of level n structures $A[n] \cong L/nL$, see [133] for the definition.

This makes it possible to prove along the lines of the above construction the existence of moduli spaces M_{K_f} over $\mathbb{Q}$ or even over $\mathbb{Z}\left[\frac{1}{m}\right]$ with m depending on K_f such that $M_{K_f}(\mathbb{C}) = X'_{K_f}$. For m one can take the product of those primes p for which the p-part of K_f is not the full group $\prod_{v|p} GL(2, \mathfrak{o}_v)$. The construction yields an algebraic stack over $\mathbb{Z}\left[\frac{1}{m}\right]$.

We thus get the following (cf. [21, 23]).

i) there is a scheme M defined over $\mathbb{Q}$ such that $M(\mathbb{C}) = X'_{K_0}$ with $K_0 = \prod_{v \text{ finite}} GL(2, \mathfrak{o}_v)$,

ii) for every compact open K_f in G_f there is a scheme X'_{K_f} defined over $\mathbb{Q}$ such that

a) if K_f, L_f are compact open in G_f and $gK_fg^{-1} \subseteq L_f$ for some $g \in G_f$ then there is a morphism $I_{K,L}(g)\colon X'_{K_f} \to X'_{L_f}$ with $I_{L,N}(h)I_{K,L}(g) = I_{K,N}(gh)$ and $I_{K,K}(g) = \mathrm{id}$ (see I.7), all defined over $\mathbb{Q}$,

b) if K_f is a normal subgroup of L_f then L_f/K_f acts on X'_{K_f} via I and $I_{K,L}(1)\colon X'_{K_f}/(L_f/K_f) \cong X'_{L_f}$.

Properties ii) a) and b) are summarized by saying that $G(\mathbb{A}^f)$ *acts continuously* on X'.

The whole galaxy of schemes X'_{K_f} is called a model over $\mathbb{Q}$ in the terminology of Deligne, see [21].

All connected components of the scheme X'_{K_f} can be compactified in a minimal way after Baily-Borel. We denote these compactifications by X_{K_f}.

(4.1) *Remark.* The connected components of X_{K_f} are in general not defined over $\mathbb{Q}$ but over a well-determined abelian extension of $\mathbb{Q}$. We refer to [21] for these fields of definition. For $K_f = K_0$ all components are defined over $\mathbb{Q}$.

We want to describe the set of connected components of the X_{K_f}. Let us write $T = R_{K/\mathbb{Q}}\mathbb{G}_m$. The determinant is then a morphism $G \to T$.

(4.2) Lemma. *The determinant induces bijections*

$$\pi_0(X_{K_f}(\mathbb{C})) \leftrightarrow \pi_0(T(\mathbb{Q})\backslash T(\mathbb{A})/\det(K_f))$$

and these give in the limit a bijection

$$\pi_0(X(\mathbb{C})) \leftrightarrow \pi_0(\mathsf{T}(\mathbb{Q})\backslash \mathsf{T}(\mathbb{A}))\,.$$

Proof (cf. [21]). We have $\pi_0(\mathsf{G}(\mathbb{Q})\backslash \mathsf{G}(\mathbb{A})/\mathsf{K}_f) = \mathsf{G}(\mathbb{Q})\backslash \mathsf{G}(\mathbb{A})/\mathsf{G}(\mathbb{R})^0 \times \mathsf{K}_f$ and the determinant induces a bijection $\pi_0(\mathsf{G}(\mathbb{Q})\backslash \mathsf{G}(\mathbb{A})/\mathsf{K}_f) \leftrightarrow \pi_0(\mathsf{T}(\mathbb{Q})\backslash \mathsf{T}(\mathbb{A})/\det \mathsf{K}_f)$ and $\pi_0(\mathsf{G}(\mathbb{Q})\backslash \mathsf{G}(\mathbb{A})) \leftrightarrow \pi_0(\mathsf{T}(\mathbb{Q})\backslash \mathsf{T}(\mathbb{A}))$. Since $\pi_0(\mathsf{K}_\infty) = 1$ we get

$$\begin{aligned}\pi_0(X_{\mathsf{K}_f}(\mathbb{C})) \leftrightarrow\ & \pi_0(\mathsf{T}(\mathbb{Q})\backslash \mathsf{T}(\mathbb{A})/\det(\mathsf{K}_\infty \mathsf{K}_f)) \\ & \qquad \| \\ & \pi_0(\mathsf{T}(\mathbb{Q})\backslash \mathsf{T}(\mathbb{A})/\det \mathsf{K}_f)\end{aligned}$$

and

$$\pi_0(X(\mathbb{C})) \leftrightarrow \pi_0(\mathsf{T}(\mathbb{Q})\backslash \mathsf{T}(\mathbb{A}))\,. \qquad \square$$

The X_{K_f} are defined over $\mathbb{Q}$ and we thus get a natural action of $\mathrm{Gal}(\bar{\mathbb{Q}}/\mathbb{Q})$ on X_{K_f} and on X. In particular we get an action on the set of connected components $\pi_0(X_{\mathsf{K}_f})$ and on $\pi_0(X)$. Since this action commutes with that of $\mathsf{G}(\mathbb{A}^f)$ on $\pi_0(X)$ it follows that the action of an element $\sigma \in \mathrm{Gal}(\bar{\mathbb{Q}}/\mathbb{Q})$ on $\pi_0(X)$ is given by $\lambda(\sigma) \in \pi_0(\mathsf{G}(\mathbb{Q})\backslash \mathsf{G}(\mathbb{A}))$. The theory of canonical models of Shimura and Deligne describes the map λ.

(4.3) Proposition (Shimura, Deligne [21]). *The action of* $\mathrm{Gal}(\bar{\mathbb{Q}}/\mathbb{Q})$ *on the set of connected components of* $X(\mathbb{C})$ *is given by*

$$\mathrm{Gal}(\mathbb{Q}^{\mathrm{ab}}/\mathbb{Q}) \xrightarrow{\sim} \pi_0(\mathbb{A}_{\mathbb{Q}}^*/\mathbb{Q}^*) \to \pi_0(\mathbb{A}_K^*/K^*) = \pi_0(\mathsf{G}(\mathbb{Q})\backslash \mathsf{G}(\mathbb{A}))\,,$$

where the second arrow is induced by the inclusion $\mathbb{Q} \to K$ *and the first is the reciprocity isomorphism of class field theory.*

For the proof we refer to [21]. One studies the Galois action by looking at the points that represent abelian varieties with complex multiplication. They are defined over precisely determined number fields and one can deduce how $\mathrm{Gal}(\bar{\mathbb{Q}}/\mathbb{Q})$ acts.

XI. The Tate Conjectures for Hilbert Modular Surfaces

Once one has models for Hilbert modular varieties over (rings of integers of) number fields one can start investigating the arithmetic properties of Hilbert modular varieties. The wealth of results (and conjectures) for the moduli spaces of elliptic curves suggests that unexpected treasures lie buried in a largely unexplored territory.

The results of Harder, Langlands and Rapoport on the Tate conjectures for Hilbert modular surfaces form the first breakthrough in this field and they constitute the main theme of this chapter. Tate's first conjecture describes algebraic cycles as Galois invariant cohomology classes, while the second one predicts the rank of the group of algebraic cycles as the order of the pole of a certain L-function at a critical point.

In the case of a Hilbert modular variety X associated to a totally real field K the interesting part of the middle cohomology breaks up into pieces corresponding to cuspidal automorphic representations π of weight 2—in classical terms to modular forms of weight 2 which are eigenforms for the action of the Hecke algebra. Each such π defines a 2^n-dimensional piece with $n = [K:\mathbb{Q}]$. If the Hilbert modular variety is defined over $\mathbb{Q}$ one finds a natural 2^n-dimensional ℓ-adic representation r of the Galois group $\mathrm{Gal}(\bar{\mathbb{Q}}/\mathbb{Q})$ with coefficients in a field k_π.

It is one of the standard conjectures in the theory of automorphic representations that for each π as above there exists a 2-dimensional Galois representation τ of $\mathrm{Gal}(\bar{\mathbb{Q}}/K)$ such that the L-functions $L(\pi, s)$ and $L(\tau, s)$ agree. If $K = \mathbb{Q}$ one can use the Jacobian of a modular curve and the Eichler-Shimura relation to construct τ.

In that case the part of the first cohomology group of a modular curve X determined by a Hecke eigenform $f = \sum a(n)q^n$ and its conjugates $f^\alpha = \sum a(n)^\alpha q^n$ can be identified with the first cohomology group of an abelian variety A, namely an isogeny factor of the Jacobian of X and τ is the natural representation of $\mathrm{Gal}(\bar{\mathbb{Q}}/\mathbb{Q})$ on $H^1(A)$. In higher dimensions it seems difficult to construct τ. Oda conjectured that the part of the middle cohomology of a Hilbert modular variety X defined by a cuspidal automorphic representation and its conjugates can be described by the first cohomology of an abelian variety defined over K. To be more precise, Oda conjectured that for each π there exists an abelian variety A defined over K with real multiplication by k_π such that

$$\bigoplus_{\alpha: k_\pi \to \mathbb{C}} H^n(X)(\pi^\alpha) = \bigotimes_{\sigma: K \to \mathbb{C}} H^1(A^\sigma) \qquad (1)$$

as Hodge structures; here $H^n(X)(\pi^\alpha)$ is the piece of the cohomology of X associated to π^α and A^σ is the complex abelian variety obtained from A by $\sigma: K \to \mathbb{C}$ and the tensor product is taken over k_π. (Note that $H^1(A^\sigma)$ is a k_π-module of rank 2.) The existence of an abelian variety A over K such that (1) holds as motives would offer a reasonable explanation for the existence of the Galois representation. It is tempting to conjecture conversely, that for each abelian variety defined over K with real multiplication by a totally real field F (i.e. $F \to \operatorname{End}_K^\circ(A)$) we can find a Hilbert modular variety and a cuspidal automorphic representation π for GL_2/K such that (1) holds.

This seems to fit in the frame work of the Langlands conjectures. Whatever the truth is in these matters, these ideas can serve as an excellent guiding principle as we shall see.

In the real quadratic case Oda showed the existence of complex abelian varieties A, A' such that $\bigoplus H^2(X)(\pi^\alpha) = H^1(A) \otimes H^1(A')$ as k_π-Hodge structures. As we saw in Chapter VI this puts already strong restrictions on the Hodge cycles in $H^2(X)(\pi)$. The attack of Harder, Langlands and Rapoport on the Tate conjectures uses the same idea in étale cohomology. The intersection pairing on the 4-dimensional space $H^2(X, \mathbb{Q}_\ell)(\pi)$ obtained from Poincaré duality defines a (projective) quadric in $\mathbb{P}^3$. A non-singular quadric over a field with a rational point over that field is isomorphic to $\mathbb{P}^1 \times \mathbb{P}^1$ over a quadratic extension of that field. This idea can be used to split $H^2(X, \bar{\mathbb{Q}}_\ell)(\pi)$ as $V \otimes V'$ as a Galois module. This gives us strong restrictions on the Tate cycles. Although this splitting does not yield the existence of τ it points strongly in that direction. This splitting is one main ingredient for the proof of the first conjecture of Tate. Another main ingredient (for both the first and the second conjecture of Tate) is the shape of the L-function of a Hilbert modular surface. It was determined by Brylinski and Labesse and independently by Harder, Langlands and Rapoport.

Of crucial importance is furthermore the fact that we know lots of algebraic cycles: the cycles defined by the skew-hermitian curves. They are called Hirzebruch-Zagier cycles. The other main ingredient of the proof of the second conjecture (over abelian extensions of $\mathbb{Q}$) is an adelic version of the result of Asai on the order of the pole of certain L-fuctions of automorphic forms.

The contents of this chapter are the following. Section 1 is introductory and states the Tate conjectures. In Section 2 we give (without proof) the L-function of a Hilbert modular variety. Section 3 explains one of the main ideas of the proof—the splitting of the Galois representation on the part of the cohomology defined by an automorphic form. In Section 4 we give the results of Harder, Langlands and Rapoport on the Tate conjectures as complemented by the work of Klingenberg and Murty and Ramakrishnan.

1. Hodge and Tate Cycles

Here we stick to the real quadratic case and use the following notation: $\mathsf{G} = \mathsf{R}_{K/\mathbb{Q}}\mathrm{GL}(2)$, K_f: compact open subgroup of $\mathsf{G}(\mathbb{A}^f)$,

$$X'_{\mathsf{K}_f} = \mathsf{G}(\mathbb{Q})\backslash \mathsf{G}(\mathbb{A})/\mathsf{K}_f \mathsf{K}_\infty ,$$
$$X' = \varprojlim_{\mathsf{K}_f} X'_{\mathsf{K}_f}$$

X_{K_f} (resp. X) Baily-Borel compactification of X'_{K_f} (resp. X'), Z_{K_f}: minimal resolution of cusp singularities of X_{K_f}, Y_{K_f}: minimal resolution of all singularities.

We shall use several cohomology theories:

a) Betti-cohomology,

b) De Rham-cohomology,

c) étale cohomology.

We refer to [25] for a comparison of a), b) and c). Implicitly, we shall use intersection cohomology too.

Our main interest is of course H^2. We denote by $\mathbb{H}(X_{\mathsf{K}_f})$ the image of compactly supported cohomology:

$$\mathbb{H}^2(X_{\mathsf{K}_f}) = \operatorname{Im}(H_c^2(X'_{\mathsf{K}_f}) \to H^2(X_{\mathsf{K}_f})) .$$

To justify notation, $\mathbb{H}^2$ can be interpreted as intersection cohomology, see [16].

a) The classical Betti-cohomology $H_B^*(X_{\mathsf{K}_f}) = H_B^*(X_{\mathsf{K}_f}(\mathbb{C}), \mathbb{Q})$ was already used in Chapter VI. According to Deligne there is a mixed Hodge structure on $H_B^2(X'_{\mathsf{K}_f}) \otimes \mathbb{C}$. We saw in Chapter VI that this gives a pure Hodge structure on $\mathbb{H}_B^2(X_{\mathsf{K}_f}) \otimes \mathbb{C}$:

$$\mathbb{H}_B^2(X_{\mathsf{K}_f}) \otimes \mathbb{C} = \bigoplus_{p+q=2} H^{p,q}(X_{\mathsf{K}_f}) .$$

b) The algebraic De Rham cohomology is defined by algebraic differential forms defined over a field $F \subset \mathbb{C}$: $H_{DR}^*(/F)$. For Y compact Kähler $H_{DR}^m(Y/F)$ is provided with a Hodge filtration $F^p H_{DR}(/F)$, $p \geqslant 0$.

There is an isomorphism which compares a) and b):

$$H_B^2 \otimes \mathbb{C} \xrightarrow{\sim} H_{DR}^2(/\mathbb{C}) \tag{1}$$

compatible with Hodge filtrations.

Although X'_{K_f} is not compact we have a Hodge filtration on $\mathbb{H}_{DR}^2(X_{\mathsf{K}_f})$. Indeed, one either uses the properties of intersection cohomology or the fact that the image of $H_{DR}^2(Z_{\mathsf{K}_f}/\mathbb{C})$ in $H_{DR}^2(X'_{\mathsf{K}_f}/\mathbb{C})$ coincides with $\mathbb{H}_{DR}^2(X_{\mathsf{K}_f}/\mathbb{C})$ and under (1) also with $\mathbb{H}_B^2(X_{\mathsf{K}_f})$. The Hodge filtration on $\mathbb{H}_B^2$ is given explicitly in Chapter VI.

c) If Y is a scheme over $\mathbb{Q}$, then for each rational prime ℓ there is étale cohomology $H_{\text{et}}^*(Y \otimes \bar{\mathbb{Q}}, -)$. (Of course, we mean $Y \times_{\operatorname{Spec}(\mathbb{Q})} \operatorname{Spec}(\bar{\mathbb{Q}})$ when we write $Y \otimes \bar{\mathbb{Q}}$.) We can put the groups $H_{\text{et}}^*(Y \otimes \bar{\mathbb{Q}}, \mathbb{Q}_\ell)$ together in a restricted direct product w.r.t. $H_{\text{et}}^*(Y \otimes \bar{\mathbb{Q}}, \mathbb{Z}_\ell)$:

$$H_{\text{et}}^*(Y \otimes \bar{\mathbb{Q}}) = \varprojlim_n H_{\text{et}}^*(Y \otimes \bar{\mathbb{Q}}, \mathbb{Z}/n\mathbb{Z}) \otimes \mathbb{Q} .$$

For $\bar{\mathbb{Q}} \subset \mathbb{C}$ we have by the proper base change theorem

$$H_{\text{et}}^*(Y \otimes \bar{\mathbb{Q}}) \xrightarrow{\sim} H_{\text{et}}^*(Y \otimes \mathbb{C}) .$$

The important point about étale cohomology is that $\operatorname{Gal}(\bar{\mathbb{Q}}/\mathbb{Q})$ acts in a natural way on it (via the action on $\bar{\mathbb{Q}}$).

We have a comparison of a) and c) via

$$\mathbb{H}_B^2(X_{K_f}) \otimes \mathbb{A}^f \simeq \mathbb{H}_{et}^2(X_{K_f} \otimes \bar{\mathbb{Q}}). \tag{2}$$

We shall take the contributions of all places of $\mathbb{Q}$ together in

$$\mathbb{H}_{\mathbb{A}}^2(X_{K_f}) := \mathbb{H}_{DR}^2(X_{K_f}/\mathbb{C}) \oplus \mathbb{H}_{et}^2(X_{K_f} \otimes \mathbb{C}).$$

Then (1) and (2) combine to an isomorphism

$$\mathbb{H}_B^2(X_{K_f}) \otimes_{\mathbb{Q}} (\mathbb{C} \times \mathbb{A}^f) \simeq \mathbb{H}_{\mathbb{A}}^2(X_{K_f}). \tag{3}$$

If $K_f \subset K'_f$, both compact open subgroups of G_f, we have a finite morphism $X_{K_f} \to X_{K'_f}$ inducing $\mathbb{H}_{\mathbb{A}}^2(X_{K_f}) \to \mathbb{H}_{\mathbb{A}}^2(X_{K'_f})$. This is an injection and we can take the injective limit

$$\varinjlim_{K_f} \mathbb{H}_{\mathbb{A}}^2(X_{K_f}) =: \mathbb{H}_{\mathbb{A}}^2(X).$$

The Galois group $\mathrm{Gal}(\bar{\mathbb{Q}}/\mathbb{Q})$ still acts on $\mathbb{H}_{et}^2(X \otimes \bar{\mathbb{Q}})$.

Before we look at cycle class maps we must make some remarks about Tate twists. We refer to Milne's paper in [25] for more explanation. We put

$$\mathbb{Q}_B(1) = 2\pi i \mathbb{Q}$$

$$\mathbb{Q}_{DR}(1) = \mathbb{Q}$$

$$\mathbb{Q}_{et}(1) = \mathbb{A}^f(1) = \left(\varprojlim_m \mu_m\right) \otimes_{\mathbb{Z}} \mathbb{Q}$$

$$H_B^n(Y)(m) = H_B^n(Y) \otimes_{\mathbb{Q}} (2\pi i)^m \mathbb{Q}$$

$$H_{DR}^n(Y)(m) = H_{DR}^n(Y)$$

$$H_{et}^n(Y)(m) = H_{et}^n(Y) \otimes_{\mathbb{A}^f} (\mathbb{A}^f(1)) \otimes m.$$

These twists can be defined also for negative m. The comparison isomorphisms are compatible with "twisting".

The representation of $\mathrm{Gal}(\bar{\mathbb{Q}}/\mathbb{Q})$ on roots of unity gives a 1-dimensional Galois representation by

$$\zeta^\sigma = \zeta^{r(\sigma)}$$

By Artin reciprocity this defines a character of the idele class group

$$\alpha\colon \mathbb{A}_{\mathbb{Q}}^*/\mathbb{Q}^* \to \mathbb{C}^* \qquad x \to |x| = \prod_v |x|_v$$

called the *Tate character*.

In each of these theories we have Chern class maps. If Y is a smooth projective surface over $\mathbb{Q}$ we have isomorphisms

$$\mathrm{Pic}(Y(\mathbb{C})) \to H^1(Y(\mathbb{C}), \mathcal{O}_Y^*)$$

$$\mathrm{Pic}(Y \otimes \bar{\mathbb{Q}}) \to H_{et}^1(Y \otimes \bar{\mathbb{Q}}, \mathbb{G}_m),$$

the first is $\mathrm{Gal}(\mathbb{C}/\mathbb{R})$-invariant and the second $\mathrm{Gal}(\bar{\mathbb{Q}}/\mathbb{Q})$-invariant. Furthermore,

$$\mathrm{Pic}(Y/\mathbb{Q}) = \mathrm{Pic}(Y \otimes \bar{\mathbb{Q}})^{\mathrm{Gal}(\bar{\mathbb{Q}}/\mathbb{Q})}.$$

The exact sequences

$$0 \to 2\pi i\mathbb{Z} \to \mathcal{O}_{Y^{\mathrm{an}}} \to \mathcal{O}^*_{Y^{\mathrm{an}}} \to 0$$

$$1 \to \mathbb{U}_{l^n} \to \mathbb{G}_m \to \mathbb{G}_m \to 1 \qquad (n \geqslant 1)$$

define Chern class maps

$$\mathrm{Pic}(Y^{\mathrm{an}}) \to H^1(Y^{\mathrm{an}}, \mathcal{O}^*_{Y^{\mathrm{an}}}) \to H^2_B(Y(\mathbb{C}), 2\pi i\mathbb{Z}) \hookrightarrow H^2_B(Y)(1)$$

$$\mathrm{Pic}(Y \otimes \bar{\mathbb{Q}}) \to H^1_{\mathrm{et}}(Y \otimes \bar{\mathbb{Q}}, \mathbb{G}_m) \to H^2_{\mathrm{et}}(Y \otimes \bar{\mathbb{Q}}, \hat{\mathbb{Z}}(1)) \hookrightarrow H^2_{\mathrm{et}}(Y)(1),$$

denoted by c_B and c_{et}, respectively. The first is $\mathrm{Gal}(\mathbb{C}/\mathbb{R})$ equivariant and the second is $\mathrm{Gal}(\bar{\mathbb{Q}}/\mathbb{Q})$-equivariant. They are compatible with the comparison isomorphism (2), see [25], p. 21. The image of $\mathrm{Pic}(Y^{\mathrm{an}})$ lands in the Hodge $(1,1)$-part, while $\mathrm{Pic}(Y/\mathbb{Q})$ lands in the $\mathrm{Gal}(\bar{\mathbb{Q}}/\mathbb{Q})$-invariant part of $H^2_{\mathrm{et}}(Y)(1)$. Similarly, we have a Chern class c_{DR} in $H^2_{DR}(Y/\mathbb{C})$, again compatible with c_B and c_{et} under (1) and (2). So we have a class $c_{\mathbb{A}} = (c_{DR}, c_{\mathrm{et}}) \in H^2_{\mathbb{A}}(Y)(1)$. The subspace of $H^2_{\mathrm{et}}(Y, \mathbb{Q}_\ell)(1)$ (or $H^2_B(Y)(1)$ or $H^2_{\mathbb{A}}(Y)(1)$) generated by the c_{et} (resp. c_{DR}, $c_{\mathbb{A}}$) is called the space of algebraic cycles.

Two famous conjectures characterize algebraic cycles: the Hodge conjecture and a conjecture of Tate. For codimension 1 cycles the Hodge conjecture is true by the Lefschetz-criterion.

(1.1) Definition. A cohomology class $a \in H^2_B(Y)(1)$ is called a *Hodge cycle* if
i) a is in the image of $H^2_B(Y, \mathbb{Q})(1) \to H^2_B(Y)(1) \otimes \mathbb{C}$
ii) a is of Hodge type $(1,1)$.

(1.2) Lefschetz-criterion (cf. [48]). *The Hodge cycles are exactly the algebraic cycles, i.e. they make up* $c_B(\mathrm{Pic}(Y^{\mathrm{an}})) \otimes \mathbb{Q}$.

(1.3) Definition. Let E be a number field. A cohomology class $a \in H^2_{\mathrm{et}}(Y \otimes \bar{\mathbb{Q}})(1)$ is called a *Tate cycle* (over E) if it belongs to the $\mathrm{Gal}(\bar{\mathbb{Q}}/E)$-invariant part of $H^2_{\mathrm{et}}(Y \otimes \bar{\mathbb{Q}})(1)$.

(1.4) Tate conjecture 1 (cf. [161]). *The algebraic cycles on $Y \otimes E$ are the Tate cycles over E.*

The group of Hodge cycles on Y^{an} is denoted by $\mathrm{Ho}(Y^{\mathrm{an}})$, the group of Tate cycles over E by $\mathrm{Ta}(Y \otimes E)$ or Ta_E.

We now apply this to Z_{K_f}. If Z_{K_f} is not smooth then we replace K_f by a finite index normal subgroup K'_f and consider $\mathsf{K}_f/\mathsf{K}'_f$-invariant cohomology of $Z_{\mathsf{K}'_f}$.

The Tate conjecture for Z_{K_f} is equivalent to:

(1.5) Tate conjecture 1 for X'_{K_f}. *The algebraic cycles on X'_{K_f} over E are exactly the $\mathrm{Gal}(\bar{\mathbb{Q}}/E)$-invariant elements of $\mathbb{H}^2_{\mathrm{et}}(X_{\mathsf{K}_f})(1)$.*

The equivalence follows from

$$\mathbb{H}^2_{\mathrm{et}}(X_{\mathsf{K}_f} \otimes \bar{\mathbb{Q}}) = j^* H^2_{\mathrm{et}}(Z_{\mathsf{K}_f} \otimes \bar{\mathbb{Q}}),$$

cf. Ch. VI. So we can write (with D the resolution divisor at ∞)

$$H^2_{\mathbb{A}}(Z_{K_f}) \cong \mathbb{H}^2_{\mathbb{A}}(X_{K_f}) \oplus H^2_{D,\mathbb{A}}(Z_{K_f}).$$

Since $H^2_{D,\mathbb{A}}(Z_{K_f})$ is generated by the classes of the resolution curves we find that (1.5) holds iff the Tate conjecture holds for Z_{K_f}. Since

$$H^2_{\text{et}}(X'_{K_f} \otimes \bar{\mathbb{Q}})/\mathbb{H}^2_{\text{et}}(X_{K_f} \otimes \bar{\mathbb{Q}}) \simeq \bigoplus \mathbb{A}^f(-2)$$

(cf. [22]) one has

$$H^2(X'_{K_f} \otimes \bar{\mathbb{Q}})(1)^{\text{Gal}(\bar{\mathbb{Q}}/E)} = \mathbb{H}^2(X_{K_f} \otimes \bar{\mathbb{Q}})(1)^{\text{Gal}(\bar{\mathbb{Q}}/E)}$$

which makes it possible to rephrase (1.5) as: the algebraic cycles on X'_{K_f} are exactly the $\text{Gal}(\bar{\mathbb{Q}}/E)$-invariant elements of $H^2(X'_{K_f} \otimes \bar{\mathbb{Q}})(1)$.

Tate formulated a second conjecture that gives the rank of the space of algebraic cycles. (We restrict to the case where Y is a smooth projective surface over $\mathbb{Q}$.) Let E be a number field.

(1.6) Tate conjecture 2. *The rank of the space of algebraic cycles on Y which are defined over E is the order of the pole of the L-function $L^{(2)}_Q(Y/E,s)$ of Y over E in $s = 2$.*

Here

$$L^{(2)}_Q(Y/E,s) = \prod_{\mathfrak{p} \notin Q} \det(1 - \mathrm{N}\mathfrak{p}^{-s} F_{\mathfrak{p}} | H^2(Y \otimes \bar{\mathbb{Q}}, \mathbb{Q}_\ell))^{-1}$$

with Q a finite set of "bad" primes of E and with $F_{\mathfrak{p}}$ the geometric Frobenius at $\mathfrak{p}$. Note that the order to the pole is independent of Q.

The conjectures of Tate (1.4) (or (1.5)) and (1.6) will be the subject of Section 4.

We have lots of algebraic cycles: besides those coming from the resolutions of the singularities we have those defined by the components of the T_N. All of these are defined over cyclotomic fields. It will turn out that we do not need all of them to generate all algebraic cycles defined over $\mathbb{Q}^{\text{ab}}$; the non-compact ones suffice. To define these in the adelic setting we consider the algebraic group $\mathsf{H} = \mathrm{GL}(2)/\mathbb{Q}$ and the diagonal embedding $\mathsf{H} \to \mathsf{G}$ induced by $\mathbb{Q} \hookrightarrow K$. It induces the diagonal embedding

$$\begin{array}{ccc} \mathsf{H}(\mathbb{R})^0/\mathsf{L}_\infty & \to & \mathsf{G}(\mathbb{R})^0/\mathsf{K}_\infty \quad \text{with } \mathsf{L}_\infty = \mathsf{K}_\infty \cap \mathsf{H}(\mathbb{R}). \\ \| & & \| \\ \mathfrak{H} & & \mathfrak{H}^2 \end{array}$$

For a fixed compact open subgroup K_f of $\mathsf{G}(\mathbb{A}^f)$ we get

$$\mathsf{H}(\mathbb{Q})\backslash\mathsf{H}(\mathbb{A})/\mathsf{L}_f\mathsf{L}_\infty \to \mathsf{G}(\mathbb{Q})\backslash\mathsf{G}(\mathbb{A})/\mathsf{K}_f\mathsf{K}_\infty$$

with $\mathsf{L}_f = \mathsf{K}_f \cap \mathsf{H}(\mathbb{A}^f)$ and an associated morphism defined over $\mathbb{Q}$ from a curve to a surface

$$d\colon X_{\mathsf{L}_f} \to X_{\mathsf{K}_f}.$$

Now X_{L_f} is not connected in general; its connected components are in 1-1 correspondence with $\pi_0(\mathbb{Q}^*\backslash\mathbb{A}^*_{\mathbb{Q}}/\det \mathsf{L}_f)$. The components are all defined over abelian extensions of $\mathbb{Q}$, see [21].

(1.7) Definition. The elements of the $\mathsf{G}(\mathbb{A}^f)$-invariant subgroup of $\mathrm{Pic}(Y_{\mathsf{K}_f} \otimes \bar{\mathbb{Q}})$ generated by the divisor classes of the components of $d(X_{\mathsf{L}_f})$ are called Hirzebruch-Zagier cycles. Their images in $\mathbb{H}^2(X_{\mathsf{K}_f})(1)$ are called Hirzebruch-Zagier classes. They span a subspace $\mathbb{T}_{HZ}$ in $\mathbb{H}^2(X_{\mathsf{K}_f})(1)$. We shall write $\mathbb{T}^0_{HZ} = \mathbb{T}_{HZ} \cap \mathbb{H}^2_{\mathrm{sp}}$.

We have an action of $\mathrm{Gal}(\mathbb{Q}^{\mathrm{ab}}/\mathbb{Q})$ on $\mathbb{T}_{HZ}$ and $\mathbb{T}^0_{HZ}$ and we can decompose

$$\mathbb{T}_{HZ} = \bigoplus_{\omega} \mathbb{T}_{HZ}(\omega)$$

according to the characters of $\mathrm{Gal}(\mathbb{Q}^{\mathrm{ab}}/\mathbb{Q})$.

2. Decomposition of the Cohomology and L-Series

We decompose the cohomology $\mathbb{H}^2(X_{\mathsf{K}_f})$ under the action of the Hecke algebra. We recall that

$$\mathbb{H}^2_{\mathbb{A}}(Y_{\mathsf{K}_f}) = \mathbb{H}^2_{\mathbb{A}}(X_{\mathsf{K}_f}) \oplus \left(\bigoplus_e (\mathbb{C} \times \mathbb{A}^f(-1)\right), \tag{1}$$

where e runs over the irreducible components of the resolutions of the cusp and quotient singularities.

Any $g \in \mathsf{G}(\mathbb{A}^f)$ defines a correspondence: let $\mathsf{K}'_f = \mathsf{K}_f \cap g\mathsf{K}_f g^{-1}$; we have a morphism

$$X_{\mathsf{K}'_f} \xrightarrow{(I_1, I_g)} X_{\mathsf{K}_f} \times X_{\mathsf{K}_f},$$

where I_1 (resp. I_g) is the identity (resp. induced by right multiplication by g). These correspondences are defined over $\mathbb{Q}$ (since X_{K}, X_{K_f}, I_1 and I_g are) and define operators on $\mathbb{H}^2(X_{\mathsf{K}_f})$ as we saw in Chapter VI.1. We thus find an action of $\mathsf{G}(\mathbb{A}^f)$ on $\mathbb{H}^2(X_{\mathsf{K}_f})$.

Recall from Chapter VI that

$$\mathbb{H}^2(X_{\mathsf{K}_f}) = \mathbb{H}^2_{\mathrm{sp}}(X_{\mathsf{K}_f}) \oplus W(X_{\mathsf{K}_f}), \tag{2}$$

with $W_B(X_{\mathsf{K}_f})$ the subspace generated by the classes of the Chern forms $\dfrac{dx_i \wedge dy_i}{y_i^2}$ on all components of X_{K_f}. $\mathbb{H}^2_{\mathrm{sp}}$ is called the *cuspidal cohomology* and is described by cusp forms of weight 2. (We showed this in Chapter VI for the Betti-cohomology, but all this holds for the $\mathbb{A}$-cohomology.) The decomposition (2) is compatible with taking the limit $\varprojlim_{\mathsf{K}_f}$, so it holds for $\mathbb{H}^2(X)$. The action of $\mathsf{G}(\mathbb{A}^f)$ respects the decomposition (2) (sends cusp forms to cusp forms and Chern forms to Chern forms).

Since $\mathbb{H}^2_{\mathrm{sp}}$ is described by cusp forms of weight 2 we can decompose it according to the Hecke eigenforms. In the terminology of automorphic representations (see e.g. [43]) we have:

(2.1) Theorem. *Under the action of* $\mathsf{G}(\mathbb{A}^f)$ *we have the splitting in isotypic components*

$$\mathbb{H}^2_{\mathbb{A},\mathrm{sp}}(X) = \bigoplus_{\pi} \mathbb{H}^2_{\mathbb{A},\mathrm{sp}}(X)(\pi),$$

where π *runs over the irreducible cuspidal automorphic representations of* $\mathsf{G}(\mathbb{A})$ *whose archimedean components lie in the discrete series of weight two. Each such* π *occurs only once.*

(2.2) *Remark.* Each π in (2.1) corresponds to a holomorphic cusp form of weight 2, a new form of a suitable level. The theory of automorphic representations tells us that for each π in the right hand side of (2.1) there is a K_f such that $\mathbb{H}^2_{\mathrm{sp}}(X)(\pi)^{\mathsf{K}_f}$ is an isotypic component of $\mathbb{H}^2_{\mathrm{sp}}(X_{\mathsf{K}_f})$. This component is defined over a number field k_π (in classical terms the field generated by the Fourier coefficients). However, we do not have a Multiplicity-One-Principle on finite levels. In general the components $\mathbb{H}^2_{\mathrm{sp}}(X_{\mathsf{K}_f})(\pi)$ occur with a multiplicity $m(\mathsf{K}_f, \pi)$. But given π there is a K_f such that $m(\mathsf{K}_f, \pi) = 1$ (new form).

(2.3) *Remark.* The essential point of the proof of (2.1) is that $\mathbb{H}^2$ can be described by L^2-cohomology. The Zucker conjecture asserts more generally that the intersection cohomology of the Baily-Borel compactification of a quotient of a bounded symmetric domain by an arithmetic group is described by L^2-cohomology. In the Hilbert modular case this is true, see [179].

Proof of (2.1). The first statement follows immediately from our results in Chapter VI and a good dictionary between the language of classical modular forms and that of automorphic representations, see e.g. [43]. The second one is the so-called 'Multiplicity-One-Principle' of [114]. □

(2.4) Notation. $D_{\mathbb{A}}(\pi) := \mathbb{H}^2_{\mathbb{A},\mathrm{sp}}(X_{\mathsf{K}_f})(\pi)$.

This is an irreducible $k_\pi \otimes \mathsf{G}(\mathbb{A}^f)$-module. (Similarly, $D_{\mathrm{et}}(\pi) = \mathbb{H}^2_{\mathrm{et},\mathrm{sp}}(X_{\mathsf{K}_f})(\pi)$, etc.) Recall from Ch. VI that $D_B(\pi) \otimes \mathbb{C}$ is spanned by $\omega_f = (2\pi i)^2 f\, dz_1 \wedge dz_2, \varepsilon_1^* \omega_f, \varepsilon_2^* \omega_f$ and $\bar{\omega}_f$, where f is a suitable modular form of weight 2, an eigenform for the Hecke algebra.

We now can split the group of Tate cycles:

$$\mathrm{Ta}_E(X) = \bigoplus_{\pi} \mathrm{Ta}_E(X)(\pi)$$

Similarly, we can split the space of Hirzebruch-Zagier classes further:

$$\mathbb{T}^0_{HZ} = \bigoplus_{\omega, \pi} \mathbb{T}^0_{HZ}(\pi, \omega),$$

where ω runs over the characters of $\mathrm{Gal}(\bar{\mathbb{Q}}/\mathbb{Q})$.

Since the action of $\mathsf{G}(\mathbb{A}^f)$ commutes with that of $\mathrm{Gal}(\bar{\mathbb{Q}}/\mathbb{Q})$, we have a representation:

$$\mathrm{Gal}(\bar{\mathbb{Q}}/\mathbb{Q}) \to \mathrm{GL}(D_{\mathrm{et}}(\pi)). \tag{2}$$

This representation has coefficients in k_π. It is a compatible system of ℓ-adic representations, see [143]. To such a system one can associate an L-series (cf. [145, 24]). Let p be a prime and let $\ell \neq p$. We consider

$$P_p(T) = (\det(1 - F_p T | D_\ell(\pi)^{I_p}) , \tag{3}$$

with F_p a geometric Frobenius and I_p the inertia group at p. Then $P_p(T)$ does not depend on ℓ and we define

$$L(D(\pi), s) = \prod_p P_p(p^{-s})^{-1} . \tag{4}$$

If p is a good prime (X_{K_f} has good reduction at p) then $D(\pi)^{I_p} = D(\pi)$ and $L(D(\pi), s)$ is a factor of the Hasse-Weil zeta function of X_{K_f}. For each $\sigma\colon k_\pi \to \mathbb{C}$ we have a $L(D(\pi^\sigma), s)$, so we can view (4) as a $k_\pi \otimes \mathbb{C}$-valued function associated to $\bigoplus_\sigma D(\pi^\sigma)$.

For the good primes one can apply the Weil-conjectures proved by Deligne:

$$P_p(T) = \prod_{i=1}^{4[k_\pi:\mathbb{Q}]} (1 - \alpha_i T)$$

with algebraic numbers α_i of absolute value p. It follows then that (4) converges for $\mathrm{Re}(s) > 2$. (Serre tells us also which Euler factors one should take at the infinite primes: $\Gamma_{\mathbb{R}}^{h^{1,+}}(s-1)\Gamma_{\mathbb{R}}^{h^{1,-}}(s)\Gamma_{\mathbb{C}}^{h^{0,2}}(s)$ where $\Gamma_{\mathbb{R}}(s) = \pi^{-s/2}\Gamma(s/2)$, $\Gamma_{\mathbb{C}}(s) = 2(2\pi)^{-s}\Gamma(s)$, and where $h^{1,+}$ (resp. $h^{1,-}$) is the dimension of the $+1$ (resp. -1) eigenspace of F_∞ (induced by complex conjugation on $X_{\mathrm{K}_f}(\mathbb{C})$) on $H^{1,1}$.)

One expects that $L(D(\pi), s)$ can be expressed in terms of the L-series of a modular form. As we saw in Chapter VI.4 one can associate several L-series to a Hilbert modular form: if f has Fourier coefficients $c(\mathfrak{m}, f)$ then we have e.g.:

i) $L(f, s) = \sum_{\mathfrak{m}} c(\mathfrak{m}, f)\mathrm{N}(\mathfrak{m})^{-s} \quad \left(\mathrm{Re}(s) > \frac{k}{2} + 1\right)$

ii) $\zeta(2s - 2k + 2)\sum_{m} c((m), f)m^{-s} \quad (\mathrm{Re}(s) > k + 1)$

iii) $\zeta_K(2s - 2k + 2)\sum_{\mathfrak{m}} c(\mathfrak{m}, f)c(\mathfrak{m}', f)\mathrm{N}(\mathfrak{m})^{-s} \quad (\mathrm{Re}(s) > k + 1)$.

The first one is also denoted by $L(\pi, s)$ if π is the automorphic representation corresponding to f.

The result of Asai (VI (4.8)) and the conjectural existence of τ (see Introduction of this Chapter) suggest that ii) gives the L-function we are looking for. (Compare the Euler factors given in VI.4.) Langlands defined these L-functions in the framework of automorphic representations. In Langlands terminology Asai's L-function is the one associated to the representation t of the L-group ${}^LG = \mathrm{GL}(2, \mathbb{C}) \times \mathrm{GL}(2, \mathbb{C}) \times \mathrm{Gal}(\bar{\mathbb{Q}}/\mathbb{Q})$ with t the extension of the natural representation (cf. [14])

$$\mathrm{GL}(2, \mathbb{C}) \times \mathrm{GL}(2, \mathbb{C}) \to \mathrm{GL}(\mathbb{C}^2 \otimes \mathbb{C}^2)$$

defined by

$$t(1 \times 1 \times \sigma)(u \otimes v) = \begin{cases} u \otimes v & \text{if } \sigma|K = \mathrm{id} \\ v \otimes u & \text{if } \sigma|K \neq \mathrm{id} . \end{cases}$$

We shall denote the Euler factors of this L-function by $L_p(s, \pi, t)$.

The main result of this section connects this L-function associated to π with the L-function associated to the Galois representation

$$r\colon \mathrm{Gal}(\bar{\mathbb{Q}}/\mathbb{Q}) \to \mathrm{GL}(D_{\mathrm{et}}(\pi)).$$

(2.5) Theorem (Brylinski-Labesse [16], cf. also [61]). *Let π be an admissible automorphic representation which defines a summand $D(\pi)$ of $\mathbb{H}^2_{\mathbb{A},\mathrm{sp}}(X)$. Then*

$$L_p(D(\pi), s) = L_p(s-1, \pi, t)$$

for almost all primes p (primes not occurring in the level of π and in the discriminant of K).

We do not prove this theorem here, but refer to [16]. A similar result in the compact quotient case was proved by Langlands. The additional problem here is to prove that the contributions of the cusps vanish.

There are two rather different cases of automorphic representations that occur in the decomposition (2.1): those that are of CM-type and those that are not. Here CM-type is defined as follows.

(2.6) Definition. An admissible cuspidal representation π of weight 2 is called of type CM if there exists a character $\chi\colon \mathbb{A}_K^*/K^* \to \mathbb{C}^*$ such that $\pi \otimes \chi \cong \pi$.

If π is of CM-type then it is rather special and quite explicitly known. In the classical case it corresponds to a theta function associated to a grössencharacter. One knows the following (see e.g. [43]): χ is quadratic and the field $M = M(\chi)$ corresponding to χ is a CM-field of degree 4 containing K. Moreover, there exists a Hecke character $\Omega\colon \mathbb{A}_M^*/M^* \to \mathbb{C}^*$ (not factoring through the norm $\mathsf{N}_{M/K}$) such that π is the automorphic representation $\pi(\Omega)$ associated to Ω, see [43, Ch. 7]. The infinity type of Ω is as follows: there are two embeddings $\tau, \tau'\colon M \to \mathbb{C}$ which extend the two embeddings $\sigma, \sigma'\colon K \to \mathbb{R}$ such that the type of Ω is $\Omega_\infty\colon \mathbb{C} \times \mathbb{C} \to \mathbb{C}$, $(z, z') \to zz'$. Furthermore, we have a 2-dimensional Galois representation r_1 of $\mathrm{Gal}(\bar{\mathbb{Q}}/K)$,

$$r_1 = \mathrm{Ind}_{\mathrm{Gal}(\bar{\mathbb{Q}}/M(\chi))}^{\mathrm{Gal}(\bar{\mathbb{Q}}/K)} \Omega$$

such that $L(\pi, s) = L(r_1, s)$.

Finally, we analyze the W-part of the cohomology. The invariant forms $\dfrac{dz_i \wedge d\bar{z}_i}{y_i^2}$ on $\mathfrak{H}^2 = \mathsf{G}(\mathbb{R})^0/\mathsf{K}_\infty$ define the Chern forms of two line bundles on each connected component of X. These generate $W_{\mathbb{A}}(1)$. Since $\mathsf{G}(\mathbb{A}^f)$ acts on the set of components via the determinant $W_{\mathbb{A}}$ breaks up in isotypic components $W_{\mathbb{A}}(\pi)$ which correspond to grössencharacters $\chi_\pi = \nu \cdot \det$ with ν a character of $\mathbb{A}_K^*/K^*$ trivial on $\mathsf{G}(\mathbb{R})^0$, i.e. ν is of finite order. The action of $\mathrm{Gal}(\bar{\mathbb{Q}}/\mathbb{Q})$ preserves these isotypic components.

(2.7) Proposition. *As a $\mathrm{Gal}(\bar{\mathbb{Q}}/\mathbb{Q})$-module $W_{\mathbb{A}}(\pi)$ is isomorphic to*

$$\chi_\pi(-1) \oplus \varepsilon_{K/\mathbb{Q}} \chi_\pi(-1),$$

where v (restricted to $\mathbb{A}_{\mathbb{Q}}^*/\mathbb{Q}^*$) *is viewed as a Galois character by class field theory and* $\varepsilon_{K/\mathbb{Q}}$ *is the quadratic character corresponding to* $K/\mathbb{Q}$.

Proof. The line bundle L whose sections are the modular forms of weight 2 is defined over $\mathbb{Q}$. One deduces that $\mathrm{Gal}(\bar{\mathbb{Q}}/\mathbb{Q})$ acts on

$$\sum_{\alpha \in \pi_0(\mathbb{A}_K^*/K^*)} v^{-1}(\alpha)(w_1^\alpha + w_2^\alpha)$$

with w_i^α the class corresponding to $-(dz_i \wedge d\bar{z}_i)/y_i^2$ on the component of X determined by α via the character $\chi_\pi = v\det$.

That it acts on $\sum v^{-1}(\alpha)(w_1^\alpha - w_2^\alpha)$ by $\varepsilon_{K/\mathbb{Q}}\chi_\pi$ follows from the fact that over K the bundle $\mathrm{Tr}_{K/\mathbb{Q}}\mathrm{t}^\vee$ (see X.3) can be decomposed as $\mathrm{t}_1^\vee \oplus \mathrm{t}_2^\vee$ (corresponding to the factors of automorphy $(c^{(i)}t_i + d^{(i)})$, $i = 1, 2$). □

3. Splitting up the Galois Representation

We start with some heuristic considerations. Recall Oda's conjecture given in the Introduction of Chapter VI. It says that if f is a Hecke eigenform of weight 2 on $\mathrm{SL}(2, \mathfrak{o})$ and if $H^2(M_f)$ is the part of the cohomology of X_{K_0} associated to f and its companions, then there exists an abelian variety of dimension $[k_f : \mathbb{Q}]$ (with k_f the field of Fourier coefficients of f) defined over our real quadratic field K such that

$$H^2(M_f) = H^1(A) \otimes_{k_f} H^1(A^\rho) \quad (\rho\text{: conjug. of } K/\mathbb{Q}) \tag{1}$$

as k_f-Hodge structures. In the terminology of automorphic representations we have

$$H^2(M_f) = \bigoplus_{\sigma: k_\pi \to \mathbb{C}} H_B^2(X_{\mathrm{K}_0}, \mathbb{C})(\pi^\sigma) \quad (k_\pi = k_f)$$

and the summands correspond to the companions of f.

Now let

$$r\colon \mathrm{Gal}(\bar{\mathbb{Q}}/\mathbb{Q}) \to \mathrm{GL}(\mathbb{H}_{\mathrm{et}}^2(X_{\mathrm{K}_f})(\pi)) \tag{2}$$

be the natural Galois representation. Then (1) suggests the existence of a Galois representation

$$\tau\colon \mathrm{Gal}(\bar{\mathbb{Q}}/K) \to \mathrm{GL}(V_{\mathrm{et}}(\pi))$$

with $V_{\mathrm{et}}(\pi)$ a rank 2 module over $k_\pi \otimes \mathbb{A}^f$ such that

$$r|_{\mathrm{Gal}(\bar{\mathbb{Q}}/K)} = \tau \otimes \tau^\rho\,. \tag{3}$$

It seems difficult to construct τ. As we shall see, what one can show is that there exist for every prime ℓ irreducible representations

$$r_{1,\ell}, r_{2,\ell}\colon \mathrm{Gal}(\bar{\mathbb{Q}}/K) \to \mathrm{GL}(Y_\ell)\,,$$

where Y_ℓ is a $\bar{\mathbb{Q}}_\ell$-vector space of dimension 2 such that

$$r_\ell|_{\mathrm{Gal}(\bar{\mathbb{Q}}/K)} = r_{1,\ell} \otimes r_{2,\ell} \tag{4}$$

with $r_\ell\colon \mathrm{Gal}(\bar{\mathbb{Q}}/\mathbb{Q}) \to \mathrm{GL}(V_\ell(\pi) \otimes \bar{\mathbb{Q}}_\ell)$.

The idea behind the possibility of splitting r_ℓ as in (4) is the following. Poincaré duality induces a Galois equivariant pairing Q on $D_\ell(\pi) = \mathbb{H}^2_{\mathrm{et}}(X_{\mathrm{K}_f} \otimes \bar{\mathbb{Q}}, \mathbb{Q}_\ell)(\pi)$. The 2-dimensional quadric $Q = 0$ in $\mathbb{P}(D_\ell \otimes \bar{\mathbb{Q}}_\ell)$ is isomorphic to $\mathbb{P}^1 \times \mathbb{P}^1$ and therefore the orthogonal group $\{\alpha \in \mathrm{End}(D_\ell)\colon Q(\alpha x, \alpha y) = \lambda(\alpha) Q(x, y)\}$ is close to $\mathrm{GL}(2) \times \mathrm{GL}(2)$.

The conjectural existence of τ or the actual existence of $r_{1,\ell}$ and $r_{2,\ell}$ have strong implications for the Tate cycles $\mathrm{Ta}_E(\pi)$. Indeed,

$$\begin{aligned}\mathrm{Ta}_E(\pi) &= (V \otimes V^\rho)^{\mathrm{Gal}(\bar{\mathbb{Q}}/E)} \quad \text{with } V = V_{\mathrm{et}}(\pi) \text{ or } Y_\ell(\pi) \\ &= \mathrm{Hom}(V^\rho \otimes V^\vee)^{\mathrm{Gal}(\bar{\mathbb{Q}}/E)}\end{aligned}$$

where $^\vee$ denotes the contragredient representation and Hom denotes Galois module homomorphisms. This implies immediately

$$\mathrm{rank}\, \mathrm{Ta}_E(\pi) \leqslant 2\,.$$

More precisely,

i) if $V^\rho \not\cong V^\vee$ then $\mathrm{rank}\, \mathrm{Ta}_E(\pi) = 0$,

ii) if $V^\rho \cong V^\vee$ then since $V^\vee \cong V \otimes \omega_\pi^{-1}$ with ω_π the central character of π (cf. [43]), we have

$$\begin{aligned}\mathrm{Ta}_{\bar{\mathbb{Q}}}(\pi) &= \mathrm{Hom}(V^\rho, V^\vee) \\ &= (\mathrm{Sym}^2 V \otimes \omega_\pi^{-1}) \oplus (\textstyle\bigwedge^2 V) \otimes \omega_\pi^{-1} \\ &= (\mathrm{Sym}^2 V \otimes \omega_\pi^{-1}) \oplus \mathbb{A}^f(-1)\end{aligned}$$

if we know that $\det r_{i,\ell} = \omega_\pi \cdot \alpha^{-1}$. But

$$\begin{aligned}(\mathrm{Sym}^2 V^\rho) \otimes \omega_\pi^{-\rho} &= (\mathrm{Sym}^2(V \otimes \omega_\pi^{-1})) \otimes \omega_\pi \\ &= (\mathrm{Sym}^2 V) \otimes \omega_\pi^{-1}\,,\end{aligned}$$

so the summands are $\mathrm{Gal}(\bar{\mathbb{Q}}/\mathbb{Q})$-invariant; the cycle corresponding to $\mathbb{A}^f(-1)$ is defined over $\mathbb{Q}^{ab}$.

If $\mathrm{rank}\, \mathrm{Ta}(\pi) = 2$ then the Galois module V admits non-trivial endomorphisms.

Note that all this is the Galois-module analogue of (8.2) of Chapter VI where we used the splitting of Hodge modules (Thm. (7.1)). In Chapter VI the existence of the abelian variety A could be proved for the case that the eigenform f is a Doi-Naganuma lift of a modular form in one variable. As we shall see in this case the existence of τ can be proved too. Another case where the existence of τ can be proved is the case of complex multiplication.

We formalize the above in the following conjecture which is suggested by Langlands' philosophy (cf. [61], cf. [123], [131]) (cf. also Remark (3.10)).

(3.1) Conjecture. *For every cuspidal automorphic representation π of weight 2 of $\mathrm{G}(\mathbb{A})$ there exists a module V of rank 2 over $\mathbb{Q}(\pi) \otimes \mathbb{A}^f_K$ giving rise to a system of λ-adic representations*

$$\tau\colon \mathrm{Gal}(\bar{\mathbb{Q}}/K) \to \mathrm{GL}(V)$$

such that

$$L(\pi, s) = L(\tau, s)$$

(*and such that for*

$$r\colon \mathrm{Gal}(\bar{\mathbb{Q}}/\mathbb{Q}) \to \mathrm{GL}(\mathbb{H}^2_{\mathrm{et}}(X)(\pi))$$

we have

$$r|_{\mathrm{Gal}(\bar{\mathbb{Q}}/K)} = \tau \otimes \tau^{\rho})) .$$

Now we turn to the construction of $r_{1,\ell}$ and $r_{2,\ell}$. The starting point is the natural pairing ("Poincaré duality")

$$H^2_{\mathrm{et}}(Z_{\mathsf{K}_f} \otimes \bar{\mathbb{Q}})(1) \times H^2_{\mathrm{et}}(Z_{\mathsf{K}_f} \otimes \bar{\mathbb{Q}})(1) \to H^4_{\mathrm{et}}(Z_{\mathsf{K}_f} \otimes \bar{\mathbb{Q}})(2)$$

which induces

$$\mathbb{H}^2_{\mathrm{et}}(X_{\mathsf{K}_f} \otimes \bar{\mathbb{Q}})(1) \times \mathbb{H}^2_{\mathrm{et}}(X_{\mathsf{K}_f} \otimes \bar{\mathbb{Q}})(1) \to \mathbb{H}^4_{\mathrm{et}}(X_{\mathsf{K}_f} \otimes \bar{\mathbb{Q}})(2) .$$

This pairing is equivariant for the action of $\mathsf{G}(\mathbb{A}^f)$, so we split into isotypic components

$$\mathbb{H}^2_{\mathrm{et}}(\pi)(1) \times \mathbb{H}^2_{\mathrm{et}}(\pi)(1) \to \mathbb{H}^4_{\mathrm{et}}(\pi)(2) \tag{5}$$

The surface X_{K_f} need not be connected, but since $\mathbb{H}^2_{\mathrm{et}}(\pi)$ is an irreducible $\mathsf{G}(\mathbb{A}^f)$-module Schur's lemma implies that $\mathbb{H}^4_{\mathrm{et}}(\pi)$ is of rank 1 over $\mathbb{A}^f$. In the right-hand side of (5) we should write $\omega_\pi \cdot \det$ instead of π since $\mathsf{G}(\mathbb{A}^f)$ acts on $\mathbb{H}^4_{\mathrm{et}}(\pi)$ as $\omega_\pi \cdot \det$. This follows from the fact that

$$\pi^\vee = \pi \otimes \omega_\pi^{-1}$$

with $\pi^\vee$ the contragredient representation and ω_π the central character of π.

The Galois action on $\mathbb{H}^4_{\mathrm{et}}(X_{\mathsf{K}_f} \otimes \bar{\mathbb{Q}})(\pi)$ is as follows:

(3.2) Lemma. *Let* $\chi_\pi = \omega_\pi | \mathbb{A}^*_{\mathbb{Q}}$. *If we view* χ_π *as a* 1-*dimensional Galois representation via the identification of* $\mathrm{Gal}(\mathbb{Q}^{\mathrm{ab}}/\mathbb{Q})$ *with* $\pi_0(\mathbb{A}^*_{\mathbb{Q}}/\mathbb{Q}^*)$ *then*

$$\mathbb{H}^4_{\mathrm{et}}(X_{\mathsf{K}_f} \otimes \bar{\mathbb{Q}})(\pi) = \chi_\pi(-2)$$

as $\mathrm{Gal}(\bar{\mathbb{Q}}/\mathbb{Q})$-*module.*

Proof. The components of X_{K_f} define elements of $\mathbb{H}^4_{\mathrm{et}}(2)$. The Galois action on the set of components is given by

$$\pi_0(\mathbb{A}^*_{\mathbb{Q}}/\mathbb{Q}^*) \to \pi_0(\mathbb{A}^*_K/K^*) \to \pi_0(X_{\mathsf{K}_f} \otimes \bar{\mathbb{Q}})$$

by X (4.3). □

For shortness we shall use the following notation:

$$D_{\mathrm{et}}(\pi) := \mathbb{H}^2_{\mathrm{et}}(X_{\mathsf{K}_f} \otimes \bar{\mathbb{Q}})(\pi)$$

$$D_\ell(\pi) := \mathbb{H}^2_{\mathrm{et}}(X_{\mathsf{K}_f} \otimes \bar{\mathbb{Q}}, \mathbb{Q}_\ell)(\pi) .$$

By (5) we have a Galois equivariant pairing

$$Q\colon D_{\mathrm{et}}(\pi) \times D_{\mathrm{et}}(\pi) \to \mathbb{A}^f(\alpha^{-2}\chi_\pi)\,,$$

satisfying

$$Q(r(\sigma)x, r(\sigma)y) = (\chi_\pi \alpha^{-2})(\sigma) Q(x, y)\,,$$

with r the natural representation $\mathrm{Gal}(\bar{\mathbb{Q}}/\mathbb{Q}) \to D_{\mathrm{et}}(\pi)$. Thus we have a homomorphism

$$r\colon \mathrm{Gal}(\bar{\mathbb{Q}}/\mathbb{Q}) \to \mathrm{GO}(D_{\mathrm{et}}(\pi), Q) := \{g \in \mathrm{GL}(D_{\mathrm{et}}(\pi))\colon Q(gx, gy) = \lambda(g) Q(x, y)\}\,,$$

into the general orthogonal group of Q. The factor λ is called the multiplier.

Harder, Langlands and Rapoport show in [61] that Q can be used to split r, i.e. they construct for every prime ℓ representations

$$r_{1,\ell}, r_{2,\ell}\colon \mathrm{Gal}(\bar{\mathbb{Q}}/K) \to \mathrm{GL}(Y_\ell)\,,$$

where Y_ℓ is a $\bar{\mathbb{Q}}_\ell$-vector space of dimension 2 such that

$$r_\ell | \mathrm{Gal}(\bar{\mathbb{Q}}/K) = r_{1,\ell} \otimes r_{2,\ell}\,.$$

The idea is that a 2-dimensional quadric $Q = 0$ in $\mathbb{P}(D_\ell \otimes \bar{\mathbb{Q}}_\ell)$ (over an algebraically closed field) is isomorphic to $\mathbb{P}^1 \times \mathbb{P}^1$; so $\mathrm{GO}(D_\ell \otimes \bar{\mathbb{Q}}_\ell, Q)$ is not too far from $\mathrm{GL}(2) \times \mathrm{GL}(2)$.

(3.3) Intermezzo. The orthogonal group $\mathrm{GO}(E, Q)$.

Let Y be a 2-dimensional vector space over an algebraically closed field k and let $Y \times Y \to k$ be a non-degenerate pairing. Using this we can identify Y and $Y^\vee$, hence $Y \otimes Y$ with $Y \otimes Y^\vee = \mathrm{End}(Y)$. Put $E = \mathrm{End}(Y)$. For $\phi \in E$ transposition $\phi \mapsto {}^t\phi$ with respect to this pairing is given by $x \otimes y \mapsto -y \otimes x$. On E we have the quadratic form

$$Q(\phi, \psi) = \mathrm{Tr}(\phi\, {}^t\psi)\,.$$

The group

$$\mathrm{GO}(E, Q) = \{\alpha \in \mathrm{End}(E)\colon Q(\alpha x, \alpha y) = \lambda(\alpha) Q(x, y)\}$$

is not connected. Indeed, there is an exact sequence

$$1 \to \mathrm{SGO}(E, Q) \to \mathrm{GO}(E, Q) \xrightarrow{d} \{+1, -1\} \to 1\,,$$

where $d(\alpha) = \det(\alpha)/\lambda(\alpha)^2$ is the similitude norm. The group $\mathrm{SGO}(E, Q)$ is the connected component of 1 and it consists of those elements that preserve the "rulings" of $\mathbb{P}^1 \times \mathbb{P}^1$ (quadric defined by pairing in $\mathbb{P}^3_k$).

(3.4) Lemma. *There is a commutative diagram*

$$\begin{array}{ccc} \mathrm{GL}(Y) \times \mathrm{GL}(Y) & \xrightarrow{p} & \mathrm{SGO}(E, Q) \\ & {\scriptstyle \det \times \det} \searrow \quad \swarrow {\scriptstyle \lambda} & \\ & k^* & \end{array}$$

with $p(\alpha,\beta)$ *the transformation* $(\phi \mapsto \alpha\phi{}^t\beta)$. *The kernel of* p *is* k^* *embedded in* $\mathrm{GL}(2) \times \mathrm{GL}(2)$ *via* $x \to (x, x^{-1})$.

Reprise. Over an algebraically closed field all non-singular quadrics are the same so we may take $k = \bar{\mathbb{Q}}_\ell$ and identify $(D_\ell(\pi) \otimes \bar{\mathbb{Q}}_\ell, Q)$ with $(E, \mathrm{Tr}\,\phi{}^t\psi)$.

(3.5) Proposition. *There is a commutative diagram*

$$\begin{array}{ccccccccc} 1 & \longrightarrow & \mathrm{Gal}(\bar{\mathbb{Q}}/K) & \longrightarrow & \mathrm{Gal}(\bar{\mathbb{Q}}/\mathbb{Q}) & \longrightarrow & \mathrm{Gal}(K/\mathbb{Q}) & \longrightarrow & 1 \\ & & \downarrow & & \downarrow r & & \downarrow \cong & & \\ 1 & \longrightarrow & \mathrm{SGO}(E,Q) & \longrightarrow & \mathrm{GO}(E,Q) & \xrightarrow{d} & \{+1,-1\} & \longrightarrow & 1 \end{array}$$

Proof. Choose $\sigma \in \mathrm{Gal}(\bar{\mathbb{Q}}/\mathbb{Q})$ mapping to the generator of $\mathrm{Gal}(K/\mathbb{Q})$. We use the information about r stored in the L-function. If $p = \mathfrak{p}\mathfrak{p}'$ is a prime that splits then

$$r(F_p) \sim \begin{bmatrix} \alpha_{\mathfrak{p}}\alpha_{\mathfrak{p}'} & & & \\ & \alpha_{\mathfrak{p}}\beta_{\mathfrak{p}'} & & \\ & & \alpha_{\mathfrak{p}'}\beta_{\mathfrak{p}} & \\ & & & \beta_{\mathfrak{p}}\beta_{\mathfrak{p}'} \end{bmatrix}$$

see (2.5), Interchanging $\mathfrak{p}$ and $\mathfrak{p}'$ gives a factor -1 in d. □

To split r_ℓ we lift it to $\mathrm{GL}(Y) \times \mathrm{GL}(Y)$ as follows. *Since the prime* ℓ *is fixed we drop the index* ℓ.

(3.6) Proposition. 1) *There exists a lift*

$$\begin{array}{ccc} & & \mathrm{GL}(Y) \times \mathrm{GL}(Y) \\ & \overset{(r_1,r_2)}{\nearrow} & \downarrow p \\ \mathrm{Gal}(\bar{\mathbb{Q}}/K) & \longrightarrow & \mathrm{SGO}(E,Q) \end{array}$$

such that $\det r_1 = \alpha_K^{-1}\mu_1$, $\det r_2 = \alpha_K^{-1}\mu_2$ *with* μ_1, μ_2 *Galois characters of finite order,*

2) r_1, r_2 *are unique up to twisting* $r_1 \to r_1 \otimes \mu$, $r_2 \to r_2 \otimes \mu^{-1}$ *with* μ *a Galois character of finite order,*

3) *for some choice of* μ *as in* 2) *we have* $r_2(\sigma) = r_1(\sigma^\rho)$ *for all* $\sigma \in \mathrm{Gal}(\bar{\mathbb{Q}}/K)$, ρ *a representative of the generator of* $\mathrm{Gal}(K/\mathbb{Q})$.

Proof. We twist with α on D_ℓ such that $Q\colon D_\ell \times D_\ell \to \bar{\mathbb{Q}}_\ell^*$, (with trivial Galois action on an open subgroup $\mathrm{Gal}(\bar{\mathbb{Q}}/L)$.) The exact sequence

$$1 \to \bar{\mathbb{Q}}_\ell^* \to \mathrm{GL}(Y) \times \mathrm{GL}(Y) \to \mathrm{SGO}(E,Q) \to 1$$

induces an exact sequence

$$\begin{aligned} 1 &\to \mathrm{Hom}(\mathrm{Gal}(\bar{\mathbb{Q}}/L), \bar{\mathbb{Q}}_\ell^*) \to \mathrm{Hom}(\mathrm{Gal}(\bar{\mathbb{Q}}/L), \mathrm{GL} \times \mathrm{GL}) \\ &\to \mathrm{Hom}(\mathrm{Gal}(\bar{\mathbb{Q}}/L), \mathrm{SGO}) \to H^2(\mathrm{Gal}(\bar{\mathbb{Q}}/L), \bar{\mathbb{Q}}_\ell^*)\,. \end{aligned} \tag{6}$$

By a theorem of Tate (see [180], § 6) $H^2(\mathrm{Gal}(\bar{\mathbb{Q}}/L), \bar{\mathbb{Q}}_\ell^*) = 0$, hence we can lift, even to SL × SL Twisting back yields 1).

For 2) the exact sequence (6) suffices. As to 3) take $R: \phi \to -{}^t\phi$ as generator of $\mathrm{GO}(E, Q) \bmod \mathrm{SGO}$. Clearly, $r(\rho)$ is conjugate to

$$(1 \times g)R \quad \text{for some } g \in \mathrm{GL}(2).$$

Hence

$$\begin{aligned} r(\rho^2) &= p((1,g)R \cdot (1,g)R) \\ &= p(g,g) \end{aligned}$$

but on the other hand

$$r(\rho^2) = p(r_1(\rho^2), r_2(\rho^2)),$$

so

$$g = \zeta r_1(\rho^2) = \zeta^{-1} r_2(\rho^2) \quad \text{for some non-zero } \zeta. \tag{7}$$

Furthermore,

$$\begin{aligned} p(r_1(\sigma^\rho), r_2(\sigma^\rho)) &= p(r_1(\rho\sigma\rho^{-1}), r_2(\rho\sigma\rho^{-1})) \\ &= p((1,g)R(r_1(\sigma), r_2(\sigma))R^{-1}(1,g^{-1})) \\ &= p((1,g)(r_2(\sigma), r_1(\sigma))(1,g^{-1})) \\ &= p(r_2(\sigma), g r_1(\sigma) g^{-1}), \end{aligned}$$

hence taking into account the kernel of p we find

$$r_2(\sigma) = r_1(\sigma^\rho)\mu(\sigma) \tag{8}$$

$$r_1(\sigma) = g^{-1} r_2(\sigma^\rho) g \mu(\sigma)^{-1} \tag{9}$$

for some character μ of finite order on $\mathrm{Gal}(\bar{\mathbb{Q}}/K)$. Substitution of $\sigma = \rho^2$ in (8) and (9) gives in view of (7)

$$g = \pm\mu(\rho) r_1(\rho^2) = \pm\mu(\rho)^{-1} r_2(\rho^2).$$

Using this identity substitution of (9) into (8) gives

$$\mu(\sigma^\rho) = \mu(\sigma) \quad \text{for all } \sigma \in \mathrm{Gal}(\bar{\mathbb{Q}}/K).$$

We claim: $\mu = \eta \cdot \eta^\rho$ for some character η. (Proof: Let $A = \mathbb{Z}/2\mathbb{Z}$. Then $\hat{H}^0(A, \mathrm{Hom}(\mathbb{A}_K^*/K^*, \mathbb{Q}/\mathbb{Z}))$ = dual of $\hat{H}^{-1}(A, \mathbb{A}_K^*/K^*)$ and since A is cyclic this is $\hat{H}^1(A, \mathbb{A}_K^*/K^*)$ which vanishes by class field theory.) If we take $r_1' = r_1 \otimes \eta$, $r_2' = r_2 \otimes \eta$ we have

$$r_1'(\sigma^\rho) = r_1(\sigma^\rho)\eta(\sigma^\rho) = r_2(\sigma)\eta(\sigma)^{-1} = r_2'(\sigma). \qquad \square$$

(3.7) Theorem. i) *r_1 and r_2 are irreducible.*

ii) *If π is not of CM-type then r_1, r_2 remain irreducible when restricted to any open subgroup* $\mathrm{Gal}(\bar{\mathbb{Q}}/L) \subset \mathrm{Gal}(\bar{\mathbb{Q}}/K)$.

Proof. i) One uses Ribet's arguments, cf. [136]. Suppose r_1 (and hence r_2) is reducible. Then after suitable conjugation we have

$$r_1(\sigma) = \begin{pmatrix} \chi_1(\sigma) & * \\ 0 & \chi_2(\sigma) \end{pmatrix} \quad \text{for all } \sigma \in \mathrm{Gal}(\bar{\mathbb{Q}}/K) \tag{10}$$

with characters χ_1, χ_2. One considers the "semi-simplication" $\chi_1 \oplus \chi_2$ of r_1. By a theorem of Serre and Lang (cf. Serre [143], Ch. III) there are characters γ_i of finite order such that

$$\chi_i = \alpha_K^{n_i}\gamma_i\,,$$

where α_K is the ℓ-adic cyclotomic character of K. We have:

$$\det r_1 = \alpha_K^{-2}\mu_1 \quad \text{by (3.6), 1)}$$

while on the other hand

$$\det r_1 = \alpha_K^{n_1+n_2}\gamma_1\gamma_2$$

From the fact that the eigenvalues of a Frobenius element all have the same absolute value it follows that $n_1 = n_2$. This contradicts $n_1 + n_2 = -1$ and proves *i*).

Alternative proof: one shows that this gives a contradiction by looking at the exterior product of $r_K = r|\mathrm{Gal}(\bar{\mathbb{Q}}/K)$: We have

$$\bigwedge^2 r_K = \left(\mathrm{Sym}^2 r_1 \otimes \bigwedge^2 r_2\right) \oplus \left(\mathrm{Sym}^2 r_2 \otimes \bigwedge^2 r_1\right).$$

We now twist with $(\omega_\pi \cdot \mathsf{N}_{K/\mathbb{Q}})^{-1}$ to get rid of the determinants $\bigwedge^2 r_1$ and $\bigwedge^2 r_2$. If eigenvalues at a uniformizing parameter $\mathfrak{p}$ are $\alpha_\mathfrak{p}, \beta_\mathfrak{p}$ for r_1 then for $\mathrm{Sym}^2 r_1$: $\alpha_\mathfrak{p}^2, \alpha_\mathfrak{p}\beta_\mathfrak{p}, \beta_\mathfrak{p}^2$. Twisting gives eigenvalues $(\alpha_\mathfrak{p}/\beta_\mathfrak{p}, 1, \beta_\mathfrak{p}/\alpha_\mathfrak{p})$. Using $\chi_i = \alpha_K^{n_i}\gamma_i$ and (10) we thus have

$$L\left(\bigwedge^2 r_K, s\right) = L(s, \chi_1/\chi_2)L_{K/\mathbb{Q}}(s)L(s, \chi_2/\chi_1)\cdot L(s, \chi_1'/\chi_2')L_{K/\mathbb{Q}}(s)L(s, \chi_2'/\chi_1')\,,$$

where χ_1', χ_2' play the role of χ_1, χ_2 for r_2. Now $L_{K/\mathbb{Q}}(s)$ has a simple pole at $s = 1$ while $L(s, \chi_1/\chi_2)$, etc. have no pole at $s = 1$ since χ_1/χ_2 is of finite order.

On the other hand, using the conjectural τ and $\bigwedge^2 E = \mathrm{Sym}^2 Y \oplus \mathrm{Sym}^2 Y$ one sees that $L(\bigwedge^2 r_K, s)$ must be $L(\mathrm{Sym}^2(\pi), s)L(\mathrm{Sym}^2(\pi^\rho), s)$, where $L(\mathrm{Sym}^2(\pi), s)$ is that L-function associated to π whose Euler factor $L_\mathfrak{p}$ at $\mathfrak{p}$ is

$$[(1 - \alpha_\mathfrak{p}^2\mathsf{N}(\mathfrak{p})^{-s})(1 - \alpha_\mathfrak{p}\beta_\mathfrak{p}\mathsf{N}(\mathfrak{p})^{-s})(1 - \beta_\mathfrak{p}^2\mathsf{N}(\mathfrak{p})^{-s})]\,.$$

Such L-functions are studied in [44]. Comparing Euler factors at the good primes yields:

$$L_\mathfrak{p}\left(\bigwedge^2 r_K \otimes (\omega_\pi \mathsf{N}_{K/\mathbb{Q}})^{-1}, s\right) = L_\mathfrak{p}(\mathrm{Sym}^2(\pi), s)L_\mathfrak{p}(\mathrm{Sym}^2(\pi^\rho), s)\,.$$

From [44] it can be deduced that the product of the L-functions $L(\mathrm{Sym}^2(\pi), s)L(\mathrm{Sym}^2(\pi^\rho), s)$ has no poles. This contradiction finishes the proof of i).

ii) The argument here is similar but more complicated. We refer to [61], 4.5.4 for a proof. □

(3.8) Corollary. *The representation r is semi-simple.*

Proof. Suppose it is not. Then $r(\mathrm{Gal}(\bar{\mathbb{Q}}/K))$ has a proper invariant subspace and therefore is contained in a parabolic subgroup of $\mathrm{GO}(E, Q)$. But then r_1, r_2 have images contained in parabolic subgroups of $\mathrm{GL}(Y)$, hence are reducible. □

We derive some consequences. Consider the Tate cycles defined over $\mathbb{Q}^{\mathrm{ab}}$. Then $\mathrm{Gal}(\bar{\mathbb{Q}}/\mathbb{Q})$ acts via characters and we can decompose $\mathrm{Ta}_{\bar{\mathbb{Q}}}(\pi)^{\mathrm{Gal}(\bar{\mathbb{Q}}/\mathbb{Q}^{\mathrm{ab}})} = \bigoplus_{\omega} \mathrm{Ta}(\pi, \omega)$. Each $\mathrm{Ta}(\pi, \omega)$ is defined over a field $k(\pi) \otimes \mathbb{Q}(\omega)$.

Often we shall use the following lemma.

(3.9) Twisting Lemma. *Let ω be a character of $\mathrm{Gal}(\bar{\mathbb{Q}}/\mathbb{Q})$ of finite order viewed as a character of $\mathbb{A}_{\mathbb{Q}}^*/\mathbb{Q}^*$. If ω' is any extension of ω to $\mathbb{A}_K^*/K^*$ then*

$$D(\pi \otimes \omega') = D(\pi) \otimes \omega$$

and

$$L_Q(D(\pi \otimes \omega'), s) = L_Q(D(\pi) \otimes \omega, s),$$

where the Q means: up to those Euler factors L_p with $p|Q$ and Q is suitably chosen.

(3.10) Corollary. *Let π be a cuspidal automorphic representation. Then* i) $\mathrm{rank}\,\mathrm{Ta}(\pi) \leqslant 2$,

ii) *If π is not of CM-type then* $\mathrm{rank}_{\mathbb{A}_f} \mathrm{Ta}(\pi) \leqslant 1$ *and all Tate cycles in* $\mathrm{Ta}(\pi)$ *are defined over* $\mathbb{Q}^{\mathrm{ab}}$.

iii) *For any character ω of* $\mathrm{Gal}(\mathbb{Q}^{\mathrm{ab}}/\mathbb{Q})$ *we have* $\mathrm{rank}\,\mathrm{Ta}(\pi, \omega) \leqslant 1$.

Proof. As explained in the beginning of this section i) follows immediately from the irreducibility; the same holds for ii) and iii). □

Before we proceed we analyze the proof of Theorem (3.7). The reason why we used the alternating square $\bigwedge^2 r_K$ is the fact that we do not have the conjectural τ but only r_1 and r_2. In classical terms this means the following. Let f be a classical Hecke eigenform of weight 2 whose Dirichlet series has Euler factors

$$1 - a(\mathfrak{p}_i)p^{-s} + \mathrm{N}(\mathfrak{p}_i)^{1-2s} \qquad i = 1, 2$$

for a splitting prime $p = \mathfrak{p}_1\mathfrak{p}_2$. We write formally

$$X^2 - a(\mathfrak{p}_i)X + \mathrm{N}\mathfrak{p}_i = (X - \alpha_i)(X - \beta_i).$$

Then the Euler factor of the L-series of r at p is the inverse of

$$(1 - \alpha_1\alpha_2 p^{-s})(1 - \alpha_1\beta_2 p^{-s})(1 - \alpha_2\beta_1 p^{-s})(1 - \beta_1\beta_2 p^{-s}).$$

We then know (from $r = r_1 \otimes r_2$) that r_1 *or* r_2 has eigenvalues α_1, β_1, but we cannot decide which one. By taking the alternating square we avoid the choice.

(3.11) *Remark 1.* Suppose there exists a representation $\tau\colon \mathrm{Gal}(\bar{\mathbb{Q}}/K) \to Y_\ell$ where $\dim_{\mathbb{Q}_\ell} Y_\ell = 2$ such that $L(\pi, s) = L(\tau, s)$. Then we can construct a representation

$$s\colon \mathrm{Gal}(\bar{\mathbb{Q}}/\mathbb{Q}) \to \mathrm{GO}(E, Q)$$

by putting

$$s(\sigma) = (\tau(\sigma), \tau(\sigma^\rho)) \qquad \sigma \in \mathrm{Gal}(\bar{\mathbb{Q}}/K)$$
$$s(\rho) = (1, g)R\,,$$

where ρ is a fixed element in $\mathrm{Gal}(\bar{\mathbb{Q}}/\mathbb{Q})$ mapping to the generator of $\mathrm{Gal}(K/\mathbb{Q})$ and $g = \tau(\rho^2)$, R as the proof of (5.6). We then have:

i) τ is irreducible,

ii) s is equivalent to r.

Proof of i): as in (3.7) but we can avoid the alternating square since τ is available, cf. above remarks.

ii) s and r are semi-simple since r_1 and τ are irreducible. By what wc know about the L-series for all Frobenii they have the same traces, hence by [143, (2.2)] they are equivalent. □

So we see r_1 (or r_2) is a good candidate for τ and the condition in (3.1) that $r_K = \tau \otimes \tau^\rho$ is superfluous.

(3.12) *Remark 2.* In (3.11) we have two choices for g: $g = \pm\tau(\rho^2)$. But if $\mathrm{Tr}\,\tau(\rho^2) \neq 0$ then the sign is determined by

$$\mathrm{Tr}\, s(\rho) = \mathrm{Tr}\, g = \pm \mathrm{Tr}\, \tau(\rho^2)\,.$$

In two cases τ is known to exist:

(3.13) Theorem. *If π is cuspidal automorphic then $\tau(\pi)$ exists if* i) *π is a lift of a $\pi^0_{\mathbb{Q}}$ (see* (4.6)), ii) *π is of CM-type.*

Proof. i) If $\pi^0_{\mathbb{Q}}$ corresponds to a Hecke eigenform f of weight 2 then f determines an (isogeny) factor of the Jacobian of a modular curve. The Tate module of this factor gives the representation $\tau(\pi^0_{\mathbb{Q}})$ of $\mathrm{Gal}(\bar{\mathbb{Q}}/\mathbb{Q})$. Restrict to $\mathrm{Gal}(\bar{\mathbb{Q}}/K)$ to get π. The Eichler-Shimura relation and the theory of lifting gives the relation $L(\pi, s) = L(\tau, s)$.

ii) If π is of CM-type there is a grössencharacter $\psi\colon \mathbb{A}^*_L/L^* \to \bar{\mathbb{Q}}^*_\ell$ of type A_0 where L is a totally imaginary quadratic extension of K. We can view ψ as $\psi'\colon \mathrm{Gal}(\bar{\mathbb{Q}}/L) \to \bar{\mathbb{Q}}^*_\ell$ and we let τ be the induced representation from $\mathrm{Gal}(\bar{\mathbb{Q}}/L)$ to $\mathrm{Gal}(\bar{\mathbb{Q}}/K)$ of ψ', see (2.6). □

4. The Tate Conjectures

(4.1) Theorem. *Let K_f be a compact open subgroup of $\mathsf{G}(\mathbb{A}^f)$ and let E be a number field. Then the algebraic cycles on the Hilbert modular surface $Y_{\mathsf{K}_f} \otimes E$ are the Tate cycles on Y_{K_f} over E.*

We outline the steps of the proof given below. Using the decomposition of the cohomology of Section 2 we reduce the discussion to a part $D(\pi)$ of $H^2(Y_{\mathsf{K}_f})$. We know $\mathrm{Ta}(\pi)$ has rank $\leqslant 2$ and this rank is 1 if π is not of CM-type by (3.10). Suppose that $\mathrm{rk}\,\mathrm{Ta}_{\bar{\mathbb{Q}}}(\pi) = 1$. Then $\mathrm{Gal}(\bar{\mathbb{Q}}/\mathbb{Q})$ acts by a character and the classes in $\mathrm{Ta}_{\bar{\mathbb{Q}}}(\pi)$ are defined over $\mathbb{Q}^{\mathrm{ab}}$. We shall prove (referred to later as steps 1, 2, 3):

1) if $\mathrm{rank}\,\mathrm{Ta}(\pi, \omega) = 1$ then the Hilbert modular form corresponding to π and ω is up to twisting a lift from a one variable modular form; to be more precise, then there exists a character v' of finite order on $\mathbb{A}_K^*/K$ such that $\pi \otimes \omega' v'$ (with ω' an extension of ω to $\mathbb{A}_K^*/K$) is a lift (base change from $\mathbb{Q}$ to K).

2) if $\pi \otimes \omega' v'$ is a base change then $\mathrm{Ta}(\pi, \omega)$ is generated by an algebraic cycle. We prove this by applying the Lefschetz criterion: the cohomology class is rational and of type (1, 1).

3) for π of CM-type we also use the Lefschetz criterion to prove that all Tate cycles are Hodge cycles.

Theorem (4.1) raises the question what these cycles are. The answer is:

(4.2) Theorem. *If* $\mathrm{rank}\,\mathrm{Ta}(\pi, \omega) = 1$ *then* $\mathrm{Ta}(\pi, \omega)$ *is generated by a Hirzebruch-Zagier cycle.*

If π is of CM-type ($\pi \otimes \chi = \pi$) and $\chi \neq \chi^\rho$ then $\mathrm{Ta}_{\bar{\mathbb{Q}}}(\pi)$ contains an algebraic cycle defined over a non-abelian extension of $\mathbb{Q}$. This cannot be Hirzebruch-Zagier cycle. One does not have geometric constructions for these cycles. Looking at the analogue of an isogeny factor of a jacobian of a modular curve with CM it is doubtful whether one can find such geometric cycles at all.

As to the second conjecture of Tate one has:

(4.3) Theorem. *Let* K_f *be a compact open subgroup of* $\mathsf{G}(\mathbb{A}^f)$. *If E is a number field contained in* $\mathbb{Q}^{\mathrm{ab}}$ *then the rank of the space of algebraic cycles on the Hilbert modular surface* Y_{K_f} *which are defined over E equals the order of the pole of the L-function* $L_Q^{(2)}(Y_{\mathsf{K}_f}/E, s)$ *at* $s = 2$.

If π is of CM-type one can do better.

(4.4) Theorem. *If* π *is of CM-type then the rank of the space of algebraic cycles in* $D(\pi)(1)$ *defined over a number field E equals the order of the pole of the L-function* $L_Q(D(\pi)/E, s)$ *at* $s = 2$.

Theorem (4.1), (4.2) and (4.3) were proved by Harder, Langlands and Rapoport for abelian extensions of $\mathbb{Q}$. The CM-type case was treated by Klingenberg and independently by Murty and Ramakrishnan see [61], [96] and [118]. We shall prove (4.1) and (4.4).

As to Theorems (4.2) and (4.3) it is illuminating to see the proof of (4.2) and (4.3) in the special case where $\mathsf{K}_f = \mathsf{K}_0$, $E = \mathbb{Q}$ and where the class number h^+ of K equals 1. In this case the CM-type does not occur. The L-function $L(D(\pi), s)$ is (apart from finitely many Euler factors) of the form $G(f, s)$ (see VI.4), the L-function studied by Asai for a Hecke eigenform f. We showed in VI.4:

(4.5). The following are equivalent: (for f of weight 2)

i) $G(f, s)$ has a simple pole at $s = 2$,

ii) $f = f^{\rho}$,

iii) f is in the image of the Doi-Naganuma lifting,

iv) $\int_{T_1} \eta_f \neq 0$.

Recall that $G(f, s)$ is the L-function we called $L(s, \pi, t)$ in the preceding section. Using (4.1) we see that algebraic cycles are precisely the Tate cycles. If $\mathrm{Ta}(\pi)$ has rank 1 then π is a lift. So if f corresponds to π then by (4.5) $\int_{T_1} \eta_f \neq 0$; therefore there is a Hirzebruch-Zagier cycle in $\mathrm{Ta}(\pi)$ and by (4.5) again $G(f, s)$ has a simple pole at $s = 2$. Conversely, if $G(f, s)$ has a simple pole at $s = 2$ then T_1 is not orthogonal to $\mathrm{Ta}(\pi)$. The twisting lemma now enables us to extend the result to all abelian extensions of $\mathbb{Q}$. In general one needs an adelic version of (4.5) for the proof of (4.2) and (4.3). We refer to [61] for a proof.

We now start the proof of (4.1).

Consider first the W-part of $\mathbb{H}^2$. Each $W(\pi)$ is well-known:

$$W(\pi) = \chi_\pi(-1) \oplus \chi_\pi(-1)\varepsilon_{K/\mathbb{Q}} \, .$$

So we have Tate cycles in $W(\pi) \otimes \omega$ if and only if $\chi_\pi \omega = 1$ or $\chi_\pi \omega = \varepsilon_{K/\mathbb{Q}}$. We already know that $W(\pi)$ consists of Hodge cycles, so these Tate cycles are algebraic. The L-function of $W(\pi)$ is

$$L(s-1, \chi_\pi) L(s-1, \chi_\pi \varepsilon_{K/\mathbb{Q}}) \, .$$

So we can verify Tate's conjectures immediately for this case.

Since the differential forms w_1, w_2 give the same value when integrated over a component of T_N we see that $W(\pi)$ contains at most one Hirzebruch-Zagier cycle. Indeed, it contains exactly one such cycle; it corresponds to the factor $\chi_\pi(-1)$ i.e. $W(\pi) \otimes \omega$ contains this cycle if and only if $\chi_\pi \omega = 1$.

Step 1

(4.6) Theorem. *Suppose* $\operatorname{rank} \mathrm{Ta}(\pi, \omega) = 1$. *Then there exists a cuspidal automorphic form* $\pi^0_{\mathbb{Q}}$ *on* $\mathrm{GL}(2, \mathbb{Q})$, *a character* v' *of finite order on* $\mathbb{A}^*_K / K^*$ *and an extension* ω' *of* ω *to* $\mathbb{A}^*_K / K^*$ *such that*

i) $\pi \otimes \omega' v'$ *is the base change of* $\pi^0_{\mathbb{Q}}$ *from* $\mathbb{Q}$ *to* K,

ii) *the central character* $\omega_{\pi^0_{\mathbb{Q}}}$ *of* $\pi^0_{\mathbb{Q}}$ *satisfies*

$$\omega_{\pi^0_{\mathbb{Q}}} = v \varepsilon_{K/\mathbb{Q}}$$

with $v = v' | \mathbb{A}^*_{\mathbb{Q}}$ *and with* $\varepsilon_{K/\mathbb{Q}}$ *the character associated to* $K/\mathbb{Q}$.

Proof. We fix a prime ℓ and work with $\mathbb{Q}_\ell$ (or $\bar{\mathbb{Q}}_\ell$)-coefficients. We identify $D_\ell(\pi)$ with Y_ℓ as above. If $\mathrm{Ta}(\pi, \omega) \neq 0$ then there is an isomorphism $Y_\ell^\vee \cong Y_\ell^\rho$, hence a $\phi \colon Y_\ell \cong Y_\ell$ with $r_1(\sigma) \phi {}^t r_1(\sigma) = \alpha^{-1} \omega(\sigma) \phi$. By twisting with ω (i.e. replacing π by $\pi \otimes \omega'$) we may assume $\omega = 1$. Then $Q(\phi, \phi) = \mathrm{Tr}(\phi {}^t \phi) = 2 \det \phi \neq 0$ and

$$Q(r(\sigma)\phi, r(\sigma)\phi) = \alpha^{-2}(\sigma) Q(\phi, \phi) \, .$$

So $\lambda(\sigma) = \chi_\pi \alpha^{-2}(\sigma)$. Hence using $\chi_\pi = \omega_\pi | \mathbb{A}_\mathbb{Q}^*$ we find

$$\omega_\pi | \mathbb{A}_\mathbb{Q}^* = 1 . \tag{1}$$

We shall show that for almost all places v of K:

$$(\pi_v \otimes \omega')^\rho \quad \text{is equivalent to } (\pi_v \otimes \omega') \otimes \omega_\pi^{-1} . \tag{2}$$

Proof of this. Suppose v belongs to an inert prime $p = \mathfrak{p}$. Then conjugation by ρ has no effect and $\omega_\pi^{-1}(p) = 1$. Next, if v belongs to a splitting prime $p = \mathfrak{p}\mathfrak{p}'$ and if the Euler factor at $\mathfrak{p}$ of $L(\pi, s)$ is the inverse of $(X - \alpha_\mathfrak{p})(X - \beta_\mathfrak{p})$ then

$$r(F_p) \sim \begin{bmatrix} \alpha_\mathfrak{p}\alpha_{\mathfrak{p}'} & & & 0 \\ & \alpha_\mathfrak{p}\beta_{\mathfrak{p}'} & & \\ & & \alpha_{\mathfrak{p}'}\beta_\mathfrak{p} & \\ 0 & & & \beta_\mathfrak{p}\beta_{\mathfrak{p}'} \end{bmatrix} .$$

Let $x_\mathfrak{p}$ (resp. $x_{\mathfrak{p}'}$) be a local parameter at $\mathfrak{p}$ (resp. $\mathfrak{p}'$). We have:

$$\alpha_\mathfrak{p}\beta_\mathfrak{p} = \omega_\pi(x_\mathfrak{p}) \tag{3}$$

$$\alpha_{\mathfrak{p}'}\beta_{\mathfrak{p}'} = \omega_\pi(x_{\mathfrak{p}'})$$

$$\alpha_\mathfrak{p}\beta_\mathfrak{p}\alpha_{\mathfrak{p}'}\beta_{\mathfrak{p}'} = \omega_\pi(x_\mathfrak{p}x_{\mathfrak{p}'}) = \omega_\pi(p) = 1 . \tag{4}$$

Since $\dim \mathrm{Ta}(\pi, 1) = 1$ one of the eigenvalues, say $\alpha_\mathfrak{p}\alpha_{\mathfrak{p}'}$, equals 1, i.e. assume

$$\alpha_\mathfrak{p}\alpha_{\mathfrak{p}'} = 1 \tag{5}$$

By (4) then

$$\beta_\mathfrak{p}\beta_{\mathfrak{p}'} = 1 .$$

One finds

$$\begin{aligned} \alpha_{\mathfrak{p}'} &= \alpha_\mathfrak{p}^{-1} && \text{by (5)} \\ &= \beta_\mathfrak{p}\omega_\pi^{-1}(x_\mathfrak{p}) && \text{by (3)} \end{aligned}$$

and similarly

$$\beta_{\mathfrak{p}'} = \alpha_\mathfrak{p}\omega_\pi^{-1}(x_\mathfrak{p})$$

which proves (2).

The strong Multiplicity One Theorem of [93] tells us that we can dispense with the remaining prime:

$$\pi^\rho \sim \pi \otimes \omega_\pi^{-1} .$$

Since $\omega_\pi | \mathbb{A}_\mathbb{Q}^* = 1$ and ω_π lies in the kernel of the norm map

$$\omega_\pi = v'/(v')^\rho$$

for some $v'\colon \mathbb{A}_K^*/K^* \to \bar{\mathbb{Q}}_\ell$. Then one has

$$(\pi \otimes v')^\rho \sim \pi \otimes v' .$$

This invariance under conjugation by ρ shows that $\pi \otimes \nu'$ is a lift (base change) from $\mathbb{Q}$ to K, see [107]. This proves i).

For ii) we compute ω_π:

$$\begin{aligned}\omega_\pi &= (\omega_{\pi^0_{\mathbb{Q}}} \mathrm{N}_{K/\mathbb{Q}}) \cdot \nu'^{-2} \\ &= \nu'^{\rho}/\nu'\end{aligned}$$

hence

$$\omega_{\pi^0_{\mathbb{Q}}} = \nu \cdot \eta \quad \text{with } \eta = 1 \text{ or } \eta = \varepsilon_{K/\mathbb{Q}} . \tag{6}$$

To decide whether $\eta = 1$ or $\eta = \varepsilon_{K/\mathbb{Q}}$ we use the representation τ of $\mathrm{Gal}(\bar{\mathbb{Q}}/K)$ which is known to exist here, see Thm. (3.11). Indeed, $\pi^0_{\mathbb{Q}}$ defines a representation τ_0 of $\mathrm{Gal}(\bar{\mathbb{Q}}/\mathbb{Q})$ and we take $\tau = \tau_0|_{\mathrm{Gal}(\bar{\mathbb{Q}}/K)} \otimes (\nu')^{-1}$. We then have

$$\begin{aligned}\det \tau_0(\sigma) &= \alpha^{-1}(\sigma)\omega_\pi(\sigma)\cdot(\nu')^2 \quad (\text{for all } \sigma \in \mathrm{Gal}(\bar{\mathbb{Q}}/K)) \\ &= \alpha^{-1}(\sigma)\cdot \nu'^{-1}\cdot(\nu')^{\rho}\cdot(\nu')^2 \\ &= \alpha^{-1}(\sigma)(\nu'\cdot\nu'^{\rho}) .\end{aligned} \tag{7}$$

Recall now that according to our assumption we have a non-zero $\phi \in \mathrm{End}(Y)$ with

$$r(\sigma)\phi = \alpha^{-1}(\sigma)\phi \quad \text{for all } \sigma \in \mathrm{Gal}(\bar{\mathbb{Q}}/K) . \tag{8}$$

We claim that this is equivalent with

$$\tau_0(\sigma^\rho) = \phi^{-1}\tau_0(\sigma)\phi \quad \text{for all } \sigma \in \mathrm{Gal}(\bar{\mathbb{Q}}/K) . \tag{9}$$

Indeed, (8) is equivalent to

$$(\nu'(\sigma)\nu'(\sigma^\rho))^{-1}\tau_0(\sigma)\phi\, {}^t\tau_0(\sigma^\rho) = \alpha^{-1}(\sigma)\phi ,$$

so (by multiplication on the right with $\tau_0(\sigma^\rho)$) to

$$(\alpha(\sigma)(\nu'(\sigma)\nu'(\sigma^\rho))^{-1} \det \tau_0(\sigma^\rho))\tau_0(\sigma)\phi = \phi\tau_0(\sigma^\rho) .$$

Using (7) and $\det \tau_0(\sigma^\rho) = \det \tau_0(\sigma)$ the equivalence of (8) and (9) follows. Since τ is irreducible we have

$$\tau_0(\rho)^{-1} = \lambda\phi \quad \text{with } \lambda \in \bar{\mathbb{Q}}_\ell^* .$$

We may assume $\lambda = 1$. Then we have immediately

$$r(\rho)\phi = \pm\alpha^{-1}(\rho)\phi . \tag{10}$$

We have to show that the plus sign holds in (10) if and only if $\eta = \varepsilon_{K/\mathbb{Q}}$. But we have:

$$\begin{aligned}r(\rho)\tau_0(\rho)^{-1} &= ((1,g)R)\tau_0(\rho)^{-1} \quad \text{see (3.11)} \\ &= -{}^t\tau_0(\rho)^{-1}\cdot{}^t\tau_0(\rho^2)\cdot \nu'(\rho^2)^{-1} \\ &= -{}^t(\tau_0(\rho^2)\tau_0(\rho^{-1}))\cdot \nu'(\rho^2)^{-1} \\ &= -\tau_0(\rho^{-1})\det \tau_0(\rho)\cdot \nu'(\rho^2)^{-1} \\ &= -\alpha^{-1}(\rho)\eta(\rho)\phi\end{aligned}$$

since
$$\begin{aligned}\det \tau_0(\rho) &= \alpha^{-1}(\rho)\omega_\pi(\rho)\\ &= \alpha^{-1}(\rho)\eta(\rho)\nu(\rho)\nu'(\rho^2)^{-1}\\ &= \alpha^{-1}(\rho)\eta(\rho)\,.\end{aligned}$$

So $-\eta(\rho) = +1$. □

Step 2. We use the Lefschetz-criterion, cf. also Ch. VI (8.4). Suppose $\pi \otimes \omega'\nu'$ is a lift. We replace $\pi \otimes \omega'\nu'$ by π and consider $\bigoplus_\sigma \mathbb{H}^2_{DR}(X)(\pi^\sigma)$ where σ runs through $\sigma\colon k_\pi \to \mathbb{C}$. Conjugation ρ of $K/\mathbb{Q}$ induces an involution τ. It acts as -1 on the $++$ and $--$ eigenspaces of $\pi_0(\mathrm{G}(\mathbb{R}))$ and interchanges the $+-$ and $-+$ parts. Consequently, the subspace where τ acts as $+1$ has ℓ_π-rank 1 and consists of $(1,1)$-type classes. Hence it consists of algebraic cycles. □

Step 3. Analysis of the CM-case after [118].

Recall that π is called of CM-type if and only if there is a quadratic character χ of $\mathbb{A}_K^*/K^*$ such that $\pi \otimes \chi \cong \pi$. We have $\chi_\infty(x) = \mathrm{sgn}(x_\infty)\mathrm{sgn}(x'_\infty)$. By class field theory this character defines a quartic extension $M(\chi)$ of $\mathbb{Q}$ which is a CM-field containing K. There exists a grössencharacter Ω of type A_0 such that the associated 2-dimensional Galois representation $\tau = \tau(\pi)$ is given by

$$\tau = \mathrm{Ind}^{\mathrm{Gal}(\bar{\mathbb{Q}}/K)}_{\mathrm{Gal}(\bar{\mathbb{Q}}/M)}\,\Omega\,, \tag{11}$$

where we view Ω as a 1-dimensional Galois representation, cf (3.11).

Now we outline the arguments of Step 3.

i) By analyzing the infinity type of Ω one shows: rank $\mathrm{Ta}_{\bar{\mathbb{Q}}}(\pi) = 2$ if and only if $M(\chi)$ is the compositum of two quadratic fields.

ii) One uses χ to define a non-trivial automorphism $R(\chi)^*$ of $D_B(\pi) \otimes \mathbb{C}$ which preserves the rational structure and which acts on the $(0,2)$ and $(2,0)$ part (resp. on the $(1,1)$ part) of $D_B(\pi) \otimes \mathbb{C}$ by multiplication by G (resp. $-G$), where G is a certain Gauss sum. One deduces that $R(\chi)^*$ acts by the anti-diagonal matrix with entries $G\omega_{ss'}/\omega_{-s-s'}$ $(s, s' \in \{+1, -1\})$.

iii) By ii) one finds $\mathrm{rank}_{\ell_\pi}\mathrm{Ho}(\pi) = \begin{cases}0 & \text{if } G \notin \mathbb{Q}\\ 2 & \text{if } G \in \mathbb{Q}\end{cases}$. By the Lefschetz criterion the Hodge cycles are algebraic.

iv) One shows that $G \in \mathbb{Q}$ if and only if $M(\chi)$ is the compositum of two quadratic fields.

Combination of i) and iii), iv) gives $\mathrm{Ta}_{\bar{\mathbb{Q}}}(\pi) = \mathrm{Ho}(\pi)$ for π of CM-type.

Now we give the details.

(4.7) Proposition. *If π is of CM-type, then* $\mathrm{rank}\,\mathrm{Ta}_{\bar{\mathbb{Q}}}(\pi) = 0$ *or* 2. *Moreover,* $\mathrm{rank}_{\bar{\mathbb{Q}}}(\pi) = 2$ *if and only if $M(\chi)$ is the compositum of two quadratic fields.*

Proof. We write $M = M(\chi)$. Let $\rho \in \mathrm{Gal}(\bar{\mathbb{Q}}/\mathbb{Q})$ be such that $\rho \neq \mathrm{id}$ on K and let M^ρ be the conjugate field extension. Then

$$\tau^\rho = \mathrm{Ind}^{\mathrm{Gal}(\bar{\mathbb{Q}}/K)}_{\mathrm{Gal}(\bar{\mathbb{Q}}/M)}\,\Omega^\rho\,,$$

where

$$\Omega^\rho \colon M^\rho \xrightarrow{\rho^{-1}} M \xrightarrow{\Omega} \mathbb{C}^* .$$

Since τ (resp. τ^ρ) are direct sums of the restrictions of Ω and $\bar{\Omega}$ (resp. Ω^ρ and $\bar{\Omega}^\rho$) to MM^ρ we have

$$r|_{\mathrm{Gal}(\bar{\mathbb{Q}}/MM^\rho)} = \Omega\Omega^\rho \oplus \Omega\bar{\Omega}^\rho \oplus \bar{\Omega}\Omega^\rho \oplus \overline{\Omega\Omega^\rho} . \tag{12}$$

Here $\bar{\Omega}$, $\bar{\Omega}^\rho$ are obtained from Ω, Ω^ρ by complex conjugation of M and M^ρ. We check which of these four characters is of type α^{-1} at infinity, i.e. of type 'norm'. To be precise, we assume that Ω_∞ is given by $\mathbb{C} \times \mathbb{C} \to \mathbb{C}$ $(z, z') \to zz'$ and (μ, μ'): $L \to \mathbb{C} \times \mathbb{C}$, where μ, μ' are two fixed embeddings $L \to \mathbb{C}$ which extend $\sigma, \sigma' \colon K \to \mathbb{R}$. If $M \neq M^\rho$ then none of our four characters can have the infinity type of α^{-1}. So if $M \neq M^\rho$ then $\mathrm{rank}_{\bar{\mathbb{Q}}}(\pi) = 0$.

Assume now that $M = M^\rho$. Then M is either cyclic of degree 4 over $\mathbb{Q}$ or else is the compositum of two quadratic fields. If M is cyclic and ϕ generates $\mathrm{Gal}(M/\mathbb{Q})$ then ϕ^2 is complex conjugation and if (μ, μ') corresponds to (say) $(1, \phi^3)$ then none of the four characters is of type 'norm', hence rank $\mathrm{Ta}_{\bar{\mathbb{Q}}}(\pi) = 0$ again. So we may assume that M is the compositum of two quadratic fields and that we have a diagram

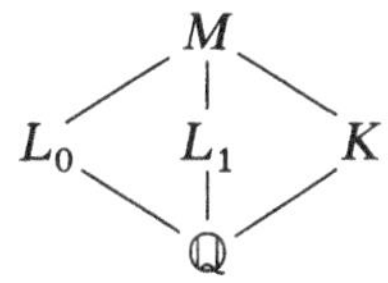

with L_0, L_1 imaginary quadratic fields and with $\rho | L_0 = \mathrm{id}$. We fix an embedding $\mu_0 \colon L_0 \to \mathbb{C}$. Then $\sigma, \sigma' \colon K \to \mathbb{R}$ induce two embeddings $M \to \mathbb{C}$ and we assume that these are μ, μ'. Then $\Omega\bar{\Omega}$, $\Omega^\rho\bar{\Omega}$, $\Omega\bar{\Omega}^\rho$ and $\Omega^\rho\bar{\Omega}^\rho$ are the only ones with infinity type α^{-1}. So exactly two of the four characters in (12), namely $\Omega^\rho\bar{\Omega}$ and $\Omega\bar{\Omega}^\rho$, are of type α^{-1} at infinity. We note that complex conjugation interchanges $\Omega^\rho\bar{\Omega}$ and $\Omega\bar{\Omega}^\rho$ and that they are invariant under $\mathrm{Gal}(M/L_1)$. So we can extend them to characters of $\mathrm{Gal}(\bar{\mathbb{Q}}/L_1)$. Since they are of type A_0 there exists a character γ of finite order on $\mathrm{Gal}(\bar{\mathbb{Q}}/L_1)$ such that

$$\Omega\bar{\Omega}^\rho \oplus \bar{\Omega}\Omega^\rho = \mathrm{Ind}_{\mathrm{Gal}(\bar{\mathbb{Q}}/L_1)}^{\mathrm{Gal}(\bar{\mathbb{Q}}/\mathbb{Q})} \alpha^{-1}\gamma .$$

Over the field extension $L_1(\gamma)$ corresponding to γ the space spanned by $\Omega\bar{\Omega}^\rho$ and $\bar{\Omega}\Omega^\rho$ is a space of Tate classes. □

(4.8) *Note.* Actually, we showed: if π is of CM-type, then rank $\mathrm{Ta}_{\bar{\mathbb{Q}}}(\pi) \neq 0$ if and only if rank $\mathrm{Ta}_{\bar{\mathbb{Q}}}(\pi) = 2$.

For ii) we define a correspondence $R(\mu)$ for any character μ of finite order on $\mathbb{A}_K^*/K^*$ as introduced by Hida. Let $\mathfrak{c}$ be the conductor of μ and let $K_\mathfrak{c} = \prod_{\mathfrak{p}|\mathfrak{c}} K_\mathfrak{p}$, where $K_\mathfrak{p}$ is the completion of K at the prime $\mathfrak{p}$. Let $\mathfrak{o}_\mathfrak{c}$ be the ring of integers of $K_\mathfrak{c}$ and choose a complete set R of representatives for $\mathfrak{c}^{-1}\mathfrak{o}_\mathfrak{c} \bmod \mathfrak{o}_\mathfrak{c}$. We assume that $D(\pi)$ occurs in $\mathbb{H}^2(X_{K_f})$ with multiplicity 1. We write

$$\mathsf{G}(\mathbb{A}) = \bigcup_{j=1}^{m} \mathsf{G}(\mathbb{Q})g_j\mathsf{G}(\mathbb{R})^0\mathsf{K}_f \quad \text{(cf. I (7.4))}.$$

For $\gamma \in \mathsf{G}(\mathbb{A}^f)$ we denote by $T(\gamma)$ the Hecke correspondence, see VI.1. If $\det(\gamma) = 1$ then $T(\gamma)$ preserves the connected components of X_{K_f}. We use the notation $T_j(\gamma)$ for the restriction of $T(\gamma)$ to the component of X_{K_f} corresponding to g_j. Define

$$R(\mu) = \sum_{j=1}^{m} \sum_{r \in R} \mu_f((\det g_j)r) T_j\left(\begin{pmatrix} 1 & r \\ 0 & 1 \end{pmatrix}\right).$$

This is a correspondence on $X_{\mathsf{K}_f} \times X_{\mathsf{K}_f(\mu)}$, where $\mathsf{K}_f(\mu)$ is a certain compact open subgroup of $\mathsf{G}(\mathbb{A}^f)$ contained in K_f. (By twisting the level goes up by at most the square of $\mathfrak{c}$.) One checks:

$$R(\mu)T(\gamma) = \mu_f(\det \gamma) T(\gamma) R(\mu).$$

A correspondence $R(\mu)$ induces a homomorphism $R(\mu)^*\colon \mathbb{H}^2_{\mathbb{A}}(X_{\mathsf{K}_f}) \to \mathbb{H}^2_{\mathbb{A}}(X_{\mathsf{K}_f(\mu)})$ by the formula $x \mapsto p_{1*}(p_2^*(x) \vee \mathrm{cl}(R(\mu)))$, with p_1, p_2 the projections of $X_{\mathsf{K}_f} \times X_{\mathsf{K}_f(\mu)}$ on its factors and $\vee\, \mathrm{cl}(R(\mu))$ the cup product with the cycle class of $R(\mu)$ in $\mathbb{H}^4(X_{\mathsf{K}_f} \times X_{\mathsf{K}_f(\mu)})(2)$.

If π is of CM-type with $\pi \otimes \chi \cong \pi$ then $R(\chi)^*$ induces an automorphism

$$R(\chi)^*\colon D_{\mathbb{A}}(\pi) \to D_{\mathbb{A}}(\pi). \tag{13}$$

We define a Gauss sum associated to χ and to an additive character Ψ on $\mathbb{A}_K$ trivial on K. We let $m_\mathfrak{p}$ (resp. $n_\mathfrak{p}$) be the order of the conductor of χ (resp. Ψ) at $\mathfrak{p}$. (For Ψ this means: Ψ is trivial on $\pi_\mathfrak{p}^{n_\mathfrak{p}}$ but not on $\pi_\mathfrak{p}^{n_\mathfrak{p}-1}$.) The Gauss sum $G(\chi, \Psi)$ is defined as a product

$$G(\chi, \Psi) = \prod_{\mathfrak{p}|\mathfrak{c}} G(\chi_\mathfrak{p}, \Psi_\mathfrak{p})$$

with

$$G(\chi_\mathfrak{p}, \Psi_\mathfrak{p}) = \sum \chi_\mathfrak{p}(x) \Psi_\mathfrak{p}(x\pi_\mathfrak{p}^{-n_\mathfrak{p}-m_\mathfrak{p}}),$$

where the sum is over $\mathfrak{o}_\mathfrak{p}^*/(1 + \pi_\mathfrak{p}^{m_\mathfrak{p}}\mathfrak{o}_\mathfrak{p})$. Then the following holds:

$$G(\chi, \Psi) = \pm\sqrt{\mathsf{N}(\mathfrak{c})} \quad (\in \mathbb{R}).$$

(4.9) Proposition. *Let π be of CM-type. Then there exists an additive character Ψ of $\mathbb{A}_K/K$ such that for every $\sigma\colon \mathfrak{k}_\pi \to \mathbb{C}$ the automorphism $R(\chi)^*$ acts on the* (0, 2) *and* (2, 0)*-part of $D_B(\pi^\sigma) \otimes \mathbb{C}$ by $G(\sigma\chi, \Psi)$ and on the* (1, 1)*-part by* $-G(\sigma\chi, \Psi)$.

Sketch of Proof. Associate to a Hecke eigenform $f = (f_1, \ldots, f_{h^+})$ (corresponding to π) the holomorphic 2-form ω_f which is on the j-th component given by $(2\pi i)^2 f_j\, dz_1 \wedge dz_2$. The action of $R(\chi)^*$ can be computed directly with the help of the Fourier expansion and this gives the eigenvalue on the (0, 2) plus (2, 0)-piece. Moreover, we have

$$R(\chi)^* \varepsilon_i^* = -\varepsilon_i^* R(\chi)^* \qquad i = 1, 2$$

since if we write

$$\varepsilon_1 = \left(\begin{pmatrix} -1 & 0 \\ 0 & 1 \end{pmatrix}, \begin{pmatrix} 1 & 0 \\ 0 & 1 \end{pmatrix}\right)$$

then $\mathsf{G}(\mathbb{Q})g_j\varepsilon_1\mathsf{G}(\mathbb{R})^0\mathsf{K}_f = \mathsf{G}(\mathbb{Q})g_{k(j)}\mathsf{G}(\mathbb{R})^0\mathsf{K}_f$ with $\chi_f(\det g_{k(j)}) = -\chi_f(\det g_j)$. This proves the result. We refer to [118] for an adelic version of the proof. □

By (4.9) and VI (6.6) we know the action of $R(\chi)^*$ on the basis $e^\sigma_{ss'}$: (Note that $G(\sigma\chi, \Psi) = G(\chi, \Psi)$.)

$$R(\mu)^* e^\sigma_{ss'} = G\frac{\omega^\sigma_{ss'}}{\omega^\sigma_{-s-s'}} e^\sigma_{-s-s'} .$$

(4.10) Theorem ([118]). *For π of CM-type there exist numbers $\alpha, \beta \in \mathscr{k}_\pi$ such that*

$$\begin{aligned} \omega^\sigma_{++} &= \alpha^\sigma G\omega^\sigma_{--} \\ \omega^\sigma_{+-} &= \beta^\sigma G\omega^\sigma_{-+} \end{aligned} \qquad (\text{for all } \sigma\colon \mathscr{k}_\pi \to \mathbb{C})$$

where G is a certain Gauss sum $G(\chi, \Psi)$.

Proof. $R(\chi)^*$ is an algebraic correspondence, so it preserves the $\mathbb{Q}$-structure of $\bigoplus_\sigma D(\pi^\sigma)$. Hence the matrix of $R(\chi)^*$ w.r.t. $e^\sigma_{ss'}$ is the σ-th conjugate of a matrix in $\mathscr{k}_\pi$. □

(4.11) Corollary. *We have*: $\operatorname{rank}_{\mathscr{k}_\pi} \operatorname{Ho}(\pi) = \begin{cases} 0 & G \notin \mathbb{Q} \\ 2 & G \in \mathbb{Q} . \end{cases}$

iv) We have for $G = G(\chi, \Psi)$: $G \in \mathbb{R}$, $G^2 = \mathsf{N}_{K/\mathbb{Q}}(\mathfrak{c})$. On the other hand we have the relation

$$D_{M/\mathbb{Q}} = \mathsf{N}_{K/\mathbb{Q}}(\mathfrak{c}) D^2_{K/\mathbb{Q}} . \quad (D_{A/B}\text{: discriminant of } A/B)$$

Therefore, $N_{K/\mathbb{Q}}(\mathfrak{c})$ is a square in $\mathbb{Q}$ if and only if $D_{M/\mathbb{Q}}$ is a square.

(4.12) Lemma. *$D_{M/\mathbb{Q}}$ is a square if and only if M is the compositum of two quadratic fields.*

Proof. We leave this as an exercise to the reader.

We analyze the situation a little further. Let π be of CM-type and suppose that it occurs with multiplicity one on X_{K_f}. We write

$$\mathsf{G}(\mathbb{A}) = \coprod_{j=1}^{m} \mathsf{G}(\mathbb{Q})g_j\mathsf{G}(\mathbb{R})^0\mathsf{K}_f$$

as in I.7. Then $X'_{\mathsf{K}_f} = \coprod_j \Gamma_j \backslash \mathfrak{H}^2$. If $c \in \mathbb{H}^2(X_{\mathsf{K}_f})(\pi)$ we can write

$$c = \sum_j c_j \quad \textit{with } c_j \in \mathbb{H}^2(X_{\Gamma_j}) .$$

We claim that for a CM-form the $(0, 2) + (2, 0)$-part and the $(1, 1)$-part live on only half of the components. To be precise:

(4.13) Proposition. *Let π be of CM-type with $\pi \otimes \chi = \pi$. Let c be a non-zero element of $\mathbb{H}^{p,q}(X_{K_f})$ with $p+q=2$. Then c_j vanishes for all j with $\chi(\det g_j) = -1$ (resp. $\chi(\det g_j) = 1$) if $(p,q) \neq (1,1)$ (resp. $= (1,1)$).*

Proof. We have $R(\chi)^* T(g)^* = \chi(g) T(g)^* R(\chi)^*$. This implies that the Fourier coefficients of f_j in $f = (f_1, \dots, f_m)$ vanish if $\chi \det(g_j) = -1$. So the associated (0, 2)-forms vanish on these components. By using the action of ε_1 and ε_2 the result follows. □

We do not give the proofs of Theorems (4.2) and (4.3). From the short proof in the special case given above one sees that one needs a generalization of the equivalences (4.5) iii) ↔ iv) and i) ↔ iii).

We give the proof of Theorem (4.4).

Proof of (4.4). Let $\phi = \mathrm{Ind}_{\mathrm{Gal}(\bar{\mathbb{Q}}/K)}^{\mathrm{Gal}(\bar{\mathbb{Q}}/\mathbb{Q})} \tau = \mathrm{Ind}_{\mathrm{Gal}(\bar{\mathbb{Q}}/M)}^{\mathrm{Gal}(\bar{\mathbb{Q}}/\mathbb{Q})} \Omega$ with τ the 2-dimensional representation used to split r (see (3.11)). Then

$$\bigwedge^2 \phi = (\Omega\Omega^\rho \oplus \Omega\bar{\Omega}^\rho \oplus \bar{\Omega}\Omega^\rho \oplus \overline{\Omega\Omega^\rho}) \oplus \mathrm{Ind}_{\mathrm{Gal}(\bar{\mathbb{Q}}/K)}^{\mathrm{Gal}(\bar{\mathbb{Q}}/\mathbb{Q})} \omega_\pi$$

(with ω_π the central character of π)

$$= D(\pi)(1) \oplus \mathrm{Ind}_{\mathrm{Gal}(\bar{\mathbb{Q}}/K)}^{\mathrm{Gal}(\bar{\mathbb{Q}}/\mathbb{Q})} \omega_\pi$$

Therefore, $D(\pi)(1)$ is a sum (in the representation ring of finite dimensional representations) of $\mathrm{Gal}(\bar{\mathbb{Q}}/E)$ of monomial representations (i.e. representations induced by 1-dimensional representations of an open subgroup). Then by well-known properties of the L-functions of monomial representations one has:

order of pole at $s = 2$ of $L(D(\pi)/E, s)$ = multiplicity of the trivial representation in $D(\pi)(1)/E$.

But on the other hand we know that the right hand side of this equals the rank of the group of Tate cycles. □

Table 1. Elliptic Fixed Points

We use the following notation: $A = \mathrm{N}(\mathfrak{a})$ with $\mathfrak{a}$ an integral ideal, a_n (resp. a_n^+, a_n^-): the number of fixed points of order n (resp. of type $(n,1,1)$, $(n,1,-1)$); $\to 2n^{\pm}$: the number of fixed points of order n that become of type $(2n,1,\pm)$ for $\Gamma_2(\mathfrak{o}\oplus\mathfrak{a})$; $h(\Delta)$ the usual class number of primitive positive definite binary quadratic forms of discriminant Δ; w: a square-free divisor of D, $\tilde{w}$ the square-free part of D/w.

I. Fixed points of $\Gamma(\mathfrak{o}\oplus\mathfrak{a})$ for $D > 12$

a) Fixed point of order 2

D	a_2	$\to 4^+$	$\to 4^-$
$D \equiv 1 \pmod 4$	$h(-4D)$	0	0
$D \equiv 0 \pmod 8$	$3h(-D)$	$\frac{1}{2}h(-D)$	$\frac{1}{2}h(-D)$
$D/4 \equiv 3 \pmod 8$ $A \equiv 1 \pmod 4$	$10h(-D/4)$	$h(-D/4)$	$3h(-D/4)$
$D/4 \equiv 3 \pmod 8$ $A \equiv 3 \pmod 4$	$10h(-D/4)$	$3h(-D/4)$	$h(-D/4)$
$D/4 \equiv 7 \pmod 8$ $A \equiv 1 \pmod 4$	$4h(-D/4)$	$2h(-D/4)$	0
$D/4 \equiv 7 \pmod 8$ $A \equiv 3 \pmod 4$	$4h(-D/4)$	0	$2h(-D/4)$

b) Fixed points of order 3

D	a_3^+	a_3^-	$\to 6^+$	$\to 6^-$
$D \not\equiv 0 \pmod 3$	$\frac{1}{2}h(-3D)$	$\frac{1}{2}h(-3D)$	0	0
$D/3 \equiv 1 \pmod 3$ $A \equiv 1 \pmod 3$	$4h(-D/3)$	$h(-D/3)$	0	$h(-D/3)$
$D/3 \equiv 1 \pmod 3$ $A \equiv 2 \pmod 3$	$h(-D/3)$	$4h(-D/3)$	$h(-D/3)$	0
$D/3 \equiv 2 \pmod 3$ $A \equiv 1 \pmod 3$	$3h(-D/3)$	0	$h(-D/3)$	0
$D/3 \equiv 2 \pmod 3$ $A \equiv 2 \pmod 3$	0	$3h(-D/3)$	0	$h(-D/3)$

II. Additional fixed points of $\Gamma_w(\mathfrak{o}\oplus\mathfrak{a})$ for $D > 12$

D	w	a_2^{new}
$D \equiv 1 \pmod 8$	$w \equiv 3 \pmod 8$, $w, \tilde{w} \neq 3$	$16h(-w)h(-\tilde{w})$
$D \equiv 1 \pmod 8$	$w \equiv 7 \pmod 8$	$4h(-w)h(-\tilde{w})$
$D \equiv 5 \pmod 8$	$w \equiv 3 \pmod 4$, $w, \tilde{w} \neq 3$	$4h(-w)h(-\tilde{w})$
$D \equiv 1 \pmod 4$	$w \equiv 1 \pmod 4$	$h(-4w)h(-4\tilde{w})$
$D \equiv 1 \pmod 8$	$w = 3$	$5h(-D/3)$

D	w	a_2^{new}
$D \equiv 1 \pmod 8$	$w = 3$	$h(-D/3)$
$D \equiv 5 \pmod 8$	$w \equiv 3 \pmod 8$, $w \neq 3$	$10h(-w)h(-4\tilde{w})$
$D \equiv 0 \pmod 8$	$w \equiv 7 \pmod 8$	$4h(-w)h(-4\tilde{w})$
$D \equiv 0 \pmod 8$	$w \equiv 1 \pmod 4$	$3h(-4w)h(-4\tilde{w})$
$D \equiv 0 \pmod 8$	$w = 3$	$3h(-D/3)$
$D \equiv 4 \pmod 8$	$w \equiv 3 \pmod 8$, $w \neq 3$	$10h(-w)h(-4\tilde{w})$
$D \equiv 4 \pmod 8$	$w \equiv 7 \pmod 8$	$4h(-w)h(-4\tilde{w})$
$D \equiv 4 \pmod 8$	$w \equiv 2 \pmod 4$	$h(-4w)h(-4\tilde{w})$
$D \equiv 4 \pmod 8$	$w = 3$	$3h(-D/3)$

III. *The special cases* D = 5, 8, 12

D = 5. $\Gamma = \mathrm{SL}(2, \mathfrak{o})$ — $a_2 = 2$, $a_3^+ = 1$, $a_3^- = 1$, $a_5 = 2$ (of type (5; 2, 1) and type (5; 3, 1))

D = 8. $\Gamma = \mathrm{SL}(2, \mathfrak{o})$ — $a_2 = 2$ (forms: [1, 0, 8] and [2, 0, 8])
$a_3^+ = 1$ (form [1, 2, 7]), $a_3^- = 1$ (form [8, 8, 8])
$a_4^+ = 1$ (form [2, 4, 6]), $a_4^- = 1$ (form [4, 0, 4])

D = 12. $\Gamma = \mathrm{SL}(2, \mathfrak{o})$ — $a_2 = 3$ (forms: [1, 0, 12] twice, [6, 0, 6] once)
$a_3^+ = 2$ (form [1, 2, 10]), $a_3^- = 0$
$a_6^- = 1$ (form [4, 4, 4])

$\Gamma = \mathrm{SL}(\mathfrak{o} \oplus (\sqrt{3}))$ — Use Remark (7.5) of Ch. V

$\Gamma = \Gamma_2(\mathfrak{o} \oplus \mathfrak{o})$ — $a_2^{\text{old}} = 1$ (form [1, 0, 12])
$a_2^{\text{new}} = 2$ (forms [10, 8, 4] and [6, 0, 12])
$a_4^- = 1$ (form [6, 0, 6])
$a_3^+ = 1$ (form [1, 2, 10])
$a_{12} = 1$ (type (12; 1, 5), form [4, 4, 4])

$\Gamma = \Gamma_2(\mathfrak{o} \oplus (\sqrt{3}))$ — Use Remark (7.5) of Ch. V

Table 2. Numerical Invariants

The table gives the following data: a discriminant D, the sign s of a fundamental unit of $\mathbb{Q}(\sqrt{D})$, a genus γ in the notation of page 3, the number a_2 of quotient singularities of order 2 of $X_\gamma(D)$, the number a_3^+ of quotient singularities of type $(3,1,1)$ of $X_\gamma(D)$, the number a_3^- of quotient singularities of type $(3,1,2)$, the total length l of the resolution cycles of the cusps on $Y_\gamma(D)$, the Euler number e of $Y_\gamma(D)$, the arithmetic genus χ of $Y_\gamma(D)$, the Chern number c_1^2 of $Y_\gamma^\circ(D)$ (just a point if $Y_\gamma(D)$ is rational) and furthermore, if $D \leqslant 12$ or if D is prime, the dimension HZ of the space of Hirzebruch-Zagier cycles, the Picard number ρ and the Lefschetz number L of $Y_\gamma(D)$. See Table 1 for the additional quotient singularities if $D \leqslant 12$.

D	s	γ	a_2	a_3^+	a_3^-	l	e	χ	c_1^2	HZ	ρ	L
5	−	+	2	1	1	1	14	1	·	1	12	0
8	−	+	2	1	1	2	15	1	·	1	13	0
12	+	++	3	2	0	2	16	1	·	1	14	0
12	+	−−	3	0	2	4	16	1	·	1	14	0
13	−	+	2	2	2	3	15	1	·	1	13	0
17	−	+	4	1	1	5	16	1	·	1	14	0
21	+	++	4	4	1	2	18	1	·			
21	+	−−	4	1	4	6	25	2	0			
24	+	++	6	3	0	4	19	1	·			
24	+	−−	6	0	3	8	26	2	0			
28	+	++	4	2	2	4	20	1	·			
28	+	−−	4	2	2	10	26	2	0			
29	−	+	6	3	3	5	28	2	0	2	23	1
33	+	++	4	3	0	8	21	1	·			
33	+	−−	4	0	3	12	28	2	0			
37	−	+	2	4	4	7	29	2	0	2	24	1
40	−	++	6	2	2	12	32	2	0			
40	−	−−	6	2	2	8	28	2	0			
41	−	+	8	1	1	11	30	2	0	2	25	1
44	+	++	10	2	2	6	32	2	0			
44	+	−−	10	2	2	12	38	3	0			
53	−	+	6	5	5	7	40	3	0	3	32	2

D	s	γ	a_2	a_3^+	a_3^-	l	e	χ	c_1^2	HZ	ρ	L
56	+	++	12	2	2	4	34	2	0			
56	+	−−	12	2	2	16	46	4	2			
57	+	++	4	4	1	14	34	2	0			
57	+	−−	4	1	4	18	41	3	0			
60	+	+++	8	6	0	4	30	1				
60	+	−−+	8	0	6	24	56	5	4			
60	+	−+−	8	6	0	12	38	3	0			
60	+	+−−	8	0	6	8	40	3	0			
61	−	+	6	4	4	11	41	3	0	3	33	2
65	−	++	8	2	2	18	44	3	0			
65	−	−−	8	2	2	14	40	3	0			
69	+	++	8	9	0	4	35	2	0			
69	+	−−	8	0	9	16	56	5	4			
73	−	+	4	2	2	21	43	3	0	3	35	2
76	+	++	10	2	2	14	44	3	0			
76	+	−−	10	2	2	20	50	4	2			
77	+	++	8	6	6	2	44	3	0			
77	+	−−	8	6	6	14	56	5	4			
85	−	++	4	6	6	18	56	4	0			
85	−	−−	4	6	6	10	48	4	2			
88	+	++	6	4	4	12	46	3	0			
88	+	−−	6	4	4	24	58	5	4			
89	−	+	12	1	1	21	52	4	2	4	41	3
92	+	++	12	4	4	4	46	3	0			
92	+	−−	12	4	4	22	64	6	8			
93	+	++	4	12	3	6	46	3	0			
93	+	−−	4	3	12	18	67	6	6			
97	−	+	4	2	2	27	53	4	2	4	42	3
101	−	+	14	5	5	11	60	5	4	5	46	4
104	−	++	18	2	2	20	64	5	4			
104	−	−−	18	2	2	12	56	5	6			
105	+	+++	8	6	0	12	46	2	0			
105	+	−−+	8	0	6	16	56	4	0			
105	+	−+−	8	0	6	44	84	8	12			
105	+	+−−	8	6	0	32	66	6	8			
109	−	+	6	6	6	17	61	5	4	5	47	4
113	−	+	8	3	3	23	60	5	6	5	46	4
120	+	+++	12	8	2	8	56	3	0			
120	+	−−+	12	2	8	40	94	9	14			
120	+	−+−	12	8	2	24	72	7	12			
120	+	+−−	12	2	8	8	62	5	4			
124	+	++	12	2	2	16	56	4	2			
124	+	−−	12	2	2	34	74	7	12			
129	+	++	12	4	1	30	74	6	8			
129	+	−−	12	1	4	34	81	7	8			
133	+	++	4	8	8	12	64	5	4			
133	+	−−	4	8	8	24	76	7	10			

D	s	γ	a_2	a_3^+	a_3^-	l	e	χ	c_1^2	HZ	ρ	L
136	+	++	12	2	2	32	74	6	8			
136	+	−−	12	2	2	20	62	6	12			
137	−	+	8	3	3	27	68	6	10	6	51	5
140	+	+++	20	4	4	4	64	4	0			
140	+	−−+	20	4	4	8	68	6	8			
140	+	−+−	20	4	4	40	100	10	20			
140	+	+−−	20	4	4	20	80	8	16			
141	+	++	8	15	0	8	57	4	2			
141	+	−−	8	0	15	28	92	9	16			
145	−	++	8	2	2	40	82	7	10			
145	−	−−	8	2	2	36	78	7	12			
149	−	+	14	7	7	15	78	7	10	7	58	6
152	+	++	18	4	4	12	70	6	10			
152	+	−−	18	4	4	24	82	8	16			
156	+	+++	16	8	2	16	76	5	4			
156	+	−−+	16	2	8	48	114	11	20			
156	+	−+	16	8	2	28	88	9	20			
156	+	+−−	16	2	8	12	78	7	10			
157	−	+	6	8	8	19	77	7	12	7	57	6
161	+	++	16	2	2	14	68	5	4			
161	+	−−	16	2	2	50	104	11	28			
165	+	+++	8	16	4	4	68	4	0			
165	+	−−+	8	4	16	8	84	8	14			
165	+	−+−	8	4	16	44	120	12	24			
165	+	+−−	8	16	4	16	80	8	16			
168	+	+++	12	12	0	8	64	4	4			
168	+	−−+	12	0	12	16	84	8	16			
168	+	−+−	12	12	0	24	80	8	18			
168	+	+−−	12	0	12	48	116	12	28			
172	+	++	10	6	6	26	88	8	16			
172	+	−−	10	6	6	32	94	9	18			
173	−	+	14	9	9	13	86	8	14	8	63	7
177	+	++	4	9	0	32	79	7	16			
177	+	−−	4	0	9	44	100	10	24			
181	−	+	10	6	6	25	85	8	16	8	62	7
184	+	++	12	4	4	16	76	6	8			
184	+	−−	12	4	4	52	112	12	32			
185	−	++	16	2	2	38	96	9	20			
185	−	−−	16	2	2	30	88	9	24			
188	+	++	20	4	4	4	70	6	12			
188	+	−−	20	4	4	34	100	11	32			
193	−	+	4	4	4	47	103	10	24	8	72	11
197	−	+	10	11	11	15	94	9	18	9	68	8
201	+	++	12	4	1	42	102	10	28			
201	+	−−	12	1	4	46	109	11	28			
204	+	+++	20	12	0	28	104	8	12			
204	+	−−+	20	0	12	56	144	14	28			

D	s	γ	a_2	a_3^+	a_3^-	l	e	χ	c_1^2	HZ	ρ	L
204	+	− + −	20	12	0	20	96	10	26			
204	+	+ − −	20	0	12	24	112	12	32			
205	+	+ +	8	10	10	32	110	10	18			
205	+	− −	8	10	10	20	98	10	24			
209	+	+ +	20	2	2	32	102	10	26			
209	+	− −	20	2	2	44	114	12	34			
213	+	+ +	8	21	0	4	71	6	12			
213	+	− −	8	0	21	32	120	13	36			
217	+	+ +	8	4	4	32	100	9	20			
217	+	− −	8	4	4	68	136	15	46			
220	+	+ + +	16	4	4	16	88	7	12			
220	+	− − +	16	4	4	32	104	11	32			
220	+	− + −	16	4	4	64	136	15	44			
220	+	+ − −	16	4	4	32	104	11	32			
221	+	+ +	16	8	8	28	108	10	20			
221	+	− −	16	8	8	12	92	10	28			
229	−	+	10	6	6	29	97	10	28	10	68	9
232	−	+ +	6	6	6	40	108	11	32			
232	−	− −	6	6	6	32	100	11	36			
233	−	+	12	5	5	39	114	12	36	10	77	13
236	+	+ +	30	2	2	18	100	10	28			
236	+	− −	30	2	2	36	118	13	40			
237	+	+ +	12	20	5	10	98	9	20			
237	+	− −	12	5	20	30	133	14	36			
241	−	+	12	2	2	59	133	14	42	10	88	17
248	+	+ +	24	6	6	4	94	9	24			
248	+	− −	24	6	6	40	130	15	50			
249	+	+ +	12	9	0	48	127	13	40			
249	+	− −	12	0	9	60	148	16	48			
253	+	+ +	4	12	12	10	98	9	20			
253	+	− −	4	12	12	46	134	15	46			
257	−	+	16	3	3	39	116	13	46	11	76	14
264	+	+ + +	24	8	2	32	124	11	28			
264	+	− − +	24	2	8	16	114	13	44			
264	+	− + −	24	8	2	32	124	15	56			
264	+	+ − −	24	2	8	64	162	17	46			
265	−	+ +	8	2	2	62	136	15	52			
265	−	− −	8	2	2	58	132	15	54			
268	+	+ +	10	6	6	38	120	13	44			
268	+	− −	10	6	6	44	126	14	46			
269	−	+	22	7	7	21	112	12	36	12	77	11
273	+	+ + +	8	8	2	28	108	10	32			
273	+	− − +	8	2	8	84	170	20	72			
273	+	− + −	8	2	8	24	110	12	42			
273	+	+ − −	8	8	2	64	144	18	72			
277	−	+	6	14	14	29	133	14	40	12	90	15

D	s	γ	a_2	a_3^+	a_3^-	l	e	χ	c_1^2	HZ	ρ	L
280	+	+ + +	12	4	4	24	104	10	32			
280	+	− − +	12	4	4	40	120	14	52			
280	+	− + −	12	4	4	72	152	18	64			
280	+	+ − −	12	4	4	40	120	14	52			
281	−	+	20	3	3	49	142	16	56	12	91	19
284	+	+ +	28	4	4	12	110	11	32			
284	+	− −	28	4	4	54	152	18	64			
285	+	+ + +	16	24	0	4	100	8	16			
285	+	− − +	16	0	24	20	140	16	52			
285	+	− + −	16	0	24	60	180	20	60			
285	+	+ − −	16	24	0	12	108	12	38			
293	−	+	18	11	11	17	120	13	40	13	82	12
296		+ +	30	6	6	40	152	16	48			
296	−	− −	30	6	6	28	140	16	54			
301	+	+ +	8	8	8	30	118	13	46			
301	+	− −	8	8	8	42	130	15	52			
305	+	+ +	16	4	4	56	152	17	60			
305	+	− −	16	4	4	44	140	17	68			
309	+	+ +	12	20	5	22	126	13	40			
309	+	− −	12	5	20	42	161	18	56			
312	+	+ + +	12	18	0	8	102	9	28			
312	+	− − +	12	0	18	80	192	23	84			
312	+	− + −	12	18	0	40	134	17	70			
312	+	+ − −	12	0	18	16	128	15	56			
313	−	+	8	4	4	65	161	19	74	13	99	24
316	+	+ +	20	6	6	36	148	16	54			
316	+	− −	20	6	6	66	178	21	76			
317	−	+	10	13	13	21	126	14	46	14	85	13
321	+	+ +	20	9	0	56	167	19	72			
321	+	− −	20	0	9	68	188	22	80			
328	−	+ +	12	6	6	56	154	17	60			
328	−	− −	12	6	6	32	130	17	76			
329	+	+ +	24	4	4	24	140	15	52			
329	+	− −	24	4	4	84	200	25	100			
332	+	+ +	30	6	6	18	132	15	56			
332	+	− −	30	6	6	36	150	18	68			
337	−	+	8	6	6	71	185	22	86	14	112	29
341	+	+ +	28	8	8	6	122	13	42			
341	+	− −	28	8	8	42	158	19	70			
344	+	+ +	30	4	4	32	146	17	66			
344	+	− −	30	4	4	44	158	19	72			
345	+	+ + +	8	8	2	28	132	14	56			
345	+	− − +	8	2	8	28	138	16	64			
345	+	− + −	8	2	8	108	218	28	118			
345	+	+ − −	8	8	2	92	196	26	116			
348	+	+ + +	24	18	0	12	130	13	46			
348	+	− − +	24	0	18	72	208	25	92			

D	s	γ	a_2	a_3^+	a_3^-	l	e	χ	c_1^2	HZ	ρ	L
348	+	− + −	24	18	0	36	154	19	76			
348	+	+ − −	24	0	18	24	160	19	72			
349	−	+	14	8	8	37	143	17	66	15	91	18
353	−	+	16	3	3	47	148	19	86	15	88	22
357	+	+ + +	8	30	0	4	110	10	30			
357	+	− − +	8	0	30	68	204	24	84			
357	+	− + −	8	0	30	24	160	20	80			
357	+	+ − −	8	30	0	8	114	14	56			
364	+	+ + +	20	4	4	48	164	18	68			
364	+	− − +	20	4	4	84	200	24	92			
364	+	− + −	20	4	4	40	156	20	88			
364	+	+ − −	20	4	4	52	168	22	96			
365	−	+ +	20	14	14	38	172	18	52			
365	−	− −	20	14	14	14	148	18	68			
373	−	+	10	16	16	37	175	20	70	16	112	23
376	+	+ +	24	4	4	24	148	17	68			
376	+	− −	24	4	4	84	208	27	116			
377	+	+ +	16	4	4	56	168	21	92			
377	+	− −	16	4	4	44	156	21	100			
380	+	+ + +	32	8	8	8	148	15	48			
380	+	− − +	32	8	8	40	180	23	96			
380	+	− + −	32	8	8	80	220	27	104			
380	+	+ − −	32	8	8	16	156	19	76			
381	+	+ +	20	20	5	22	150	17	64			
381	+	− −	20	5	20	42	185	22	80			
385	+	+ + +	8	4	4	48	172	20	84			
385	+	− − +	8	4	4	92	216	28	124			
385	+	− + −	8	4	4	68	192	24	104			
385	+	+ − −	8	4	4	120	244	32	140			
389	−	+	22	11	11	31	162	19	70	17	104	20
393	+	+ +	12	15	0	56	185	23	102			
393	+	− −	12	0	15	76	220	28	120			
397	−	+	6	12	12	35	153	19	80	17	95	20
401	−	+	20	3	3	65	194	25	112	17	112	32
408	+	+ + +	12	16	4	40	164	18	72			
408	+	− − +	12	4	16	80	216	26	100			
408	+	− + −	12	16	4	32	156	22	108			
408	+	+ − −	12	4	16	24	160	22	106			
409	−	+	16	2	2	87	225	29	130	17	127	40
412	+	+ +	20	6	6	36	168	21	94			
412	+	− −	20	6	6	66	198	26	116			
413	+	+ +	20	16	16	4	152	17	60			
413	+	− −	20	16	16	40	188	23	88			
417	+	+ +	12	12	3	62	202	26	120			
417	+	− −	12	3	12	74	223	29	130			
421	−	+	10	10	10	47	175	22	94	18	106	25

D	s	γ	a_2	a_3^+	a_3^-	l	e	χ	c_1^2	HZ	ρ	L
424	−	+ +	18	6	6	68	208	26	112			
424	−	− −	18	6	6	60	200	26	116			
428	+	+ +	30	10	10	26	180	22	92			
428	+	− −	30	10	10	44	198	25	104			
429	+	+ + +	16	30	0	8	146	14	44			
429	+	− − +	16	0	30	96	264	32	120			
429	+	− + −	16	0	30	20	188	24	100			
429	+	+ − −	16	30	0	28	166	22	98			
433	−	+	12	4	4	81	225	30	142	18	124	41
437	+	+ +	20	14	14	2	144	17	68			
437	+	− −	20	14	14	38	180	23	96			
440	+	+ + +	36	4	4	8	148	17	72			
440	+	− − +	36	4	4	40	180	25	120			
440	+	− + −	36	4	4	80	220	29	128			
440	+	+ − −	36	4	4	16	156	21	100			
444	+	+ + +	32	8	2	32	180	21	92			
444	+	− − +	32	2	8	88	242	31	132			
444	+	− + −	32	8	2	60	208	29	140			
444	+	+ − −	32	2	8	20	174	23	106			
445	−	+ +	8	14	14	56	202	24	94			
445	−	− −	8	14	14	36	182	24	108			
449	−	+	20	3	3	71	216	29	138	19	120	38
453	+	+ +	12	28	7	14	158	19	80			
453	+	− −	12	7	28	42	207	26	106			
456	+	+ + +	24	18	0	48	202	23	94			
456	+	− − +	24	0	18	40	212	29	136			
456	+	− + −	24	18	0	40	194	27	132			
456	+	+ − −	24	0	18	96	268	33	132			
457	−	+	8	6	6	89	247	33	156	19	135	46
460	+	+ + +	20	4	4	20	160	20	96			
460	+	− − +	20	4	4	24	164	22	108			
460	+	− + −	20	4	4	104	244	34	164			
460	+	+ − −	20	4	4	84	224	32	160			
461	−	+	30	9	9	29	174	22	94	20	107	23
465	+	+ + +	16	12	0	36	192	24	116			
465	+	− − +	16	0	12	44	212	28	132			
465	+	− + −	16	0	12	124	292	40	188			
465	+	+ − −	16	12	0	100	256	36	178			
469	+	+ +	16	12	12	42	198	25	110			
469	+	− −	16	12	12	54	210	27	116			
472	+	+ +	18	8	8	36	190	25	120			
472	+	− −	18	8	8	72	226	31	148			
473	+	+ +	12	6	6	56	202	28	142			
473	+	− −	12	6	6	68	214	30	150			
476	+	+ + +	40	4	4	8	168	18	68			
476	+	− − +	40	4	4	116	276	36	156			
476	+	− + −	40	4	4	32	192	28	144			

D	s	γ	a_2	a_3^+	a_3^-	l	e	χ	c_1^2	HZ	ρ	L
476	+	+ − −	40	4	4	20	180	26	132			
481	−	+ +	16	4	4	106	282	37	172			
481	−	− −	16	4	4	90	266	37	182			
485	−	+ +	20	14	14	46	204	24	92			
485	−	− −	20	14	14	18	176	24	112			
488	−	+ +	30	6	6	44	192	25	116			
488	−	− −	30	6	6	24	172	25	130			
489	+	+ +	20	4	1	82	246	34	172			
489	+	− −	20	1	4	86	253	35	172			
492	+	+ + +	20	24	0	44	204	24	104			
492	+	− − +	20	0	24	88	272	34	140			
492	+	− + −	20	24	0	20	180	26	134			
492	+	+ − −	20	0	24	40	224	32	160			
493	−	+ +	12	14	14	50	208	26	112			
493	−	− −	12	14	14	30	188	26	126			
497	+	+ +	24	6	6	18	188	25	124			
497	+	− −	24	6	6	102	272	39	196			

Bibliography

1. Altman, A., Kleiman S.: Introduction to Grothendieck Duality theory. Lecture Notes in Math. *146*. Springer-Verlag 1970.
2. Arthur, J.: The Selberg trace formula for groups of F-rank one. Ann. of Math. *100*, 326–385 (1974).
3. Artin, M.: Algebraization of formal moduli. I in: Global Analysis. Princeton Univ. Press 1969. II Ann. of Math. *99*, 88–135 (1970).
4. Artin, M.: The implicit function theorem in algebraic geometry. In: Algebraic Geometry. Oxford, Bombay, 1969.
5. Asai, T.: On the Fourier coefficients of automorphic forms at various cusps and some applications to Rankin's convolution. J. Math. Soc. Japan *28* (1), 48–61 (1976).
6. Asai, T.: On certain Dirichlet series associated with Hilbert modular forms and Rankin's method. Math. Ann. *226*, 81–94 (1977).
7. Atiyah, M., Donnelly, H., Singer, I. M.: Eta invariants, signature defects of cusps and the values of L-functions. Ann of Math. *118*, 131–177 (1983).
8. Ash, A., Mumford, D., Rapoport, M., Tai, Y.: Smooth compactification of locally symmetric varieties. Math. Sci. Press, Brookline, Mass. 1975.
9. Baily, W., Borel, A.: Compactification of arithmetic quotients of bounded symmetric domains. Ann. of Math. *84*, 442–528 (1966).
10. Barth, W., Peters, C., Van de Ven, A.: Compact complex surfaces. Ergebnisse der Math. 4. Springer-Verlag 1985.
11. Bassendowski, D.: Klassifikation Hilbertscher Modulflächen zur symmetrischen Hurwitz-Maass-Erweiterung. Bonner Math. Schriften *163* (1985).
12. Beauville, A.: L'application canonique pour les surfaces de type général Inv. Math. *55*, 121–140 (1979).
13. Blumenthal, O.: Über Modulfunktionen von mehreren Veränderlichen. Math. Ann. *56*, 509–548 (1903) and *58*, 497–527 (1904).
14. Borel, A.: Automorphic L-functions. In: Proc. Symp. Pure Math AMS *33*, vol 2, 27–62 (1979).
15. Borewicz, S. I., Shafarevich, I. R.: Zahlentheorie. Birkhäuser Verlag, Basel 1966.
16. Brylinski, J.-L., Labesse, J.-P.: Cohomologie d'intersection et fonctions L de certaines variétés de Shimura. Ann. Sci. E.N.S. 4ème série, *17*, 361–412 (1984)
17. Chern, S. S.: On the curvatura integra in a Riemannian manifold. Ann. of Math. *46* (4), 647–684 (1945).
18. Cohen, H.: Variations sur un thème de Siegel et Hecke. Acta Arithm. *30*, 63–93 (1976).
19. Cohen, H.: A lifting of modular forms in one variable to Hilbert modular forms in two variables. In: Modular Functions of One Variable VI. Lecture Notes in Math. *627*, 175–196. Springer-Verlag 1976.
20. Deligne, P.: Formes modulaires et représentations ℓ-adiques. Sém. Bourbaki 1968/1969, exp. 355, 139–172. Lecture Notes in Math. *179*. Springer-Verlag 1971.
21. Deligne, P.: Travaux de Shimura. Sém. Bourbaki 1970/71, exp. 389. Lecture Notes in Math. *244*. Springer-Verlag 1971.
22. Deligne, P.: Théorie de Hodge II, III. Publ. Math. I.H.E.S. *40*, 5–58 (1971) *44*, 5–77 (1974)
23. Deligne, P.: Variétés de Shimura: Interprétation modulaire et technique de construction de modèles canoniques. Proc. Symp. Pure Math. AMS *33*, vol. 2, 247–290 (1979).

24. Deligne, P.: Valeurs de fonctions L et périodes d'integrales. Proc. Symp. Pure Math AMS *33*, vol 2, 313–346 (1979)
25. Deligne, P., Milne, J., Ogus, A., Shih, K.: Hodge cycles, Motives and Shimura varieties. Lecture Notes in Math. *900*. Springer-Verlag 1982.
26. Deligne, P., Mumford, D.: The irreducibility of the space of curves of given genus. Publ. Math. I.H.E.S. *36*, 75–100 (1969).
27. Doi, K., Naganuma, H.: On the functional equation of certain Dirichlet series. Invent. Math. *9*, 1–14 (1969).
28. Ehlers, F.: Eine Klasse komplexer Mannigfaltigkeiten und die Auflösung einiger isolierter Singularitäten. Math. Ann *218*, 127–156 (1975).
29. Faltings, G.: Arithmetic varieties and rigidity. Sém. Delange-Pisot-Poitou 1984.
30. Faltings, G.: Arithmetische Kompaktifizierung des Modulraumes der abelschen Varietäten. In: Arbeitstagung Bonn 1984. Lecture Notes in Math. *1111*, 321–383. Springer-Verlag 1985.
31. Faltings, G.: Notes of a course given in Princeton 1985.
32. Franke, H.-G.: Kurven in Hilbertschen Modulflächen und Humbertsche Flächen im Siegelraum. Bonner Math. Schriften *104* (1978).
33. Freitag, E.: Über die Struktur der Funktionenkörper zu hyperabelschen Gruppen I, II. J.f.d.r.u.a. Math. *247*, 97–117 (1971); *254*, 1–16 (1972).
34. Freitag, E.: Eine Bemerkung zur Theorie der Hilbertschen Modulmannigfaltigkeiten hoher Stufe. Math. Zeitschrift *171*, 27–35 (1980).
35. Freitag, E., Schneider, V.: Bemerkung zu einem Satz von J. Igusa und W. Hammond. Math. Zeitschrift *102*, 9–16 (1967).
36. van der Geer, G.: Hilbert modular forms for the field $\mathbb{Q}(\sqrt{6})$. Math. Ann. *233*, 163–179 (1978).
37. van der Geer, G.: Minimal models for Hilbert modular surfaces of principal congruence subgroups. Topology *18*, 29–39 (1979).
38. van der Geer, G.: On the geometry of a Siegel modular threefold. Math. Ann. *260*, 317–350 (1982).
39. van der Geer, G.: Note on abelian schemes of level three. To appear.
40. van der Geer, G., Ueno, K.: Families of abelian surfaces with real multiplication over Hilbert modular surfaces. Nagoya Math. Journal *88*, 17–53 (1982)
41. van der Geer, G., Van de Ven, A.: On the minimality of certain Hilbert modular surfaces. In: Complex Analysis and Algebraic Geometry. Iwanami Shoten and Cambridge University Press 1977 137–150.
42. van der Geer, G., Zagier, D.: The Hilbert modular group for the field $\mathbb{Q}(\sqrt{13})$. Invent. Math. *42*, 93–133 (1977).
43. Gelbart, S.: Automorphic forms on adele groups. Ann. of Math Studies *83*. Princeton Univ. Press 1975.
44. Gelbart, S., Jacquet, H.: A relation between automorphic forms on GL(2) and GL(3). Ann. Sci. E.N.S. 4ème série t. *11*, 471–542 (1978).
45. Gérardin, P., Labesse, J.-P.: The solution to the base change problem for GL(2) (following Langlands, Saito, Shintani). Proc. Symp. Pure Math. AMS *33*, vol. 2, 115–133 (1979).
46. Giraud, J.: Surfaces d'Hilbert-Blumenthal. In: Surfaces algébriques. Lecture Notes in Math. 868 (1981).
47. Grauert, H.: Über Modifikationen und exzeptionelle analytische Mengen. Math. Ann. *146*, 331–368 (1962).
48. Griffiths, P., Harris, J.: Principles of algebraic geometry. John Wiley & Sons, New York 1978.
49. Gross, B., Zagier, D.: Heegner points and derivatives of *L*-series. Invent. Math. *84* 225–320 (1986).
50. Grothendieck, A.: Groupes de Barsotti-Tate et cristaux. Actes du Congres Int. Math. Nice 1970, t. 1, 431–436.
51. Gundlach, K-B.: Some new results in the theory of Hilbert's modular group. In: Contributions to function theory. Tata Institute Bombay, 1960, 165–180.
52. Gundlach, K-B.: Die Bestimmung der Funktionen zur Hilbertschen Modulargruppe des Zahlkörpers $\mathbb{Q}(\sqrt{5})$. Math. Ann. *152*, 226–256 (1963).
53. Gundlach, K-B.: Die Bestimmung der Funktionen zu einigen Hilbertschen Modulgruppen. J. f. d. r. u. a. Math. *220*, 109–153 (1965).

54. Gundlach, K-B.: Die Berechnung von Zetafunktionen mit Vorzeichencharakter an der Stelle 1. Acta Arithm. *24*, 201–221 (1973).
55. Gundlach, K-B.: Poincarésche und Eisensteinsche Reihen zur Hilbertschen Modulgruppe. Math. Zeit. *64*, 339–352 (1956)
56. Hammond, W.: The modular groups of Hilbert and Siegel. Amer. J. of Math. *88*, 497–516 (1966).
57. Hammond, W.: Chern numbers of 2-dimensional Satake compactifications. J. London Math. Soc. *14* (2), 65–70 (1976).
58. Hammond, W.: The two actions of the Hilbert modular group. Amer. J. of Math. *99*, 389–392 (1977).
59. Hammond, W., Hirzebruch, F.: *L*-series, modular embeddings and signatures. Math. Ann. *204*, 263–270 (1973)
60. Harder, G.: A Gauss-Bonnet formula for discrete arithmetically defined groups. Ann. Sci. E.N.S. 4, 409–455 (1971)
61. Harder, G., Langlands, R., Rapoport, M.: Algebraische Zykeln auf Hilbert Blumenthal-Flächen. J. f. d. r. u. a. Maths. *336*, 53–120 (1986).
62. Hausmann, W.: Kurven auf Hilbertschen Modulflächen. Bonner Math. Schriften *123* (1980).
63. Hausmann, W.: The fixed points of the symmetric Hilbert modular group of a real quadratic field with arbitrary discriminant. Math. Ann. *260*, 31–50 (1982).
64. Hecke. E.: Höhere Modulfunktionen und ihre Anwendung auf die Zahlentheorie. Math. Ann. *71*, 1–37 (1912) (= Werke p. 21.)
65. Hecke, E.: Analytische Funktionen und algebraische Zahlen. Abhandl. Math. Sem. Hamburg *3*, 213–236 (1924) (= Werke p. 381).
66. Hecke, E.: Über ein Fundamentalproblem aus der Theorie der elliptischen Modulfunktionen. Abhandl. Math. Sem Hamburg *6*, 235–257 (1928) (= Werke p. 525).
67. Hecke, E.: Vorlesungen über Zahlentheorie. Chelsea 1946, Springer-Verlag 1981.
68. Hejhal, D.: The Selberg trace formula for PSL(2, $\mathbb{R}$), 1. Lecture Notes in Math. *548*. Springer-Verlag 1976.
69. Helling, H.: Bestimmung der Kommensurabilitätsklasse der Hilbertschen Modulgruppe. Math. Zeit. *92*, 269–280 (1966).
70. Hermann, C. F.: Thetareihen und modulare Spitzenformen zu den Hilbertschen Modulgruppen reell-quadratischer Körper. Math. Ann *277*, 327–344 (1987).
71. Hirzebruch, F.: Über vierdimensionale Riemannsche Flächen mehrdeutiger analytischer Funktionen von zwei komplexen Veränderlichen. Math. Ann. *126*, 1–22 (1953).
72. Hirzebruch, F.: Topological methods in algebraic geometry. Grundlehren *131*. Springer-Verlag 1966.
73. Hirzebruch, F.: Automorphe Formen und der Satz von Riemann-Roch. In: Symp. Int. Top. Alg. 1956 Universidad de Mexico. (1958), 129–144.
74. Hirzebruch, F.: Hilbert modular surfaces. L'Ens. Math. *71*, 183–281 (1973).
75. Hirzebruch, F.: The Hilbert modular group, resolution of the singularities at the cusps and related problems. Sém. Bourbaki 1970/71, exp. 396. In: Lecture Notes in Math. *244*. Springer-Verlag (1971).
76. Hirzebruch, F.: The Hilbert modular group for the field $\mathbb{Q}(\sqrt{5})$ and the cubic diagonal surface of Clebsch and Klein. Usp. Mat. Nauk *31*, 153–166 (1976) (in Russian) (Russian Math. Surveys *31* (5), 96–110 (176).
77. Hirzebruch, F.: The ring of Hilbert modular forms for real quadratic fields of small discriminant. In: Modular Functions of One Variable VI. Lecture Notes in Math. *627*, 287–324. Springer-Verlag 1976.
78. Hirzebruch, F.: Modulflächen und Modulkurven zur symmetrischen Hilbertschen Modulgruppe. Ann. Sci. E.N.S. *11*, 101–166 (1978).
79. Hirzebruch, F.: The canonical map for certain Hilbert modular surfaces. In: Proc. Chern Symp. 1979. Springer-Verlag 1981.
80. Hirzebruch, F.: Überlagerungen der projektiven Ebene und Hilbertsche Modulflächen. L'Ens. Math. *24*, 63–78 (1978).
81. Hirzebruch, F., van der Geer, G.: Lectures on Hilbert modular surfaces. Les Presses de l'Univ. de Montréal 1981.
82. Hirzebruch, F., Van de Ven, A.: Hilbert modular surfaces and the classification of algebraic surfaces. Invent. Math. *23*, 1–29 (1974).

83. Hirzebruch, F., Van de Ven, A.: Minimal Hilbert modular surfaces with $p_g = 3$ and $K^2 = 2$. Amer. J. of Math. *101*, 132–148 (1979).
84. Hirzebruch, F., Zagier, D.: The Atiyah-Singer theorem and elementary number theory. Math. Lecture Series *3*. Publish or Perish, Boston 1974.
85. Hirzebruch, F., Zagier. D.: Intersection numbers of curves on Hilbert modular surfaces and modular forms of Nebentypus. Invent. Math. *36*, 57–113 (1976).
86. Hirzebruch, F., Zagier, D.: Classification of Hilbert modular surfaces. In: Complex Analysis and Algebraic Geometry. Iwanami Shoten and Cambridge Univ. Press 1977, 43–77.
87. Horikawa, E.: Algebraic surfaces of general type with small c_1^2. I, II, III, IV. Ann. of Math. *104*, 357–387 (1976); Invent. Math. *37*, 121–155 (1976); Invent. Math. *47*, 209–248 (1978); Invent. Math. *50*, 103–128 (1979).
88. Humbert, G.: Sur les fonctionnes abéliennes singulières. I, II, III. J. Math. Pures Appl. serie 5, t. V, 233–350 (1899); t. VI, 279–386 (1900); t. VII, 97–123 (1901).
89. Igusa, J-I.: A desingularization problem in the theory of Siegel modular functions. Math. Ann. *168*, 228–260 (1967).
90. Igusa, J-I.: On the ring of modular forms of degree two over $\mathbb{Z}$. Amer. J. of Math. *101*, 132–148 (1979).
91. Iyanaga, K.: Arithmetic of special unitary groups and their symplectic representations. J. Fac. of Sci. Tokyo *15*, 35–69 (1968).
92. Jacobowitz, R.: Hermitian forms over local fields. Amer. J. of Math. *84*, 441–465 (1962).
93. Jacquet, H., Langlands, R.: Automorphic forms on GL(2). Lecture Notes in Math. *114*. Springer-Verlag 1970.
94. Katz, N., Mazur, B.: Arithmetic moduli of elliptic curves. Ann. of Math. Studies *108*. Princeton Univ. Press 1985.
95. Klingen, H.: Über die Werte der Dedekindschen Zetafunktion. Math. Ann. *145*, 165–272 (1962).
96. Klingenberg, C.: Die Tate-Vermutungen für Hilbert-Blumenthal-Flächen. Invent. Math. *89*, 291–318 (1987).
97. Kloosterman, H.: Thetareihen in total-reellen algebraischen Zahlkörpern. Math. Ann. *103*, 279–299 (1930).
98. Knöller, F.: Beispiele dreidimensionaler Hilbertscher Mannigfaltigkeiten von allgemeinem Typ. Manuscr. Math. *37*, 135–161 (1982).
99. Knöller, W.: Über die Plurigeschlechter Hilbertscher Modulmannigfaltigkeiten. Math. Ann. *264*, 413–422 (1983).
100. Koehl, J.: Algebraische Zykel auf Hilbertschen Modulflächen. Dissertation, Bonn 1987.
101. Kohnen, W.: Modular forms of half-integral weight on $\Gamma_0(4)$. Math. Ann. *248*, 249–266 (1980).
102. Koll, F-J.: Die elliptischen Fixpunkte und die Spitzen der diskreten Erweiterungen der Hilbertschen Modulgruppe. Bonner Math. Schriften *84* (1976).
103. Kudla, S.: Intersection numbers for quotients of the complex ball and Hilbert modular forms. Invent. Math. *47*, 189–208 (1978).
104. Kudla, S.: Relations between automorphic forms produced by theta functions. In. Modular Functions of One Variable VI, 277–286, Lecture Notes in Math. 627. Springer-Verlag 1976.
105. Kudla, S., Millson, J.: Geodesic cycles and the Weil representation. I. Quotients of hyperbolic space and Siegel modular forms. Comp. Math. *45*, 207–271 (1982).
106. Lang, S.: Algebra. Addison-Wesley, Reading, Mass. 1965 (6th pr. 1974).
107. Langlands, R.: Base change for GL(2). Ann. of Math. Studies *96*. Princeton Univ. Press 1980.
108. Laufer, H. B.: Normal 2-dimensional singularities. Ann. of Math. Studies *71*. Princeton Univ. Press 1971.
109. Lion, G., Vergne, M.: The Weil representation, Maslov index and theta series. Progress in Math. *6*. Birkhäuser Verlag 1980.
110. Maass, H.: Über Gruppen von hyperabelschen Transformationen. Sitzungsber. Heidelb. Akad Wiss., 3–26 (1940).
111. Maass, H.: Über die Erweiterungsfähigkeit der Hilbertschen Modulgruppe. Math. Zeit. *51*, 255–261 (1948).
112. Meyer, C.: Die Berechnung der Klassenzahl abelscher Körper über quadratischen Zahlkörpern. Berlin 1957.

113. Meyer, W., Szcech, R.: Uber eine topologische und eine zahlentheoretische Anwendung von Hirzebruch's Spitzenauflösung. Math. Ann. *240*, 69–96 (1979).
114. Miyake, T.: On automorphic forms on GL(2) and Hecke operators. Ann. of Math. *94*, 174–189 (1971).
115. Mumford, D.: Abelian varieties. Oxford University Press 1970.
116. Mumford, D.: An analytic construction of degenerating abelian varieties over complete local rings. Comp. Math. *24*, 239–272 (1972)
117. Mumford, D.: Hirzebruch's proportionality in the non-compact case. Invent. Math. *42*, 239–272 (1977).
118. Murty, V. K., Ramakrishnan. D.: Period relations and the Tate conjecture for Hilbert modular surfaces. Invent. Math. *89*, 319–346 (1987).
119. Naganuma, H.: On the coincidence of two Dirichlet series associated with cusp forms of Hecke's "Neben"-type and Hilbert modular forms over a real quadratic field. J. Math. Soc. Japan 25, 547–555 (1973).
120. Nagaoka, S.: On the ring of Hilbert modular forms over $\mathbb{Z}$. J. Math. Soc. Japan *35*, 589–608 (1983) + Correction.
121. Oda, Tadao: The first de Rham cohomology group and Dieudonné modules. Ann. Sci. E.N.S. *2*, 63–135 (1969).
122. Oda, T.: On modular forms associated with indefinite quadratic forms of signature $(2, n-2)$. Math. Ann. *231*, 97–144 (1977).
123. Oda, T.: Periods of Hilbert modular surfaces. Proc. Japan Acad. *57* Ser A (8), 415–419.
124. Oda, T.: Periods of Hilbert modular surfaces. Progress in Math. *19*, Birkhäuser Verlag 1982.
125. Oda, T.: Hodge structures of Shimura varieties attached to the unit groups of quaternion algebras. In: Galois groups and their representations. Adv. Studies in Pure Math. *2*, 15–36 (1983).
126. Ogg, A.: Hyperelliptic modular curves. Bull. Soc. Math. France *102*, 449–462 (1974).
127. Perron. O.: Die Lehre von den Kettenbrüchen. Leipzig & Berlin 1913.
128. Prasad, G.: Strong approximation for semi-simple groups over function fields. Ann. of Math. *105*, 553–572 (1977).
129. Prestel, A.: Die elliptischen Fixpunkte der Hilbertschen Modulgruppen. Math. Ann. *177*, 181–209 (1968).
130. Prestel, A.: Die Fixpunkte der symmetrischen Hilbertschen Modulgruppe zu einem reell-quadratischen Körper mit Primzahldiskriminante. Math. Ann. *200*, 123–139 (1973).
131. Ramakrishnan, D.: Arithmetic of Hilbert-Blumenthal surfaces. Proc. CMS conference on number theory, 7 (1987).
132. Ramakrishnan, D.: Valeurs de fonctions L des surfaces d'Hilbert-Blumenthal en $s = 1$. C. R. Acad. Sci. Paris t. 301. Série I, N° 18 809–912 (1985).
133. Rapoport, M.: Compactifications de l'espace de modules de Hilbert-Blumenthal Comp. Math. *36*, 255–335 (197).
134. Raynaud, M.: Variétés abéliennes et géometrie rigide. Actes du Congres Int. Math. Nice, t. I, 437–477.
135. Resnikoff, H. L.: On the graded ring of Hilbert modular forms associated with $\mathbb{Q}(\sqrt{5})$. Math. Ann. *208*, 161–170 (1974).
136. Ribet, K.: Galois representations attached to eigenforms with Nebentypus. In: Modular Functions of One Variable VI. Lecture Notes in Math. *627*, 18–52. Springer-Verlag 1977.
137. Rosati, M.: Su una generazione del gruppo simplettico modulare di Siegel Conforto $\Gamma(n, \Delta)$. Rend. Sem. Mat. Padova *60*, 229–236 (1978).
138. Saito, H.: Automorphic forms and algebraic extensions of number fields. Kinikuniya Bookstore. Tokyo 1975.
139. Saito, H.: On the representation of $SL(2, \mathbb{F}_q)$ in the space of Hilbert modular forms. J. Math. Univ. Kyoto *15*, 101–128 (1975).
140. Schlessinger, M.: Functors of Artin rings. Trans. AMS *130*, 205–222 (1968)
141. Sczech, R.: Zur Summation von L-Reihen. Bonner Math. Schriften 141 (1982).
142. Selberg, A.: Recent developments in the theory of discontinuous groups of motions of symmetric spaces. In: Lecture Notes in Math. *118*, Springer-Verlag (1970).

143. Serre, J-P.: Abelian *l*-adic representations. Benjamin, New York 1968.
144. Serre, J-P.: Le problème des groupes de congruence pour SL_2. Ann. of Math. *92*, 489–527 (1970).
145. Serre, J-P.: Facteurs locaux des fonctions zêta des variétés algébriques. Sém. Delange-Pisot-Poitou 1969/70, N° 19 (also: Oeuvres II).
146. Shimura, G.: Arithmetic of unitary groups. Ann. of Math. *79*, 369–409 (1964).
147. Shimura, G.: On elliptic curves with complex multiplication as factors of the Jacobians of modular function fields. Nagoya Math. J. *43*, 199–208 (1971).
148. Shimura, G.: Class fields over real quadratic fields and Hecke operators. Ann. of Math. *95*, 130–190 (1972).
149. Shimura, G.: On the factors of a Jacobian variety of a modular function field. J. Math. Soc. Japan *25*, 523–544 (1973).
150. Shimura, G.: Introduction to the arithmetic theory of automorphic functions. Princeton Univ. Press 1971.
151. Shimura, G.: On the periods of modular forms. Math. Ann. *229*, 211–221 (1977).
152. Shimura, G.: The special values of the zeta functions associated with Hilbert modular forms. Duke Math, J. *45*, 637–679 (1978).
153. Shimizu, H.: On discontinuous groups operating on a product of upper half planes. Ann. of Math. *77*, 33–71 (1963).
154. Shintani, T.: On the evaluation of zeta functions of totally real algebraic number fields at non-positive integers. J. Fac. Sci. Univ. Tokyo, Sect. IA, *23*, 393–417 (1976).
155. Shintani, T.: On liftings of holomorphic cusp forms. Proc. Symp. Pure Math. AMS *33*, vol. 2, 97–110 (1979).
156. Siegel, C. L.: Lectures on advanced analytic number theory. Tata Institute Bombay, 1961, 1965.
157. Siegel, C. L.: The volume of the fundamental domain for some infinite groups. Trans. AMS *39*, 209–218 (1936). Correction in: Zur Bestimmung des Fundamentalbereiches der unimodularen Gruppe. Math. Ann. *137*, 427–432 (1959).
158. Siegel, C. L.: Über die analytische Theorie der quadratischen Formen. Ann. of Math. *38*, 212–291 (1937).
159. Siegel, C. L.: Berechnung von Zetafunktionen an ganzzahligen Stellen. Nachr. Akad. Wiss. Götttingen *10*, 87–102 (1969).
160. Svarcman, O. V.: Simple-connectedness of the factor space of the Hilbert modular group. Funct. Anal. *8*, 99–100 (1974).
161. Tate, J.: Algebraic cycles and the poles of zeta functions. In: Arithmetic algebraic geometry, 93–110. Harper and Row 1965.
162. Tate, J.: Rigid analytic spaces. Invent. Math. *12*, 257–289 (1971).
163. Thomas, E., Vasquez, A.: Rings of Hilbert modular forms. Comp. Math. *48*, 139–165 (1983).
164. Thomas, E., Vasquez, A.: On the resolution of cusp singularities and the Shintani decomposition in totally real cubic fields. Math. Ann. *247*, 1–20 (1980).
165. Tong, Y.: Weighted intersection numbers on Hilbert modular surfaces. Comp. Math. *38*, 299–310 (1979).
166. Vaserstein, L.: On the group SL(2) over Dedekind rings of arithmetic type. Mat. Sbornik *89* (1972) (= Math. USSR Sbornik *18*, 321–325 (1972)).
167. Vignéras, M-F.: Invariants numériques des groupes modulaires de Hilbert. Math. Ann. *224*, 189–215 (1976).
168. Vignéras, M-F.: Série thêta des formes quadratiques indéfinies. In: Modular Functions of One Variable VI, 227–240. Lecture Notes in Math. *627*. Springer Verlag 1976.
169. Waterhouse, W., Milne, J.: Abelian varieties over finite fields. Proc. Symp. Pure Math. AMS *20* (1969).
170. Weil. A.: Sur certaines groupes d'opérateurs unitaires. Acta Math. *111*, 143–211 (1964).
171. Yangbo Ye: Kloosterman integrals and base change. Preprint 1986.
172. Zagier, D.: Modular forms associated to real quadratic fields. Invent. Math. *30*, 1–46 (1975).
173. Zagier, D.: On the values at negative integers of the zeta function of a real quadratic field. L'Ens. Math. *22* 55–95 (1976).

174. Zagier, D.: Modular forms whose Fourier coefficients involve zeta functions of quadratic fields. In: Modular Functions of One Variable VI, 105–169. Lecture Notes in Math. *627*. Springer-Verlag 1976.
175. Zagier, D.: Valeurs des fonctions zêta des corps quadratiques réels aux entiers négatifs. Astérisque *41–42*, 135–151 (1977).
176. Zagier, D.: The Rankin-Selberg method for automorphic functions which are not of rapid decay. J. Fac. Sci. Tokyo *28*, 415–437 (1981).
177. Zagier, D.: Nombres de classes et fractions continues. Astérisque 24–25 (1975).
178. Zagier, D.: Zetafunktionen und Quadratische Körper. Springer-Verlag 1980.
179. Zucker, S.: L_2-cohomology of warped products and arithmetic groups. Invent. Math. *70*, 169–218 (1982).
180. Serre, J-P.: Modular forms of weight one and Galois representations. In: Algebraic Number Fields, ed. A. Fröhlich. Acad. Press, 1977.
181. Watson, G. L.: Integral quadratic forms. Cambridge Tracts *51*, 1960.

List of Notations

Index

Ergebnisse der Mathematik und ihrer Grenzgebiete, 3. Folge

A Series of Modern Surveys in Mathematics

Volume 13: **E. Freitag, R. Kiehl**

Etale Cohomology and the Weil Conjecture

With an Historical Introduction by J. A. Dieudonné
Translated from the German Manuscript

1987. XVIII, 317 pages. ISBN 3-540-12175-7

Contents: Introduction to the History of the Weil Conjecture (by J. A. Dieudonné). - Introduction to Etale Cohomology Theory. - Rationality of Weil ξ-Functions. - The Monodromy Theory of Lefschetz Pencils. - Deligne's Proof of the Weil Conjecture. - Appendices. - Bibliography. - Notation Index. - Subject Index.

Volume 14: **M. R. Goresky, R. D. MacPherson**

Stratified Morse Theory

1987. Approx. 300 pages. ISBN 3-540-17300-5

Contents: Introduction: Stratified Morse Theory. Topology of complex analytic varieties and the Lefschetz hyperplane theorem. - Morse Theory of Whitney Stratified Spaces: Whitney stratifications and subanalytic sets. Morse functions and nondepraved critical points. Dramatis Personae and the main theorem. Moving the wall. Fringed sets. Absence of characteristic convectors: Lemmas for moving the wall. Local, normal, and tangential Morse data are well defined. Proof of the main theorem. Relative Morse theory. Nonproper Morse functions. Relative Morse theory of nonproper functions. - Normal Morse data of two Morse functions. - Morse Theory of Complex Analytic Varieties: Introduction. Statement of Results. Normal Morse data for complex analytic spaces. Homotopy Type of the Morse Data. Morse theory of the complex link. Proofs of the Main Theorems. Morse Theory and Intersection Homology. Connectivity theorems for q-defective pairs. Counterexamples. - Complementa of Affine Subspaces: Introduction. Statement of Results. Geometry of the order complex. Morse theory of $\mathbb{R}^n$. Proofs of theorems B, C, and D. - Examples. - Bibliography. - Index.

Volume 15: **T. Oda**

Convex Bodies and Algebraic Geometry

An Introduction to the Theory of Toric Varieties

1987. 42 figures. VIII, 212 pages.
ISBN 3-540-17600-4

Contents: Fans and Toric Varieties. - Integral Convex Polytopes and Toric Projective Varieties. - Toric Varieties and Holomorphic Differential Forms. - Applications. - Geometry of Convex Sets. - References. - Subject Index.

Volume 16: **van der Geer**

Hilbert Modular Surfaces

1987. 39 figures. IX, 291 pages.
ISBN 3-540-17601-2

Contents: Hilbert's Modular Group. - Resolution of the Cusp Singularities. - Local Invariants. - Global Invariants. - Modular Curves on Modular Surfaces. - The Cohomology of Hilbert Modular Surfaces. - The Classification of Hilbert Modular Surfaces. - Examples of Hilbert Modular Surfaces. - Humbert Surfaces. - Moduli of Abelian Schemes with Real Multiplication. - The Tate Conjectures for Hilbert Modular Surfaces. - Tables. - Bibliography. - Index. - List of Notations.

Forthcoming title:

G. A. Margulis

Discrete Subgroups of Lie Groups

ISBN 3-540-12179-X

Springer-Verlag
Berlin Heidelberg New York
London Paris Tokyo

Springer

GPSR Compliance
The European Union's (EU) General Product Safety Regulation (GPSR) is a set of rules that requires consumer products to be safe and our obligations to ensure this.

If you have any concerns about our products, you can contact us on

ProductSafety@springernature.com

In case Publisher is established outside the EU, the EU authorized representative is:

Springer Nature Customer Service Center GmbH
Europaplatz 3
69115 Heidelberg, Germany

www.ingramcontent.com/pod-product-compliance
Lightning Source LLC
Chambersburg PA
CBHW081926120926
51338CB00004B/26
* 9 7 8 3 5 4 0 1 7 6 0 1 5 *